CONFIDENTIAL FREQUENCY LIST

FIFTH EDITION

Oliver P. Ferrell

GILFER ASSOCIATES, INC

P.O. Box 239 52 Park Avenue Park Ridge, NJ 07656 USA

ISBN: 0-914542-10-9

Library of Congress Catalog Card: 82-80199

FIFTH EDITION

Printed in the United States of America

Contents

To my son Mike

With love and great admiration
for your exceptional courage

Foreword

Acclaim for the re-design of the 4th edition of my CONFIDENTIAL FREQUENCY LIST was far greater than any compiler/editor could anticipate. To those dozens of readers and users that wrote me about the 4th edition – and possibly did not receive an acknowledgement – my belated THANK YOU.

The 5th edition has been greatly expanded in terms of number of frequencies and stations, ancilliary remarks and a wholly new section of the book devoted to a reverse listing by callsign, location, service, mode and frequency(ies). The number of "details" entered in the 5th edition exceeds 100,000 items concerning 8500 frequencies and stations. Handling this magnitude of information was only possible with the aid of a minicomputer word processor.

In the past 20 years I have been witness to numerous changes in the use of the high frequency (HF) spectrum. Two-way HF communication has become more affordable greatly expanding the "usage base". Stationary satellite relays have siphoned off a majority of the RTTY press broadcasters and many of the large radiotelephone links between continents. However, it has also become apparent that there is a vulnerability in satellite communications – especially for the military and ships at sea. Diminution of the number of HF circuits due to the switchover to satellites has not taken place – nor can it or will it take place. Shortwave – regardless of its modest shortcomings – is here to stay.

As electronic technology made HF communications more affordable so did the advent of new general-coverage (0.5-30.0 MHz) direct frequency readout receivers. Within the past 4 years at least a half dozen very effective general coverage receivers have been made available to the public. Most of them are reasonably good single-sideband (SSB) receivers. Simultaneously, integrated circuits have been used in developing new demodulators to print out CW (or RTTY) on a billboard screen, video display or paper rolls. We can look forward to major improvements in receiving equipment within the near future: frequency memory and recall expansion, scanning of frequencies for activity, easier and more positive tuning, better antennas, filters and totally foolproof CW demodulators.

Having been directly or indirectly involved in HF (and VHF) reception since 1934, I still enjoy twisting that dial and looking for the new and unexpected. Tuning the frequencies between the International Broadcast and Radio Amateur bands provides a daily on-going challenge I find difficult to resist. It is a rare day that I do not intercept at least two new stations or frequencies. In the HF spectrum you can find it all: emergencies and distress calls, hurricane and typhoon warnings, embassy messages, news, war games, ordinary and out-of-the-ordinary radio telephone conversations, etc. The variety seems almost endless. I hope you enjoy using this new CONFIDENTIAL FREQUENCY LIST as much as I have had assembling the tabulations.

Oliver P. Ferrell

February 1982

Anyone divulging the content of any non-broadcast radio transmission is in violation of Sec. 605 of the Communications Act of 1934 (USA) or Article 17 of the 1938 Radio Regulations adopted by the International Telecommunication Union. Similar disclosure prohibitions are in the laws of numerous countries throughout the world.

Why a CFL?

The CONFIDENTIAL FREQUENCY LIST is an index of "unusual" radio stations that can be heard in the shortwave spectrum. It is the only widely circulated publication of its kind and contains information not otherwise available to owners of shortwave receivers. This edition of the CONFIDENTIAL FREQUENCY LIST is a distillation of the enormous International Frequency List prepared by the International Telecommunications Union (ITU), plus numerous other specialized sources and considerable monitoring by the author and collaborators.

Curious shortwave listeners cannot help but wonder about the number and variety of signals they can hear on frequencies between the International Broadcasting Bands and the radio amateur bands. Some of the signals are nothing more than clicks and grinding sounds (FAX), others might sound like a musical typewriter (RTTY), or every once in a while a snatch of a conversation can be heard when the BFO was accidentally switched on (SSB) and then there's the unforgettable "dahdidahdit dahdahdidah" of CW.

This natural curiosity has grown in the past 20 years as shortwave receivers became easier to tune and specialized "International Broadcast" receivers (Squires-Sanders IBS, Drake SW-4A & SPR-4, Allied SX-190, etc.) gave way to "general coverage" receivers tuning the whole HF spectrum from 1.6 to 30.0 MHz. Simultaneously, the necessity to "learn the code" vanished as IC technology enabled the construction of CW readers. The world of tuning between the International Broadcast Bands and radio amateur is now open to one and all. And, it's a fascinating experience that adds a new dimension to shortwave listening.

The CONFIDENTIAL FREQUENCY LIST is primarily assembled for a listener in North America. Thus, greater emphasis—especially on the frequencies below 6 MHz—is placed on "things" most likely to be heard in that part of the world, albeit this includes many stations from Australasia and the Far East. As the radio frequency increases, the list becomes worldwide. The CFL also "assumes" that the listener has a reasonably good idea of the frequency to which he is tuned and recognized the transmission format (CW, LSB, etc.).

Musical Chairs The ebb and flow of HF assignments reflects the continuing interest and demand for HF communication links. As a result there is never a "good time" to publish a new edition of the CONFIDENTIAL FREQUENCY LIST—something is always about to happen. In the preceding edition of this book and its companion—GUIDE TO RTTY FREQUENCIES—we dealt with last-minute changes in callsign blocks, deletions of complete HF networks, disappearance of countries (plus many name changes) and the 5th edition is no different—the aeronautical HF services are playing musical chairs.

As this is being written, aeronautical HF circuits are beginning a massive, frequency shift. Some nets are moving a few kiloHertz, others are moving 50 to 200 kHz, while a third group is moving from one band to an entirely different band—several megaHertz away. The logic behind all of these changes is beyond the scope of this book, although it can be said that it is one of the results of the 1979 World Administrative Radio Conference (WARC) of which we will hear more in the years to come.

In addition to the frequency shifts, aeronatucial circuits (aircraft and ground stations) are expected to switch over to single-sideband transmissions using either a partially or a fully suppressed carrier, A3H and USB, respectively. Prior to February 1982, aeronautical circuits could be either AM or A3H, although some USB circuits were operational.

The magnitude of these changes is staggering and involves thousands of aircraft and hundreds of ground stations. Fortunately, the changeover takes place in several steps: some changes take place in 1982, some in 1983 and the remainder in 1984.

At this writing it is impossible to foresee how smoothly, or what difficulties may arise with the changes planned for 1982 . Most of the frequency shifts are in the 8, 10, 11, 13 and 17 MHz bands. Shifts in the lower frequency bands are planned for 1983. In the CONFIDENTIAL FREQUENCY LIST projected frequency shifts are identified by the section symbol "§" and show the "before" and "after" frequencies.

Although not wholly pertinent to the aeronautical changeover, the sub-heads in the CFL tabulation by frequency show the block assignments where there will be major shifts in the makeup of the HF spectrum. These are identified as "Post-WARC". Implementation of these massive changes are not expected until 1985, and beyond.

Letter Beacons In the 4th edition of the CFL little attention was drawn to the single CW letter "beacons". Isolated instances of a radio signal copied while transmitting the single letter K or W were included in the main tabulation ("K"—4002.5; "W"—10,698.0, etc.). What purpose these beacons served, where the transmitters were located and who "owned" them was not identified.

Four years later many of the very same letter beacons—or so it appears—are still functioning. The number of intercepted beacons has tripled—either due to the increased number of beacons on the air or better receivers and observational techniques of CFL collaborators. Although the purpose of these beacons remain shrouded in mystery some insight has been gained from recent observations:

1. Some letter beacons operate on parallel frequencies.

2. Letter beacons are most frequently reported by observers in Australasia, the Far East and North America.

3. Letter beacons have a distinctly patterned on/off time.

4. There is a slight variation in the transmitting frequency of the same letter beacon over a period of time.

5. The repetition rate of the transmitted letter varies with time and frequency; some letters are transmitted every 2 or 3 seconds and some are transmitted every 6 or 7 seconds, or any interval in between. There is no immediately obvious frequency to rep-rate correlation.

6. Some letter beacons began transmitting data bursts in 1980.

The signal strength of most letter beacon signals is an indication that transmitters radiating 25-100 watts are being used. Also, the times that letter beacons are heard favors a transmitter functioning during the hours of darkness—most likely in the South Pacific Ocean.

All-in-all, the mystery remains "why" a single CW letter beacon and "why" does an unattended transmitter change frequency and repetition rate? Possibly these three factors are the information to be broadcast—frequency shift conveying one channel of information and rep rate conveying a second information channel. Changing the CW letter could conceivably provide a third information channel. Even the times of audibility (when the transmitting site is known) renders valuable information on radio wave propagation. The possibility that letter beacons are actually transponders—transmitters that are electronically coupled to receivers and react to "query" signals—cannot be excluded.

Inquires made by the author to various US Government agencies and funded research foundations sheds no light on the letter beacon mystery. The ITU files contain no leads, even though beacons are HF radio signals and should be registered. Who, what, where and why remain unanswered.

Those "NUMBERS" Stations Oldest and as much a mystery as every, the so-called "number" or "spy" (rapidly going out of fashion for lack of evidence) stations did not lessen their activities in 1981-82. If nothing else, the NUMBERS stations are going modern with increasing usage of SSB transmitters on many new frequencies.

The listener in North America that has never heard a NUMBERS station isn't really trying. Although on the air for only 10 to 15 minutes, the chances of intercepting a Spanish speaking woman announcer carefully enunciating a list of numbers: "ocho seis uno dos cuatro-ocho seis uno dos cuatro" is extraordinarily high. Listeners in Europe are favored with German numbers and those in the Far East by Korean.

A wholly new trend in this mysterious category have been the "LETTERS" stations. First observed in 1981, LETTERS stations usually have a woman announcer reciting a random English language list of unconnected letters—unlike the NUMBERS stations that almost inevitably use groups of 3, 4 or 5 figures. Transmissions end with plain text announcements, "End of messages—End of transmission". Preliminary analysis of the timing and frequencies are indicative of locations in Europe.

The CFL collaborators reported on approximately 80 NUMBERS frequencies in the 12-month period prior to publication. Due to the transient nature of these broadcasts, plus the inconsistencies of frequency usage, NUMBERS stations have not been emphasized in this edition of CFL. A few general examples are shown in the Tabulation by Frequency, but even these stations may have moved 20-50 kHz away by the time this book is in circulation.

Notes Made While Compiling US Government Public Law 87-795 enables foreign governments to operate HF transmitters from their respective Washington embassies. All of these stations have the assigned prefix "KNY". New guidelines for assigning radio frequencies to foreign embassies was approved in late 1981. It includes a provision for embassies to operate satellite earth stations. Most KNY stations are expected to close HF circuits in 1982/83. . . . . Beware of unsubstantiated claims regarding callsigns and assignees. In the course of "massaging" this edition of CFL prior to final pagination the French EMBASSY CW station "P6Z" sorted out as being in the Parisian North Korean Embassy. Unless all foreign government embassies in Paris have the same P6Z callsign (very unlikely) any claim that this is North Korean is extremely unlikely. The original P6Z was employed by the French Embassy for communications with Tokyo. . . . . The People's Republic of China inundated the IFRB in 1980 by filing thousands of frequencies and callsigns. The PRC now leads all other countries in having its IFRB/ITU files up to date. Meanwhile, the Soviet Union continues to have thousands of pre-World War II frequencies and callsigns on file and has not deleted any sizeable number of circuits in the past decade.

Military tactical nets often occupy frequencies allocated to maritime, aeronautical, etc. services in complete disregard of the interference they are creating. Such nets use a variety of alpha/numeric callsigns and are usually impossible to identify. Fortunately, activity is limited to hours or one or two days when occupancy returns to normal. . . . . The very strong CW signal calling itself "A4I" would attract any listeners immediate attention (14,894.1 and 24,769.7 kHz). Is it an embassy? Is it in the Middle East—as reported by some Australian listeners, or in South America as suspected by the author due to the times he has heard this mysterious signal.

What do "KILO PAPA ALFA 2" (on 8925.0 kHz) and "MIKE INDIA WHISKEY 2" (on 12,747.0 kHz) have in common? A strong USB signal with a woman announcer repeating either phrase for several minutes (over a period of several weeks at various times of the day) was heard in Europe and North America in 1981. Oddly enough, in 1980 the same frequencies were occupied by a strong CW signal that signed as "J3R". The possibility that any of these stations are military appears remote—the announcing is too professional. . . . . Also consistently heard virtually worldwide has been the CW signal of "EC3Y". It appears to alternate between about 8158.0 and about 13,582.0. Never does anything but rattle off a string of "VVV DE EC3Y VVV DE EC3Y VVV DE EC3Y" and close down in about 15 minutes.

There is always something interesting taking place. On February 2nd, 1982, the author copied this NUKO broadcast from NAR:

> WARNING NO 57 NICARAGUA. MARINERS ARE ADVISED TO AVOID CARIBBEAN PORTS AND WATERS OF NICARAGUA UNTIL FURTHER NOTICE. THE NICARAGUAN GOVERNMENT HAS RESTRICTED TRAVEL TO THE EASTERN PORTION OF THE COUNTRY. THERE HAVE BEEN CASES OF FOREIGN FLAG VESSELS BOARDED BY NICARAGUAN AUTHORITIES AND RELEASED AFTER A FEW DAYS. U.S. EMBASSY CONSULAR ACCESS TO DETAINED U.S. CITIZENS HAS BEEN DIFFICULT AND SUBJECT TO DELAYS.

Polish HF stations began changing over to new callsigns in 1980. As this edition was being prepared, most HF point-to-point stations were employing new callsigns. Coastal Polish stations at Gydnia and Szczecin began changeover in 1981, but some stations were still being heard in early 1982 with old calls. When possible both new and old callsigns are shown.

Using the CFL

FREQUENCY All frequencies listed in the CONFIDENTIAL FREQUENCY LIST are in kiloHertz. For the few AM stations and several thousand CW stations the frequency listed is zero beat (exact) and is either ITU registered or averaged from monitoring observations. All SSB (both upper and lower) frequencies are those of the missing "carrier" and are about 1.5 kHz high or lower than "assigned" frequencies. These published frequencies are also derived from official sources, or averaged from monitoring.

It is necessary to stress to the CFL user that certain receivers (such as the Sony ICF-6800W and Yaesu FRG-7700) do not reflect an SSB offset in their digital readout systems. The operator must make a mental compensation of 1 or 2 kHz, up or down, to determine the carrier frequency. A few receivers (JRC NRD-505 and Panasonic RF-9000) require retuning of the "KHZ' dial to optimize SSB signal recovery and indicate the true carrier frequency. In the Drake R7A the Passband Tuning (PBT) is adjusted to accomplish the same purpose. Lastly, receivers such as the JRC NRD-515 automatically track digital readout and carrier frequencies. Although it is common federal and international practice to show both the assigned and carrier frequencies of SSB stations, it is not done in the CFL in order to eliminate unnecessary confusion. Again, ALL LSB or USB frequencies are carrier frequencies.

Theoretically, all HF radio circuits are to be registered with the International Frequency Registration Board of the ITU. In practice about three dozen nations do attempt to file with the IFRB those frequencies, callsigns, locations, power, hours of operation, beam headings, etc. that do not affect their own internal security. Probably about five times as many nations file with the IFRB whatever material their resources permit. The remainder of the ITU signatories may, or may not, file.

All countries are also expected to delete from the IFRB files those stations and frequencies that have been shut down. Deletions lag literally decades behind current filings and the IFRB/ITU station lists show frequencies and callsigns that were registered pre World War II.

As a rough approximation, about 50% of all radio stations shown in this CFL edition have the ability to shift frequency up or down 5 to 10 kHz from that listed. Frequency shifting is very common in the Coastal Station CW Bands to avoid interference from stations on the same channel. A deviation of 0.5 to 2.0 kHz in the Coastal Bands is not uncommon. Those stations remaining fixed in frequency include those whose services are of a broadcast nature (aeronautical, Coastal SSB Duplex, etc.)—they must stay rooted to their frequency—someone listening for an announcement cannot go tuning around to find the station.

MODE The abbreviations or acronyms used in this column are those most easily recognized by the majority of users. If a user has any doubt, exact interpretations appear in the ABBREVIATIONS section of the CFL. Stations, nets and frequencies in use are shown in the mode most frequently monitored. Dual notations (CW/USB) show that both modes have been monitored.

As in the 4th edition, the use of "ISB" remains purposefully ambiguous and should be viewed as any of the following: (A) Two independent sidebands each carrying its own modulation or traffic content. VOA feeders are a good example of two different languages being broadcast by the same Greenville transmitter on the same carrier frequency. (B) A station that changes from upper to lowersideband, or vice versa, to avoid interference, or has no apparent preference for LSB or USB and has been monitored in both modes during the preparation of this CFL edition. (C) A combination of voice and RTTY.

This edition introduces the word "Data" to reflect a transmission format of high speed computer language. On some USAF SAC circuits, data burst transmissions are supplementing a voice check-in.

The combination "CW/RTTY" or "CW/TOR" is indicative that the station operates in either mode and has been so monitored during the preparation of the CFL. Coastal stations will frequently idle on CW before switching over to TOR transmissions. Further details on RTTY and TOR will appear in GUIDE TO RTTY FREQUENCIES to be published in 1982.

CALL Virtually every HF radio station involved in international circuits has been assigned a callsign. These callsigns usually comprise an internationally accepted country designator and a national or federal suffix to further identify the station within its own country. Unfortunately, adherence to this basic precept varies from extreme to extreme and is reflected in the CONFIDENTIAL FREQUENCY LIST.

Slowly developing nations of the Third World may lack the financing and personnel necessary for licensing and frequency management. In these instances filing new station assignments with the IFRB may take 3 or 4 years. Also when the licensing authority is known to be lax many small users "appropriate" frequencies and whatever identifiers they desire. A typical early morning tuning session in the eastern half of the USA will reveal Spanish-speaking nets on: 4460.0 (LSB); 4510.0 (USB); 4555.0 (LSB); 4650.0 (USB); 5006.5 (USB); 5215.0 (USB); 5324.5 (LSB); 5396.0 (LSB); 5838.0 (USB); 5870.0 (USB); etc. Similar activity can be found between 6.7-7.0 MHz and 7.4-7.7 MHz*. Only one of these nets have filed callsign or frequency details with the IFRB.

Many HF stations rarely— and in some instance never—use callsigns. These stations may substitute a place name, a place name plus a service identifier (or even contraction of the two) or a service identifier only. Military tactical callsigns may be randomly selected

*It is safe to assume that most of the equipment used in these nets are "converted" radio amateur transceivers.

nouns, numeric, alphabetical, or alpha-numeric. Such callsigns are usually intended to be for short-term use and are not included in the CONFIDENTIAL FREQUENCY LIST, albeit some tactical callsigns have been in continuous use for years.

Coastal stations (FC) are notorious for not transmitting the proper callsign to match the frequency in use. There are numerous reasons for this discrepancy including: operator error, correct callsign tape misplaced or broken, simultaneous operation on several frequencies, etc. Users of the CFL can expect to find occasional deviation of the received callsign vis-a-vis the properly assigned callsign.

A dagger (†) preceding the callsign indicates that monitoring has not confirmed frequent operation of the station in question. As far as can be determined the "license" is still in effect and equipment is standing by, but actual operation cannot be confirmed.

LOCATION As the title implies, the principal task of this column is to identify the city, state or province and country associated with the callsign. The usual format is to show the city, the transmitting site (in parenthesis), state or province—to aid in further identification—and finally the country.

There are several variations to this scheme and this column may also be used to identify a particular station—or frequency in use—by slogan, acronym or even an arbitrary "title". Thus, most Coastal CW stations have the title "Radio" appended to their city (Slidell Radio, Murmansk Radio, etc.), while aeronautical stations are frequently referred to as "Aeradio" (New York Aeradio, Sydney Aeradio, etc.). Similarly, most naval stations—regardless of nationality—are referred to as "Naval Radio", but some aeronautical stations may be called "Air" ("Metz Air", "Bampton Air", etc.). Canadian Forces aeronautical stations use their own phrases—which have been also used in this CFL—such as, "Halifax Military", "St. Johns Military", etc. An exception to these general rules is the US Air Force whose stations frequently employ he word "Airways" ("Scott Airways", "Lajes Airways", etc.), but which are identified here by callsign.

Acronyms and abbreviated identifiers, such as, "COMMSTA Portsmouth" for the announced, "Portsmouth Communications Station" (NMN of the USCG), or the more complex "NAVCOMCEN" for those South African Naval stations in the NAVal COMmunications CENter at Silvermine.

Also appearing in this column are the words "unknown", "several" and "numerous", occasionally followed by a few qualifying remarks. Frequencies in use by stations at unidentifiable locations are simply shown as "unknown". If there is evidence that a particular country or continent may encompass the transmitting site, it is added to the "unknown" as "probably in North America", "probably in the Far East", etc. To indicate a net of 2 or 3 stations, the word "several" is employed, "numerous" may be taken to mean that the net has more than 4 or 5 active participants.

A descriptive phrase may appear in this column to identify a net of stations involved in one specific task. Examples of this usage are "Northwest Pacific LORAN Net", "Atlantic Domestic Emergency Net", etc. These phrases are then followed by a listing of the stations known to be active, or assigned to the net.

Actual location names appearing in this column are mostly derived from the ITU, US Department of State and numerous other official sources. Place names have been "Americanized" and there will be many disagreements with other lists. A good example would be the city in Netherlands that Americans call "The Hague", but in some lists would be shown as "s'Gravenhage" or the shorter version (sic) "Gravenhage". Similarly, there is clean division between those sources that call the country "Viet Nam" as opposed to those that call it "Vietnam" (we have used the latter).

One of the principal objectives in this column was to provide consistency within the framework described above.

SERVICE The two letter groups in this column are used by the ITU to quickly identify the character of a particular station's operation. Interpretation is quite simple: FA—Fixed Aeronautical (ground station); FC—Fixed Coastal; FS—Fixed Station involved only in life-saving activities; FX—Fixed Ground Station; MA—aircraft and MS—motor vessel. Further details in the ABBREVIATIONS Section.

POWER There are only two sources for information on the output power of HF stations—contact with the station itself, or blind dependence on the figures filed with the ITU. Obviously, the former would be the ideal method of obtaining this information, albeit thoroughly and totally impractical. Power figures from the ITU—if their limitations are taken into account—do have the advantage of becoming reasonably accurate due to mass averaging. In other words, overstatements about power are offset by stations that are using more powerful equipment than their ITU filings would indicate. Power figures should be regarded as "relative" and mostly useful when comparing similar stations operating in the same Service classification.

The above assumption is particularly true in the Coastal Station Bands (FC) where antenna patterns are frequently non-directional. The possibility of intercepting a station with a claimed 10 kW transmitter is substantially greater than if the same station were using a 1 or 2 kW transmitter. Stations with transmitters of less than 500 watts are only likely to be heard when conditions are just "right". The advantage of having this information at hand in the CONFIDENTIAL FREQUENCY LIST is to aid the user to identify (or possibly seek) new stations.

The author recognizes that the usefulness of this column is declining and it may be eliminated in future editions. A caveat we must always respect is the ever-increasing use of beam or directional antennas. Beams have been used by point-to-point radiotelephone links for many years, but they have started appearing in the Coastal Station Bands and can confuse or confound the listener. Recently the author observed a sudden decline in the signal strength of a station only 1200 kM distant—a telephone call to the Station Manager quickly revealed that a log—aperiodic beam antenna had just been installed and it was pointed at right angles to New Jersey—the radiated signal had dropped 20 dB in that direction.

REMARKS Identifying radio stations by frequency and callsign are only part of the information the CONFIDENTIAL FREQUENCY LIST user requires. What is the purpose of a particular station, to whom does it belong, what is the nature of its activity, where are its alternative frequencies (if any) are also questions that need answering. This is the intent of the REMARKS column.

To keep within the constraints of page size it is necessary to make use of abbreviations and acronyms, plus drawing upon the users ability to rationalize. Although interpreting this column is straight forward some guidelines will help explain the format:

> Primary information appears as an abbreviation, acronym or phrase. For example, USCG indicates that the station or channel is used by the United States Coast Guard ... AMVER means that this Coast Station accepts ship

position reports as part of the AMVER program ... A "Feeder" relays a broadcast signal to a remote transmitting site and EMBASSY—when it appears alone—indicates that the country of origin is that shown under LOCATION.

TIME SIGNALS identifies stations with scheduled broadcasts (sometimes shown in this column) described in detail in the LIST OF TIME SIGNAL STATIONS—available from GILFER Associates ... "Coded Tfc" refers to transmissions of alpha/numeric codes that are NOT meteo. These stations may be military, commercial, embassies, etc ... "Net Control" indicates a station that appears to be coordinating the activities of more than 3 other stations in a net ... "Telcom" refers to two-way radiotelephone communications—usually commercial.

Frequencies and stations identified as RCAF, USAF, USN, etc. are assigned to the appropriate military service (see ABBREVIATIONS) ... NATO and Soviet military identifiers—used in addition to or in place of an assigned call—will also appear in this column: e.g., K13A, LCMP-2, 4LA, etc ... Numerous schedules of weather, FAX and NAVAREA broadcasts are itemized and the source (NOAA, "Reported", etc.) is usually identified.

Both parentheses and brackets appear in this column. Generally, information enclosed in parentheses is of prime importance and information enclosed in brackets is of secondary importance, or may be the author's interpolation or explanation. ... An exception is in the MARITIME USB DUPLEX bands where the frequency in the parentheses refers to the ship USB answerback channel.

A virgule (/) is used in this column as either a conjunction or preposition to represent "and", "by", "for", "in", "of" or "to". Thus, NUMBERS/YL/SS may be taken to mean, "This is a NUMBERS broadcast in the Spanish language by a woman".

Most other REMARKS can be "understood" by combining the information contained in the LOCATION column with that of the REMARKS column.

Acknowledgments

Preparation of the CONFIDENTIAL FREQUENCY LIST was facilitated by the co-operation of dozens of contributors. I am particularly grateful to Albert H. Planchon for his special assistance and encouragement.

Valuable and very useful information is acknowledged from Arthur Andrews, Joseph Ambrose, Clif Brown, Anna L. Case, Robert H. French, Frank Gardiner, B.J. Grasser, Ian Kerr, Thomas Kneitel, Roger Legge, Lawrence Magne, Alex Moore, T.E. Muszynski, Spence Naylor, Fred Osterman, Thomas Reppert and David L. Walcutt.

Information supplied by many other contributors cannot be acknowledged in print due to restrictions--in some countries--on HF radio monitoring.

SPECIAL NOTATION

At the request of various USA and foreign government agencies, research organizations and universities, the tabular content of the CONFIDENTIAL FREQUENCY LIST--sans the reverse callsign listing and verification/substantiation proprietary information--is available on 5 1/4" floppy disks. These disks are updated monthly and contain additions and deletions not in the printed tabulation. Disks are compatible with the Lanier "No Problem" word processor. Other formats may be made available should the demand warrant. Please request price per set on your letterhead.

Confidential Frequency List

kHz	Mode	Call	Location	Service	kW	Remarks

4000.0 - 4063.0 kHz -- FIXED SERVICE (Worldwide)

kHz	Mode	Call	Location	Service	kW	Remarks
4002.0	CW	YRR4	Bucharest, Rumania	FX	0.8	Meteo
4005.0	CW	"K"	unknown	RC	....	Beacon

The band between 4003.0 and 4048.0 kHz is utilized in the USA by radio amateurs in a communications program called MARS (Military Affilated Radio System). Arranged geographically in networks, there are hundreds of stations and 30-40 frequencies (mostly SSB) involved. These MARS stations are used for short-haul traffic and are not too easily identified. MARS channels involved in long-haul traffic are identified elsewhere by frequency & service.

kHz	Mode	Call	Location	Service	kW	Remarks
4020.0	CW	†OVG	Frederikshaven Naval Radio, Denmark	FX	5.0	Danish Marine
4037.5	FAX	SMA4	Stockholm (Norrkoping), Sweden	AX	2.5	Meteo
			(0040, 0740 & 0930)			
4040.0	CW	UCY2	Astrakhan Radio, USSR	FC	5.0	Caspian Sea
	USB	....	Barrigada (Agana NAS), Guam	FX	0.1	USCG/USN
			[Emergency joint communications--"PAPA" channel]			
4047.5	FAX	FTE4	Paris (St. Assise), France	FX	10.0	Meteo
4048.5	USB	†....	North Atlantic (East Coast) LORAN Net	FX	....	USCG (Includes: NMA7, Jupiter, FL; NMP32, Nantucket, MA; NMF33, Caribou, ME; NMN73, Carolina Beach [Cape Fear], NC; NOL, "Bermuda Monitor") [Alt: 5313.0]
4048.5	USB	....	West Indies LORAN Net	FX	....	USCG (Includes: NMA, Miami, FL; NMA4, San Salvador [Watling Is.], Bahamas; NMA5, South Caicos, Bahamas; NMA7, Jupiter, FL & NMR, San Juan, PR) [Alt: 5422.5]
4050.0	ISB	....	Hawii & Central Pacific LORAN Net	FX	....	USCG (Includes: NMO, Lualualei, HI; NRO, Johnston Is.; NRO5, Upolu Point, HI & NRO7, Kure Is. [Ocean Is.], HI) [Alt: 5063.0]

4063.0 - 4438.0 kHz -- MARITIME MOBILE (Worldwide)

4063.0 USB – 4140.5 USB

The band between these frequencies is divided into 26 channels spaced 3.1 kHz apart and assigned numbers from 401 to 426. These are ship transmitting channels and are paired with Coastal USB stations operating between 4357.4 and 4434.9 kHz.

kHz	Mode	Call	Location	Service	kW	Remarks
4063.0	USB	WCM	Pittsburgh, PA	FC/MS	0.8	Inland Waterway
4069.2	USB	....	several, on/offshore NS/Nfld., Canada	FC/MS	..	Canadian Fishing
4075.0	CW	UBN	Jdanov Radio, Ukrainian SSR	FC	5.0	
4087.8	USB	WJG	Memphis, TN	FC/MS	0.8	Inland Waterway
4115.7	USB	WFN	Louisville, KY	FC/MS	0.8	Inland Waterway
4125.0	AM/USB		INTERNATIONAL SHIP STATION CALLING FREQUENCY			
	USB	KCI95	Cold Bay Radio, AK	FC	0.8	NOAA reports
			wx broadcasts at 0800 & 2200			
	AM	KCI98	King Salmon Radio, AK	FC	1.0	NOAA reports
			wx broadcasts at 0200 & 2000			
	AM	KGD58	Annette Radio, AK	FC	1.0	NOAA reports
			wx broadcasts at 1000 & 1500			

kHz	Mode	Call	Location	Service	kW	Remarks
	USB	KGD91	Yakutat Radio, AK	FC	0.8	NOAA reports wx broadcasts at 0330 & 1600
	USB	NOJ	COMMSTA Kodiak, AK	FC	5.0	USCG/Wx broadcasts at 0703 & 1903
	USB	WBH29	Kodiak Radio, AK	FC	0.8	NOAA reports wx broadcasts at 0400 & 1100
	USB	....	Atlantic & Pacific Voice Circuits	FC/MS	....	USCG (Includes NMC, San Francisco, CA; NMG, New Orleans, LA & NMO, Honolulu, HI)
	USB	....	Coast & Ship Simplex Frequency	FC/MS	....	Inland & intercoastal waterway traffic. Employed worldwide by hundreds of base stations and various types of ships. Frequently heard in North America: KDR, St. Louis, MO; KJG, New Orleans, LA; KMB, Houston, TX; KXQ, Houston, TX; KDL, Houston, TX; WGE, Baton Rouge, LA; WDJ, Morgan City, LA; etc.
4134.6	USB	ZLB	Awarua Radio, New Zealand	FC	1.0	AMVER
	USB	ZLD	Auckland Radio, New Zealand	FC	1.0	AMVER/NOAA reports wx broadcasts at 0318, 0918, 1518 & 2118
	USB	ZLW	Wellington Radio, New Zealand	FC	1.0	AMVER
	USB	....	Atlantic & Pacific Voice Circuits	FC/MS	....	USCG (Includes: NMA, Miami, FL; NMC, San Francisco, CA; NMF, Boston, MA; NMG, New Orleans, LA; NMN, Portsmouth, VA; NMO, Honolulu, HI; NMR, San Juan, PR; NOJ, Kodiak, AK & NRV, Barrigada, Guam
4143.6	USB	....	Coast & Ship Simplex Frequency	FC/MS	....	Inland & intercoastal waterway traffic. Employed worldwide by hundreds of base stations and various types of ships. Frequently heard in North America: KDL, Houston, TX; KDR, St Louis, MO; KMB, Houston, TX; WEC, Norfolk, VA; etc.
4180.0 | | 4187.2	CW		The band between these two frequencies is divided into 18 channels each 400 Hz wide and assigned numbers from 1 to 18. These are ship transmitting channels. The most commonly used channels are 5 and 6, centered on 4181.8 and 4187.0 kHz, respectively.			
4220.0	CW	ZAD2	Durres Radio, Albania	FC	1.0	
	CW	LZW2	Varna Radio, Bulgaria	FC	5.0	
	CW	YIR	Basrah Control, Iraq	FC	1.0	
4220.5	CW	C6N	Nassau Radio, Bahamas	FC	0.5	
	CW/TOR	DAF	Norddeich Radio, GFR	FC	5.0	
4221.0	CW	ODR9	Beirut Radio, Lebanon	FC	1.0	
4221.2	CW	GYU	Gibraltar Naval Radio, Gibraltar	FC	2.0	British Navy
4223.0	CW	SVD2	Athens Radio, Greece	FC	2.5	
	CW	UGH2	Juzno-Sakhalinsk Radio, USSR	FC	5.0	
	CW	UCW4	Leningrad Radio, USSR	FC	....	
4223.5	CW	9HD	Malta Radio, Malta	FC	1.0	
4223.9	CW	ZRQ2	Cape (Simonstown) Naval Radio, RSA	FC	....	RSA Navy Reported wx broadcasts at 1015 & 1800
4225.0	CW	LPC43	Ushuaia Radio, Argentina	FC	1.0	
	CW	JCT	Choshi Radio, Japan	FC	5.0	
	CW	XSJ	Zhanjiang Radio, PRC	FC	1.0	
4225.1	CW	CLA3	Havana Radio, Cuba	FC	5.0	
4225.2	CW	XFM	Manzanillo Radio, Mexico	FC	0.5	
4226.0	CW	FFD	St. Denis Radio, Reunion	FC	2.0	
4226.5	CW	STP	Port Sudan Radio, Sudan	FC	1.5	
4226.8	CW	SXA3	Spata Attikis Naval Radio, Greece	FC	5.0	Greek Navy/NATO
4228.0	CW	XSP	Shantou Radio, PRC	FC	0.5	
	CW	GKR4	Wick Radio, England	FC	0.3	
	CW	KFS	San Francisco Radio, CA	FC	10.0	
4228.5	CW	VIM	Melbourne Radio, Vic., Australia	FC	1.0	NOAA reports wx broadcast at 0948
	CW	VII	Thursday Island, Qld., Australia	FC	1.0	NOAA reports wx broadcasts at 0018 & 0948
4229.0	CW	VIP	Perth Radio, WA, Australia	FC	1.0	AMVER
4229.3	CW	PBC94	Goeree Naval Radio, Netherlands	FC	....	(N13A) Dutch Navy

kHz	Mode	Call	Location	Service	kW	Remarks
4229.5	CW	VIT	Townsville Radio, Qld., Australia	FC	1.0	NOAA reports wx broadcasts at 0930 & 2348
	CW	UNM3	Klaipeda Radio, Lithuanian SSR	FC	5.0	
4230.0	CW	UFB	Odessa Radio, Ukrainian SSR	FC	5.0	
4230.5	CW	VIB	Brisbane Radio, Qld., Australia	FC	1.0	NOAA reports wx broadcast at 0948
4231.0	CW	PPJ	Juncao Radio, Brazil	FC	1.0	
4232.0	CW	HWN	Paris (Houilles) Naval R., France	FC	2.0	French Navy
	CW	6WW	Dakar Naval Radio, Senegal	FC	2.0	French Navy
4232.5	CW	5RS4	Tamatave Radio, Madagascar	FC	2.0	
4233.0	CW	A4M	Muscat Radio, Oman	FC	1.0	
	CW	UKX	Nakhodka Radio, USSR	FC	5.0	
4233.5	CW	Y5M	Ruegen Radio, GDR	FC	5.0	
4234.0	CW/RTTY	3SB	Datong Naval Radio, PRC	FC	5.0	PRC Navy
	CW	CTV4	Monsanto Naval Radio, Portugal	FC	3.0	Port. Navy NOAA reports wx broadcasts at 0800 & 2000
4235.0	CW	VAI	Vancouver CG Radio, BC, Canada	FC	5.0	AMVER/NOAA reports wx broadcasts at 0130, 0530 & 1730
	CW	CLA2	Havana (Cojimar) Radio, Cuba	FC	5.0	
	CW	ICB	Genoa P.T. Radio, Italy	FC	1.0	
	CW/RTTY	CTP	Oeiras Naval Radio, Portugal	FC	1.5	Port. Navy/NATO
	FAX	CTV4	Monsanto Naval Radio, Portugal	FC	3.0	Port. Navy NOAA reports operation 0635-2200
	CW	HLP2	Pusan Radio, South Korea	FC	1.0	
	CW	UFL	Vladivostok Radio, USSR	FC	10.0	
4236.0	CW	UDH	Riga Radio, Latvian SSR	FC	5.0	
	CW	JYO	Aqaba Radio, Jordan	FC	1.0	
4236.5	CW	VFF	Frobisher CG Radio, NWT, Canada	FC	1.0	AMVER
4237.0	CW	PKN	Balikpapan Radio, Indonesia	FC	0.3	
	CW	PKG	Banjarmasin Radio, Indonesia	FC	0.3	
	CW	PKD	Surabaja Radio, Indonesia	FC	1.0	
	CW	VTP4	Vishakhapatnam Naval Radio, India	FC	1.0	Indian Navy
	CW	4XO	Haifa Radio, Israel	FC	2.5	
4238.0	CW	WCC	Chatham Radio, MA	FC	10.0	
4239.0	CW	HAR	Budapest Naval Radio, Hungary	FC	2.5	
4240.0	CW	HPN6Ø	Canal (Puerto Armuelles) R., Panama	FC	1.0	
	CW	ZSD43	Durban Radio, RSA	FC	3.5	
4240.5	CW	Y5M	Ruegen Radio, GDR	FC	5.0	
4241.0	CW	5AL	Tobruk Radio, Libya	FC	0.5	
	CW	LGW	Rogaland Radio, Norway	FC	5.0	
	CW	UKA	Vladivostok Radio, USSR	FC	10.0	Reported NAVAREA broadcast at 0030
4241.4	CW	4XZ	Haifa Naval Radio, Israel	FC	....	Israel Navy
4242.0	CW	Y5M	Ruegen Radio, GDR	FC	5.0	
4242.5	CW	ZSD4	Durban Radio, RSA	FC	3.5	NOAA reports wx broadcasts at 0905 & 1705
4243.0	CW	UVD	Magadan 1 Radio, USSR	FC	5.0	Okhotsk Sea
4244.0	CW	PPR	Rio Radio, Brazil	FC	1.0	TIME SIGNALS NOAA reports wx broadcasts at 0400, 1000 & 2020
	CW/TOR	DAL	Norddeich Radio, GFR	FC	5.0	
	CW	URL	Sevastopol Radio, Ukrainian SSR	FC	....	
4244.2	CW	GYA	Whitehall (London) Naval R., England	FC	1.0	British Navy
4245.0	CW	VIS53	Sydney Radio, NSW, Australia	FC	3.5	AMVER
	CW	UFN	Novorossiysk Radio, USSR	FC	1.0	
4246.2	CW	GYA	Whitehall (London) Naval R., England	FC	1.0	British Navy
4246.5	CW/RTTY	ZRH2	Cape (Fisantekraal) Naval R., RSA	FC	5.0	RSA Navy
4247.0	CW	P2R	Rabaul Radio, Papua New Guinea	FC	1.0	Reported wx broadcasts at 0000 & 1000
	CW	OBC3	Callao Radio, Peru	FC	1.0	
	CW	KPH	San Francisco Radio, CA	FC	5.0	
4247.5	CW	PPL	Belem Radio, Brazil	FC	1.0	
4247.8	FAX	GZZ2	Whitehall (Northwood) Naval R., Eng.	FC	1.0	British Navy (0450-1525)
4248.0	CW	ZRQ2	Cape (Simonstown) Naval Radio, RSA	FC	5.0	RSA Navy Reported wx broadcasts at 1015 & 1800

MARITIME MOBILE (con.)

kHz	Mode	Call	Location	Service	kW	Remarks
4250.0	FAX	LOK	Orcadas Radio, S. Orkney Islands (Local winter 0300, 1700 & 1915, only)	FC	....	Meteo
	CW	LZW2	Varna Radio, Bulgaria	FC	5.0	
	CW	JCK	Kobe Radio, Japan	FC	2.0	
	CW	XFL	Mazatlan Radio, Mexico	FC	0.5	
	CW	PCH95	Scheveningen Radio, Netherlands	FC	3.0	AMVER
	CW	ZLP2	Irirangi Naval Radio, New Zealand	FC	10.0	RNZ Navy
4251.0	CW	PPJ	Juncao Radio, Brazil	FC	1.0	NOAA reports wx broadcasts at 0130, 0730 & 1900
	CW	3VP	La Skhirra Radio, Tunesia	FC	0.3	
4251.6	CW	GKC2	Portishead Radio, England	FC	5.0	
4253.0	CW	TAH	Istanbul Radio, Turkey	FC	3.0	
	CW	UXN	Arkhangelsk Radio, USSR	FC	....	
4255.0	CW	CFH	Maritime Command Radio, Halifax, NS, Canada (Canadian Forces)	FC	10.0	(C13L) NOAA reports wx broadcasts at 0200, 0630, 1400 & 1800/Seasonal ice warning broadcasts at 0130 & 1430
	CW	UBF5	Leningrad Radio, USSR	FC	5.0	
	CW	UMV	Murmansk Radio, USSR	FC	....	
4255.3	CW	PZN25	Paramaribo Radio, Suriname	FC	1.0	
4255.6	CW	VIR	Rockhampton Radio, Qld., Australia	FC	1.0	AMVER/NOAA reports wx broadcast at 0048
4256.0	CW	CCM	Magallenes Radionaval, Chile	FC	....	Chilean Navy
	CW	XSK4	Haimen Radio, PRC	FC	0.5	
	CW	GKD2	Portishead Radio, England	FC	5.0	
	CW	KLC	Galveston Radio, TX	FC	3.5	NOAA reports wx broadcasts at 0530 & 1130
4256.5	CW	WLO	Mobile Radio, AL	FC	1.5	NOAA reports wx forecasts at 1300 & 2300
4259.0	CW	HEB	Berne Radio, Switzerland	FC	5.0	
4260.0	CW	TBA2	Izmir Naval Radio, Turkey	FC	1.0	[aka TCC21] Turkist Navy/NATO
	CW	UDK2	Murmansk Radio, USSR	FC	5.0	
4260.4	CW	ZLO2	Irirangi Naval Radio, New Zealand	FC	10.0	RNZ Navy
4261.0	CW	ZSC45	Cape Town Radio, RSA	FC	10.0	
4262.0	CW	LPD62	Gen. Pacheco Radio, Argentina	FC	5.0	NOAA reports EE wx broadcasts at 0130 & 1615
	CW	LZL2	Bourgas Radio, Bulgaria	FC	1.0	
	CW	J2A4	Djibouti Radio, Djibouti	FC	2.5	
	CW	9GA	Takoradi Radio, Ghana	FC	5.0	
	CW	DZW	Manila Radio, Philippines	FC	....	
	CW	SAG2	Goteborg Radio, Sweden	FC	5.0	AMVER
4263.0	CW	FFP2	Fort de France Radio, Martinique	FC	5.0	NOAA reports wx broadcasts at 0130, 0300, 0530 & 1300
4264.0	CW	RCV	Moscow Naval Radio, USSR	FC	....	Russian Navy
4265.0	CW	PPL	Belem Radio, Brazil	FC	1.0	NOAA reports wx broadcasts at 0130, 0800 & 1930
	CW	DAM	Norddeich Radio, GFR	FC	5.0	TIME SIGNALS (2355-0006, winter only)/Tfc
	CW	UBN	Jdanov Radio, Ukrainian SSR	FC	5.0	Reported wx broadcasts at 0500 & 1700
4266.0	CW	XSR	Haikow Radio, PRC	FC	1.0	
4268.0	CW	LPD68	Gen. Pacheco Radio, Argentina	FC	5.0	AMVER
	CW/FAX	CKN	Vancouver Forces Radio, BC, Canada	FC	....	(C13E) NAWS Canadian Forces
	CW/TOR	GKG2	Portishead Radio, England	FC	5.0	
	CW	VTG4	Bombay Radio Naval, India	FC	2.0	Indian Navy Reported wx broadcasts at 0130, 0500, 0900 & 1500
	CW	DZE	Mandaluyong Radio, Philippines	FC	0.4	
	CW	WCC	Chatham Radio, MA	FC	10.0	
4271.0	FAX	CFH	Maritime Command Radio, Halifax, NS, Canada (Canadian Forces)	FC	5.0	Meteo/NOAA re-reports seasonal ice warning pics at 0000 & 2200, plus various test charts, surface analysis, wave analysis, depictions, etc.

kHz	Mode	Call	Location	Service	kW	Remarks
	CW	CCV	Valparaiso Naval Radio, Chile	FC	....	Chilean Navy
	CW	XSG	Shanghai Radio, PRC	FC	1.0	
	CW	XFS	Tampico Radio, Mexico	FC	0.5	
	CW	UBE2	Petropavlovsk Radio, USSR	FC	5.0	
4272.0	CW	OFJ8	Helsinki Radio, Finland	FC	5.0	
4272.5	CW	VIA	Adelaide Radio, SA, Australia	FC	1.0	NOAA reports wx broadcasts at 0018 & 1018
	CW	VID	Darwin Radio, NT, Australia	FC	1.0	
4273.0	CW	HLF	Seoul Radio, South Korea	FC	3.5	
4274.0	CW	LPD74	Gen. Pacheco Radio, Argentina	FC	5.0	
	CW	GKB2	Portishead Radio, England	FC	5.0	
	CW	DZF	Manila (Bacoor) Radio, Philippines	FC	....	Reported wx broadcasts at 1600 & 2200
	CW	XSU	Yantai Radio, PRC	FC	0.2	
	CW	KFS	San Francisco Radio, CA	FC	10.0	
	CW	WPD	Tampa Radio, FL	FC	5.0	NOAA reports wx broadcasts at 1420 & 2320
4274.2	FAX	JFA	Chuo Gyogyo (Matsudo) Radio, Japan	FC	3.0	Meteo
4275.0	CW	UFN	Novorossiysk Radio, USSR	FC	....	
4275.5	CW	EAC	Cadiz (Tarifa) Radio, Spain	FC	1.0	
4276.0	CW	JNA	Tokyo Naval Radio, Japan	FC	....	JSDA/Reported NAVAREA broadcasts at 0005, 0405, 0805 & 1205
	CW	DXD	Manila (Jolo) Radio, Philippines	FC	0.4	
4277.0	CW	VRT	Bermuda (Hamilton) Radio, Bermuda	FC	1.0	
	CW	ZLB2	Awarua Radio, New Zealand	FC	2.5	AMVER
	CW	UJO3	Izmail Radio, Ukrainian SSR	FC	2.0	
4277.1	CW	HKB	Barranquilla Radio, Colombia	FC	2.5	
4278.0	CW	DZP	Manila (Novaliches) R., Philippines	FC	2.5	Reported wx broadcasts at 0300 & 1100
	CW	CTP93	Oeiras Naval Radio, Portugal	FC	....	Port. Navy/NATO
4278.5	CW	JFH	Hamajima Gyogyo Radio, Japan	FC	0.5	
4279.0	CW/RTTY	EBA	Madrid Radionaval, Spain	FC	1.0	Spanish Navy
4280.0	CW	PPO	Olinda Radio, Brazil	FC	1.0	
	CW	MTI	Plymouth Naval Radio, England	FC	1.0	British Navy
	CW	IDQ	Rome Naval Radio, Italy	FC	5.0	Italian Navy NOAA reports wx forecasts at 0500, 0830, 1800 & 2100
	CW	PJK34	Suffisant Dorp Naval Radio, Curacao	FC	1.0	Dutch Navy
	CW	CWM	Montevideo Armada Radio, Uruguay	FC	1.0	Uruguay Navy
4280.1	CW	PBC3	Goeree Naval Radio, Netherlands	FC	1.0	Dutch Navy
4282.7	CW	DHJ59	Sengwarden Naval Radio, GFR	FC	0.4	GFR Navy/NATO
4283.0	CW	UQK2	Riga Radio, Latvian SSR	FC	5.0	
	CW	ZSJ2	NAVCOMCEN (Silvermine) Cape R., RSA	FC	5.0	RSA Navy/AMVER
	CW	CQL	Principe Radio, Sao Tome e Principe	FC	0.5	
	CW	KOK	Los Angeles Radio, CA	FC	7.5	
	CW	UKW3	Korsakov Radio, USSR	FC	1.0	
4283.1	CW	XSV	Tianjin Radio, PRC	FC	1.0	
4284.0	CW	A9M	Bahrain Radio, Bahrain	FC	2.0	
	CW	DZR	Manila Radio, Philippines	FC	5.0	Reported wx broadcast at 0120
4285.0	CW	VCS	Halifax CG Radio, NS, Canada	FC	1.0	AMVER
	CW	UAH	Tallinn Radio, Estonian SSR	FC	5.0	Reported wx broadcasts at 0400 & 2040
4285.3	CW	XFU	Veracruz Radio, Mexico	FC	0.5	
4286.0	CW	VHP2	COMMSTA Canberra, ACT, Australia	FC	5.0	(A13B) RA Navy NOAA reports wx broadcasts at 0130 & 0530
	CW	GKA2	Portishead Radio, England	FC	5.0	NOAA reports wx broadcasts at 0930 & 2130/NAVAREA 0730 & 1730
	CW	VWC	Calcutta Radio, India	FC	2.5	TIME SIGNALS NOAA reports wx forecasts at 0918 & 1818
4287.0	CW	FJY4	St. Paul et Amsterdam Radio, French Antarctic	FC	0.3	
4288.0	CW	7TA2	Algiers Radio, Algeria	FC	5.0	
	CW	XSQ	Guangzhou Radio, PRC	FC	1.0	
	CW	DZZ	Manila Radio, Philippines	FC	....	
4289.0	CW	PWZ	Rio de Janeiro Naval Radio, Brazil	FC	5.0	Brazil Navy/NOAA reports wx broadcasts at 0030, 0630 & 1730

MARITIME MOBILE (con.)

kHz	Mode	Call	Location	Service	kW	Remarks
	CW	4XZ	Haifa Naval Radio, Israel	FC	1.0	Israel Navy
	CW	C9L2	Maputo Radio, Mozambique	FC	0.5	
	CW/RTTY	NUW	Whidbey Island Radio, WA	FC	5.0	USN/NOAA
4289.1	CW	TFA	Reykjavik Radio, Iceland	FC	1.5	
4290.0	CW	DZO	Manila (Bulacan) Radio, Philippines	FC	2.0	NOAA reports
			wx broadcasts at 1030 & 2330			
	CW	UCW4	Leningrad Radio, USSR	FC	....	
4290.5	CW	OST22	Oostende Radio, Belgium	FC	5.0	
4291.0	CW	TIM	Limon Radio, Costa Rica	FC	10.0	
	CW	ZSC46	Cape Town Radio, RSA	FC	10.0	TIME SIGNALS
			(0755-0800 & 1655-1700, only) NOAA reports wx broad-casts at 1030 & 2330			
4291.9	CW	CUB28	Funchal Radio, Maderia Islands	FC	5.0	
4292.0	CW	XYR6	Rangoon Radio, Burma	FC	1.0	
	CW	D4D	Praia de Cabo Radio, Cape Verde	FC	1.0	
	CW	†EQZ	Abadan Radio, Iran	FC	0.5	
	CW	IAR24	Rome P.T. Radio, Italy	FC	5.0	NOAA reports
			wx broadcasts at 0050, 0700, 1250 & 1900			
	CW	XFA	Acapulco Radio, Mexico	FC	0.5	
	CW	XFP	Chetumal Radio, Mexico	FC	0.5	
	CW	WOE	Lantana Radio, FL	FC	5.0	
	CW	CUG32	Sao Miguel Radio, Azores	FC	5.0	
4292.1	CW	CUL4	Lisbon Radio, Portugal	FC	5.0	
4292.5	CW	OMP2	Prague Radio, Czechoslovakia	FC	1.0	
4293.9	CW	SXA34	Spata Attikis Naval Radio, Greece	FC	5.0	Greek Navy/NATO
4294.0	CW	WNU31	Slidell Radio, LA	FC	2.0	
	CW	DZG	Manila (Las Pinas) Radio, Philippines	FC	5.0	
4295.0	CW	PKE	Amboina Radio, Indonesia	FC	1.0	
	CW	PKB	Belawan Radio, Indonesia	FC	1.0	
	CW	PKF	Makassar Radio, Indonesia	FC	1.0	
	CW	XSJ	Zhanjiang Radio, PRC	FC	1.0	
	CW	6VA3	Dakar Radio, Senegal	FC	4.0	
	CW	UHK	Batumi Radio, USSR	FC	5.0	
4295.5	CW	OMP2	Prague Radio, Czechoslovakia	FC	1.0	
4296.0	FAX	NOJ	COMMSTA Kodiak, AK	FC	5.0	USCG/Meteo
			[Pics at various times: analysis, prognosis, outlook, surface temp, ice, fisheries aids, etc.]			
	CW	DZK	Manila (Bulacan) Radio, Philippines	FC	1.0	
	CW	EBC	Cadiz Naval Radio, Spain	FC	1.0	Spanish Navy
4297.0	CW	TIM	Limon Radio, Costa Rica	FC	10.0	
4297.9	CW	PPO	Olinda Radio, Brazil	FC	1.0	NOAA reports
			wx broadcasts at 0100, 0600, 1000, 1200, 1500 & 2100			
4298.0	CW	OST2	Osteende Radio, Belgium	FC	5.0	
4298.0	CW	CCV	Valparaiso Naval Radio, Chile	FC	....	Chilean Navy
			NOAA reports wx forecasts at 0110 & 1330/TIME SIGNALS			
	CW	FJA41	Mahina Radio, Tahiti	FC	1.0	
	CW	JMC2	Tokyo Radio, Japan	FC	4.0	NOAA reports
			wx broadcasts at 0318, 0918, 1518 & 2118			
4299.0	CW	9KK2	Kuwait Radio, Kuwait	FC	5.0	
4299.5	CW	SAB2	Goteborg Radio, Sweden	FC	5.0	
4299.7	CW	FLK	unknown	FC	....	clg WGP
4300.0	USB	....	several, on/offshore NS/Nfld., Canada	FC/MS	..	Canadian Fishing
4300.2	CW	GLP	Whitehall (London) Naval R., England	FC	....	British Navy
4301.0	CW	VWM	Madras Radio, India	FC	2.5	
4302.0	CW	ZAD2	Durres Radio, Albania	FC	1.0	
	CW	A9M	Bahrain Radio, Bahrain	FC	2.0	
4303.0	CW	EBK	Las Palmas Radionaval, Canary Is.	FC	....	Reported
			NAVAREA 0050 & 2050			
	CW	OXZ2	Lyngby Radio, Denmark	FC	5.0	
4304.0	CW	LSO5	Buenos Aires Radio, Argentina	FS	....	Prefectura Naval
			Reported wx broadcasts at 0210, 1210 & 1810/NAVAREA at 0030, 1530 & 2100			
	CW	VHR2	Darwin Naval Radio, NT, Australia	FC	10.0	RA Navy
	CW	JFA	Chuo Gyogyo (Matsudo) Radio, Japan	FC	3.0	

kHz	Mode	Call	Location	Service	kW	Remarks
4305.0	CW	XSZ	Dairen Radio, PRC	FC	....	
	CW	HZW	Khafji Radio, Saudi Arabia	FC	2.0	
	CW	ZCB4	unknown	FC	....	
4305.2	CW	6WW	Dakar Naval Radio, Senegal	FC	10.0	French Navy
4305.8	CW	MTI	Plymouth Naval Radio, England	FC	1.0	British Navy
4306.0	CW	SPE21	Szczecin Radio, Poland	FC	5.0	
	CW	CTU4	Monsanto Naval Radio, Portugal	FC	5.0	Port. Navy
4307.0	CW	D3L28	Luanda Naval Radio, Angola	FC	6.0	
	CW	GYA	Whitehall (London) Naval R., England	FC	5.0	British Navy
	CW	C9C3	Maputo Naval Radio, Mozambique	FC	3.0	
	CW	UMV	Murmansk Radio, USSR	FC	....	Reported wx broadcasts at 0630 & 1700
4308.0	CW	CTV	Monsanto Naval Radio, Portugal	FC	....	Port. Navy
	CW	HLG	Seoul Radio, South Korea	FC	3.5	
4308.2	CW/TOR	DAN	Norddeich Radio, GFR	FC	5.0	
4310.0	CW	MTI	Plymouth Naval Radio, England	FC	1.0	British Navy
	CW	WNU41	Slidell Radio, LA	FC	2.0	
4311.3	CW	EBA	Madrid Radionaval, Spain	FC	0.5	(EBCQ) Spanish Navy
4312.0	CW	OSN44	Oostende Naval Radio, Belgium	FC	1.0	Belgium Navy
4312.9	CW	FUG3	La Regine (Castelnaudry), France	FC	5.0	French Navy
4313.0	CW	†NPM	Lualualei Naval Radio, HI	FC	10.0	USN
	CW	9VG33	Singapore Radio, Singapore	FC	3.0	
4314.0	CW	DZN	Manila (Novotas) Radio, Philippines	FC	5.0	NOAA reports wx forecasts at 0200 & 1400
4315.0	CW	URD	Leningrad Radio, USSR	FC	....	
	CW	XVG22	Haiphong Radio, Vietnam	FC	1.0	
4316.0	CW	GKM2	Portishead Radio, England	FC	5.0	
	CW	VWB	Bombay Radio, India	FC	1.0	
	FAX	JJC	Tokyo Radio, Japan	FC	5.0	Meteo/Nx/NTM
	CW	DYV	Manila (Iloilo) Radio, Philippines	FC	0.4	
	CW	WLC	Rogers City Radio, MI	FC	0.3	
4317.0	CW	ZSC33	Cape Town Radio, RSA	FC	10.0	
4317.5	CW	GKI2	Portishead Radio, England	FC	5.0	
4318.0	CW	DZY	Manila (Malabon) Radio, Philippines	FC	....	
	CW	DZU	Manila (Pasig) Radio, Philippines	FC	1.0	
4319.0	CW	OXZ21	Lyngby Radio, Denmark	FC	5.0	
	CW	SPH23	Gydnia Radio, Poland	FC	5.0	[aka SPI2]
4319.1	CW	XSG	Shanghai Radio, PRC	FC	1.0	
4320.0	CW	IAR4	Rome P.T. Radio, Italy	FC	5.0	NOAA reports wx forecast at 0050
4321.0	CW	†GYA	Whitehall (London) Naval R., England	FC	5.0	British Navy
4322.0	CW	CBM2	Punta Arenas Radiomaritima, Chile	FC	0.7	
	CW	9VG54	Singapore Radio, Singapore	FC	3.0	NOAA reports wx broadcasts at 0120 & 1320
	CW	WPA	Port Arthur Radio, TX	FC	5.0	
	CW	YVL	Puerto Cabello Radio, Venezuela	FC	0.6	
	CW	YVM	Puerto Ordaz Radio, Venezuela	FC	0.6	
4323.0	CW	VIC	Carnarvon Radio, WA, Australia	FC	1.0	NOAA reports wx broadcasts at 0148 & 1130
	CW	UFH	Petropavlovsk Radio, USSR	FC	5.0	
4323.5	CW	YQI4	Constanta Radio, Rumania	FC	1.0	
4323.6	CW	VIE	Esperance Radio, WA, Australia	FC	1.0	NOAA reports wx broadcasts at 0018 & 1118
4324.0	CW	DZH	Manila Radio, Philippines	FC	0.3	Reported wx broadcasts at 1100 & 2320
4324.9	CW	HKC	Buenaventura Radio, Colombia	FC	0.5	
4325.0	CW	SUP	Port Said Radio, Egypt	FC	2.0	
	CW/RTTY	FUB	Paris (Houilles) Naval Radio, France	FC	1.0	French Navy
	CW	LFW	Rogaland Radio, Norway	FC	5.0	AMVER
	CW	OBQ5	Iquitos Radio, Peru	FC	0.5	
4325.1	CW	AQP2	Karachi Naval Radio, Pakistan	FC	5.0	Pakistan Navy
4326.5	CW	GKJ2	Portishead Radio, England	FC	5.0	
	CW	JYO	Aqaba Radio, Jordan	FC	1.0	

kHz	Mode	Call	Location	Service	kW	Remarks
4328.0	CW	FFL2	St. Lys Radio, France	FC	5.0	
	CW	JOS	Nagasaki Radio, Japan	FC	3.0	
	CW	DZM	Manila (Bulacan) Radio, Philippines	FC	10.0	
4330.0	CW	DZJ	Manila (Bulacan) Radio, Philippines	FC	2.5	Reported wx broadcasts at 0300 & 1000
4331.0	CW	WCC	Chatham Radio, MA	FC	10.0	
4332.5	CW	JCK	Kobe Radio, Japan	FC	1.0	
4334.0	CW	LSA2	Boca Radio, Argentina	FC	2.0	
	CW	PJC	Curacao (Willemstad) Radio, Curacao	FC	2.5	
4335.0	CW	UDH	Riga Radio, Latvian SSR	FC	....	
4336.0	CW	GKK2	Portishead Radio, England	FC	5.0	
	CW	DZI	Manila (Bacoor) Radio, Philippines	FC	1.0	Reported wx broadcast at 0000
4337.0	CW	SPH27	Gydnia Radio, Poland	FC	5.0	[aka SPA2]
	CW	....	numerous	FA/FC/FX	..	USCG [Command Net Calling Frequency]
4338.0	CW	HWN	Paris (Houilles) Naval R., France	FC	1.0	French Navy
	CW	DYM	Cebu Radio, Philippines	FC	0.4	
4340.0	CW	9GX	Tema Radio, Ghana	FC	5.0	
4340.1	CW	XSQ	Guangzhou Radio, PRC	FC	5.0	
4342.0	CW	XUK4	Kompong Som-Ville Radio, Cambodia	FC	0.3	
4342.5	CW	IRM2	CIRM, Rome Radio, Italy	FC	5.0	Medico/AMVER
4342.7	CW	WSL	Amagansett Radio, New York	FC	4.5	
4343.0	CW	TUA3	Abidjan Radio, Ivory Coast	FC	4.0	
	CW	SVB2	Athens Radio, Greece	FC	2.5	
	CW	XSL	Fuzhou Radio, PRC	FC	0.4	
	CW	UNM2	Klaipeda Radio, Lithuanian SSR	FC	5.0	
	CW	YJM4	Port-Vila Radio, Vanuata	FC	0.5	
4344.1	FAX	†NMC	COMMSTA San Francisco, CA	FC	2.0	USCG/Meteo
4344.5	CW	OMC	Bratislava Radio, Czechoslovakia	FC	1.0	
	CW/TOR	GKS2	Portishead Radio, England	FC	15.0	
4345.0	CW	UFM3	Nevelsk Radio, USSR	FC	5.0	
4346.0	CW	XFU2	Veracruz Radio, Mexico	FC	1.0	
	CW	YUR	Rijeka Radio, Yugoslavia	FC	5.0	
	CW	WMH	Baltimore Radio, MD	FC	0.8	
	CW	NMC	COMMSTA San Francisco, CA	FC	4.0	USCG/NOAA reports wx forecast at 0630
	CW	CWA	Cerrito Radio, Uruguay	FC	10.0	Reported wx broadcasts at 0000, 1400 & 1900
4347.5	CW	5BA	Cyprus (Nicosia) Radio, Cyprus	FC	5.0	
4349.0	CW	CBV	Valparaiso Radiomaritima, Chile	FC	0.3	AMVER
	CW	JCS	Choshi Radio, Japan	FC	3.0	
	CW	S7Q	Seychelles Radio, Seychelles	FC	1.0	
	CW	EAD	Aranjuez radio, Spain	FC	3.5	
	CW	UXN	Arkhangelsk Radio, USSR	FC	5.0	
	CW	KLB	Seattle Radio, WA	FC	3.5	
4350.0	CW/TOR	9VG74	Singapore Radio, Singapore	FC	10.0	
4350.5	CW/TOR	GKE2	Portishead Radio, England	FC	5.0	
4351.0	CW	ODR9	Beirut Radio, Lebanon	FC	1.0	
	CW/TOR	LGW2	Rogaland Radio, Norway	FC	5.0	[aka LGB]
4351.9	CW/TOR	WLO	Mobile Radio, AL	FC	5.0	
4352.0	CW	YVG	La Guaira Radio, Venezuela	FC	0.3	
4356.1	CW/TOR	KPH	San Francisco Radio, CA	FC	5.0	

4357.4 - 4434.9 kHz — MARITIME COAST USB DUPLEX (Worldwide)

kHz	Mode	Call	Location	Service	kW	Remarks
4357.4	USB	SAG	Goteborg Radio, Sweden	FC	5.0	CH#401 (4063.0)
	USB	ZSV8	Walvis Bay Radio, RSA	FC	10.0	" "
	USB	KMI	San Francisco Radio, CA	FC	10.0	" " [NOAA reports wx broadcasts at 0000,0600 & 1500]
4358.0	A2/CW	C9C4	Maputo Naval Radio, Mozambique	FX	3.0	TIME SIGNALS
4360.5	USB	SPC23	Gdynia Radio, Poland	FC	10.0	CH#402 (4066.1)
4363.6	USB	CFH	Maritime Command Radio, Halifax, NS, Canada (Canadian Forces)	FC	10.0	CH#403 (4069.2)
	USB	VFC	Cambridge Bay CG Radio, NWT, Canada	FC	1.0	CH#403 (4069.2)

MARITIME COAST USB DUPLEX (con.)

kHz	Mode	Call	Location	Service	kW	Remarks
	USB	WOM	Miami Radio, FL	FC	10.0	" "
			NOAA reports wx forecast at 1230			
4366.7	USB	KWJ91	Anchorage Radio, AK	FC	....	CH#404 (4072.3)
	USB	....	numerous, on/offshore Alaska	FC/MS	....	
	USB	FFL21	St. Lys Radio, France	FC	10.0	CH#404 (4072.3)
	USB	KGN	Delcambre Radio, LA	FC	1.0	" "
4369.8	USB	....	numerous, on/offshore Alaska	FC/MS	....	
	USB	PPL	Belem Radio, Brazil	FC	1.0	CH#405 (4075.4)
	USB	FFL23	St. Lys Radio, France	FC	10.0	" "
	USB	PCG21	Scheveningen Radio Netherlands	FC	10.0	" "
	USB	9VG60	Singapore Radio, Singapore	FC	10.0	" "
	USB	ZSC25	Cape Town Radio, RSA	FC	5.0	" "
	USB	WAK	New Orleans Radio, LA	FC	1.0	" "
4372.9	USB	GTK26	Portishead Radio, England	FC	10.0	CH#406 (4078.5)
	USB	EHY	Pozuelo del Rey Radio, Spain	FC	10.0	" "
4373.0	USB	....	numerous, east & west USA coast	FA/FC/MS	..	USN
4376.0	USB	VFU2	Cape Dorset CG Radio, NWT, Canada	FC	1.0	CH#407 (4081.6)
	USB	VOK	Cartwright CG Radio, Nfld., Canada	FC	1.0	" "
	USB	VAP	Churchill CG Radio, Man., Canada	FC	1.0	CH#407 (4081.6)
	USB	VAW	Killinek CG Radio, NWT, Canada	FC	1.0	" "
	USB	VFF	Frobisher Bay CG Radio, NWT, Canada	FC	5.0	" "
			[NOAA reports wx broadcasts at 0140, 1040, 1340 & 2240]			
	USB	VFR	Resolute Bay CG Radio, NWT, Canada	FC	1.0	CH#407 (4081.6)
	USB	VFU	Coral Harbor CG Radio, NWT, Canada	FC	1.0	" "
	USB	JBO	Tokyo Radio, Japan	FC	10.0	" "
	USB	LGN4	Rogaland Radio, Norway	FC	10.0	" "
	USB	EHY	Pozuelo del Rey Radio, Spain	FC	10.0	" "
	USB	ZSD37	Durban Radio, RSA	FC	5.0	" "
			[NOAA reports wx forecast at 1703]			
	USB	NMC	COMMSTA San Francisco, CA	FC	5.0	USCG (This frequency also used by USCG Coast Stations NMF, Boston, MA & NMN, Portsmouth, VA) [Used for special law enforcement activities when required]
4379.1	USB	VFZ	Goose Bay CG Radio, Nfld., Canada	FC	1.0	CH#408 (4084.7)
	USB	HEB14	Berne Radio, Switzerland	FC	10.0	" "
	USB	KVK	Miami Radio, FL	FC	1.0	" "
	USB	....	Rijeka Radio, Yugoslavia	FC	7.5	" "
4380.5	CW	CCS	Santiago Naval Radio, Chile	FC	....	Chilean Navy
4382.2	USB	LGN5	Rogaland Radio, Norway	FC	10.0	CH#409 (4087.8)
	USB	WBL	Buffalo Radio, NY	FC	1.0	" "
	USB	WLC	Rogers City Radio, MI	FC	1.0	" "
	USB	WMI	Lorain Radio, OH	FC	1.0	" "
	USB	WCM	Pittsburgh, PA	FC/MS	0.8	Inland Waterway
	USB	WJG	Memphis, TN	FC/MS	0.8	Inland Waterway
4385.3	USB	VAI	Vancouver CG Radio, BC, Canada	FC	1.0	CH#410 (4090.9)
	USB	PCG23	Scheveningen Radio, Netherlands	FC	10.0	" "
	USB	WOO	New York (Ocean Gate, NJ) Radio, NY	FC	10.0	" "
			[NOAA reports wx forecasts at 0100 & 1300]			
4388.4	USB	EHY	Pozuelo del Ray Radio, Spain	FC	10.0	CH#411 (4094.0)
	USB	WOO	New York (Ocean Gate, NJ) Radio, NY	FC	10.0	" "
			[NOAA reports wx forecast at 1900]			
4391.5	USB	IAR	Rome Radio 4, Italy	FC	7.0	CH#412 (4097.1)
	USB	WOM	Miami Radio, FL	FC	10.0	" "
			[NOAA reports wx forecast at 1330]			
4394.6	USB	CTH	Horta Naval Radio, Azores	FC	2.5	Port. Navy CH#413 (4100.2)
	USB	VCS	Halifax CG Radio, NS, Canada	FC	5.0	" "
4397.7	USB	....	numerous, on/offshore Alaska	FC/MS	....	
	USB	WLO	Mobile Radio, AL	FC	10.0	CH#413 (4100.2)
			[NOAA reports wx broadcasts at 0000, 1200 & 1800]			
4400.8	USB	9MG42	Penang Radio, Malaysia	FC	5.0	CH#415 (4106.4)
	USB	....	numerous, worldwide	FC	....	USN/CH#415 (4106.4)
4403.9	USB	....	numerous, on/offshore Alaska	FC/MS	....	
	USB	FFL22	St. Lys Radio, France	FC	10.0	CH#416 (4109.5)

kHz	Mode	Call	Location	Service	kW	Remarks
	USB	EHY	Pozuelo del Rey Radio, Spain	FC	10.0	" "
	USB	KMI	San Francisco Radio, CA	FC	10.0	" "
	USB	WOO	New York (Ocean Gate, NJ) Radio, NY	FC	10.0	" "
4407.0	USB	VIS	Sydney Radio, NSW, Australia	FC	5.0	CH#417 (4112.6)
	A3H	P2R	Rabaul Radio, Papua New Guinea	FC	1.0	" "
						[NOAA reports wx broadcasts at 0233 & 2348]
	USB	KMI	San Francisco Radio, CA	FC	10.0	" "
	USB	WOM	Miami Radio, FL	FC	10.0	CH#417 (4112.6)
						[NOAA reports wx forecast at 2230]
4410.1	USB	CFW	Vancouver Radio, BC, Canada	FC	1.0	CH#418 (4115.7)
	USB	VCS	Halifax CG Radio, NS, Canada	FC	1.0	" "
	USB	KQM	Honolulu Radio, HI	FC	1.0	" "
	USB	WBL	Buffalo Radio, NY	FC	1.0	" "
	USB	WLC	Rogers City Radio, MI	FC	1.0	" "
	USB	WMI	Lorain Radio, OH	FC	1.0	" "
	USB	WCM	Pittsburgh, PA	FC/MS	0.8	Inland Waterway
	USB	WFN	Louisville, KY	FC/MS	0.8	" "
	USB	WGK	St. Louis, MO	FC/MS	0.8	" "
4413.2	USB	PPL	Belem Radio, Brazil	FC	1.0	CH#419 (4118.6)
	USB	FFL24	St. Lys Radio, France	FC	10.0	" "
	USB	PCG22	Scheveningen Radio, Netherlands	FC	10.0	" "
	USB	WLO	Mobile Radio, AL	FC	10.0	" "
						[NOAA reports wx forecast at 1200]
4416.3	USB	....	Goteborg Radio, Sweden	FC	5.0	CH#420 (4121.9)
4419.0	USB	WAK	New Orleans Radio, LA	FC	....	NOAA reports wx broadcasts at 0500 & 1400
4419.4	USB	....	INTERNATIONAL COAST STATION CALLING FREQUENCY			CH#421 (4125.0)
	USB	....	numerous, coastal waters, worldwide	FC/MS	...	Simplex (Used for communications with oil rigs)
4422.5	USB	....	numerous, on/offshore Alaska	FC/MS	....	
	USB	CKN	Vancouver Forces Radio, BC, Canada (Canadian Forces)	FC	10.0	CH#422 (4128.1)
	USB	COB	Havana Radio, Cuba	FC	2.5	" "
	USB	COB41	Cienfuegos Radio, Cuba	FC	1.0	" "
	USB	WOO	New York (Ocean Gate, NJ) Radio, NY	FC	10.0	" "
4425.6	USB	....	numerous, on/offshore Alaska	FC/MS	....	
	USB	WOM	Miami Radio, FL	FC	10.0	CH#423 (4131.2)
						[NOAA reports wx forecast at 2330]
4428.7	USB	VIA	Adelaide Radio, SA, Australia	FC	1.0	CH#424 (4134.2)
						[NOAA reports wx forecasts at 0845 & 2303]
	USB	VIB	Brisbane Radio, Qld., Australia	FC	1.0	CH#424 (4134.3)
						[NOAA reports wx broadcasts at 0318 & 2333]
	USB	VIO	Broome Radio, WA, Australia	FC	1.0	CH#424 (4134.3)
						AMVER [NOAA reports wx broadcasts at 0803 & 2333]
	USB	VIC	Carnarvon Radio, WA, Australia	FC	1.0	CH#424 (4134.3)
						AMVER [NOAA reports wx forecasts at 0418, 0503 & 2318]
	USB	VID	Darwin Radio, NT, Australia	FC	1.0	CH#424 (4134.3)
						AMVER [NOAA reports wx broadcasts at 0348, 0733 & 2233]
	USB	VIE	Esperance Radio, WA, Australia	FC	1.0	CH#424 (4134.3)
						[NOAA reports wx broadcasts at 0103 & 0833]
	USB	VIH	Hobart Radio, Tasmania, Australia	FC	1.0	CH#424 (4134.3)
						[NOAA reports wx forecasts at 0303, 0718 & 2218]
	USB	VII	Thursday Island, Qld., Australia	FC	1.0	CH#424 (4134.3)
						AMVER [NOAA reports wx forecasts at 0248, 0648 & 2303]
	USB	VIM	Melbourne Radio, Vic., Australia	FC	1.0	CH#424 (4134.3)
						[NOAA reports wx broadcast at 0730]
	USB	VIP	Perth Radio, WA, Australia	FC	1.0	CH#424 (4134.3)
						AMVER [NOAA reports wx broadcasts at 0103, 0503 & 0903]
	USB	VIR	Rockhampton Radio, Qld., Australia	FC	1.0	CH#424 (4134.3)
						AMVER [NOAA reports wx forecasts at 0218, 0633 & 2248]
	USB	VIS	Sydney Radio, NSW, Australia	FC	1.0	CH#424 (4134.3)
						[NOAA reports wx forecasts at 0203, 0703 & 2203]
	USB	VIT	Townsville Radio, Qld., Australia	FC	1.0	CH#424 (4134.3)
						[NOAA reports wx forecasts at 0333, 0748 & 2133]

kHz	Mode	Call	Location	Service	kW	Remarks
	USB	....	Goteborg Radio, Sweden	FC	5.0	CH#424 (4134.3)
	USB	HEB24	Berne Radio, Switzerland	FC	10.0	" "
	USB	NMC	COMMSTA San Francisco, CA	FC	10.0	USCG/AMVER
			[NOAA reports wx broadcasts at 0430 & 1030]			CH#424 (4134.3)
	USB	NMG	COMMSTA New Orleans, LA	FC	10.0	USCG/AMVER CH#424 (4134.3)
	USB	NMN	COMMSTA Portsmouth, VA	FC	10.0	USCG/AMVER
			[NOAA reports wx broadcasts at 0400 & 0530 & 1000]			CH#424 (4134.3)
	USB	....	Caribbean & Atlantic Voice Circuits	FC/MS	....	USCG (Includes:
			NMA, Miami, FL; NMF, Boston, MA & NMN,			Portsmouth, VA)
	USB	WOM	Miami Radio, FL	FC	10.0	CH#424 (4134.3)
4431.8	USB	OXZ	Lyngby Radio, Denmark	FC	10.0	CH#425 (4137.3)
	USB	JBO	Tokyo Radio, Japan	FC	2.5	" "
	USB	WWD	La Jolla Radio (Scripps), CA	FC	10.0	" "
			[NOAA reports wx forecasts--weekdays--June			through October]
4434.9	USB	JBO	Toyko Radio, Japan	FC	10.0	CH#426 (4140.5)
4435.0	USB	....	several, on/offshore NS/Nfld., Canada	FC/MS	..	Canadian Fishing

4438.0 - 4650.0 kHz -- FIXED & MOBILE SERVICE (Worldwide)

kHz	Mode	Call	Location	Service	kW	Remarks
4441.5	USB	VL5MB	Darwin, NT, Australia	FX	0.5	Telcom
4445.0	FAX	ROF73	Novosibirsk, USSR	FX	20.0	Meteo
	CW	NPO	San Miguel (Capas), Philippines	FC	5.0	USN/TIME SIGNALS
			NOAA reports wx broadcasts at 0400 & 1300			
4450.0	USB	....	numerous, northwestern USA	FX	....	USAF MARS
4454.0	USB	3DV31	Suva Radio, Fiji	FC	0.2	
4455.0	USB	....	unknown, possibly in NWT, Canada	FX	....	Net/EE
4460.0	USB	VCA318	Chipman, NB, Canada	FX	0.1	Fishing/Telcom
	USB	XLN45	Thunder Bay, Ont., Canada	FX	0.1	Telcom
4462.0	USB	XMP347	Halifax, NS, Canada	FX	0.1	Net Control
4464.5	USB	....	numerous, USA nationwide	FX	....	CAP
4465.0	ISB	†....	Holzkirchen, GFR	FX	10.0	RFE Feeder
4467.0	USB	†....	Daventry, England	FX	30.0	BBC Feeder
4467.5	USB	....	numerous, in midwest USA	FX	....	CAP
4472.5	USB	CJN911	unknown, probably in NE Canada	FX	....	Canadian Fishing
4474.0	CW	UJO5	Izmail Radio, Ukrainian SSR	FC	1.0	Tfc/Danube
4475.0	ISB	....	Holzkirchen, GFR	FX	10.0	RFE Feeder
	CW	....	unknown	FX	....	Coded Tfc
4483.5	USB	XOP654	Fort Chimo, PQ, Canada	FX	0.1	Net Control
4484.0	CW	T2A4	Funafuti, Tuvala	FX	0.1	Net Control
4487.0	USB	....	numerous, southwestern USA	FX	....	USAF MARS
4495.0	ISB	....	several, around North America	FX	....	USAF/SAC
4497.5	FAX	CKN	Vancouver Forces Radio, BC, Canada	FX	....	C. Forces/Meteo
4500.0	AM/A2	VNG	Lyndhurst, Vic., Australia	SS	10.0	TIME SIGNALS
4505.0	ISB	....	Holzkirchen, GFR	FX	10.0	RFE Feeder
	LSB	NAM	Norfolk Naval Radio, VA	FX	....	USN
4509.0	USB	....	numerous, in midwest USA	FX	....	CAP
4512.0	CW	"KKQ"	unknown	FX	....	Coded Tfc
4513.0	USB	NZJ	El Toro NAS, CA	FX	0.5	USN/Net Control
4515.0	ISB	....	several, in Nfld., Canada	FX	....	Telcom
4516.7	FAX	RHB/RHO	Khabarovsk, USSR	FX	....	Meteo
4517.0	USB	....	numerous, in midwest USA	FX	....	USAF MARS
4525.0	CW	Y3S	Berlin (Nauen), GDR	FX	5.0	TIME SIGNALS
	CW	NMO	COMMSTA Honolulu, HI	FX	10.0	USCG/NOAA re-
			ports wx broadcasts at 0100, 0400, 0700, 1300 & 2000			
4526.0	FAX	SUU36	Cairo, Egypt	FX	....	Meteo
4529.5	ISB	5TN231	Nouakchott, Mauritania	FX	6.0	Telcom
4531.0	USB	....	Aleutian LORAN Net	FX	....	USCG (Includes:
			NMJ22, Attu; NOJ, Kodiak; NRW2, St. Paul Is. & NRW3, Pt. Clarence--all AK) [Alt: 4575.0]			
4532.0	USB	CKA214	Inuvik, NWT, Canada	FX	0.5	Net Control
4532.5	CW	LMB	Bergen Radio, Norway	FX	0.5	NOAA reports
			wx broadcasts at 1350 & 2240 (April-Sept., only)			

kHz	Mode	Call	Location	Service	kW	Remarks
4533.5	CW	FDE	Villacoublay Air, France	FX	1.0	French AF
4537.0	USB	....	unknown, on/offshore NS/Nfld., Canada	FX/MS	..	Canadian Fishing
4550.0	CW/USB	....	several, in Australia	FX	1.0	Monitoring Net-
			work (Includes: VNA3, Melbourne; VNA4, Brisbane; VNA5, Adelaide; VNA6, Perth; VNA7 Hobart & VNA21, Townsville) [Alt: 5422.0 & 6810.0]			
	USB	....	numerous (ICAO CAR)	FA	....	ARINC
4550.1	USB	....	Northwest Pacific LORAN Net	FX	....	USCG (Includes:
			NRT, "Yokota Monitor", Japan; NRT2, Gesashi, Japan, NRT3, Iwo Jima; NRT9, Hokkaido, Japan; NRV, Barrigada, Guam; NRV6, Marcus Is. & Yap Island) [Alt: 5053.6]			
4553.5	USB	†....	Delano, CA	FX	50.0	VOA Feeder
4557.0	USB	....	numerous, USA nationwide	FX	....	USAF MARS
4560.0	USB	....	several, nationwide Australia	FX	1.0	Police Net (In-
			cludes: VKA, Adelaide; VKC, Melbourne; VKR, Brisbane & VKT, Hobart, Tasmania) [Alt: 5180.0 & 7660.0]			
4560.0	FAX	YMA35	Ankara, Turkey	FX	5.0	Meteo
4562.5	CW	JWT	Stavanger Naval Radio, Norway	FX	....	Norwegian Navy
4565.0	ISB	....	Holzkirchen, GFR	FX	10.0	RFE Feeder
	USB	†....	Delano, CA	FX	50.0	VOA Feeder
4571.0	USB	CHF280	Prince Albert, Sask., Canada	FX	0.5	Net Control
4572.0	USB	....	numerous, offshore North America	FX	....	Oil Rigs
4575.0	USB	....	Aleutian LORAN NET	FX	....	USCG (For list
			of stations see 4531.0) [Alt: 4531.0 & 5226.0]			
4577.0	USB	TUP43	Abidjan, Ivory Coast	FX	6.0	Telcom
4577.5	CW	....	unknown	FX	....	Wx 1130
4580.0	USB	....	numerous, southweastern USA	FX	....	USAF/MARS
4582.0	USB	....	numerous, USA nationwide	FX	....	CAP
4585.0	USB	NMJ1	COMMSTA Juneau, AK	FA	....	USCG
	USB	....	unknown--probably in Canada	FX	....	Telcom
	USB	....	numerous, USA nationwide	FX	....	CAP
	CW/RTTY	....	several, in Pacific Ocean area	FX	....	USN
4586.0	CW	DHJ59	Sengwarden Naval Radio, GFR	FX	....	GFR Navy/NATO
	CW	PAA21	The Hague Naval Radio, Netherlands	FX	....	Dutch Navy/NATO
4588.0	CW	FUG	La Regine (Castelnaudry), France	FX	....	French Navy
4590.0	USB	....	several, USA nationwide	FX	....	USAF
4593.5	USB	....	numerous, northweastern USA	FX	....	USAF MARS
4597.2	USB	†TUP45	Abidjan, Ivory Coast	FX	2.0	Telcom
4601.0	USB	ZLBC5	Campbell Island	FX	....	Meteo/Wellington
4601.4	USB	VCA294	unknown, probably NS, Canada	FX	....	Net Control
	USB	VCA737	Battle Harbor, Nfld., Canada	FX	0.1	Net Control
	USB	VCA744	Cape St. Francis, Nfld., Canada	FX	0.1	Net Control
	USB	VCC42	Dartmouth, NS, Canada	FX	0.3	Net Control
4601.5	USB	....	several, throughout France	FX	0.1	Net (Includes:
			stations in Bordeaux, Limoges, Pau, Limoges, Poitiers & Toulouse)			
4602.5	USB	....	numerous, USA nationwide	FX	....	CAP
4604.0	USB	†....	numerous, USA nationwide	FX	....	FEMA
4610.0	FAX	GFA22	Bracknell, England	FX	0.5	Meteo
					(1800-2400)	
4610.5	USB	....	numerous, offshore North America	FX	....	Oil Rigs
4613.5	USB	....	numerous, offshore North America	FX	....	Oil Rigs
4615.6	CW	IDR2	Rome Naval Radio, Italy	FX	1.0	Italian Navy
4616.0	USB	CGZ	St. Johns, Nfld., Canada	FX	1.0	Net Control
4617.5	CW	FDE4	Bourges Air, France	FX	....	French AF
4620.0	USB	VII	Thursday Island, Qld., Australia	FX	1.0	Telcom
4622.0	USB	....	unknown, probably in PQ, Canada	FX	....	Telcom/FF
4623.0	CW	NGR	Kato Souli Naval Radio, Greece	FX	3.5	USN
			NOAA reports wx broadcasts at 0030 & 1230			
4625.0	AM	VEB2	unknown, probably in Canada	FX	....	TIME SIGNALS
4626.1	CW	†KRH5Ø	London, England	FX	....	USA EMBASSY
4626.5	USB	....	numerous, USA nationwide	FX	....	CAP
4627.0	USB	VGF91	Mount Bertha, BC, Canada	FX	0.1	Net Control
4628.5	USB	....	unknown, possibly in Canada	FX	....	Telcom

kHz	Mode	Call	Location	Service	kW	Remarks
4629.8	USB	....	numerous, USA nationwide	FX	....	CAP
4630.0	CW/RTTY	SOE263	Warsaw, Poland	FX	10.0	Tfc
4631.6	USB	....	unknown, probably in PQ, Canada	FX	....	Telcom/FF
4632.5	CW	†5BP3	Nicosia, Cyprus	FX	0.5	INTERPOL
	CW	FEB	Paris (St. Martin Abbat), France	FX	3.0	"
	CW	DEB	Weisbaden, GFR	FX	2.0	"
	CW	LJP24	Oslo, Norway	FX	....	"
	CW	†SHX	Stockholm, Sweden	FX	1.5	"
	CW	HEP46	Zurich (Waltikon), Switzerland	FX	2.0	"
4634.5	USB	....	numerous, offshore North America	FX	....	Oil Rigs
4635.0	USB	VJC	Broken Hill, NSW, Australia	FX	0.3	Outpost Control
	CW	FSS267	Strasbourg, France	FX	....	CRS
4637.5	CW	FDC2Ø	Toul Air, France	FX	....	French AF
	USB	....	numerous, offshore North America	FX	....	Oil Rigs
4638.0	USB	....	numerous, western USA	FX	....	Commercial Tfc
4640.0	AM	....	unknown	FX	....	NUMBERS/SS/OM
4642.5	FAX	LMO34	Olso, Norway	FX	0.5	Meteo
4645.0	CW	FUF	Fort de France Naval R., Martinique	FX	2.0	French Navy
	CW	FUV	Djibouti Naval Radio, Djibouti	FX	2.0	French Navy

4650.0 - 4750.0 kHz — AERONAUTICAL MOBILE (Worldwide)

kHz	Mode	Call	Location	Service	kW	Remarks
4654.0	USB	....	unknown, in North America	FA	....	Telcom/EE [Alt: 8070.0]
4663.0	A3H	....	Moscow (Vnukovo) Aeradio, USSR	FA	....	VOLMET/EE (25+55)
	A3H	....	Novosibirsk (Tolmachevo) Aeradio, USSR	FA	....	VOLMET/EE (10+40)
	A3H	....	Tashkent Aeradio, Uzbek SSR	FA	....	VOLMET/EE (05+35)
4668.0	A3H	KAB8	Anchorage Aeradio, AK	FA	0.1	
	USB	WDB2	Nome Aeradio, AK	FA	1.0	
	CW/USB	†KJY74	Miami "Monitor", FL	FA	....	NOAA
4670.0	AM	....	unknown	FX	....	LETTERS/EE/SS/YL
4684.0	A3H	....	Novosibirsk (Tolmachevo) Aeradio, USSR	FA	....	VOLMET/RR
4689.0	AM/USB	....	numerous (ICAO EUR)	FA/MA	....	Airports & in-flight covering all of Europe, Middle East & Mediterrean
4696.0	AM/USB	....	numerous (ICAO SAM)	FA/MA	....	Airports & in-flight covering west coast of South America
	USB	VIS54	Sydney Aeradio, NSW, Australia	FA/MA	1.5	
	USB	WDB2	Nome Aeradio, AK	FA	1.0	
4700.0	USB	....	numerous, worldwide	FA/MA	....	USN
4704.0	CW/USB	†....	several, throughout Canada	FA/MA	1.0	RCAF (Usually CHR, "Trenton Military", Ont.; CJX, "St, Johns Military, Nfld.; DHM94, "Lahr Military", GFR & VXA, "Edmonton Military", Alta., Canada)
4704.6	FAX	....	unknown, apparently in North America	FA	....	
4705.0	CW	....	unknown, apparently in North America	FA	....	
4707.0	USB	....	reported Bampton, England	FA	....	RAF
4710.0	USB	....	reported Findhorn, Scotland	FA	....	RAF
4716.0	FAX	....	unknown, in the USA	FC	....	USN
4720.0	AM	†....	Baku, Azerbaijan SSR	FA	....	VOLMET/RR
4722.0	USB	MVU	West Drayton (Upavon) AB, England	FA	5.0	RAF (Frequent wx broadcasts)
4723.0	CW	FDEU	unknown, probably in Caribbean	FA	....	French AF
4725.0	CW	IDR	Rome Naval Radio, Italy	FA	....	Italian Navy
	USB	....	numerous, worldwide	FA/MA	....	USAF/SAC (In-frequently used--sometimes called "VICTOR" frequency)
4732.0	USB	....	numerous, worldwide	FA/MA	....	USAF
4737.0	USB	....	numerous, worldwide	FA/MA	....	USN
4739.0	USB	....	reported Bampton, England	FA	....	RAF
	CW	IBA	Naples Naval Radio, Italy	FA	....	Italian Navy/NATO
	CW	YMB	Izmir Naval Radio, Turkey	FA	....	Turkist Navy/NATO
4746.0	USB	CUW	Lajes Field, Azores	FA/MA	....	USAF
	USB	AFE8	MacDill AFB, FL	FA/MA	....	USAF
	USB	AFI2	McClellan AFB, CA	FA/MA	....	USAF (May be used for SKY KING broadcasts)

kHz	Mode	Call	Location	Service	kW	Remarks

4750.0 - 4850.0 kHz — FIXED & TROPICAL BROADCASTING—(Shared)

kHz	Mode	Call	Location	Service	kW	Remarks
4752.0	USB	....	Halifax, NS, Canada	FX/MA	....	Oil Rigs/Helio. [Alt: 5744.2]
4753.0	CW	FDE	Villacoublay Air, France	FX	1.0	French AF
	USB	HZY473	Dhahran, Saudi Arabia	FX	1.0	Control/Oil Rigs
4755.0	USB	....	Halifax, NS, Canada	FX	....	Net control
4767.0	USB	AJF7	Rhein Main AB, GFR	FX	1.0	USAF
4762.0	CW	†IDR2	Rome Naval Radio, Italy	FX	1.0	Italian Navy
4770.0	AM	....	unknown	FX	....	NUMBERS/GG/YL
4777.5	FAX	IMB31	Rome, Italy	FX	....	Meteo (cont.)
4780.0	USB	VKC	Melbourne, Vic., Australia	FX	0.5	Police
	USB	†....	numerous, USA nationwide	FX	....	FEMA
4782.0	FAX	GFE21	Bracknell, England	FX	5.0	Meteo (cont.)
4789.0	USB	CON3Ø1	Havana, Cuba	FX	0.1	Net Control
4792.0	AM	....	unknown	FX	....	NUMBERS/SS/YL
4793.0	FAX	AFA	Washington (Andrews AFB, MD), DC	FX	3.0	Meteo
4802.5	FAX	NPM	Lualualei Naval Radio, HI	FX	15.0	USN/Meteo
4810.5	CW	....	unknown	FX	....	Meteo
4813.5	USB	....	Agana (Barrigada NAS), Guam	FX	0.1	USCG/USN
			[Emergency joint communications--"SIERRA" channel]			
4825.0	USB	C6Q51	Nassau, Bahamas	FX	0.2	Net Control
4826.0	USB	VCA77	St. Anthony, Nfld., Canada	FX	0.1	Net Control
	USB	....	several, offshore North America	FX	....	Oil Rigs
4832.0	USB	ZKS	Rarotonga, Cook Islands	FX	1.0	Net Control
4837.5	CW/RTTY	OEQ35	Vienna, Austria	FX	1.0	INTERPOL
	CW/RTTY	DEB	Wiesbaden, GFR	FX	5.0	"
	CW/RTTY	SHX	Stockholm, Sweden	FX	1.0	"
4839.1	CW	CSF36	Punta Delgada, Azores	FX	1.0	Port. AF
	CW	CSF37	Funchal, Madeira Islands	FX	3.0	" "
	CW	CSF46	Horta, Azores	FX	0.5	" "
	CW	NAM	Norfolk Naval Radio, VA	FX	....	(LCMP3) USN/NUKO
4840.0	CW/RTTY	OMZ29	Prague, Czechoslovakia	FX	1.0	GDR EMBASSY
	CW/RTTY	Y7A24	Berlin, GDR	FX	....	EMBASSY

4850.0 - 4995.0 kHz — FIXED, MOBILE & TROPICAL BROADCASTING—(Shared)

kHz	Mode	Call	Location	Service	kW	Remarks
4850.0	USB	....	several, southeastern USA	FX/MS	....	USACE/USCG Liaiso
4856.0	CW	LOL	Buenos Aires, Argentina	FX	0.5	Prefectura Naval/ TIME SIGNALS
4857.5	USB	†....	Norwegian Sea & N. Atlantic LORAN Net	FX	....	USCG (Includes:
			DML, Sylt, GFR; JXL, BØ, Norway; JXP, Jan Mayen Is.; NMS Shetland Is.; OUN, Ejde, Denmark; OVY, Angissog, Greenland; TFR, Sandur, Iceland & TFR2, Reykjavik, Iceland) [Alt: 7512.5]			
4860.0	ISB	†JBA44	Tokyo, Japan	FX	10.0	Telcom
4864.5	CW	FDC9	Nancy Air, France	FX	0.5	French AF
4867.5	CW	FDG	Bordeaux Air, France	FX	0.5	French AF
4870.0	USB	....	several, in England	FX	0.5	RAF
4880.0	CW	KKN5Ø	Washington, DC	FX	....	EMBASSY
4882.0	USB	....	unknown, on/offshore NS/Nfld., Canada	FX/MS	..	Canadian Fishing
	CW	FDE4	Bourges Air, France	FX	....	French AF
4883.0	USB	XNZ583	St. Johns, Nfld., Canada	FX	0.1	Net Control
			(Includes: XNZ585, Bernard River, Nfld. & XNZ584, Tote River, Nfld.; etc.)			
	USB	....	Northwest Pacific LORAN Net	FX	....	USCG (Includes:
			NRT "Yokota Monitor", Japan; NRT3, Iwo Jima; NRV, Barrigada, Guam; NRV6, Marcus Is. & NRV7, Yap Island) [Alt: 5300.0 & 7580.0]			
4885.7	CW	N2S	unknown, probably within USA	FX	....	Coded Tfc
				[May ID as C9Ø or Z8C]		
4886.0	CW	FDC	Metz Air, France	FX	....	French AF
	CW	KKN44	Monrovia, Liberia	FX	....	USA EMBASSY

kHz	Mode	Call	Location	Service	kW	Remarks
4888.4	USB	XJP26	unknown, on/shore NS/Nfld., Canada	FX/MS	....	Canadian Fishing
4893.0	CW/RTTY	RLX	Dublin, Ireland	FX	0.5	Russian EMBASSY [Alt: 5181.0]
4897.0	CW	FAD58	Tours, France	FX	....	French Army
4900.0	CW/RTTY	†OMZ	Prague, Czechoslovakia	FX	1.0	EMBASSY
4902.0	FAX	JMB4Ø	Tokyo (Fusa), Japan	FX	0.5	Meteo
4907.5	CW	....	several, throughout France	FX	0.5	French AF
4910.0	CW	KWS78	Athens Greece	FX	....	USA EMBASSY
4918.5	CW	FDI	Aix les Milles Air, France	FX	....	French AF
4940.0	CW	FDC	Metz Air, France	FX	....	French AF
	USB	ZKAK	Auckland Aeradio, New Zealand	AX	5.0	Tfc/Rarotonga
	USB	ZKRG	Rarotonga Aeradio, Cook Islands	AX	0.1	Tfc/Auckland
	USB	....	Norfolk Island	AX	0.1	Tfc/Auckland
4942.0	CW	FDY	Orleans Air, France	FX	0.5	French AF
4955.0	CW	NPN	Barrigada (Agana NAS), Guam	FA	5.0	USN
4963.0	USB	VCB513	Halifax, NS, Canada	FX	0.1	Telcom
	USB	VCB9Ø1	Fredericton, NB, Canada	FX	0.1	Telcom
4974.0	CW	....	several, throughout France	FX	0.4	GNF
4975.0	FAX	NPN	Barrigada (Agana NAS), Guam	FC	15.0	USN/Meteo
	FAX	NAM	Norfolk Naval Radio, VA	FC	40.0	USN/Meteo
4982.0	USB	XMH356	Dartmouth, NS, Canada	FX/MS	1.0	Canadian Fishing
4992.0	USB	....	several, on/offshore NS/Nfld., Canada	FX/MS	..	Canadian Fishing
	USB	AFE71	"Cape Radio", Cocoa, FL	FX	3.0	USAF/NASA Shuttle Support [Alt: 5870.0]

4995.0 - 5003.0 kHz — STANDARD FREQUENCY & TIME SIGNALS

kHz	Mode	Call	Location	Service	kW	Remarks
4996.0	CW	RWN	Moscow, USSR	SS	5.0	TIME SIGNALS
5000.0	A2	LOL	Buenos Aires, Argentina	SS	2.0	" "
	A2	PPE	Rio de Janeiro, Brazil	SS	1.0	" "
	AM/CW	BPM	Xian, PRC	SS	15.0	" "
	A2	HD21ØA	Guayaquil, Ecuador	SS	....	" "
	A2	MSF	Teddington, England	SS	0.5	" "
	A2	WWVH	Kauai, HI	SS	10.0	" "
	A2	ATA	New Delhi, India	SS	8.0	" "
	CW	IAM	Rome, Italy	SS	1.0	" "
	A2	IBF	Turin, Italy	SS	5.0	" "
	A2	JJY	Tokyo (Sanwa, Ibaraki), Japan	SS	2.0	" "
	A2	ZUO	Pretoria, RSA	SS	4.0	" "
	A2	BSF	Chung-Li, Taiwan	SS	5.0	" "
	CW	RZH	Tashkent, USSR	SS	1.0	" "
	AM/A2	WWV	Ft. Collins, CO	SS	10.0	" "

5003.0 - 5005.0 kHz — STANDARD FREQUENCY/TIME/SPACE RESEARCH

kHz	Mode	Call	Location	Service	kW	Remarks
5004.0	A2	....	Irkutsk, USSR	SS	1.0	TIME SIGNALS

5005.0 - 5060.0 kHz — FIXED & TROPICAL BROADCASTING (Shared)

kHz	Mode	Call	Location	Service	kW	Remarks
5010.0	CW	FDY	Orleans Air, France	FX	0.5	French AF
	CW	FDY4	unknown	FX	....	"
	CW	FDY5	Aix les Milles Air, France	FX	0.5	"
	CW	OVC	Groennedal, Greenland	FX	1.2	Danish Marine
5013.5	FAX	....	unknown, probably in NA	FX	....	
5014.0	USB	GZU	Portsmouth (Petersfield), England	FX	1.0	British Naval Air
5015.0	LSB	WUB4	Baltimore, MD	FX	....	USACE Net Control (Callup Monday/1300) [Alt: 5400.0]
5016.2	CW	†CTV25	Monsanto Naval Radio, Portugal	FX	3.0	Port. Navy
5019.5	USB	....	numerous, throughout the GFR	FX	....	Police
5020.0	AM	....	unknown	FX	....	NUMBERS/SS/YL
5021.5	CW	FDE8	Doullens Air, France	FX	0.5	French AF
5023.5	AM	....	unknown	FX	....	carrier only
5026.0	USB	AFS	Offutt AFB, NE	FX	0.5	USAF/SAC [Alt: 6826.0]
5026.5	CW	FDE4	Bourges Air, France	FX	0.5	French AF
5031.5	USB	XLI334	Frobisher Bay, NWT, Canada	FX	0.1	Net Control

kHz	Mode	Call	Location	Service	kW	Remarks
5033.0	CW	....	unknown, probably in the USA	FX	....	Coded Tfc
5035.0	LSB	FJD251	Papeete, Tahiti	FX	0.1	Meteo/Control
5036.0	CW	FDC21	Drackenbronn Air, France	FX	....	French AF
5037.5	FAX	KVM7Ø	Honolulu (International), HI	FX	10.0	Meteo (2315-0106)
5055.5	USB	C6L22	Nassau, Bahamas	FX	0.2	Net Control

5060.0 - 5450.0 kHz -- FIXED/LAND MOBILE SERVICE (Worldwide)

kHz	Mode	Call	Location	Service	kW	Remarks
5060.0	USB	VCA388	Lower Lahave, NS, Canada	FX/MS	....	Canadian Fishing
	USB	CJW31	unknown, probably in NS, Canada	FX	....	
	CW	†D4B62	Sal, Cape Verde	AX	1.5	Aero Tfc
5063.0	USB	....	Hawaii & Central Pacific LORAN Net	FX	....	USCG (Includes:
			NMO, Lualualei, HI; NRO, Johnston Is.; NRO5, Upolu Point, HI & NRO7, Kure Island [Ocean Is.], HI) [Alt: 7473.0]			
5063.6	USB	....	Northwest Pacific LORAN Net	FX	....	USCG (Includes:
			NRT, "Yokota Monitor", Japan; NRT2, Gesashi, Japan; NRT3, Iwo Jima; NRT9, Hokkaido, Japan; NRV, Barrigada, Guam; NRV6, Marcus Is.; NRV7, Yap Island) [Alt: 5315.5]			
5066.0	USB	....	Mediterranean LORAN Net	FX	....	USCG (Includes:
			AOB5Ø, Estartit, Spain; NCI, Sellia Marina, Italy; NCI3, Lampedusa, Italy, NCI4, Laragabarun Turkey & NCI1Ø, Rhodes) [Alt: 5123.0]			
5066.5	USB	....	Papeete Aeradio, Tahiti	AX	0.2	Net Control
5070.0	CW/ISB	CSF461	Angra do Heroismo, Azores	FX	0.5	Port. AF
	CW/ISB	CSF462	Horta, Azores	FX	0.5	" "
	CW/ISB	CSF37	Funchal, Madeira Islands	FX	0.5	" "
5071.0	USB	NAM	Norfolk Naval Radio, VA	FC	3.0	USN
5072.5	ISB	9VF6	Singapore, Singapore	FX	8.0	Tfc/Hong Kong
5073.5	USB	....	several, around Persian Gulf	AX	....	Aero Tfc
5075.0	USB	XOU5Ø1	Eldorado, Sask., Canada	FX	0.2	Net Control
	AM	....	unknown	FX	....	NUMBERS/YL/SS
5080.0	USB	....	Channel Island Harbor Station, Navy Point Mugu, CA	FX	....	USCG/USN ["Plead Control"]
5081.0	ISB	5UR8Ø	Niamey, Niger	FX	2.0	Telcom
5083.5	USB	C6Y42	Nassau, Bahamas	FX	0.2	Net Control
5085.0	CW	ZVN3	Belem Aeradio, Brazil	AX	3.0	Aero Tfc
5090.0	FAX	....	Tashkent, Uzbek SSR	FX	....	Meteo
5093.0	FAX	LZA8	Sofia, Bulgaria	FX	....	Meteo (0430-1110)
5097.0	CW	CFH	Maritime Command Radio, Halifax, NS, Canada	FX	10.0	NAWS (Candian Forces)
5100.0	FAX	AXM32	Canberra, ACT, Australia	FX	10.0	Meteo
5110.0	CW	†KKN44	Monrovia, Liberia	FX	....	USA EMBASSY
	FX	RYP29	Khabarovsk, USSR	FX	15.0	Meteo
5112.0	CW/RTTY	4OC3	Belgrade (Makis), Yugoslavia	FX	30.0	Tfc/Moscow
5113.5	USB	†MKT	London (Stanbridge), England	FX	3.0	RAF Telcom/Tfc
5123.0	USB	....	Mediterranean LORAN Net	FX	....	USCG (Includes:
			NCI4, Karagabun, Turkey & NCI1Ø, Rhodes) [Alt: 6943.5]			
5123.5	CW	FDY	Orleans Air, France	FX	0.5	French AF
5125.0	CW	P6Z	Paris, France	FX	2.0	EMBASSY
	ISB	....	Holzkirchen, GFR	FX	10.0	RFE Feeder
5127.0	FAX	†5YE4	Nairobi, Kenya	FX	10.0	Meteo
5145.0	USB	VJN	Cairns, Qld., Australia	FX	0.3	Outpost Control
5150.3	FAX	RVO73	Moscow, USSR	FX	20.0	Meteo
5160.0	CW	FDI	Aix les Milles Air, France	FX	....	French AF
5167.0	CW	NRK	Keflavik Naval Radio, Iceland	FX	2.5	USN/NOAA reports
			wx broadcasts at 0030, 0630, 1230 & 1900/Ice at 0500 & 1700			
	CW	NAM	Norfolk Naval Radio, VA	FX	....	USN
5167.5	USB	KPI29	unknown	FX	....	
5172.5	FAX	†SMA5	Stockholm (Norrkoping), Sweden	FX	2.5	Meteo
5175.0	LSB	ZUD214	Olifantsfontein, RSA	FX	45.0	Telcom
5180.0	USB	VKA	Adelaide, SA, Australia	FX	....	Police
5181.0	CW/RTTY	RLX	Dublin, Ireland	FX	0.5	Russian EMBASSY [Alt: 4893.0 & 5744.0]

FIXED/LAND MOBILE SERVICE (con.)

kHz	Mode	Call	Location	Service	kW	Remarks
5185.0	FAX	LRO	Buenos Aires (G. Pacheco), Argentina	FX	....	Meteo
			(0300, 1600, 1900 & 2000)			
5187.0	CW	FLR26	Metz, France	FX	....	French Army
5192.0	CW	CSO	unknown	FX	....	VVV's
5204.0	CW	†OMZ	Prague, Czechoslovakia	FX	1.0	EMBASSY
5205.0	USB	KUP65	Dalap, Majuro Atoll	FX	1.0	Used on request
	USB	KUP66	Param, Ponape Island	FX	1.0	" " "
	USB	KUP67	Moen, Truk Islands	FX	1.0	" " "
	USB	KUP68	Korak, Palau Islands	FX	1.0	" " "
	USB	KUP69	Nif, Yap Islands	FX	1.0	" " "
	USB	KUP71	Kobler, Saipan, Mariana Islands	FX	1.0	" " "
	USB	KUP72	Shinaparnu, Rota Is., Mariana Islands	FX	1.0	" " "
5206.0	FAX	NGR	Kato Souli Naval Radio, Greece	FX	20.0	USN/Meteo [2000-0800]
5208.0	CW/RTTY	OEQ36	Vienna, Austria	FX	1.0	INTERPOL
	CW	FSB	Paris (St. Martin Abbat), France	FX	3.0	"
	CW	HEP52	Zurich (Waltikon), Switzerland	FX	3.0	"
5211.0	USB	†....	numerous, USA nationwide	FX	....	FEMA
5215.0	USB	CFY82	Whitehorse, YT, Canada	FX	0.4	Net Control
	USB	4Q029	Colombo (Kotugoda), Sri Lanka	FX	10.0	Telcom
5216.5	CW	WAR	Washington, DC	FX	....	US Army
5217.0	USB	C6Q89	Nassau, Bahamas	FX	0.1	Net Control
5226.0	USB	....	Aleutian LORAN Net	FX	....	USCG (Includes:
			NMJ22, Attu; NOJ, Kodiak; NRW2, St. Paul Is. & NRW3, Pt. Clarence--all AK) [Alt: 5422.0]			
	CW	JXU	BØ, Norway	FX	....	Norwegian Army
	CW/RTTY	YZG6	Belgrade (Makis), Yugoslavia	FX	30.0	Tfc/Moscow
5230.2	CW	....	unknown	FX	....	Nx/EE
5240.0	CW	FDC22	Contrexeville Air, France	FX	0.5	French AF
5250.0	CW	†OMZ	Prague, Czechoslovakia	FX	1.0	EMBASSY
5254.0	USB	ZLC7	Chatham Island	FX	0.4	Telcom/Wellington
5258.0	CW	CMU967	Santiago Naval Radio, Cuba	FX	....	Russian Navy
5263.0	CW	CNO	Casablanca, Morocco	FX	0.5	NOAA reports
			wx broadcasts at 0835 & 1935			
	CW	XVS	Ho Chi Minh-Ville, Vietnam	FX	....	NOAA reports
			wx broadcasts at 0420 & 1020			
5265.0	CW	5VA	Lome, Togo	FX	0.2	NOAA reports
			wx broadcast at 0835			
5271.0	CW	†KWS78	Athens, Greece	FX	....	USA EMBASSY
5278.5	FAX	DFE28	Bonames, GFR	FX	20.0	
5281.5	USB	VCC43	Dartmouth, NS, Canada	FX/MS	0.3	Commercial Tfc
	USB	CJR68Ø	Sable Island, NS, Canada	FX/MS	0.5	" "
	USB	....	numerous, offshore NS/Nfld., Canada	FX	....	Oil Rigs
5284.5	FAX	....	Tashkent, Uzbek SSR	FX	....	Meteo
5287.5	USB	....	Pago Pago, American Samoa	FX	....	USCG Liaison
5290.0	ISB	†REN35	Moscow, USSR	FX	20.0	R. Moscow Feeder
5295.0	ISB	....	Holzkirchen, GFR	FX	10.0	RFE Feeder
5297.6	USB	†FPF29	Paris (Le Vernet), France	FX	20.0	Telcom
5300.0	USB	VJN	Cairns, Qld., Australia	FX	0.3	Outpost Control
	CW	....	unknown	FX	....	Net
5303.5	CW	NKS	unknown	FX	....	Net Control
5305.5	CW/RTTY	FSB	Paris (St. Martin Abbat), France	FX	3.0	INTERPOL
	CW/RTTY	OEQ37	Vienna, Austria	FX	1.0	"
5305.6	CW	"D"	unknown	RC	....	Beacon
5307.0	CW	"F"	unknown	RC	....	Beacon
5308.7	CW	"L"	unknown	RC	....	Beacon
5313.0	USB	....	†North Atlantic (East Coast) LORAN Net	FX	....	USCG (Includes:
			NMA7, Jupiter, FL; NMP32, Nantucket, MA; NMF33, Caribou, ME; NMN73, Carolina Beach [Cape Fear], NC & NOL, "Bermuda Monitor") [Alt: 4048.5]			
5315.0	USB	†3XK23	Conakry, Guinea	FX	2.0	Telcom
5315.5	USB	....	Northwest Pacific LORAN Net	FX	....	USCG (Includes:
			NRT, "Yokota Montior", Japan; NRT2, Gesashi, Japan, NRT3, Iwo Jima; NRT9, Hokkaido, Japan; NRV, Barrigada, Guam: NRV6, Marcus Is.; NRV7, Yap Is.) [Alt: 5063.6 & 7836.6]			

kHz	Mode	Call	Location	Service	kW	Remarks
5317.0	USB	....	Aleutian LORAN Net	FX	....	USCG(Includes: NMJ22, Attu; NOJ, Kodiak; NRW2, St. Paul Is. & NRW3, Pt. Clarence--all AK) [Alt: 5226.0 & 5422.0]
5320.0	USB	....	Northwest Pacific LORAN Net	FX	....	USCG (Includes: NRT, "Yokota Monitor", Japan; NRT3, Iwo Jima; NRV, Barrigada, Guam; NRV6, Marcus Is. & NRV7, Yap Is.) [Alt: 7580.0]
	CW	NIK	COMMSTA Boston, MA	FX	....	USCG/NOAA reports seasonal ice broadcast at 0018
	USB	....	numerous, USA nationwide	FX	....	USCG Operations
5327.0	LSB	WIC	Tulsa, OK	FX	....	USACE
5335.0	FAX	ROF76	Novosibirsk, USSR	FX	20.0	Meteo
5350.0	USB	....	several, on/offshore NS/Nfld., Canada	FC/MS	..	Canadian Fishing
	CW	FDC9	Nancy Air, France	FX	....	French AF
5355.0	FAX	RND77	Moscow, USSR	FX	15.0	Meteo (cont.)
5357.0	ISB	CUA2Ø	Lisbon (Alfragide), Portugal	FX	5.0	Telcom
5358.0	CW	CCS	Santiago Naval Radio, Chile	FX	....	Chilean Navy
5360.0	USB	....	unknown, possibly Venezuela	FX	....	Net/SS
5362.0	CW	OFB35	Helsinki, Finland	FX	5.0	NOAA reports seasonal ice warnings at 2118
5365.0	USB	....	Halifax, NS, Canada	FX	....	Commercial Tfc
5375.0	USB	....	several, in North Sea	FX	....	Oil Rigs
5376.5	USB	....	several, around Persian Gulf	AX	....	Aero
5381.0	CW	EA3V	unknown	FX	....	VVV's
5382.0	CW	NAM	Norfolk Naval Radio, VA	FX	1.0	USN
5390.0	USB	CGD2Ø6	Alma, PQ, Canada	FX	1.0	Net Control
	LSB	....	Riyadh, Saudi Arabia	FX	....	Feeder
5392.0	CW	LBJ5	Harstad Naval Radio, Norway	FX	....	Norwegian Navy
5395.0	USB	....	unknown, numerous, possibly Mexico	FX	....	Net/SS
5398.5	USB	†5TN58	Nouakchott, Mauritania	FX	6.0	Telcom
5400.0	CW	GZU	Portsmouth (Petersfield), England	FX	....	British Naval Air
	CW	MTO	Rosyth Naval Radio, England	FX	....	" Navy
	USB	WUD	Omaha, NE	FX	....	USACE (Net Control--many stations authorized for this channel. Recently heard have included: WUD2, Buffalo, NY; WUD6, St. Paul, MN & WUD7, Rock Island, IL)
5405.0	FAX	JMJ2	Tokyo (Usui), Japan	FX	5.0	Meteo
5411.0	USB	....	numerous, east coast of USA	FC	....	USN
5415.0	CW	†FNO32	Paris (Orly), France	AX	10.0	Meteo
5418.9	CW	VMH	unknown	FX	....	VVV's
5420.5	FAX	....	unknown	FX	....	
5422.0	USB	....	Aleutian LORAN Net	FX	....	USCG (Includes: NMJ22, Attu, NOJ, Kodiak; NRW2, St. Paul Is. & NRW3, Pt. Clarence--all AK) [Alt: 5317.0 & 6812.0]
5422.5	USB	....	West Indies LORAN Net	FX	....	USCG (Includes: NMA, Miami, FL; NMA4, San Salvador [Watling Is.], Bahamas; NMA5, South Caicos, Bahamas; NMA7, Jupiter, FL & NMR, San Juan, PR) [Alt: 4048.5 & 7530.0]
	USB	....	numerous, USA nationwide	FA/FC/FX	..	USCG Operations
5425.0	USB	....	numerous, in Cook Islands	FX	....	Tfc Net
5426.0	CW	KRH5Ø	London, England	FX	....	USA EMBASSY
5426.8	CW	LBA2	Stavanger Naval Radio, Norway	FX	....	Norwegian Navy
5430.0	CW	BPM	Xian, PRC	FX	....	TIME SIGNALS
5435.0	USB	....	several, offshore N & S America	FX	....	Oil Rigs
5446.0	CW	....	Ankara, Turkey	FX	1.0	Belgium EMBASSY

5450.0 - 5480.0 kHz -- FIXED/LAND/AERONAUTICAL MOBILE (Worldwide)

kHz	Mode	Call	Location	Service	kW	Remarks
5455.0	CW	FUJ	Noumea Naval Radio, New Caledonia	FX	2.0	French Navy
	CW	FUM	Papeete Naval Radio, Tahiti	FX	2.0	French Navy
	ISB	†RBK74	Moscow, USSR	FX	20.0	R. Moscow Feeder
5456.0	USB	....	several, in England	FA/MA	....	British Navy Air
5457.0	FAX	JKC	Tokyo (Kemigawa), Japan	FX	5.0	Meteo
5461.5	CW	FDI21	Nimes Air, France	FX	....	French AF
5462.0	CW	FDI22	Narbonne Air, France	FX	0.5	French AF
5462.5	ISB	FTF46	Paris (St. Assise), France	FX	20.0	Telcom/Tfc

kHz	Mode	Call	Location	Service	kW	Remarks
5470.0	LSB	†RKU29	Moscow, USSR	FX	80.0	R. Moscow Feeder
5471.0	USB	....	reported Findhorn, Scotland	FX	....	RAF

5480.0 - 5730.0 kHz -- AERONAUTICAL MOBILE (Worldwide)

kHz	Mode	Call	Location	Service	kW	Remarks
5484.0	AM/A3H	....	numerous (ICAO AFI & CAR)	FA/MA	....	Airports & in-
			flight covering eastern Africa, Arabian Sea, western Indian Ocean, Middle East and the eastern Caribbean flights to/from New York City			
5490.0	USB	....	numerous, throughout Caribbean	FA	....	Tfc Net
5500.0	USB	....	numerous, USA nationwide	FX	....	CAP
5505.0	AM/A3H	....	numerous (ICAO AFI & CWP)	FA/MA	....	Airports & in-
			flight covering most of Africa, Arabian Sea and western Pacific--Honolulu to Ulan Bator			
			§AFI moves to 5634.0 in 1983			
	AM	†....	Brazzaville Aeradio, Congo	FA	....	VOLMET/FF
5519.0	AM/A3H	....	numerous (ICAO AFI)	FA/MA	....	Airports & in-
			flight covering Mediterranean Sea to South Africa			
	A3H	KIS7Ø	Anchorage Aeradio, AK	FA	....	VOLMET (25+55)
	AM	....	Kai Tak Aeradio (Hong Kong)	FA	1.5	TIME SIGNALS & VOLMET (15+45)
	A3H	KVM7Ø	Honolulu (International), HI	FA	1.6	VOLMET (00+30)
	A3H	JMA	Narita Aeradio, Japan	FA	1.5	" (10+40)
	USB	....	Auckland Aeradio, New Zealand	FA	5.0	" (20+50)
	A3H	KSF7Ø	Oakland Aeradio, CA	FA	1.6	" (05+35)
5525.0	FAX	BAF6	Beijing Aeradio, PRC	FA	....	Meteo
5530.0	CW/AM	....	Kargabarun LORAN Station, Turkey	FX	....	USCG
5533.0	AM	EIP	Shannon Aeradio (Rineanna), Ireland	FA	1.5	VOLMET (cont.)
5547.0	A3H	KWO3	Anchorage Aeradio, AK	FA	1.0	
5554.0	AM/A3H	....	numerous (ICAO CEP & MID)	FA/MA	....	Airports & in-
			flight covering eastern Pacific, western USA and Middle East			
5561.0	AM	....	Baghdad Aeradio, Iraq	FA	....	VOLMET (00+30)
	AM	....	Bahrain Aeradio	FA	1.0	" (10+40)
	AM	....	Cairo Aeradio, Egypt	FA	1.0	" (20+50)
	AM	....	Beirut Aeradio, Lebanon	FA	....	" (15+45)
	AM	....	Istanbul Aeradio, Turkey	FA	....	" (25+55)
	AM	....	Teheran Aeradio, Iran	FA	....	" (05+35)
5568.0	AM/A3H	....	numerous (ICAO CAR)	FA/MA	....	Airports & in-
			flight covering Caribbean & Central America			
5575.0	AM	OKL	Prague Aeradio, Czechoslovakia	FA	1.5	VOLMET (15+45)
	AM	4XL	Ben Gurion Aeradio, Israel	FA	1.0	VOLMET (05+35)
5582.0	AM/A3H	....	numerous (ICAO SAM)	FA/MA	....	Airports & in-
			flight covering virtually all of South America			
5589.0	A3H	C7P	Ocean Station Vessel 50°N & 145°W	FA	0.3	(aka "PAPA")
	AM/A3H	....	numerous (ICAO NP)	FA/MA	....	Airports & in-
			flight covering northern Pacific Ocean & Arctic regions			
	AM/A3H	....	INTERNATIONAL AIR-TO-GROUND FREQUENCY			
5603.0	AM/A3H	....	numerous (ICAO CEP & MID)	FA/MA	....	Airports & in-
			flight covering Honolulu to San Francisco and all of the Middle East			
5610.0	AM/A3H	....	numerous (ICAO NAT)	FA/MA	....	Airports & in-
			flight covering all of the North Atlantic Ocean (may be referred to as the "ALFA" net)			
5624.0	AM/A3H	....	numerous (ICAO NAT & SEA)	FA/MA	....	Airports & in-
			flight covering all of the Northern Atlantic and Arctic regions and all of Southeast Asia			
	AM/A3H	....	This frequency is internationally used for Paid In-Flight Services by commercial carriers			
5624.8	CW	....	several, in the GDR	FA	....	Aero Tfc
5625.0	CW	C7A	Ocean Station Vessel 62°N & 32°W	MS	....	[aka "ALFA"]
	CW	C7I	Ocean Station Vessel 59°N & 19°W	MS	....	[aka "INDIA"]
	CW	C7M	Ocean Station Vessel 66°N & 02°E	MS	....	[aka "MIKE"]
			(Ocean Station Vessels will discontinue CW in 1982)			

kHz	Mode	Call	Location	Service	kW	Remarks
5634.0	A3H/USB	....	§AFI moves here from 5505.0 in 1983			
5638.0	AM/A3H	....	numerous (ICAO NAT & SP)	FA/MA	....	Airports & in-flight covering all of the North Atlantic and all of the southwestern Pacific Ocean
5640.7	CW	LOD2	unknown	FA	....	
5645.0	A3H/USB	....	numerous, in the North Sea	FA/MA	....	Airports & in-flight for heliocopters to oil rigs
	USB	PHW	Amsterdam Heliport, Netherlands	FA/MA	....	Oil Rigs
5652.0	USB	VFG	Gander Aeradio, Nfld., Canada	FA	1.0	VOLMET (20+50)
	USB	WSY70	New York Aeradio, NY	FA	1.5	" (05+35)
5673.0	AM/A3H	....	numerous (ICAO NAT & SEA)	FA/MA	....	Airports & in-flight covering the North Atlantic and India to Australia §SEA moves to 6556.0 in 1983
5680.0	AM/CW/A3H		INTERNATIONAL FREQUENCY FOR SEARCH & RESCUE			
5680.0	AM/A3H	....	numerous (ICAO AFI)	FA/MA	....	Airports & in-flight covering entire western half of Africa
	USB	....	numerous (Canada, England & USA)	FA/MA	....	Used by several military services (RCAF, RAF, USAF & USN)
5688.0	USB	ZUJ	Johannesburg (Jan Smuts) Aeradio, RSA	FA	5.0	
	USB	AFE8	MacDill AFB, FL	FA/MA	....	USAF (Used for SKY KING broadcasts)
	USB	....	several, southwestern USA	FA/MA	....	In-flight tests
5690.0	USB	DHM95	"Lahr Military", GFR	FA/MA	....	RCAF [Reported wx broadcasts at +15]
	USB	....	This frequency is internationally used for Paid In-Flight Services by commercial carriers			
5692.0	USB	....	several, throughout Australia	FA/MA	....	RAAF
	USB	....	numerous, worldwide	FA/MA	....	USN/USCG
5695.5	CW/USB	JJG8	Atsugi Ichihar, Japan	FA	0.3	Tfc
5696.0	USB	....	numerous, USA nationwide	FA/MA	....	USCG/USN (Primary air/surface channel)
5700.0	USB	....	numerous, worldwide	FA/MA	....	USAF (Some code names for USAF bases--was called "BRAVO QUEBEC" frequency--now part of "GIANT TALK")
5703.0	USB	....	numerous, worldwide	FA/MA	....	RCAF & USAF
	AM	†....	Rostov Aeradio, USSR	FA	....	VOLMET/RR
5710.0	USB	....	numerous, worldwide	FA/MA	....	USAF (enroute)
5715.0	USB	AXF	"Air Force Sydney", NSW, Australia	FA/MA	....	RAAF
	USB	AXH	"Air Force Townsville", Qld., Aust.	FA/MA	5.0	RAAF
5718.0	USB	....	several, in New Zealand	FA/MA	....	RNZAF
	USB	....	numerous, worldwide	FA/MA	....	USN
5725.5	CW	....	several, throughout Italy	FA	....	Italian AF
5726.0	USB	....	several, in Antarctica	FA/MA	....	RNZAF/USN

5730.0 - 5950.0 kHz -- FIXED SERVICE (Worldwide)

kHz	Mode	Call	Location	Service	kW	Remarks
5733.0	CW	....	unknown	FX	....	Coded Tfc
5735.1	ISB	†5TN22	Nouakchott, Mauritania	FX	6.0	Telcom
5738.5	CW	NPM	Lualualei Naval Radio, HI	FC	1.0	USN
5744.0	CW/RTTY	RLX	Dublin, Ireland	FX	0.5	Russian EMBASSY [Alt: 5181.0 & 6842.0]
5744.2	USB	....	Halifax, NS, Canada	FX/MA	....	Telcom [Alt: 4752.0]
5745.0	USB	....	Aleutian LORAN Net	FX	....	USCG (Includes: NMJ22, Attu; NRW2, ST. Paul Is. & NRW3, Pt. Clarence--all AK) [Alt: 9073.0]
5745.0	ISB	†....	Greenville, NC	FX	40.0	VOA Feeder
5751.0	USB	....	several, USA nationwide	FX	....	DOE
5755.0	USB	LTU265	La Quiara, Argentina	FX	0.1	Net Control
	FAX	AXI32	Darwin, NT, Australia	FX	5.0	Meteo (0800-2100)
5765.0	CW/ISB	†CUA23	Lisbon (Alfragide), Portugal	FX	3.0	Tfc

FIXED SERVICE (con.)

kHz	Mode	Call	Location	Service	kW	Remarks
	FAX	RRRQ	Novosibirsk, USSR	FX	15.0	Meteo
	USB	NMA	COMMSTA Miami, FL	FA/FX	....	USCG
			[Frontier Guard Forces Cuba]			
5767.5	FAX	JBK3	Tokyo (Kemigawa), Japan	FX	5.0	Meteo/Nx
5773.5	USB	†TUP57	Abidjan, Ivory Coast	FX	6.0	Telcom
5780.0	CW	LMB5	Bergen, Norway	FX	0.5	NOAA reports
			wx broadcasts at 1305 & 2240			
5781.0	USB	PRW863	Belo Horizonte, Brazil	FX	0.1	Net Control
5783.5	USB	NAX	Barber's Point Naval Radio, HI	FC	3.0	USN
	USB	NQM	Sand Island Naval Radio, Midway Is.	FC	6.0	USN
5785.0	CW	....	numerous, worldwide	FC/FX	....	USN
5790.0	ISB	....	Holzkirchen, GFR	FX	10.0	RFE Feeder
5793.0	CW/RTTY	GFT26	Bracknell, England	FX	3.5	Meteo
5794.7	CW	"K"	unknown	RC	....	Beacon
5800.0	FAX	YZZ1	Belgrade, Yugoslavia	FX	1.0	Meteo
					(0700-1700)	
5806.0	USB	VCR976	Ottawa, Ont., Canada	FX	1.0	Saudi Arabia
			EMBASSY [Alt: 14,748.5]			
5810.0	CW/AM	HLL	Seoul, South Korea	FX	1.5	NOAA reports
			wx broadcasts at 0000 & 1200			
	LSB	RIQ24	Yuzhno-Sakhalinsk, USSR	FX	....	R. Moscow Feeder
	USB	AFE71	"Cape Radio", Patrick AFB, FL	FX	10.0	USAF/NASA
			Shuttle Support [Alt: 4992.0 & 6708.0]			
5811.0	CW	JXU	BØ, Norway	FX	....	Norwegian Army
5812.0	AM	....	unknown	FX	....	NUMBERS/SS/YL
5813.0	ISB	†TLZ58	Bangui, Central African Republic	FX	2.0	Telcom
5815.0	ISB	RVO72	Moscow, USSR	FX	15.0	R. Moscow Feeder
5823.0	CW	KWL9Ø	Tokyo, Japan	FX	....	USA EMBASSY
5830.0	CW/RTTY	ONN3Ø	Brussels, Belgium	FX	0.5	EMBASSY
	USB	DFE83	Bonames, GFR	FX	20.0	DW Feeder
	USB	†RWD52	Moscow, USSR	FX	20.0	R. Moscow Feeder
5842.0	ISB	†FTF84	Paris (St. Assise), France	FX	20.0	Telcom
5845.0	ISB	†....	Holzkirchen, GFR	FX	10.0	RFE Feeder
5850.0	FAX	OXT	Copenhagen (Skamleback), Denmark	AX	20.0	Meteo
					(0030-1005)	
5851.0	FAX	JKE2	Tokyo, Japan	FX	5.0	
5858.0	USB	PSG351	Salvador, Brazil	FX	0.1	Net Control
5861.0	USB	KEM8Ø	Washington, DC	FX	....	FAA HQ
			[Backup Emergency Net]			
5865.0	USB	VJN	Cairns, Qld., Australia	FX	0.3	Outpost Control
	USB	VJQ	Kalgoorlie, WA, Australia	FX	0.3	Outpost Control
5866.0	USB	ZLW	Wellington, New Zealand	FX	0.5	Telcom/Portland I.
5868.0	CW/RTTY	ONN3Ø	Brussels, Belgium	FX	0.5	EMBASSY
	CW/RTTY	OMZ28	Prague, Czechoslovakia	FX	1.0	EMBASSY
5870.0	CW	NAM	Norfolk Naval Radio, VA	FX	20.0	USN
	CW	NAR	Key West Naval Radio, FL	FX	20.0	(LCMP-2) USN
			NOAA reports wx broadcasts at 0030, 0630, 1230 & 1700			
			Seasonal ice warnings at 0500 & 1700			
5876.0	CW	DDH5	Pinneburg, GFR	FX	5.0	Meteo
5876.5	USB	XMH356	Dartmouth, NS, Canada	FX	1.0	Net Control
	CW	NAR	Key West Naval Radio, FL	FX	20.0	(LCMP-2) USN/NUKO
5882.0	USB	LRO31	Buenos Aires (G. Pacheco), Argentina	FX	2.5	Feeder
5889.8	CW	"K"	unknown	RC	....	Beacon
5890.0	ISB	†....	Holzkirchen, GFR	FX	10.0	RFE Feeder
	FAX	RBV78	Tashkent, Uzbek SSR	FX	5.0	Meteo
5893.5	USB	†RND72	Moscow, USSR	FX	20.0	Telcom
5902.5	ISB	....	Delano, CA	FX	40.0	VOA Feeder
5907.5	USB	....	several, in Alaska & Hawaii	FX	....	US Governmental
			(Includes: KWL24, Tern Island, HI; KWL47, Bethel, AK; etc.)			
5909.0	CW	BMB	Taipei, Taiwan	FX	...	NOAA reports
			wx broadcasts at 0400, 1000, 1600 & 2200			
5915.0	CW	ZLZ2Ø	Wellington (Himatangi), New Zealand	FC	5.0	NOAA reports
			wx broadcasts at 0200, 0520, 0815, 0840, 0910, 1800 2040 & 2110			
5917.0	CW	NAM	Norfolk Naval Radio, VA	FX	....	(LCMP-3) USN

kHz	Mode	Call	Location	Service	kW	Remarks
5918.5	CW	AOK	Moron de la Frontera, Spain	FX	1.0	USN/NOAA reports wx broadcasts at 0030, 0630, 1230 & 1700/Seasonal ice warnings at 0500 & 1700
5919.9	CW	"K"	unknown	RC	....	Beacon
5920.0	USB	....	Novosibirsk, USSR	FX	100.0	R. Moscow Feeder
5921.0	USB	....	numerous, in Mexico	FX	....	Net/SS
5923.5	LSB	....	numerous, western USA	FX	....	NWS Net (Infrequently used. Most recently reported stations include: KAE41, Albuquergue, NM; KGD55, Redding, CA; KGD69, Salt Lake City, UT; KHB24, Billings, MT & KME57, Portland, OR)
5930.0	USB	ZKS	Rarotonga, Cook Islands	FX	1.0	Net Control [Alt: 7390.0]
5932.5	ISB	....	Delano, CA	FX	....	VOA Feeder
5940.0	ISB	VZSY	Sydney (Llandilo), NSW, Australia	AX	1.0	Telcom/Lord Howe
5945.0	ISB	†....	Holzkirchen, GFR	FX	10.0	RFE Feeder
	FAX	LMO5	Oslo, Norway	FX	0.5	Meteo

5950.0 - 6200.0 kHz — INTERNATIONAL BROADCASTING (Worldwide)

kHz	Mode	Call	Location	Service	kW	Remarks
6100.0	AM	YVTO	Caracas, Venezuela	BC	1.0	TIME SIGNALS

6200.0 - 6525.0 kHz — MARITIME MOBILE (Worldwide)

6200.0 USB – 6215.5 USB: The band between these frequencies is divided into 6 channels spaced 3.1 kHz apart and assigned numbers from 601 to 606. These are ship transmitting channels and are paired with Coastal USB stations operating between 6506.4 and 6521.9 kHz.

kHz	Mode	Call	Location	Service	kW	Remarks
6203.0	CW	"P"	unknown	RC	....	Beacon
6209.3	USB	WGK	St. Louis, MO	FC/MS	0.8	Inland Waterway
	USB	WJG	Memphis, TN	FC/MS	0.8	Inland Waterway
6212.4	USB	WFN	Louisville, KY	FC/MS	0.8	Inland Waterway
	USB	WGK	St. Louis, MO	FC/MS	0.8	Inland Waterway
6215.5	AM/USB	....	INTERNATIONAL SHIP STATION CALLING FREQUENCY			
6215.5	USB	PJC	Curacao (Willemstad) Radio, Curacao	FC	....	NOAA reports wx broadcasts 1200 & 2300, July-November, only
6218.6	USB	....	Atlantic & Pacific Voice Circuits	FC/MS	....	USCG (Includes: NMA, Miami, FL; NMC, San Francisco, CA; NMF, Boston, MA; NMG, New Orleans, LA; NMN, Portsmouth, VA; NMO, Honolulu, HI; NMO, Long Beach, CA; NMR, San Juan, PR; NOJ, Kodiak AK & NRV, Barrigada, Guam
	USB	....	Coast & Ship Simplex Frequency	FC/MS	....	Inland & intercoastal waterway traffic. Employed worldwide by hundreds of base stations and various types of ships. Frequently heard in North America: KMB, Houston, TX; KGA595, Seattle, WA; WGW, San Juan, PR; WPE, Jacksonville, FL; etc.
6221.6	USB	....	Atlantic & Pacific Voice Circuits	FC/MS	....	USCG (See 6218.6)
	USB	....	Coast & Ship Simplex Frequency	FC/MS	....	Inland & intercoastal waterway traffic. Employed worldwide by hundreds of base stations and various types of ships. Frequently heard in North America: KXE, Des Allemands, LA; KZU, Harvy, LA; KRM, Port Arthur, TX; WNC, Tampa, FL; etc.
6264.0	CW	FUM	Papeete Naval Radio, Tahiti	FC	....	French Navy

6270.0 CW – 6280.8 CW: The band between these two frequencies is divided into 18 channels each 600 Hz wide and assigned numbers from 1 to 18. These are ship transmitting channels.. The most commonly used channels are 5 and 6, centered on 6272.7 and 6273.3 kHz, respectively.

kHz	Mode	Call	Location	Service	kW	Remarks
6326.5	CW	PBC26	Goeree Naval Radio, Netherlands	FC	....	Dutch Navy
	CW	DZJ	Manila (Bulacan) Radio, Philippines	FC	2.5	

kHz	Mode	Call	Location	Service	kW	Remarks
	CW	WNU32	Slidell Radio, LA	FC	3.0	NOAA reports wx broadcast at 0430 & 1630
6328.0	CW	OST32	Oostende Radio, Belgium	FC	5.0	
6329.0	FAX	CFH	Maritime Command Radio, Halifax, NS, Canada	FC (Canadian Forces)	5.0	Meteo/NOAA reports seasonal ice broadcasts at 0000 & 2200, plus various test charts, surface analysis, wave analysis, depictions, etc.
6330.0	CW	YIR	Basrah Control, Iraq	FC	1.0	
	CW	FJP6	Noumea Radio, New Caledonia	FC	1.0	
6331.5	CW	UMV2	Murmansk Radio, USSR	FC	25.0	Reported wx broadcasts at 0630 & 1700
6332.0	CW	JHC	Choshi Radio, Japan	FC	1.0	
6333.5	CW	VII	Thursday Island, Qld., Australia	FC	1.0	AMVER/NOAA reports wx broadcasts at 0018 & 0948
	CW	VIM	Melbourne Radio, Vic., Australia	FC	1.0	NOAA reports wx broadcasts at 0948 & 2318
	CW	VIR	Rockhampton Radio, Qld., Australia	FC	1.0	NOAA reports wx broadcast at 0048
	CW	XSZ	Dalian Radio, PRC	FC	....	NOAA reports wx broadcasts at 0050 & 1050
	CW	GKR	Wick Radio, England	FC	0.3	
	CW	9HD	Malta Radio, Malta	FC	1.0	
	CW	WCC	Chatham Radio, MA	FC	6.0	
	CW	WMH	Baltimore Radio, MD	FC	10.0	
6334.5	CW	CTH47	Horta Naval Radio, Azores	FC	....	Port. Navy Reported wx broadcasts at 2130 & 2230
	CW	UFM3	Nevelsk Radio, USSRSR	FC	5.0	
6335.5	CW	VFA	Inuvik CG Radio, NWT, Canada	FC	0.7	
	CW	ARL	Karachi Naval Radio, Pakistan	FC	10.0	Pakistan Navy
	CW	DZI	Manila (Bacoor) Radio, Philippines	FC	1.0	Reported wx broadcast at 0000
6336.1	CW	LZW3	Varna Radio, Bulgaria	FC	5.0	
6336.5	CW	ZRQ3	Cape (Simonstown) Naval Radio, RSA	FC	5.0	RSA Navy
6336.8	CW	ZLO3	Irirangi Naval Radio, New Zealand	FC	....	RNZ Navy
6337.0	CW	CLA5	Havana (Cojimar) Radio, Cuba	FC	5.0	
	CW	PKY4	Sorong Radio, Indonesia	FC	0.3	
	CW	WCC	Chatham Radio, MA	FC	10.0	
	CW	UXN	Arkhangelsk Radio, USSR	FC	25.0	
6338.5	CW	OSN46	Oostende Naval Radio, Belgium	FC	1.0	Belgium Navy
	CW	DYM	Cebu Radio, Philippines	FC	0.4	
6339.5	CW	YIR	Basrah Control, Iraq	FC	2.0	
6340.0	CW	UEK	Feodosia Radio, USSR	FC	....	
	CW	UFN	Novorossiyk Radio, USSR	FC	5.0	
6340.5	CW	9VG9	Singapore Radio, Singapore	FC	3.0	
6341.5	CW	UDH	Riga Radio, Latvian SSR	FC	5.0	
	CW	UFB	Odessa Radio, Ukrainian SSR	FC	5.0	Reported wx broadcasts at 0600 & 1330
6342.0	CW	GKK3	Portishead Radio, England	FC	5.0	
	CW	JFA	Chuo Gyogyo (Matsudo) Radio, Japan	FC	3.0	
6344.0	CW	D3L	Luanda Radio, Angola	FC	6.0	
	CW	SVB3	Athens Radio, Greece	FC	2.5	
	CW	TFA6	Reykjavik Radio, Iceland	FC	1.5	
	CW	HLF	Seoul Radio, South Korea	FC	3.5	
	CW	YJM6	Port-Vila Radio, Vanuatu	FC	0.5	
6345.0	CW	UHK	Batumi Radio, USSR	FC	5.0	
	CW	UFB	Odessa Radio, Ukrainian SSR	FC	5.0	
	CW	UAT	Moscow Radio, USSR	FC	25.0	
	CW	UFL	Vladivostok Radio, USSR	FC	10.0	
6348.0	CW	J2A6	Djibouti Radio, Djibouti	FC	2.5	
6348.0	CW	HWN	Paris (Houilles) Naval R., France	FC	2.0	French Navy
	CW	5RS	Tamatave Radio, Madagascar	FC	2.0	
	CW	FUX	Le Port Naval Radio, Reunion	FC	4.0	French Navy
	CW	KFS	San Francisco Radio, CA	FC	10.0	NOAA reports wx broadcasts at 0420, 1620 & 2300
6350.0	CW	HRW	La Ceiba Radio, Honduras	FC	....	

kHz	Mode	Call	Location	Service	kW	Remarks
6351.5	CW	VIB	Brisbane Radio, Qld., Australia	FC	1.0	NOAA reports wx broadcasts at 0948 & 2318
	CW	VAI	Vancouver CG Radio, BC, Canada	FC	5.0	NOAA reports wx broadcasts at 0130, 0530 & 1730
	CW	3BM3	Mauritius (Bigara) Radio, Mauritius	FC	4.0	NOAA reports wx broadcasts at 0130 & 0430
	CW	P2M	Port Moresby Radio, Papua New Guinea	FC	1.0	NOAA reports wx broadcasts at 0000 & 0900
	CW	P2R	Rabaul Radio, Papua New Guinea	FC	1.0	NOAA reports wx broadcasts at 0048 & 0530
	CW	CTV6	Monsanto Naval Radio, Portugal	FC	5.0	Port. Navy
	CW	WMH	Baltimore Radio, MD	FC	0.8	
	CW	YVG	La Guairia Radio, Venezuela	FC	0.3	
6352.0	CW	FUE	Brest Naval Radio, France	FC	4.0	French Navy
	CW	FUG3	La Regine (Castelnaudry), France	FC	....	" "
	CW	FUO	Toulon Naval Radio, France	FC	4.0	" "
6353.0	CW	UFB	Odessa Radio, Ukrainian SSR	FC	5.0	
6354.0	CW	XFM	Manzanillo Radio, Mexico	FC	1.0	Reported wx broadcast at 2330
	CW	URD	Leningrad Radio, USSR	FC	5.0	
6355.0	CW	OFJ6	Helsinki Radio, Finland	FC	5.0	
	CW	PKS	Pontianak Radio, Indonesia	FC	0.3	
	CW	PZN2	Paramaribo Radio, Suriname	FC	1.0	
	CW	UBN	Jdanov Radio, Ukrainian SSR	FC	10.0	
6358.4	CW	PBC26	Goeree Naval Radio, Netherlands	FC	6.0	Dutch Navy
6358.5	CW	LPW63	Bahia Blanca Radio, Argentina	FC	2.0	
	CW	Y5M	Ruegen Radio, GDR	FC	5.0	
6360.0	CW	CLS	Havana (Industria Pesquera) R., Cuba	FC	5.0	
	CW	XFQ	Salina Cruz Radio, Mexico	FC	1.0	
	CW	OBC3	Callao Radio, Peru	FC	1.0	
	CW	UJY	Kaliningrad Radio, USSR	FC	5.0	
6361.0	CW	KUQ	Pago Pago Radio, American Samoa	FC	2.5	AMVER
6362.0	CW	GYA3	Whitehall (London) Naval R., England	FC	15.0	British Navy
	CW	†EQZ	Abadan Radio, Iran	FC	0.5	
	CW/RTTY	FUX	Le Port Naval Radio, Reunion	FC	5.0	French Navy
6363.5	CW/TOR	DAF	Norddeich Radio, GFR	FC	5.0	
6364.7	CW	7CB	unknown	FC	....	
6365.0	CW	IRM4	CIRM, Rome Radio, Italy	FC	10.0	AMVER/Medico
	CW	XFS	Tampico Radio, Mexico	FC	0.5	
6365.0	CW	UCP2	Okhotsk Radio, USSR	FC	1.0	Okhotsk Sea
	CW	UVD	Magadan Radio, USSR	FC	10.0	
6365.5	CW	UGK2	Kaliningrad Radio, USSR	FC	6.0	
	CW	KFS	San Francisco Radio, CA	FC	10.0	
	CW	WPD	Tampa Radio, FL	FC	5.0	
6367.0	CW	XFU	Veracruz Radio, Mexico	FC	0.5	
6368.9	CW/TOR	GKA3	Portishead Radio, England	FC	5.0	NOAA reports wx broadcasts at 0930 & 2130
6369.0	CW	D3E	Luanda Radio, Angola	FC	2.0	
	CW	D3F	Lobito Radio, Angola	FC	2.0	
	CW	D3G	Mocamedes Radio, Angola	FC	2.0	
	CW	XST	Qingdao Radio, PRC	FC	1.0	
	CW	KLC	Galveston Radio, TX	FC	3.5	NOAA reports wx broadcasts at 0530, 1130, 1730 & 2330
	CW	UQK	Riga Radio, Latvian SSR	FC	5.0	
	CW	UFJ	Rostov Radio, USSR	FC	5.0	
6370.0	CW	URD	Leningrad Radio, USSR	FC	15.0	
	CW	UBE2	Petropavlovsk Radio, USSR	FC	10.0	NOAA reports wx broadcasts at 0900, 2000 & 2310/NAVAREA 0900 & 2200
6371.0	CW	VPS25	Cape d'Aguilar Radio, Hong Kong	FC	2.0	NOAA reports wx broadcast at 1318
	CW	UEK	Feodosia Radio, USSR	FC	....	
6371.3	CW	GYU3	Gibraltar Naval Radio, Gibraltar	FC	10.0	British Navy
6371.5	CW	UFD9	Arkhangelsk Radio, USSR	FC	15.0	(RKLM)
6372.0	CW	FDC	Metz Air, France	FX	....	French AF
	CW	ZSD44	Durban Radio, RSA	FC	3.5	

kHz	Mode	Call	Location	Service	kW	Remarks
	CW	SAG3	Goteborg Radio, Sweden	FC	5.0	AMVER
6372.5	CW	YVM	Puerto Ordaz Radio, Venezuela	FC	....	
6374.3	CW	TB02	Izmir Naval Radio, Turkey	FC	....	Turkish Navy
6375.0	CW	HAR	Budapest Naval Radio, Hungarey	FC	2.5	
	CW	UGH2	Juzno-Sakhalinsk Radio, USSR	FC	5.0	
	CW	UNM2	Klaipeda Radio, Lithuanian SSR	FC	5.0	
6376.0	CW	C6N	Nassua Radio, Bahamas	FC	0.5	Reported wx broadcasts at 0130 & 1300
	CW	IQH	Naples Radio, Italy	FC	6.0	
	CW	RIT	unknown, probably in USSR	FC	....	
88/113 –	CW	WCC	Chatham Radio, MA	FC	10.0	Tfc/UPI Nx
	CW	SPE32	Szczecin Radio, Poland	FC	5.0	[aka SPB3]
	CW	CWN	Montevideo Armada Radio, Uruguay	FC	2.0	Uruguayan Navy
6377.0	CW	Y4K	unknown	FX	....	Coded Tfc
6377.7	CW	EBK	Las Palmas Radio Naval, Canary Is.	FC	1.0	Spanish Navy
	CW	EBA	Madrid Radionaval, Spain	FC	1.0	Spanish Navy
6379.0	CW	4XZ	Haifa Naval Radio, Israel	FC	2.0	Israel Navy
6379.5	CW	8PO	Barbados Radio, Barbados	FC	1.0	
	CW	GKB3	Portishead Radio, England	FC	5.0	
	CW	ZSC21	Cape Town Radio, RSA	FC	10.0	
	CW	UBN	Jdanov Radio, Ukrainian SSR	FC	10.0	Reported wx broadcast at 0700
	CW	UBE	Petropavlovsk Radio, USSR	FC	5.0	NOAA reports wx broadcasts at 0500 & 2310
	CW	UFL	Vladivostok Radio, USSR	FC	5.0	NOAA reports wx broadcasts at 0530 & 2130
6379.9	CW	GKN3	Portishead Radio, England	FC	5.0	
6380.0	CW	UQB	Kholmsk Radio, USSR	FC	5.0	
6381.0	CW	9KK4	Kuwait Radio, Kuwait	FC	5.0	
6382.0	CW	EAD2	Aranjuez Radio, Spain	FC	3.5	
6382.5	CW	GKB3	Portishead Radio, England	FC	5.0	
6383.0	CW	TYA3	Cotonou Radio, Benin	FC	1.0	
	CW	XSV	Tianjin Radio, PRC	FC	6.0	
	CW	3XC2	Conakry Radio, Guinea	FC	0.8	
	CW	SPH31	Gdynia Radio, Poland	FC	5.0	[aka SPA3]
	CW	UFL	Vladivostok Radio, USSR	FC	5.0	Reported wx broadcast at 1500
	CW	URL	Sevastopol Radio, Ukrainian SSR	FC	....	
	CW	NMC	COMMSTA San Francisco, CA	FC	10.0	USCG/AMVER
	CW	....	numerous	FA/FC/FX	..	USCG [Command Net Calling Frequency]
6384.5	CW	CKN	Vancouver Forces Radio, BC, Canada	FC	10.0	Canadian Forces/ NAWS
6385.0	CW	SXA3	Spata Attikis Naval Radio, Greece	FC	5.0	Greek Navy/NATO
	CW	UFL	Vladivostok Radio, USSR	FC	5.0	Reported wx broadcast at 1500/NAVAREA at 1500
6386.0	CW	6VA4	Dakar Radio, Senegal	FC	4.0	NOAA reports wx broadcasts at 1000 & 2200
6386.3	CW	HKC2	Buenaventura Radio, Columbia	FC	0.5	
	CW	ZSJ3	NAVCOMCEN (Silvermine) Cape R., RSA	FC	5.0	RSA Navy/AMVER
6386.4	CW	UJ05	Izmail Radio, Ukrainian SSR	FC	5.0	
6386.5	CW	†CFH	Maritime Command Radio, Halifax, NS, Canada	FC	10.0	C. Forces
	CW	†VPC	Falkland Islands Radio, Falkland Is.	FC	2.0	
	CW	UJY	Kaliningrad Radio, USSR	FC	5.0	
6387.9	CW	FUF	Fort de France Naval R., Martinique	FC	10.0	French Navy
6388.0	CW	JFA	Chuo Gyogyo (Matsudo) Radio, Japan	FC	3.0	
	CW	EBA	Madrid Radionaval, Spain	FC	10.0	Spanish Navy
6389.0	CW	CTP94	Oeira Naval Radio, Portugal	FC	5.0	Port. Navy/NATO
6389.6	CW	WNU42	Slidell Radio, LA	FC	3.0	
6389.9	CW	XFU	Veracruz Radio, Mexico	FC	1.0	
6390.0	CW	UTA	Tallinn Radio, Estonian SSR	FC	5.0	
	CW	JYO	Aqaba Radio, Jordan	FC	1.0	
	CW	JJK	Ominato Radio, Japan	FC	1.0	
	CW	AQP2	Karachi Naval Radio, Pakistan	FC	10.0	Pakistan Navy

kHz	Mode	Call	Location	Service	kW	Remarks
	CW	XSQ	Guangzhou Radio, PRC	FC	10.0	Reported wx broadcasts at 0030 & 1230
	CW	UFB	Odessa Radio, Ukrainian SSR	FC	....	
	CW	UPB	Providenia Bukhta Radio, USSR	FC	25.0	
6390.3	CW	IDQ3	Rome Naval Radio, Italy	FC	10.0	Italian Navy NOAA reports wx broadcasts at 0500, 0830, 1800 & 2100
	CW	....	numerous, in Italy	FC	....	Italian Navy
6392.0	CW	CLS	Havana (Industria Pesquera) R., Cuba	FC	5.0	
6393.0	CW	OVG	Frederikshaven Naval Radio, Denmark	FC	5.0	Danish Marine
	CW	DZW	Manila Radio, Philippines	FC	....	
6393.5	CW	CUG33	Sao Miguel Radio, Azores	FC	2.5	
	CW	OMC	Bratislava Radio, Czechoslovakia	FC	1.0	
	CW	CUB29	Madeira Radio, Madeira Islands	FC	2.5	
	CW	3BA3	Mauritius (Bigara) Radio, Mauritius	FC	5.0	
	CW	UDK2	Murmansk Radio, USSR	FC	15.0	
	CW	ZLB3	Awarua Radio, New Zealand	FC	5.0	
	CW	CUL6	Lisbon Radio, Portugal	FC	5.0	
6394.0	CW	RIW	Khiva Naval Radio, Uzbek SSR	FC	....	Russian Navy
6395.0	CW	TBA3	Izmir Naval Radio, Turkey	FC	2.0	Turkist Navy/ NATO [aka TCA21]
6396.0	FAX	GKN3	Portishead Radio, England	FC	5.0	
6397.0	CW	GKM3	Portishead Radio, England	FC	5.0	
	CW	JMC3	Tokyo Radio, Japan	FC	4.0	NOAA reports wx broadcasts at 0318, 0648, 0918, 1148, 1518, 1748, 2118 & 2348
6397.9	CW/TOR	GKO3	Portishead Radio, England	FC	5.0	
6398.0	CW	SPH32	Gdynia Radio, Poland	FC	5.0	[aka SPH3]
6400.0	CW	UCY2	Astrakhan Radio, USSR	FC	5.0	Caspian Sea
	CW	UMV	Murmansk Radio, USSR	FC	25.0	
6400.5	CW	EDZ2	Aranjuez Radio, Spain	FC	5.0	
6402.0	CW	GKS3	Portishead Radio, England	FC	5.0	
	CW	DZE	Mandaluyong Radio, Philippines	FC	....	
6402.5	CW	LZS24	Sofia Naval Radio, Bulgaria	FC	....	
6404.0	CW	LPD44	Gen. Pacheco Radio, Argentina	FC	5.0	
	CW	PCH3Ø	Scheveningen Radio, Netherlands	FC	3.0	
6405.0	CW	LZW3	Varna Radio, Bulgaria	FC	5.0	
	CW	PJK26	Suffisant Dorf Naval Radio, Curacao	FC	6.0	Dutch Navy
	CW	UAH	Tallinn Radio, Estonian SSR	FC	5.0	Reported wx broadcast at 1030
	CW/RTTY	UVD	Magadan Radio, USSR	FC	5.0	
	CW	UFN	Novorossiysk Radio, USSR	FC	15.0	
	CW/RTTY	UFH	Petropavlovsk Radio, USSR	FC	5.0	
	CW	CCP	Puerto Montt Naval Radio, Chile	FC	....	Chilean Navy Reported wx broadcasts at 0030, 1230 & 1830
6407.5	CW	VIC	Carnarvon Radio, WA, Australia	FC	1.0	AMVER/NOAA reports wx broadcasts at 0148 & 1130
	CW	VIE	Esperance Radio, WA, Australia	FC	1.0	NOAA reports wx broadcasts at 0048 & 1148
	CW	VIO	Broome Radio, WA, Australia	FC	1.0	AMVER/NOAA reports wx broadcasts at 0118 & 1230
	CW	VIP2	Perth Radio, WA, Australia	FC	1.0	
	CW	XSL	Fuzhou Radio, PRC	FC	0.5	
	CW	GKC3	Portishead Radio, England	FC	5.0	
	CW	GYA	Whitehall (London) Naval R., England	FC	10.0	British Navy
	CW	UXN	Arkhangelsk Radio, USSR	FC	....	
6408.0	CW	DZA	Mandaluyong Radio, Philippines	FC	....	
6408.5	CW	EBA	Madrid Radionaval, Spain	FC	10.0	Spanish Navy
6409.5	CW	IAR6	Rome P.T. Radio, Italy	FC	5.0	
6410.0	CW	ARH	Gwadar Naval Radio, Pakistan	FC	6.0	Pakistan Navy
	CW	UDH	Riga Radio, Latvian SSR	FC	5.0	
	CW	UHK	Batumi Radio, USSR	FC	5.0	
6411.0	CW	LPD41	Gen. Pacheco Radio, Argentina	FC	3.0	
	CW	OST3	Oostende Radio, Belgium	FC	5.0	
	CW	SVD3	Athens Radio, Greece	FC	2.5	
	CW	5OW6	Lagos Radio, Nigeria	FC	1.0	

kHz	Mode	Call	Location	Service	kW	Remarks
	CW	DZF	Manila (Bacoor) Radio, Philippines	FC	0.4	Reported wx broadcasts at 1600 & 2200
	CW	9LL	Freetown Radio, Sierra Leone	FC	1.5	
	CW	UKA	Vladivostok Radio, USSR	FC	10.0	
	CW	KLB	Seattle Radio, WA	FC	3.5	
6411.4	CW	WOE	Lantana Radio, FL	FC	5.0	
6412.0	CW	9VG5	Singapore Radio, Singapore	FC	3.0	NOAA reports wx broadcasts at 0120 & 1320
6413.7	CW	GYA	Whitehall (London) Naval R., England	FC	5.0	British Navy
	CW	GLP	Whitehall (London) Naval R., England	FC	5.0	British Navy
6414.1	CW	GYC3	Whitehall (London) Naval R., England	FC	5.0	British Navy
	CW	ASK	Karachi Radio, Pakistan	FC	3.0	
	CW	WSL	Amagansett Radio, NY	FC	10.0	
6414.5	CW	S3D	Chittagong Radio, Bangladesh	FC	6.0	
	CW	CBM2	Magallanes (Punta Arenas) R., Chile	FC	2.0	
	CW	XSG	Shanghai Radio, PRC	FC	1.0	TIME SIGNALS NOAA reports wx broadcasts at 0306 & 0906
	CW	XFA	Acapulo Radio, Mexico	FC	0.5	
	CW	UOP	Tuapse Radio, USSR	FC	5.0	(RPLT)
6415.0	CW	7TA4	Algiers Radio, Algeria	FC	5.0	
6416.0	CW	UJQ	Kiev Radio, Ukrainian SSR	FC	5.0	
6417.0	CW	DZP	Manila (Novaliches) R., Philippines	FC	3.0	
	CW	UKX	Nakhodka Radio, USSR	FC	5.0	
6418.0	CW	VTP5	Vishakhapatnam Naval Radio, India	FC	5.0	Indian Navy
	CW	IQX	Trieste P.T. Radio, Italy	FC	2.0	
	CW	WSL	Amagansett Radio, NY	FC	10.0	
6418.2	CW	IAR26	Rome P.T. Radio, Italy	FC	7.0	
	CW	CCS	Santiago Naval Radio, Chile	FC	....	Chilean Navy
6418.5	CW	DFF41	Frankfurt (Usingen), GFR	FC	....	MARPRESS/Reported GG nx broadcast at 0118
6420.0	CW	PWZ	Rio de Janeiro Naval Radio, Brazil	FC	5.0	Brazil Navy
	CW	IRM5	CIRM, Rome Radio, Italy	FC	5.0	Medico
	CW	UJQ7	Kiev Radio, Ukrainian SSR	FC	5.0	
	CW	UNO2	Severo Kurilsk Radio, USSR	FC	....	
6421.5	CW	FFL3	St. Lys Radio, France	FC	5.0	
	CW	UJO3	Izmail Radio, Ukrainian SSR	FC	6.0	
6423.0	CW	LZL3	Bourgas Radio, Bulgaria	FC	1.0	
	CW	DZE	Mandaluyong Radio, Philippines	FC	0.4	
	CW	ZSC47	Cape Town Radio, RSA	FC	10.0	
6424.0	CW	9GA	Takoradi Radio, Ghana	FC	5.0	
6425.0	CW	CLQ	Havana (Cojimar) Radio, Cuba	FC	5.0	
	CW	SUP	Port Said Radio, Egypt	FC	2.0	
	CW	OFJ7	Helsinki Radio, Finland	FC	5.0	
	CW	ICB	Genoa P.T. Radio, Italy	FC	5.0	
	CW	URD	Leningrad Radio, USSR	FC	5.0	
6426.0	CW	DZR	Manila Radio, Philippines	FC	5.0	
6427.0	CW	UNM2	Klaipedia Radio, Lithuanian SSR	FC	5.0	
6428.5	CW	VHP3	COMMSTA Canberra, ACT, Australia	FC	5.0	RA Navy
	CW	VIX	Master Station Canberra, Australia	FC	5.0	NOAA reports wx broadcasts at 0130, 0530, 0930, 1330, 1730 & 2130
	CW	GKD3	Portishead Radio, England	FC	5.0	
	CW	Y5M	Ruegen Radio, GDR	FC	5.0	
	CW	PKM	Bitung Radio, Indonesia	FC	1.0	
	CW	PKE5	Ternate Radio, Indonesia	FC	....	
	CW	NPG	San Francisco Naval Radio, CA	FC	10.0	USN/TIME SIGNALS
6429.0	CW	DZD	Manila (Antipolo) R., Philippines	FC	0.4	Reported wx broadcasts at 0300 & 1100
6430.0	CW	CFH	Maritime Command Radio, Halifax, NS, Canada (Canadian Forces)	FC	10.0	(C13L) NOAA reports wx broadcasts at 0200, 0630, 1400 & 1800/Seasonal ice warning broadcasts at 0230 & 1430
	CW	4XO	Haifa Radio, Israel	FC	2.5	
	CW	UKJ	Astrakhan Radio, USSR	FC	1.0	Caspian Sea
	CW	UGG2	Belomorsk Radio, USSR	FC	5.0	
	CW	UKA	Vladivostok Radio, USSR	FC	10.0	Reported NAVAREA broadcast at 1730

MARITIME MOBILE (con.)

kHz	Mode	Call	Location	Service	kW	Remarks
6432.0	CW	LGU	Rogaland Radio, Norway	FC	5.0	AMVER
	CW	DZZ	Manila Radio, Philippines	FC	....	
6433.0	CW	GYA	Whitehall (London) Naval R., England	FC	5.0	British Navy
6434.0	CW	ZAD2	Durres Radio, Albania	FC	1.0	
6434.7	CW	GYU	Gibraltar Naval Radio, Gibraltar	FC	....	British Navy
6435.0	CW	CLQ	Havana (Cojimar) Radio, Cuba	FC	5.0	
	CW	ICT	Taranto Radio, Italy	FC	....	
	CW	DZO	Manila (Bulacan) Radio, Philippines	FC	2.0	NOAA reports wx broadcasts at 1030 & 2330
6435.5	CW	PWZ	Rio de Janeiro Naval Radio, Brazil	FC	....	Brazil Navy/ Reported wx broadcasts at 0030, 0630 & 1800/NAVAREA at 0200, 1330 & 2000
	CW	DAN	Norddeich Radio, GFR	FC	5.0	
	FAX	GZZ	Whitehall (Northwood) Naval R., Eng.	FC	5.0	British Navy/ Meteo
	CW	IDR3	Rome Naval Radio, Italy	FC	10.0	Italian Navy
	CW	ZLP3	Irirangi Naval Radio, New Zealand	FC	10.0	RNZ Navy
	CW	CWA	Cerrito Radio, Uruguay	FC	....	
6436.3	CW	WPA	Port Arthur Radio, TX	FC	5.0	
6439.0	CW	OXZ31	Lyngby Radio, Denmark	FC	5.0	
6439.0	CW	UKK3	Nakhodka Radio, USSR	FC	5.0	
6440.0	CW	OMC	Bratislava Radio, Czechoslovakia	FC	1.0	
	CW	OMP	Prague Radio, Czechoslovakia	FC	1.0	
	CW	UFN	Novorossiysk Radio, USSR	FC	....	
6440.8	CW	GYA	Whitehall (London) Naval R., England	FC	5.0	British Navy
6440.9	CW	DZG	Manila (Las Pinas) R., Philippines	FC	5.0	AMVER/NOAA reports wx broadcasts at 0300 & 1500
6442.5	CW	XFM2	Manzanillo Radio, Mexico	FC	1.0	
	CW	XFS3	Tampico Radio, Mexico	FC	1.0	
	CW	XFU2	Veracruz Radio, Mexico	FC	1.0	
6443.0	CW	UCW4	Leningrad Radio, USSR	FC	....	
6444.0	CW	DZK	Manila (Bulacan) Radio, Philippines	FC	4.0	
6444.5	CW	SVF3	Athens Radio, Greece	FC	5.0	
6445.0	CW	ROT2	Moscow Naval Radio, USSR	FC	25.0	Russian Navy
	CW	UFL	Vladivostok Radio, USSR	FC	5.0	
6445.5	CW	CKN	Vancouver Forces Radio, BC, Canada (Canadian Forces)	FC	10.0	NOAA reports wx broadcasts at 0030, 0430 & 1300
6446.0	CW	DZR	Manila Radio, Philippines	FC	2.0	NOAA reports wx broadcasts at 0120
	CW	WLO	Mobile Radio, AL	FC	10.0	NOAA reports wx broadcasts at 1300 & 2300
6446.8	CW	OXZ3	Lyngby Radio, Denmark	FC	5.0	
6447.0	CW	UFB	Odessa Radio, Ukrainian SSR	FC	5.0	
	CW	UFH	Petropavlosk Radio, USSR	FC	5.0	
6448.0	CW	LZW3	Varna Radio, Bulgaria	FC	5.0	
6449.5	CW	FFS3	St. Lys Radio, France	FC	5.0	
	CW	Y5M	Ruegen Radio, GDR	FC	5.0	
6450.0	CW	HSX	Songkhla Naval Radio, Thailand	FC	3.0	Royal Thai Navy
	CW	HSY	Sattahip Radio, Thailand	FC	3.0	" " "
	CW	HSZ	Bangkok Naval Radio, Thailand	FC	3.0	" " "
6451.0	CW	HLG	Seoul Radio, South Korea	FC	5.0	
6451.5	CW	SAB3	Goteborg Radio, Sweden	FC	5.0	
6452.0	CW	XSG	Shanghai Radio, PRC	FC	1.0	
6453.0	CW	PWZ	Rio de Janeiro Naval Radio, Brazil	FC	5.0	Brazil Navy
	CW	DZJ	Manila (Bulacan) Radio, Philippines	FC	....	
6454.0	CW	CLA4	Havana (Cojimar) Radio, Cuba	FC	5.0	
	CW/FAX	LOK	Orcadas Radio, South Orkney Islands	FC	0.2	NOAA reports wx broadcast at 1625 [Some seasonal FAX ice warnings]
6455.0	CW	UON	Baku Radio, Azerbaijan SSR	FC	5.0	
	CW	UDK2	Murmansk Radio, USSRSR	FC	5.0	
6456.0	CW	DAL	Norddeich Radio, GFR	FC	5.0	
6456.5	CW	PKR6	Cilacap Radio, Indonesia	FC	0.4	
	CW	PKB2Ø	Lhok Seumawe Radio, Indonesia	FC	1.0	
	CW	PKN2	Balikpapan Radio, Indonesia	FC	0.5	

kHz	Mode	Call	Location	Service	kW	Remarks
	CW	PKN7	Bontang Radio, Indonesia	FC	1.0	
	CW	PKY41	Sorong Radio, Indonesia	FC	1.0	
6457.5	CW	JOR	Nagasaki Radio, Japan	FC	5.0	
6459.0	CW	SPE31	Szczecin Radio, Poland	FC	5.0	[aka SPB4]
6460.0	CW	LSA3	Boca Radio, Argentina	FC	1.0	
	CW/RTTY	UAT	Moscow Radio, USSR	FC	25.0	
	CW	UKA	Vladivostok Radio, USSR	FC	15.0	
6463.4	CW	HKB	Barranquilla Radio, Colombia	FC	2.5	
6463.5	CW	VIA	Adelaide Radio, SA, Australia	FC	1.0	NOAA reports wx broadcasts at 0018 & 1018
	CW	VID	Darwin Radio, NT, Australia	FC	1.0	AMVER
	CW	VIT	Townsville Radio, Qld., Australia	FC	1.0	AMVER/NOAA reports wx broadcasts at 0930 & 2348
	CW	KOK	Los Angeles Radio, CA	FC	7.5	NOAA reports wx broadcasts at 1650
6464.0	CW	VIS3	Sydney Radio, NSW, Australia	FC	2.5	AMVER
6465.0	CW	UJY	Kaliningrad Radio, USSR	FC	5.0	
	CW	UKA	Vladivostok Radio, USSR	FC	5.0	
6466.0	CW	Y5M	Ruegen Radio, GDR	FC	5.0	
6467.0	CW	VTG5	Bombay Naval Radio, India	FC	5.0	Indian Navy Reported wx broadcasts at 0130, 0900 & 1500
	CW	JCS	Choshi Radio, Japan	FC	3.0	
	CW	LFU	Rogaland Radio, Norway	FC	5.0	AMVER
	CW	ARN	Jiwani Naval Radio, Pakistan	FC	5.0	Pakistan Navy
	CW	HPN6Ø	Canal (Puerto Armuelles) R., Panama	FC	1.0	
	CW	ZSC36	Cape Town Radio, RSA	FC	10.0	
6468.0	CW	DZN	Manila (Navotas) Radio, Philippines	FC	3.0	NOAA reports wx broadcasts at 0200 & 1400
6469.0	CW	HSX	Songkhla Naval Radio, Thailand	FC	1.0	Royal Thai Navy
	CW	HSY	Sattahip Naval Radio, Thailand	FC	1.0	" " "
	CW	HSZ	Bangkok Naval Radio, Thailand	FC	1.0	" " "
6469.3	CW	GKG3	Portishead Radio, England	FC	5.0	
6470.0	CW	JCX	Naha Radio, Okinawa, Japan	FC	1.0	
	CW	UXN	Arkhangelsk Radio, USSR	FC	5.0	
	CW	UQD2	Magadan Radio, USSR	FC	5.0	(RWHC)
6470.1	CW	SXA24	Spata Attikis Naval Radio, Greece	FC	....	Greek Navy/Nato
6470.5	CW	4XO	Haifa Radio, Israel	FC	2.5	
	CW	6YI	Kingston Radio, Jamaica	FC	1.0	
	CW	9YL	North Post Radio, Trinidad	FC	1.0	NOAA reports wx broadcast at 2300
6470.8	CW	GKH3	Portishead Radio, England	FC	5.0	
6472.3	CW	GKI3	Portishead Radio, England	FC	5.0	
6473.0	CW/RTTY	RNO	Moscow Radio, USSR	FC	25.0	SAAMC
6473.5	CW	YQI3	Constanta Radio, Rumania	FC	5.0	
6474.0	CW	JJF	Tokyo Naval Radio, Japan	FC	2.0	JSDA
	CW	OBQ5	Iquitos Radio, Peru	FC	0.5	
	CW	DYV	Manila (Iloilo) Radio, Philippines	FC	0.4	
6475.0	CW	UTA	Tallinn Radio, Estonian SSR	FC	....	
	CW	UNM2	Klaipeda Radio, Lithuanian SSR	FC	5.0	
	CW	UAT	Moscow Radio, USSR	FC	25.0	
	CW	YQBF	unknown	FX	....	clg XUQC
6475.5	CW	DAM	Norddeich Radio, GFR	FC	5.0	TIME SIGNALS (2355-0006 spring & summer months only)
6477.5	CW	XST	Qingdao Radio, PRC	FC	10.0	
	CW	CLQ	Havana (Cojimar) Radio, Cuba	FC	5.0	
	CW	GKJ3	Portishead Radio, England	FC	5.0	
	CW	UBN	Jdanov Radio, Ukrainian SSR	FC	5.0	
	CW	KPH	San Francisco Radio, CA	FC	5.0	
6478.0	CW	ZSC34	Cape Town Radio, RSA	FC	10.0	
6478.5	CW	SVA3	Athens Radio, Greece	FC	10.0	
	CW	9MB3	Penang Naval Radio, Malaysia	FC	....	Malaysian Navy
6479.0	CW	JYO	Aqaba Radio, Jordan	FC	0.5	
6479.8	CW	9MB2	Penang Naval Radio, Malaysia	FC	....	Malaysian Navy
6480.0	CW	9GX	Tema Radio, Ghana	FC	5.0	
	CW/RTTY	UMV	Murmansk Radio, USSR	FC	25.0	
6481.0	CW	JJJ	Maizuru Radio, Japan	FC	1.0	

kHz	Mode	Call	Location	Service	kW	Remarks
	CW/RTTY	URD	Leningrad Radio, USSR	FC	25.0	
6481.5	CW	CCS	Santiago Naval Radio, Chile	FC	....	Chilean Navy
6483.0	CW	UFB	Odessa Radio, Ukrainian SSR	FC	5.0	
6484.5	CW	XSG	Shanghai Radio, PRC	FC	10.0	
	CW	XSV	Tianjin Radio, PRC	FC	1.0	
6485.0	CW	JCU	Choshi Radio, Japan	FC	3.0	
	CW	UAH	Tallinn Radio, Estonian SSR	FC	5.0	
6486.0	CW	DZH	Manila Radio, Philippines	FC	0.3	
6487.0	CW	GXH	Thurso Naval Radio, Scotland	FC	5.0	USN
	CW	UKW3	Korsakov Sakalinsk Radio, USSR	FC	1.0	
6487.5	CW	VRT	Bermuda (Hamilton) Radio, Bermuda	FC	1.0	
6488.0	CW	D4A5	Sao Vicente Radio, Cape Verde	FC	3.0	
	CW	JJG3	Yokosuka Radio, Japan	FC	0.5	
	CW	†C9L2	Maputo Radio, Mozambique	FC	0.5	
6489.0	CW	KPH	San Francisco Radio, CA	FC	5.0	
6890.0	CW	UJE	Moscow Radio, USSR	FC	....	
6490.7	CW	†GYA	Whitehall (London) Naval R., England	FC	10.0	British Navy
6491.5	CW	VCS	Halifax CG Radio, NS, Canada	FC	1.0	AMVER
	CW	PKC	Palembang Radio, Indonesia	FC	1.0	
	CW	JOS	Nagasaki Radio, Japan	FC	3.0	
	CW	PJC	Curacao (Willemstad) Radio, Curacao	FC	2.5	NOAA reports seasonal (7/15-11/14) hurricane warnings broadcast at 0000 & 2300
	CW	TAH	Istanbul Radio, Turkey	FC	3.0	
6492.0	CW	DZM	Manila (Bulacan) Radio, Philippines	FC	10.0	
6492.3	FAX	GYJ3	Whitehall (Northwood) Naval R., Eng.	FC	....	Meteo (0450 & 1545)
6493.0	CW/RTTY	OSN6	Oostende Naval Radio, Belgium	FC	1.0	Belgium Navy
	CW	VFC	Cambridge CG Radio, NWT, Canada	FC	0.3	AMVER
	CW	VFF	Frobisher CG Radio, NWT, Canada	FC	1.0	AMVER/NOAA reports wx broadcasts at 0200, 1100, 1400 & 2300
	CW	VAI	Vancouver CG Radio, BC, Canada	FC	....	AMVER
	CW	UNM2	Klaipedia Radio, Lithuanian SSR	FC	5.0	
	CW	....	numerous, worldwide	FC	....	USN
6493.2	CW	YIR	Basrah Control, Iraq	FC	1.5	NOAA reports wx forecasts on request
6493.5	CW/RTTY	URB2	Klaipeda Radio, Lithuanian SSR	FC	5.0	
6495.0	CW/RTTY	GYA	Whitehall (London) Naval R., England	FC	....	Britsh Navy [aka GYR3]
6496.8	CW	OST32	Oostende Radio, Belgium	FC	10.0	
6498.0	CW/TOR	LGU2	Rogaland Radio, Norway	FC	5.0	
6498.5	CW	SSF	Port Said Naval Radio, Egypt (Also using this frequency: SSG, Matrouh; SSJ, Safaga & SSK, Alexandria--all Egyptian Navy)	FC	0.3	Egyptian Navy
6500.5	CW/TOR	KPH	San Francisco Radio, CA	FC	5.0	
6501.5	CW/TOR	WLO	Mobile Radio, AL	FC	5.0	
6504.0	CW/TOR	ZUD76	Olifantsfontein Radio, RSA	FC	5.0	

6506.4 - 6521.9 kHz — MARITIME COAST USB DUPLEX (Worldwide)

kHz	Mode	Call	Location	Service	kW	Remarks
6506.4	USB	NOJ	COMMSTA Kodiak, AK	FC	10.0	CH#601 (6200.0) USCG/NOAA reports wx broadcasts at 0200 & 1645
	USB	NOX	COMMSTA Adak, AK	FC	10.0	USCG CH#601 (6200.0)
	USB	NRV	COMMSTA Barrigada, Guam	FC	10.0	CH#601 (6200.0) USCG/NOAA reports wx broadcast at 1330 & 1900
	USB	NMO	COMMSTA Honolulu, HI	FC	10.0	" " USCG/NOAA reports wx broadcasts at 0545 & 1145
	USB	ZLW	Wellington Radio, New Zealand	FC	10.0	CH#601 (6200.0)
	USB	EHY	Pozuelo del Rey Radio, Spain	FC	10.0	" "
	USB	NMN	COMMSTA Portsmouth, VA	FC	10.0	" " USCG/NOAA reports wx broadcasts at 0400, 0530, 1000, 1130, 1600, 2200 & 2330
	USB	....	This frequency also used by USCG Coast Stations: NMA, Miami, FL; NMC, San Francisco, CA; NMF, Boston, MA; NMG, New Orleans, LA & NMR, San Juan, PR			

MARITIME COAST USB DUPLEX (con.)

kHz	Mode	Call	Location	Service	kW	Remarks
6509.5	USB	CFH	Maritime Command Radio, Halifax, NS, Canada (Canadian Forces)	FC	10.0	CH#602 (6203.1)
	USB	CGF	Halifax, NS, Canada	FC	1.0	" "
	USB	CGK	Vancouver, BC, Canada	FC	1.0	" "
	USB	CGZ	St. Johns, Nfld., Canada	FC	1.0	" "
	USB	PCG31	Scheveningen Radio, Netherlands	FC	10.0	" "
	USB	CUL	Lisbon Radio, Portugal	FC	5.0	" "
	USB	....	This frequency also used by NOAA Coastal Stations: KWL21, Auke Bay, AK; KWL43, King Salmon, AK; KWL39, Little Port Walter, AK; KAI, San Juan, PR; KAC, Woods Hole, MA; KAF, Atlantic Highlands, NJ; KBR, Beaufort, NC; KHW, Pascagoula, MS; KJS, Kings Point, NY; KVH, "Atlantic Marine Center", Norfolk, VA; KVK, Miami, FL; KVJ, Seattle, WA & KVR, Detroit, MI			
6512.6	USB	LPL3Ø	Gen. Pacheco Radio, Argentina	FC	10.0	CH#603 (6206.2)
	USB	VIO	Broome Radio, WA, Australia	FC	1.0	" "
	USB	VIC	Carnarvon Radio, WA, Australia	FC	1.0	" "
	USB	VID	Darwin Radio, NT, Australia	FC	1.0	" "
	USB	VII	Thursday Island, Qld., Australia	FC	1.0	" "
	USB	VIP	Perth Radio, WA, Australia	FC	1.0	" "
	USB	VIR	Rockhampton Radio, Qld., Australia	FC	1.0	" "
	USB	VIS	Sydney Radio, NSW, Australia	FC	1.0	" "
	USB	VIT	Townsville Radio, Qld., Australia	FC	1.0	" "
			(All of the above Australian Coastal Stations accept AMVER)			
	USB	....	Bermuda (Hamilton) Radio, Bermuda	FC	1.0	CH#603 (6206.2)
	USB	NMG	COMMSTA New Orleans, LA [May be used for special law enforcement activities when required]	FC	5.0	USCG CH#603 (6206.2)
6515.7	USB	P2M	Port Moresby Radio, Papua New Guinea	FC	1.0	CH#604 (6209.3) NOAA reports wx on request
	USB	KRV	Ponce Radio, PR	FC/MS	1.0	Coastal Waterway
	USB	WCM	Pittsburgh, PA	FC/MS	0.8	Inland Waterway
	USB	WJG	Memphis, TN	FC/MS	0.8	" "
6518.8	USB	VAI	Vancouver CG Radio, BC, Canada	FC	1.0	CH#605 (6212.4)
	USB	VCS	Halifax CG Radio, NS, Canada	FC	1.0	" "
	USB	....	Lyngby Radio, Denmark	FC	10.0	" "
	USB	....	Goteborg Radio, Sweden	FC	10.0	" "
	USB	....	Coast & Ship Duplex Frequency Inland & intercoastal waterway traffic--mostly North & South America. Some simplex operation with ships & oil rigs.	FC	....	" "
	USB	....	This frequency also used by USCG Coast Stations: NMA, Miami, FL; NMC, San Francisco, CA; NMF, Boston, MA; NMN, Portsmouth, VA; NMO, Honolulu, HI; NOJ, Kodiak, AK & NRV, Barrigada, Guam			
6521.9	AM/USB	....	INTERNATIONAL COAST STATION CALLING FREQUENCY			CH#606 (6215.5)
	USB	....	Coast & Ship Duplex Frequency Inland & intercoastal waterway traffic--mostly North & South America. Some simplex operation with ships & oil rigs. Frequently heard in North America: KLH, Houston TX; WNK, Greenville, MS; etc.	FC	....	CH#606 (6215.5)
	USB	....	This frequency also used by USCG Coast Stations: NMA, Miami, FL; NMC, San Francisco, CA; NMF, Boston, MA; NMG, New Orleans, LA; NMN, Portsmouth, VA; NMO, Honolulu, HI; NMQ, Long Beach, CA; NMR, San Juan, PR; NOJ, Kodiak, AK & NRV, Barrigada, Guam			

6525.0 - 6765.0 kHz -- AERONAUTICAL MOBILE (Worldwide)

kHz	Mode	Call	Location	Service	kW	Remarks
6526.0	USB	....	numerous, worldwide, operational	FA/MA	....	Airports & in-flight--particularly ARINC New York to Honolulu where phone patches are available.
6540.0	AM/A3H	....	numerous (ICAO CAR)	FA/MA	....	Airports & in-flight eastern Caribbean to New York
6550.0	ISB	....	†Beijing, PRC	FX	....	R. Beijing Feeder

kHz	Mode	Call	Location	Service	kW	Remarks
6556.0	A3H/USB	....	§SEA moves here from 5673.0 in 1983			
6561.0	AM	†....	Riga Aeradio, Latvian SSR	FA	1.0	VOLMET/RR
	AM/A3H	....	numerous (ICAO AFI & CAR)	FA/MA	....	Airports & in-flight covering all of eastern Africa, Indian Ocean to Bombay, all of the western Caribbean and Central America
6568.0	AM/A3H	....	numerous (ICAO CAR)	FA/MA	....	Airports & in-flight covering all of the Caribbean & Central America
6575.0	AM	....	Algiers Aeradio, Algeria	FA	0.4	VOLMET (00+30)
	AM	....	Dakar Aeradio, Senegal	FA	1.3	" (15+45)
	A3H	....	Kano Aeradio, Nigeria	FA	....	" (05+35)
	AM	....	Khartoum Aeradio, Sudan	FA	....	" (25+55)
6582.0	AM/A3H	....	numerous (ICAO EUR)	FA/MA	....	Airports & in-flight covering Central Europe, Mediterranean, Middle East and Soviet Union
6589.0	AM/A3H	....	numerous (ICAO AFI)	FA/MA	....	Airports & in-flight--mostly Equatorial Africa
6608.0	CW/SSB	....	numerous, mostly in USSR	FA/MA	....	USSR Air Force
6610.0	AM/A3H	....	numerous (ICAO AFI & SAT)	FA/MA	....	Airports & in-flight West Africa & across South Atlantic Ocean
	A3H	....	Khabarovsk Aeradio, USSR	FA	....	VOLMET/RR (00+30)
	A3H	....	Novosibirsk Aeradio, USSR	FA	....	" " (10+40)
	A3H	....	Tashkent Aeradio, Uzbek SSR	FA	....	" " (05+35)
6617.0	USB	NMJ22	Attu CG LORAN Station, AK	FA	....	USCG
	AM	....	Brazzaville Aeradio, Congo	FA	1.2	VOLMET (15+45)
	A3H	....	Nairobi Aeradio, Kenya	FA	....	" (05+35)
	AM	....	Antananarive Aeradio, Madagascar	FA	1.5	" (25+55)
	USB	....	Attu & St. Paul, AK	FA/MA	....	USCG [Reeve AA]
	A3H	....	Johannesburg (Jan Smuts) Aeradio, RSA	FA	....	VOLMET (00+30)
	A3H	....	Kiev Aeradio, Ukrainian SSR	FA	....	VOLMET/RR (20+50)
	AM	....	Leningrad Aeradio, USSR	FA	....	" " (05+35)
	AM	....	Moscow Aeradio, USSR	FA	....	" " (15+45)
	A3H	....	Riga Aeradio, Latvian SSR	FA	....	" " (00+30)
	A3H	....	Rostov Aeradio, USSR	FA	....	" " (10+40)
6624.0	AM/A3H	....	numerous (ICAO MID)	FA/MA	....	Airports & in-flight covering Middle East to India sub-continent
6631.0	AM/A3H	....	numerous (ICAO CWP)	FA/MA	....	Airports & in-flight covering all of the western Pacific Ocean from Honolulu to as far west as Ulan Bator
6638.0	AM/A3H	....	numerous (ICAO AFI)	FA/MA	....	Airports & in-flight covering entire western Africa
	A3H	....	Novorossiysk Aeradio, USSR	FA	....	VOLMET/RR (cont.)
6645.0	CW/USB	†KJY74	Miami "Monitor", FL	FA	....	NOAA
6666.0	AM/A3H	....	numerous (ICAO SAM)	FA/MA	....	Airports & in-flight covering entire western South America
6670.0	USB	†....	numerous, worldwide	FA/MA	....	USAF
6674.0	CW	....	unknown	FA	....	Wx 1100
6680.0	AM	....	Bombay Aeradio, India	FA	0.8	VOLMET (25+55)
	AM	....	Calcutta Aeradio, India	FA	0.8	" (05+35)
	AM	....	Karachi Aeradio, Pakistan	FA	....	" (15+45)
	AM	9VA4Ø	Singapore Aeradio	FA	1.6	" (20+50)
	A3H	VLS	Sydney Aeradio, NSW, Australia	FA	3.0	" (00+30)
	AM	....	Bangkok Aeradio, Thailand	FA	1.6	" (10+40)
6683.0	ISB	....	several, Central & North America	FA/MA	....	USAF
6685.0	CW	†....	several, in Canada	FA	....	RCAF
6686.0	CW/USB	....	numerous, in Europe & USA	FA/MA	....	USAF/USN
6690.0	USB	MQP	West Drayton (Upavon) AB, England	FA/MA	2.0	RAF
6691.0	USB	....	numerous, in Australia	FA/MA	....	RAAF
6693.0	USB	....	several, in Canada	FA/MA	....	RCAF (Particularly CJX, "St. Johns Military", Nfld. & CZW, "Halifax Military", NS)
	AM/USB	....	several, interior of the USSR	FA	....	VOLMET/RR (Includes: Salekhard Aeradio [25+55]; Sverdlovsk Aeradio [05+35]; Syktyvka Aeradio [00+30]; Yeniseysk Aeradio [10+40]; --all low/medium power and rarely heard in North America)
6697.0	USB	MKL	Pitreavie Castle, Scotland	FA	3.5	RAF

kHz	Mode	Call	Location	Service	kW	Remarks
	USB	....	numerous, east & Gulf coasts	FA/FC	....	USN (Primary coordination frequency--used by NAF, Newport, RI; NAM, Norfolk, VA; NHK, Patuxent, MD; etc.)
6701.0	USB	....	numerous, worldwide	FA/FC	....	USN
6705.0	USB	....	several, in Canada	FA/MA	....	RCAF (Usually CHR, "Trenton Military", Ont.; CJX, "St. Johns Military" Nfld.; CKN, "Vancouver Military" & VXA, "Edmonton Military", Alta.)
6706.0	CW	Y7A24	Berlin, GDR	FX	....	EMBASSY
6708.0	USB	AFE71	"Cape Radio", Patrick AFB, FL	FX	....	USAF/NASA Shuttle Support [Alt: 5810.0 & 9006.0]
6714.0	CW	†MQY	Pitreavie Castle, Scotland	FA	1.0	RAF
	CW	MQD	Plymouth, England	FA	1.0	"
	FAX	....	unknown, probably in North America	FX	.....	
6715.0	ISB	....	numerous, worldwide	FA/MA	....	RAAF, RCAF & USAF
6720.0	USB	....	numerous, worldwide	FA/MA	....	USN (Used for "OVERWORK" broadcasts [Alt: 8277.2]
6723.0	ISB	....	numerous, USA & surrounding waters	FA/MA	....	USN (Primary night frequency) [Also used by USAF]
6726.0	USB	....	several, in England	FA/MA	....	RAF
6727.0	USB	AFG37	Scott AFB, IL	FA/MA	....	USAF
6730.0	AM	....	Aktyubinsk Aeradio, Kazakh SSR	FA	....	VOLMET/RR (05+35)
	A3H	....	Alma Ata Aeradio, USSR	FA	....	" " (15+45)
	AM	....	Baku Aeradio, Azerbaijan SSR	FA	....	" " (25+55)
	AM	....	Krasnodar Aeradio, USSR	FA	....	" " (10+40)
	USB	....	Tashkent Aeradio, Uzbek SSR	FA	4.0	" " (20+50)
	AM	....	Tbilisi Aeradio, Georgian SSR	FA	....	" " (00+30)
6735.0	USB	....	numerous, western USA & Far East	FA/MA	....	USAF
6738.0	USB	....	numerous	FA/MA	....	USAF (Typically used by AIE2, Andersen AFB, Guam; AIC2, Clark AB, Philippines; AKA5, Elemendorf AFB, AK; AGA2, Hickam AFB, HI; XPH, Thule AB, Greenland; AFI2, McClellan AFB, CA; AIF8Ø, Yokota AB, Japan for MAINSAIL & SKY KING broadcasts) [Also used by the RAAF, RAF, RCAF & RNZAF]
6740.0	USB	AFE8	MacDill AFB, FL	FA/MA	....	USAF
6745.0	USB	....	several, in England	FA/MA	....	RAF
6745.5	CW	CAK	Santiago (Los Cerrillos AB), Chile	FA/MA	....	Chilean AF NOAA reports wx forecasts at 0130, 1330 & 2000
6748.0	CW	RFNV	Moscow Aeradio, USSR	FA/MA	....	Aeroflot (Mostly in-flight to/from Havana/Moscow
6750.0	USB	....	several, worldwide	FA/MA	....	USAF (Typically used by CUW, Lajes Field, Azores; AJE, Croughton, England; AFE8, MacDill AFB, FL; etc.) [Also used by RAF]
6753.0	USB	CHR	"Trenton Military", Ont., Canada	FA	4.0	RCAF/Wx (+30)
	USB	CJX	"St. Johns Military", Nfld., Canada	FA	4.0.	" " (+40)
	USB	CKN	"Vancouver Military", BC, Canada	FA	4.0	" " (+35)
	USB	VXA	"Edmonton Military", Alta., Canada	FA	4.0	" " (+20)
6756.0	ISB	....	several, worldwide	MA	....	USAF VIP A/C
6760.0	USB	....	several, in New Zealand	FA/MA	....	RNZAF
6760.8	USB	....	unknown, probably Central America	FA	....	Net/SS
6761.0	USB	....	numerous	FA/MA	....	USAF (Used for "SKY KING" broadcasts--some code names for air bases--sometimes referred to as the SAC "QUEBEC" frequency)

6765.0 - 7000.0 kHz -- FIXED SERVICE (Worldwide)

kHz	Mode	Call	Location	Service	kW	Remarks
6765.0	FAX	HSW69	Bangkok, Thailand	FX	....	Meteo (0330, 0500 & 1700)
6768.0	CW	FDY	Orleans Air, France	FX	....	French AF
6770.0	ISB	RAN78	Moscow, USSR	FX	50.0	R. Moscow Feeder
6775.0	USB	TFK	Keflavik Airport, Iceland	FX	2.0	USAF
	USB	....	unknown, probably Central America	FX	....	Net/SS
6778.5	USB	HJM273	Bogota, Colombia	FX	0.2	Net Control
6778.6	CW	....	unknown	FX	....	Tfc/SS
6785.0	USB	WUJ3	Portland, OR	FX	....	USACE

kHz	Mode	Call	Location	Service	kW	Remarks
	USB	VEW	unknown	FX	....	
6788.5	USB	VCA77	St. Anthony, Nfld., Canada	FX	0.1	Canadian Fishing
6790.0	FAX	YMA22	Ankara, Turkey	FX	....	Meteo
					(0500-1400)	
6792.0	CW	FSB54	Paris (St. Martin Abbat), France	FX	2.5	INTERPOL
	CW	50P25	Lagos, Nigeria	FX	1.5	"
	CW	LJP26	Oslo, Norway	FX	1.5	"
6796.4	LSB	ART	Rawalpindi, Pakistan	FX	10.0	Telcom/Teheran
6800.0	CW	....	unknown	FX	....	Coded Tfc
6802.0	USB	†FPG8Ø	Paris (St. Assise), France	FX	20.0	Telcom
6803.5	CW/RTTY	ONN3Ø	Brussels, Belgium	FX	0.5	EMBASSY
6805.0	CW	JJC2Ø	Tokyo, Japan	FX	0.5	Tfc
6808.0	USB	†RAT25	Moscow, USSR	FX	20.0	R. Moscow Feeder
6810.0	CW/USB	....	several, in Australia	FX	1.0	Monitoring Net-
			work (Includes: VNA3, Melbourne; VNA4, Brisbane; VNA5, Adelaide; VNA6, Perth; VNA7, Hobart & VNA21, Townsville) [Alt: 4550.0 & 9940.0]			
	USB	....	Beijing, PRC	FX	....	R. Beijing Feeder
6812.0	USB	....	Aleutian LORAN Net	FX	....	USCG (Includes:
			NMJ22, Attu; NOJ, Kodiak; NRW2, St. Paul Is. & NRW3, Pt. Clarence--all AK) [Alt: 5422.0 & 6935.0]			
6820.0	FAX	JKA2	Tokyo (Kemigawa), Japan	FX	5.0	
6825.0	USB	VJQ	Kalgoorlie, WA, Australia	FX	0.3	Outpost Control
6810.0	ISB	RBV79	Tashkent, Uzbek SSR	FX	20.0	R. Moscow Feeder
6826.0	USB	AFS	Offutt AFB, NE	FX	0.5	USAF/SAC
				[Alt: 5026.0 & 11,494.0]		
6830.0	ISB	FTG83	Paris (St. Assise), France	FX	20.0	Telcom
6831.5	LSB	VQJ366	Honiara, Solomon Islands	FX	0.5	Net Control
6833.6	USB	....	numerous, Hawaii to US east coast	FA/FC	1.0	USN (Usually
			NAM, Norfolk, VA; NAX, Barbers Pt., HI; NEL, Lakehurst NAS, NJ; NFC, Cape May NAS, NJ; NGZ, Alameda NAS, CA; NHK, Patuxent NAS, MD; NHZ, Brunswick NAS, ME; etc.)			
6840.0	USB	VJQ	Darwin, NT, Australia	FX	0.4	Net Control
	CW/RTTY	OMZ	Prague, Czechoslavkia	FX	1.0	EMBASSY
	CW	EBC	Cadiz Radionaval, Spain	FX	....	Spanish Navy
				TIME SIGNALS (1029-1055, only)		
6842.0	LSB	....	Daventry, England	FX	30.0	BBC Feeder
	CW/RTTY	RLX	Dublin, Ireland	FX	0.5	Russian EMBASSY
					[Alt: 5744.0 & 9217.0]	
6843.0	CW	†OMZ	Prague, Czechoslovakia	FX	1.0	EMBASSY
6845.0	USB	VJJ	Charleville, Qld., Australia	FX	0.3	Outpost Control
6847.5	ISB	FTG84	Paris (St. Assise), France	FX	20.0	Telcom/Abidjan
6850.0	USB	VLZ	Davis Base, Antarctica	FX	1.0	ANARE
	USB	VNJ	Casey Base, Antarctica	FX	10.0	ANARE
6852.1	CW	BZOS	unknown	FX	....	Coded Tfc
6858.5	LSB	†....	Lyndhurst, Vic., Australia	FX	30.0	R. Australia Feeder
	CW	....	several, throughout France	FX	0.5	French AF
6860.0	USB	....	unknown	FX	....	Telcom/SS
6866.5	CW	KWL9Ø	Tokyo, Japan	FX	....	USA EMBASSY
6868.5	USB	....	several, eastern USA	FX	....	USAF/SAC
6870.0	LSB	KDM5Ø	Atlanta (Hampton), GA	FX	....	FAA Net Control
			(Some use of code names, such as, "Associate", "Drag-net", etc.) [Alt: 8125.0]			
6871.0	CW	†OMZ	Prague, Czechoslovakia	FX	1.0	EMBASSY
6873.0	ISB	....	Greenville, NC	FX	50.0	VOA Feeder
6875.0	ISB	AJF7	Rhein-Main AB, GFR	FX	1.0	USAF
6878.5	CW/USB	NQM	Sand Island Naval Radio, Midway Is.	FC	3.0	USN
6880.0	FAX	RAN77	Moscow, USSR	FX	....	Meteo
6885.0	CW/LSB	†NAW11	Guantanamo, Cuba	FX	....	Telcom/Tfc
6888.5	USB	RAN76	Moscow, USSR	FX	15.0	R. Moscow Feeder
6890.0	USB	VJT	Carnarvon, WA, Australia	FX	0.3	Outpost Control
	USB	VNZ	Port Augusta, SA, Australia	FX	0.3	Outpost Control
6893.0	FAX	JKD2	Tokyo, Japan	FX	5.0	
6895.0	CW	VKV1	Cato Is., Qld., Australia	FX	0.1	

FIXED SERVICE (con.)

kHz	Mode	Call	Location	Service	kW	Remarks
	CW/RTTY	PBC92	Goeree Naval Radio, Netherlands	FX	5.0	Dutch Navy
6898.0	ISB	MKG	London (Stanbridge), England	FX	30.0	RAF Telcom/Tfc
6901.0	FAX	SMA6	Stockholm (Norrkoping), Sweden	FX	....	Meteo (cont.)
6905.0	USB	VKR	Brisbane, Qld., Australia	FX	....	Police
	CW/RTTY	OEQ38	Vienna, Austria	FX	1.0	INTERPOL
	CW/RTTY	SHX	Stockholm, Sweden	FX	1.0	"
	CW/RTTY	HEP26	Zurich (Waltikon), Switzerland	FX	3.0	"
6912.5	FAX	AFA	Washington (Andrews AFB, MD), DC	FX	3.0	Meteo
6915.0	CW/RTTY	LHB	Oslo (Jeloey), Norway	FX	25.0	
6918.5	FAX	ECA7	Madrid, Spain	AX	5.0	Meteo
						(0410 & 1555)
	LSB	OCP	Ouagadougou, Upper Volta	FX	1.5	Net Control
			[Includes stations in Benin, Ghana, Ivory Coast, Mali, Niger & Togo]			
	USB	RAT25	Moscow, USSR	FX	15.0	R. Moscow Feeder
6920.0	A3H	VJC	Broken Hill, NSW, Australia	FX	1.2	Outpost Control
6925.0	USB	VJB	Derby, WA, Australia	FX	0.3	Outpost Control
6925.4	CW	KKN5Ø	Washington, DC	FX	....	EMBASSY
6934.5	LSB	A2P	Gaborone, Botswana	FX	0.4	Net Control
6935.0	USB	....	Aleutian LORAN Net	FX	....	USCG (For list of stations see 6812.0)
6937.0	CW	....	several, in Queensland, Australia	FX	....	Meteo/Net (Includes: VKV2, Frederick; VKV3, Marion & VKV5, Flinders)
6943.5	USB	....	Mediterranean LORAN Net	FX	....	USCG (Includes: NCI, Sellia Marina, Italy & NCI3, Lampedusa, Italy) [Alt: 5066.0 & 7441.0]
6945.0	USB	VKF	Wyndham, WA, Australia	FX	0.3	Outpost Control
	USB	....	Aleutian LORAN Net	FX	....	USCG (For list of stations see 6812.0 or 7441.0)
	USB	TJF25	Douala, Cameroon	FX	6.0	Telcom
	USB	T2U	Funafuti, Tuvala	FX	0.2	Net Control
6946.0	CW	CKN	Vancouver Forces Radio, BC, Canada	FX	10.0	(C13E) C. Forces NOAA reports wx forecasts at 0030, 0430 & 1330
	FAX	CKN	Vancouver Forces Radio, BC, Canada	FX	10.0	C. Forces/Meteo
6946.2	FAX	†CLN53	Havana (Bauta), Cuba	FX	30.0	
6950.0	USB	VJD	Alice Spring, NT, Australia	FX	0.3	Outpost Control
	FAX	RBQ74	Alma Ata, USSR	FX	20.0	Meteo
6957.0	ISB	NAW25	Guantanamo, Cuba	FX	....	Telcom
6960.0	USB	VJT	Carnavon, WA, Australia	FX	0.3	Outpost Control
	USB	VKJ	Meekatharra, WA, Australia	FX	0.3	" "
	USB	VKL	Port Hedland, WA, Australia	FX	0.3	" "
6965.0	USB	VJI	Mt. Isa, Qld., Australia	FX	0.3	Outpost Control
	CW	YMH3	Bandirma, Turkey	FX	0.1	NOAA reports wx broadcasts at 0450, 0650, 1030, 1710, 1850 & 2230
	CW	YMY	Samsun, Turkey	FX	....	Meteo
6968.5	USB	NAR	Key West NAS, FL	FA/FC	1.0	USN
	USB	NAW	Guantanamo, Cuba	FC	1.0	USN
	USB	NAX	Barbers Point Naval Radio, HI	FA/FC	1.0	USN
6970.0	ISB	VNV	Sydney (Doonside), NSW, Australia	FX	30.0	Telcom/Norfolk
	ISB	....	Holzkirchen, GFR	FX	10.0	RFE Feeder
6975.0	USB	VJY	Darwin, NT, Australia	FX	0.5	Outpost Control
6975.0	USB	DFF97	Frankfurt, GFR	FX	20.0	DW Feeder
6976.4	LSB	AQY286	Karachi, Pakistan	FX	1.0	Net Control
6977.5	LSB	....	numerous, throughout western USA	FX	....	NOAA Net
						[Infrequently used]
6986.0	LSB	....	numerous, throughout Cuba	FX	....	Net
6987.0	CW/RTTY	ONN3Ø	Brussels, Belgium	FX	0.5	EMBASSY
6987.5	USB	RAT23	Moscow, USSR	FX	20.0	R. Moscow Feeder
6995.0	ISB	MKG	London (Stanbridge), England	FX	30.0	RAF Telcom/Tfc
	ISB	....	Holzkirchen, GFR	FX	10.0	RFE Feeder
6996.0	ISB	SAR88	Stockholm, Sweden	FX	1.0	Inter. RED CROSS
	USB	....	several, western USA & Far East	FX	....	USAF MARS
6997.5	LSB	....	several, southeastern USA	FX	....	NOAA
6998.5	CW	HBC88	Geneva, Switzerland	FX	0.8	Inter. RED CROSS
6999.5	LSB	ONY52	unknown	FX	....	Belgium Army/ NATO

kHz	Mode	Call	Location	Service	kW	Remarks

7000.0 - 7100.0 kHz -- AMATEUR RADIO (Worldwide)

7100.0 - 7300.0 kHz -- AMATEUR RADIO/INTERNATIONAL BROADCASTING (Shared)

7300.0 - 8100.0 kHz -- FIXED SERVICE (Worldwide)

kHz	Mode	Call	Location	Service	kW	Remarks
7302.1	CW	IDR3	Rome Naval Radio, Italy	FX	15.0	Italian Navy
7302.0	USB	....	numerous, south & central USA	FX	....	USAF MARS
7305.0	FAX	JMH2	Tokyo, Japan	FX	5.0	Meteo
	USB	....	numerous, southeastern USA	FX	....	USAF MARS
	FSK	....	unknown, probably in North America	FX	....	
7306.5	ISB	5TA14	Nouadhibou, Mauritania	FX	6.0	Telcom/Paris
7307.0	USB	....	several, throughout Australia	FX	....	Outpost Net
	USB	6YF21	Kingston (Coopers Hill), Jamaica	FX	1.0	Telcom
7308.0	CW	....	several, worldwide	FX	....	French Navy
7310.0	USB	VL6BA	Hammersley, WA, Australia	FX	0.5	Commercial Tfc
7312.0	USB	VL6NX	Wanneroo, WA, Australia	FX	1.0	Telcom
7313.5	USB	....	numerous, USA nationwide	FX	....	USAF MARS
7318.5	USB	PUZ4	Brasilia, Brazil	FX	1.0	Governmental
7320.0	ISB	AFH3	Albrook AFS, Panama	FX	3.0	USAF
					[Alt: 9473.0]	
7324.0	USB	....	numerous, northeastern USA	FX	....	USAF MARS
7327.5	FAX	JAE27	Tokyo, Japan	FX	10.0	
7329.0	USB	....	numerous	FX	....	USAF MARS
7330.1	USB	XBD959	unknown, probably in Mexico	FX	....	Tfc/SS
7335.0	A3H	CHU	Ottawa, Ont., Canada	SS	10.0	TIME SIGNALS
7336.0	USB	PRX347	Brasilia, Brazil	FX	0.1	Net Control
7345.0	USB	....	numerous, USA nationwide	FX	....	USN MARS
7349.5	USB	CTH27	Horta Naval Radio, Azores	FX	5.0	Port. Navy
7350.0	LSB	†XDD212	Mexico City, DF, Mexico	FX	10.0	Telcom
7351.0	CW	CTH41	Horta Naval Radio, Azores	FC	3.0	Port. Navy
			NOAA reports wx broadcast at 2130			
7353.0	CW	†CTV27	Monsanto Naval Radio, Portugal	FC	3.0	Port. Navy
7370.0	ISB/RTTY	FTH37	Paris (St.Assise), France	FX	20.0	
	FAX	†JKE5	Tokyo (Usui), Japan	FX	5.0	
7373.0	USB	....	Bloemfontein, RSA	FX	1.0	Net Control
						[Alt: 7655.0]
7374.0	CW	EEQ	Madrid, Spain	FX	1.0	Net Control
7377.5	USB	†....	North Atlantic (East Coast) LORAN Net	FX	....	USCG (Includes: NMA7, Jupiter, FL; NMP32, Nantucket, MA; NMF33, Caribou, ME; NMN73, Carolina Beach [Cape Fear], NC; & NOL, "Bermuda Monitor"--daytime, only)
7389.5	LSB	YVL5	Caracas, Venezuela	FX	0.5	Telcom
7390.0	USB	ZKS	Rarotonga, Cook Islands	FX	1.0	Net Control
					[Alt: 5930.0 & 9095.0]	
7392.0	CW	....	unknown	FX	....	Coded Tfc
	LSB	....	Novosibirsk, USSR	FX	....	R. Moscow Feeder
7394.0	USB	....	numerous, worldwide	FX	....	USN MARS
7395.0	FAX	HSW64	Bangkok, Thailand	FX	3.0	Meteo
				(0330, 0500 & 1700)		
7398.0	USB	LOL	Buenos Aires, Argentina	FX	1.0	Prefectura Naval
7400.0	USB	CGD432	Kenora, Ont., Canada	FX	0.4	Net Control
7401.0	CW/RTTY	OEQ39	Vienna, Austria	FX	1.0	INTERPOL
	CW	ONA2Ø	Brussels, Belgium	FX	3.0	"
	CW/RTTY	FSB	Paris (St. Martin Abbat), France	FX	3.0	"
	CW	DEB	Wiesbaden, GFR	FX	....	"
	CW	HEP74	Zurich (Waltikon), Switzerland	FX	3.0	"
7405.0	FAX	ATP57	New Delhi, India	FX	15.0	Meteo
					(1430-0200)	
7407.0	CW	FDC7	Strasbourg Air, France	FX	....	French AF
7408.5	USB	†RAN72	Moscow, USSR	FX	20.0	R. Moscow Feeder
7417.0	USB	....	Daventry, England	FX	30.0	BBC Feeder
7422.0	USB	VL4SX	Birdsville, Qld., Australia	FX	0.1	Outpost Control
	USB	VL5LM	Alice Springs, NT, Australia	FX	0.1	" "
	USB	VL6OF	Giles, WA, Australia	FX	0.1	" "
7427.5	ISB	ZEN32	Cape d'Aguilar, Hong Kong	FX	10.0	Telcom/Calcutta

FIXED SERVICE (con.)

kHz	Mode	Call	Location	Service	kW	Remarks
7428.0	CW	FTH42	Paris (St. Assise), France	FX	20.0	TIME SIGNALS/Tfc
	CW	NDT6	Totsuka Naval Radio, Japan	FC	5.0	USN
	CW	NDT	Yokosuka Naval Radio, Japan	FX	1.0	USN
7430.0	ISB	†NAW26	Guantanamo, Cuba	FX	....	Telcom
	ISB	†ZPB74	Asuncion, Paraguay	FX	10.0	Telcom/Uruguay
7434.0	CW	KWS78	Athens, Greece	FX	....	USA EMBASSY
7440.0	ISB	†....	Holzkirchen, GFR	FX	10.0	RFE Feeder
	ISB	RVF53	Dushambe, USSR	FX	20.0	R. Moscow Feeder
7441.0	USB	....	Mediterranean LORAN Net	FX	....	USCG (Includes:
			AOB5Ø, Estartit, Spain; NCI, Sellia Marina, Italy & NCI3, Lampedusa, Italy) [Alt:6943.5 & 8021.5]			
	USB	....	Aleutian LORAN Net	FX	....	USCG (Includes:
			NMJ22, Attu; NOJ, Kodiak; NRW2, St. Paul Is. & NRW3, Pt. Clarence--all AK) [Alt: 6945.0 & 10,368.5]			
7445.0	CW	VCS838	Ottawa, Ont., Canada	FX	....	Yugoslav EMBASSY [Alt: 8172.0]
7457.0	FAX	RHB/RHO	Khabarovsk, USSR	FX	....	Meteo
	USB	....	numerous, southwestern USA	FX	....	USAF MARS
7465.0	USB	VJN	Cairns, Qld., Australia	FX	0.3	Outpost Control
	CW	LBJ7	Harstad Naval Radio, Norway	FX	....	Norwegian Navy
7470.0	CW	†KKN5Ø	Washington, DC	FX	10.0	EMBASSY
	LSB	RLG71	Yuzhno-Sakhalinsk, USSR	FX	....	R. Moscow Feeder
7473.0	USB	....	Hawaii & Central Pacific LORAN Net	FX	....	USCG (Includes:
			NMO, Lualualei, HI; NRO, Johnston Is.; NRO5, Upolu Point, HI & NRO7, Kure Is. [Ocean Is.], HI) [Alt: 9303.0]			
7475.0	FAX	RHB/RHO	Khabarovsk, USSR	FX	....	Meteo
	FAX	RMP44	Petropavlo Kam, USSR	FX	20.0	Meteo
	USB	KDM5Ø	Atlanta (Hampton), GA	FX	....	FAA Net Control
			(Also heard have been KDM95, Gulfport, MS; KKU4Ø, Kansas City, MO; etc.) (Alt: 6870.0 & 8125.0]			
7483.5	ISB	†MKE	Akrotiri, Cyprus	FX	10.0	RAF Telcom/Tfc
7493.0	LSB	....	Yuzhno-Sakhalinsk, USSR	FX	....	R. Moscow Feeder
7498.5	LSB	†5TN228	Nouakchott, Mauritania	FX	6.0	Telcom/Paris
7498.6	USB	....	numerous, mostly eastern USA	FA/FC	1.0	USN
7500.0	AM	VNG	Lyndhurst, Vic., Australia	SS	10.0	TIME SIGNALS
7505.5	CW	GXH	Thurso Naval Radio, Scotland	FX	....	USN/NOAA re-
			ports wx broadcasts at 0630 & 1900			
	CW	NAM	Norfolk Naval Radio, VA	FX	....	USN
7507.0	USB	NMR	COMMSTA San Juan, PR	FX	....	USCG/USN
			Hurricane Warning Net ["PAPA" Channel]			
7508.0	FAX	ZRO2	Pretoria, RSA	FX	3.5	Meteo
7512.5	USB	†....	Norwegian Sea & N. Atlantic LORAN Net		....	USCG (Includes:
			DML, Sylt, GFR; JXL, BØ, Norway, JXP, Jan Mayen Is.; NMS, Shetland Is.; OUN, Ejde, Denmark; OVY, Anqissog, Greenland; TFR, Sandur, Iceland & TFR2, Reykjavik, Iceland) [Alt: 4857.5 & 10,337.5]			
7515.0	CW	JMB2	Tokyo, Japan	FX	2.0	Meteo
7518.0	ISB	†IRE25	Rome, Italy	FX	10.0	Telcom
7519.0	USB	....	Daventry, England	FX	30.0	BBC Feeder
7527.0	USB	....	This frequency is used by the US Customs Service for supplemental communications. Callsigns are usually code names and have included: "Homeplate"; Hightide"; Slingshot"; "Omaha"; etc.			
7528.6	USB	....	Atlantic Domestic Emergency Net	FX	....	USCG (Includes:
			NMA, Miami, FL; NMF, Boston, MA; NMG, New Orleans, LA; NMN, Portsmouth, VA; NMR, San Juan, PR & NOZ, Elizabeth City, NC--plus transportables) [Alt: 4046.6 & 9125.6]			
	USB	....	Pacific Area Domestic Emergency Net	FX	....	USCG (Includes:
			NMC, San Francisco, CA; NMO, Honolulu, HI; NMQ, Long Beach, CA; NOJ, Kodiak, AK & NRV, Barrigada, Guam--plus transportables) [Alt: 4046.6 & 9125.6]			
7530.0	USB	....	West Indies LORAN Net	FX	....	USCG (Includes:
			NMA, Miami, FL; NMA4 San Salvador [Watling Is.], Bahamas; NMA5, South Caicos, Bahamas; NMA7, Jupiter, FL & NMR, San Juan, PR) [Alt: 5422.5 & 8083.5]			
	FAX	RKIC	Moscow, USSR	FX	15.0	TASS Nx

kHz	Mode	Call	Location	Service	kW	Remarks
7532.0	CW	OGX	Helsinki, Finland	FX	1.0	INTERPOL
	CW/RTTY	DEB	Wiesbaden, GFR	FX	5.0	"
	CW	8UF75	New Delhi, India	FX	5.0	" [Alt: 8006.5]
	CW/RTTY	JPA55	Nagoya (Komaki), Japan	FX	10.0	INTERPOL
	CW	LJP26	Oslo, Norway	FX	....	"
	CW	SHX	Stockholm, Sweden	FX	1.0	"
	USB	†9HC32A	Malta, Malta	FX	10.0	Telcom
7535.0	FAX	AXI33	Darwin, NT, Australia	FX	5.0	Meteo (0800-2100)
7536.9	CW	XUQC	unknown	FX	....	Coded Tfc
7537.0	CW/RTTY	†ONN3Ø	Brussels, Belgium	FX	1.0	EMBASSY
7537.6	CW/FAX	RLI9	Kalinkovitch, USSR	FX	20.0	
7539.0	CW	ART	Rawalpindi. Pakistan	FX	1.0	Tfc/Ankara
7540.0	USB	....	several, USA nationwide	FX	....	USAF
7541.5	ISB	†ROK28	Moscow, USSR	FX	25.0	R. Moscow Feeder
7548.5	USB	†RNN58	Moscow, USSR	FX	15.0	R. Moscow Feeder
7553.5	USB	HVC	Vatican City, Vatican	FX	10.0	Telcom
7558.5	USB	6VK27	Dakar, Senegal	FX	6.0	Telcom
7565.0	USB	VJC	Broken Hill, NSW, Australia	FX	0.3	Outpost Control
	CW	FIT75	Paris, France	FX	0.5	
7570.0	CW	OZL4Ø	Angmagssalik, Greenland	FX	4.0	NOAA reports wx forecasts at 0705, 1305 & 1905
7571.8	LSB	AJE	Croughton (Barford), England	FX	4.0	AFRTS Feeder
7576.0	CW/USB	....	numerous, throughout France	FX	0.3	GDF [Most stations have the prefix FVA]
7577.0	CW	†RIW	Khiva Naval Radio, Uzbek SSR	FX	....	Russian Navy
7580.0	USB	NRV	COMMSTA Barrigada, Guam	FX	3.0	USCG
	USB	NRT	COMMSTA Totsuka, Japan	FX	1.0	USCG
7584.0	USB	....	Daventry, England	FX	30.0	BBC Feeder
7587.5	FAX	6VY41	Dakar, Senegal	FX	5.0	Meteo (2000-0830)
7593.5	USB	....	numerous, USA nationwide	FA/FC	....	USN
7600.0	AM	HD21ØA	Guayaquil, Ecuador	SS	....	TIME SIGNALS
	CW	ZLZ22	Wellington (Himatangi), New Zealand	FX	5.0	NOAA reports wx broadcasts at 0200, 0805, 0815. 0840, 0910, 1400, 2005, 2015, 2040 & 2110
7605.0	CW	FUB	Paris (Houilles) Naval R., France	FX	10.0	French Navy
	CW	FUJ	Noumea Naval Radio, New Caledonia	FX	1.0	" "
	CW	FUM	Papeete Naval Radio, Tahiti	FX	2.5	" "
	USB	SAM	Stockholm, Sweden	FX	1.2	EMBASSY
	USB	SAM38	Moscow, USSR	FX	1.0	Swedish EMBASSY
7606.0	USB	WEH97	New York, NY	FX	30.0	Telcom
7624.0	CW	†OMZ	Prague, Czechoslovakia	FX	1.0	EMBASSY
7630.0	ISB	JBE67	Tokyo, Japan	FX	10.0	Nx/Tfc
7632.0	USB/RTTY	AIR	Andrews AFB, MD	FX	3.0	USAF/Tfc AJF7
7633.9	CW	KKN44	Monrovia, Liberia	FX	....	USA EMBASSY
7635.0	USB	....	numerous, southwestern USA	FX	....	CAP
7645.0	CW	KWS78	Athens, Greece	FX	....	USA EMBASSY
	FAX	NPN	Barrigada (Agana NAS), Guam	FX	30.0	USN/Meteo
7647.5	CW/RTTY	ONN3Ø	Brussels, Belgium	FX	0.5	EMBASSY [Alt: 7537.0 & 9910.0]
	CW/RTTY	OMZ29	Bratislava, Czechoslovakia	FX	1.0	GDR EMBASSY [Alt: 5868.0 & 9078.0]
7650.0	CW	ONN38	Brussels, Belgium	FX	0.1	Swiss EMBASSY
7651.0	USB	LTC31	Buenos Aires, Argentina	FX	0.5	Net Control
	LSB	....	Greenville, NC	FX	50.0	VOA Feeder
7652.0	CW	KKN44	Monrovia, Liberia	FX	....	USA EMBASSY
7655.0	ISB	....	Bloemfontein, RSA	FX	1.0	Net Control [Alt: 7373.0]
7656.0	CW	"W"	unknown	FX	....	Beacon
7660.0	USB	....	numerous, in Australia	FX	1.0	Police (Includes: VKA, Adelaide; VKC, Melbourne; VKI, Perth; VKM, Darwin; VKR, Brisbane; VKT, Hobart & VKX, Canberra) [Alt: 4560.0 & 10,505.0]
7662.0	CW	KWL9Ø	Tokyo, Japan	FX	....	USA EMBASSY

kHz	Mode	Call	Location	Service	kW	Remarks
7665.0	CW	†UGE2	Bellingshausen USSR Base, Antarctica	FX	....	SAAM
	CW/RTTY	CUA33	Lisbon (Alfragide), Portugal	FX	10.0	Tfc
7670.0	FAX	RCC76	Moscow, USSR	FX	15.0	Meteo
7677.0	ISB	†FTH67	Paris (St. Assise), France	FX	20.0	Telcom
7685.0	USB	....	Daventry, England	FX	30.0	BBC Feeder
7686.0	FAX	....	unknown, probably in North America	FX	....	
7688.0	ISB	†FTH68	Paris (St. Assise), France	FX	20.0	Telcom
7691.0	USB	....	Funafuti, Tuvalu	FX	0.2	Police
7693.0	CW	ONY27	Rouveroy, Belgium	FX	....	Belgium Army
7694.0	CW	SOH269	Warsaw, Poland	FX	20.0	Tfc
7698.5	USB	PCW	The Hague, Netherlands	FX	1.2	EMBASSY
	USB	PCW2	Jerusalem, Israel	FX	1.0	Dutch EMBASSY
7700.0	ISB	XTA77	Ouagadougou, Upper Volta	FX	20.0	Telcom/Tfc
	USB	....	several, USA nationwide	FX	....	DOE
			[Uses call prefix "KRF" or "KRM" for fixed stations and "Lobo" for mobiles]			
7705.0	CW	NAM	Norfolk Naval Radio, VA	FX	....	USN
7706.0	CW	AOK	Moron de la Frontera, Spain	FX	5.0	USN/NOAA re-
			ports wx broadcasts at 0030, 0630, 1230 & 1900/Seasonal ice warnings at 0500 & 1700			
7710.0	FAX	VFF	Frobisher CG Radio, NWT, Canada	FX	2.5	Meteo (Seasonal)
7717.5	USB	....	Norwegian Sea & North Atlantic LORAN Net	FX	..	USCG (Includes:
			JXL, BØ, Norway; JXP, Jan Mayen Is.; NMS, Shetland Is.; OUN, Ejde, Denmark & OVY, Angissog, Greenland) [Alt: 7512.5 & 10,337.5]			
7719.0	CW	†KNY23	Washington, DC	FX	1.0	Czech EMBASSY
7722.5	CW	FN043	Paris (Orly), France	AX	3.0	Tfc/Brazzaville
7724.0	CW	KRH5Ø	London, England	FX	....	USA EMBASSY
7732.5	CW	SMA7	Stockholm (Norrkoping), Sweden	FX	2.5	NOAA reports
			wx broadcasts at 0949, 1148 & 2048--plus seasonal Baltic Sea ice warnings			
7739.9	ISB	9HC35	Malta, Malta	FX	6.0	Telcom/Tfc
7740.0	USB	ZKS	Rarotonga, Cook Islands	AX	1.0	Aero Net
	ISB	9RE77	Lubumbashi, Zaire	FX	0.8	Net Control
7745.0	ISB	†NAW27	Guantanamo, Cuba	FX	....	Telcom/Tfc
7750.0	FAX	RAW78	Moscow, USSR	FX	15.0	Meteo (cont.)
7756.5	LSB	†NAW12	Guantanamo, Cuba	FX	....	Telcom/Tfc
7759.5	FAX	RGH77	Irkutsk, USSR	FX	15.0	
7768.5	USB	....	Greenville, NC	FX	50.0	VOA Feeder
7770.0	USB	LSM397	Buenos Aires, Argentina	FX	0.4	Net Control
	ISB	†FTH77	Paris (St. Assise), France	FX	20.0	Telcom
	FAX	KVM7Ø	Honolulu (International), HI	AX	10.0	Meteo
					(0653-0816)	
7771.5	USB	GYU	Gibraltar Naval Radio, Gibraltar	FX	....	British Navy
7778.2	CW	LMB7	Bergen, Norway	FC	0.5	NOAA reports
				broadcasts at 1305 & 2240		
7785.0	ISB	VJU	Norfolk Island	FX	15.0	Telcom/Sydney
7788.0	CW	GFT27	Bracknell, England	FX	3.5	Meteo
7799.0	USB/RTTY	TFK	Keflavik Airport, Iceland	FX	2.0	USAF
7810.0	ISB	†TJF78	Douala, Cameroon	FX	2.0	Telcom
7811.5	A3H	AUZ38	Balasore, India	FX	2.5	Telcom/Karnal
7812.0	CW/RTTY	Y7B32	Belgrade, Yugoslavia	FX	....	GDR EMBASSY
7812.5	CW	ONN3Ø	Brussels, Belgium	FX	0.5	EMBASSY
7816.5	CW	FDI	Aix les Milles Air, France	FX	....	French AF
7819.9	CW	4MO	unknown	FX	....	
7823.5	USB	TUA47	Abidjan, Ivory Coast	FX	6.0	Telcom
7830.0	CW	KKN44	Monrovia, Liberia	FX	....	USA EMBASSY
7830.3	CW	KAP	unknown	FX	....	EE/Nx
7832.0	CW	†FSB69	Paris (St. Martin Abbat), France	FX	3.0	INTERPOL
7836.6	USB	....	Northwest Pacific LORAN Net	FX	....	USCG (For list
			of stations see 5315.5 & 7918.5)			
7840.0	ISB	†ZKS34	Rarotonga, Cook Islands	FX	10.0	Telcom/Wellington
7845.0	CW	SOH289	Warsaw, Poland	FX	20.0	Tfc
7846.0	ISB	†....	Daventry, England	FX	30.0	BBC Feeder
7858.0	ISB	CUW	Lajes Field, Azores	FX	10.0	USAF

kHz	Mode	Call	Location	Service	kW	Remarks
7859.0	CW	ZBP	Pitcairn Island	FX	0.7	Tfc/Suva
7861.0	USB	CJR68Ø	Sable Island, NS, Canada	FX	0.5	Telcom
	USB	VCC43	Dartmouth, NS, Canada	FX	0.3	Telcom
7863.0	CW	NKT	Cherry Point NAS, NC	FX	....	USN
7875.0	FAX	....	unknown, probably in North America	FX	....	
7879.0	CW	ONY27	Rouveroy, Belgium	FX	....	Belgium Army
7880.0	CW/USB	OEC44	Tel Aviv, Israel	FX	1.0	Austrian EMBASSY [Alt: 10,298.0]
7880.0	FAX	....	unknown, probably in the Far East	FX	....	
7893.5	USB	....	numerous, Hawaii to US east coast	FA/FC	1.0	USN (Usually NAM, Norfolk, VA; NAX, Barbers Pt., HI; NEL Lakehurst NAS, NJ; NFC, Cape May NAS, NJ; NGZ, Alameda NAS, CA; NHK, Patuxent NAS, MD; NHZ, Brunswick NAS, ME; etc.) [Alt: 6833.6]
7894.0	CW/USB	OEC61	Rome, Italy	FX	1.0	Austrian EMBASSY [Alt: 10,425.5]
7894.0	CW	SOH289	Warsaw, Poland	FX	20.0	Tfc
7895.0	CW	†FUF	Fort de France Naval R., Martinique	FX	10.0	French Navy
7906.0	CW	FSB7Ø	Paris (St. Martin Abbat), France	FX	3.0	INTERPOL
7918.5	USB	....	Northwest Pacific LORAN Net	FX	....	USCG (Includes: NRT, "Yokota Monitor", Japan; NRT2, Gesashi, Japan; NRT3, Iwo Jima; NRT9, Hokkaido, Japan; NRV, Barrigada, Guam; NRV6, Marcus Is.; NRV7, Yap Island) [Alt: 7836.6 & 8063.6]
7920.0	CW	EBB	Ferrol de Caudillo Radionaval, Spain	FX	1.0	Spanish Navy
	CW	EBC	Cadiz Radionaval, Spain	FX	1.0	Spanish Navy
7922.5	CW/RTTY	VLV	Mawson Base, Antarctica	FX	5.0	ANARE
	CW/RTTY	VNJ	Casey Base, Antarctica	FX	10.0	"
7925.0	LSB	ROK22	Moscow, USSR	FX	20.0	R. Moscow Feeder
7926.0	CW	EBJ	Palma Naval Radio, Mallorca, Spain	FX	1.0	Spanish Navy
7930.0	CW	FDC6	Luxevil Air, France	FX	....	French AF
7935.0	CW	†COY895	Havana Naval Radio, Cuba	FX	....	Russian Navy
7937.5	CW	HGX39	New Delhi, India	FX	1.0	Hungarian EMBASSY
7938.0	USB	AKA	Elemendorf AFB, AK	FX	1.0	USAF
7940.0	USB	....	Recife Aeradio, Brazil	AX	2.5	Tfc/Dakar
7953.0	USB	CFW	Vancouver, BC, Canada	FX	....	Telcom
7954.0	CW	"K"	unknown	FX	....	Beacon
7955.0	CW	HSA/HSJ	Bangkok, Thailand	FC	....	NOAA reports wx broadcasts at 0150 & 0750
7960.0	USB	"CP"	unknown, probably in Australia	FX	....	"Channel 3"
7960.0	CW/RTTY	†OMZ	Prague, Czechoslovakia	FX	1.0	EMBASSY
7963.6	USB	....	numerous, worldwide	FA	....	USN
7969.0	CW/RTTY	SHX	Stockholm, Sweden	FX	1.5	INTERPOL
7973.0	USB	....	Daventry, England	FX	30.0	BBC Feeder
7980.0	ISB	VNV3Ø	Sydney (Doonside), NSW, Australia	FX	1.0	Telcom/Casey
7984.0	ISB	ODF98	Beirut, Lebanon	FX	10.0	Telcom/Tfc
7991.0	LSB	....	Daventry, England	FX	30.0	BBC Feeder
7997.0	CW/RTTY	SOH299	Warsaw, Poland	FX	10.0	Tfc
7998.0	LSB	VL5MB	Darwin, WA, Australia	FX	0.5	Tfc/Grooteyland
8000.0	CW	FUB	Paris (Houilles) Naval R., France	FX	2.5	French Navy
	AM/CW	JJY	Tokyo (Sanwa, Ibaraki), Japan	SS	3.0	TIME SIGNALS
8003.5	USB	RCD33	Moscow, USSR	FX	20.0	R. Moscow Feeder
8006.2	ISB	†HSD85	Bangkok, Thailand	AX	7.5	
8006.5	CW	8UF75	New Delhi, India	FX	5.0	INTERPOL [Alt: 7532.0]
	CW/RTTY	JPA22	Tokyo, Japan	FX	5.0	INTERPOL
8008.0	CW/RTTY	OMZ29	Bratislava, Czechoslovakia	FX	1.0	GDR EMBASSY
8010.0	FAX	CUA37	Lisbon (Alfragide), Portugal	FX	10.0	
	CW/RTTY	....	numerous, worldwide	FX	....	USN
8015.0	ISB	5VH8Ø	Lome, Togo	FX	6.0	Telcom
8017.4	CW	JJC2Ø	Tokyo, Japan	FX	1.0	Police
8018.0	FAX	OFB28	Helsinki, Finland	FX	5.0	Meteo (0040, 0740 & 0930)
8020.0	CW	OMC2	Bratislava, Czechoslovakia	FX	1.0	Tfc/Izmail
8020.5	USB	C8B53	Maputo, Mozambique	FX	1.0	Net Control

kHz	Mode	Call	Location	Service	kW	Remarks
8021.5	USB	....	Mediterranean LORAN Net	FX	....	USCG (Includes NCI, Sellia Marina, Italy, NCI3, Lampedusa, Italy & NCI4, Karagabun, Turkey) [Alt: 5066.0 & 9474.0]
8022.5	ISB/RTTY	FTI2	Paris (St. Assise), France	FX	20.0	Meteo/Telcom
8030.0	CW	LOL3	Buenos Aires, Argentina	FX	3.0	Prefectura Naval/ TIME SIGNALS (0100, 1300 & 2100, only)
	USB	....	Alice Springs, NT, Australia	FX	0.5	Outpost Control
8035.0	ISB/RTTY	9VE27	Singapore, Singapore	AX	2.5	Tfc
8037.0	ISB	IRF4Ø	Rome (Torrenova), Italy	FX	10.0	Telcom
	USB	CXL21	Montevideo, Uruguay	FX	2.5	Telcom
8038.0	CW/RTTY	OEQ41	Vienna, Austria	FX	1.0	INTERPOL
	CW/RTTY	ONA2Ø	Brussels, Belgium	FX	3.0	"
	CW	FSB	Paris (St. Martin Abbat), France	FX	3.0	"
	CW/RTTY	DEB	Wiesbaden, GFR	FX	5.0	" [Alt: 8097.5]
	CW/RTTY	SHX	Stockholm, Sweden	FX	1.0	INTERPOL
	CW	HEP88	Zurich (Waltikon), Switzerland	FX	3.0	"
8040.0	FAX	GFA23	Bracknell, England	FX	2.0	Meteo (cont.)
8045.0	CW/RTTY	OEQ41	Vienna, Austria	FX	1.0	INTERPOL
	CW/RTTY	ONA2Ø	Brussels, Belgium	FX	3.0	"
	CW/RTTY	DEB	Weisbaden, GFR	FX	5.0	"
	CW/RTTY	HEP85	Zurich (Waltikon), Switzerland	FX	3.0	"
8052.0	LSB	AFA	Washington (Andrews AFB, MD), DC	FX	10.0	USAF [Alt: 18,650.0]
8053.0	LSB	VL5MR	Grooteyland, NT, Australia	FX	0.5	Telcom/Darwin
8055.0	USB	ZKY	Palmerston North, New Zealand	FX	5.0	Telcom/Nadi
8057.5	FAX	LMO8	Oslo, Norway	FX	0.5	Meteo
8058.0	USB	RAW71	Mosocw, USSR	FX	20.0	Telcom
8063.5	ISB	HBX58	Geneva, Switzerland	FX	40.0	Telcom/Tfc
8063.6	USB	....	Northwest Pacific LORAN Net	FX	....	USCG (For list of stations see 7918.5 or 9223.6)
8065.6	USB	....	unknown, probably in North America	FX	....	Telcom
8067.0	USB	KWY43	Kodiak, AK	FX	0.2	Net Control
8072.5	CW	EEQ	Madrid, Spain	FX	1.0	Net Control
8075.0	ISB	FTI9	Paris (St. Assise), France	FX	20.0	Telcom/Dakar
8077.5	FAX	SMA8	Stockholm (Norrkoping), Sweden	FX	2.5	Meteo (cont.)
8080.0	ISB	....	several, Java & S. China Seas	FX	1.0	Oil Rigs
	FAX	NAM	Norfolk Naval Radio, VA	FX	2.5	USN/Meteo
8081.0	CW	SOI2Ø8	Warsaw, Poland	FX	20.0	Nx/Tfc
8083.5	USB	....	West Indies LORAN Net	FX	....	USCG (Includes: NMA, Miami, FL; NMA4, San Salvador [Watling Is.], Bahamas; NMA5, South Caicos, Bahamas; NMA7, Jupiter, FL & NMR, San Juan, PR--daytime, only) [Alt: 7530.0 & 12,150.0]
8088.0	FAX	JKE3	Tokyo, Japan	FX	5.0	
8090.0	CW	NAM	Norfolk Naval Radio, VA	FX	1.0	(LCMP-3) USN/NUKO NOAA reports wx broadcasts at 0030, 0630, 1230 & 1900/ Seasonal ice warnings broadcast at 1000 & 2300
	CW/RTTY	NGD	McMurdo Station, Antarctica	FX	3.0	USN
	CW	NRK	Keflavik Naval Radio, Iceland	FX	....	USN
8090.4	CW	YCN2	Pontianak, Indonesia	FX	5.0	
8092.0	USB	†AJO	Adana, Turkey	FX	0.5	USAF
8097.5	CW/RTTY	OEQ43	Vienna, Austria	FX	1.0	INTERPOL
	CW/RTTY	ONA2Ø	Brussels, Belgium	FX	3.0	"
	CW/RTTY	DEB	Wiesbaden, GFR	FX	5.0	" [Alt: 8038.0]
8098.5	USB	†....	Kabul, Afghanistan	FX	5.0	Telcom
	USB	C6L22	Nassau, Bahamas	FX	0.2	Net Control
8100.0	FAX	NGR	Kato Souli Naval Radio, Greece	FX	15.0	USN
8101.0	USB	....	several, throughout Pacific Area	FX	....	USAF (Usually AIE, Andersen AFB, Guam; AIF8Ø, Yokota AB, Japan; AID, Misawa AB, Japan & AFH39, March AFB, CA. Was referred to as "ALFA PAPA ONE" frequency, now part of "GIANT TALK")
8110.0	CW	....	unknown, probably in Far East	FX	....	Coded Tfc
8111.0	USB	....	Mediterranean LORAN Net	FX	....	USCG (Includes

FIXED SERVICE (con.)

kHz	Mode	Call	Location	Service	kW	Remarks
			AOB5Ø, Estartit, Spain; NCI, Sellia Marina, Italy; NCI3 Lampedusa, Italy; NCI4, Kargabun, Turkey & NCI1Ø, Rhodes)			
8117.0	CW	BMB	Taipei, Taiwan	FX	....	NOAA reports wx broadcasts at 1600 & 2200
8120.0	FAX	BAF36	Beijing, PRC	AX	7.0	Meteo
8125.0	USB	RDZ79	Moscow, USSR	FX	15.0	R. Moscow Feeder
	USB	KDM5Ø	Atlanta (Hampton), GA	FX	....	FAA Net Control
			(Recently reported included: KDM47, Fort Worth, TX; KDN49, Atlanta, GA; KEM8Ø, Washington, DC; "Dragnet"; etc.) [Alt: 7475.0 & 13,630.0]			
8132.0	CW	LMT	Tromso Naval Radio, Norway	FX	0.3	Norwegian Navy
			[Net control of meteo stations located at Hopen, Jan Mayen & Stavanger]			
8133.0	CW/RTTY	SOI213	Warsaw, Poland	FX	10.0	Tfc/PAP Nx
8135.0	USB/RTTY	CZW	"Halifax Military", NS, Canada	FX	1.0	Net Control (Canadian Forces)
8136.5	CW	"U"	unknown	RC	....	Beacon
8144.1	CW	"K"	unknown	RC	....	Beacon
8146.6	FAX	IMB54	Rome, Italy	AX	1.0	Italian AF/ Meteo (cont.)
8148.0	CW/USB	OVG8	Frederikshaven Naval Radio, Denmark	FX	5.0	Danish Marine
	CW/USB	OVC	Groennedal, Greenland	FX	5.0	" "
8149.0	USB	P28BM	Port Moresby, Papua New Guinea	FX	1.0	Net Control
8150.0	USB	VJC	Broken Hill, NSW, Australia	FX	....	Outpost Control
	CW	NRV	COMMSTA Barrigada, Guam	FX	....	USCG/Reported wx broadcast at 1300
8156.0	USB	C6Q9Ø	Nassau, Bahamas	FX	0.1	Net Control
8158.5	CW	EC3Y	unknown	FX	....	
8165.0	USB	VNZ	Port Augusta, SA, Australia	FX	0.3	RFDS
	USB	....	unknown, possibly Central America	FX	....	Net/EE/SS
8167.5	CW	LQB9	Buenos Aires (San Martin), Argentina	FX	5.0	TIME SIGNALS
8172.0	CW/RTTY	VCS838	Ottawa, Ont., Canada	FX	5.0	Yugoslav EMBASSY [Alt: 7445.0 & 10,780.0]
8175.0	LSB	....	unknown	FX	....	Telcom/FF
8176.5	USB	ONN27	Brussels, Belgium	FX	1.0	EMBASSY
8182.0	CW	....	unknown, probably in Cuba	FX	....	Coded Tfc
8185.0	FAX/ISB	FPI88	Paris, France	FX	10.0	Meteo
8188.5	USB	†NAW	Guantanamo, Cuba	FX	2.0	Telcom
8190.0	CW/RTTY	....	numerous, mostly eastern USA	FC/FX	....	USN
8192.4	CW/RTTY	SOI219	Warsaw, Poland	AX	2.0	
8194.0	CW	XVS37	Ho Chi Minh-Ville, Vietnam	FX	5.0	NOAA reports wx broadcast at 1620
8195.0	CW	TBA3	Izmir Naval Radio, Turkey	FX	....	Turkist Navy

8195.0 - 8815.0 kHz — MARITIME MOBILE (Worldwide)

kHz	Mode	Call	Location	Service	kW	Remarks
8195.0 – 8288.0	USB		The band between these frequencies is divided into 31 channels spaced 3.1 kHz apart and assigned numbers from 801 to 831. These are ship transmitting channels and are paired with Coastal USB stations operating between 8718.9 and 8811.9 kHz			
8201.0	USB	WJG	Memphis, TN	FC/MS	0.8	Inland Waterway
8213.6	USB	WCM	Pittsburgh, PA	FC/MS	0.8	Inland Waterway
8257.0	AM/USB		INTERNATIONAL SHIP STATION CALLING FREQUENCY			
8277.2	USB	....	numerous, around USA coastline	FC/MS	....	USN (Used for "OVERWORK" broadcasts) [Alt: 6720.0 & 13,181.0
8291.1	USB	....	Atlantic & Pacific Voice Circuits	FC/MS	....	USCG (Includes:
			NMA, Miami, FL; NMC, San Francisco, CA; NMF, Boston, MA; NMG, New Orleans, LA; NMN, Portsmouth, VA; NMO, Honolulu, HI; NMQ, Long Beach, CA; NMR, San Juan, PR; NOJ, Kodiak, AK & NRV, Barrigada, Guam			

kHz	Mode	Call	Location	Service	kW	Remarks
	USB	....	Coast & Ship Simplex Frequency	FC/MS	....	Inland & inter-coastal waterway traffic. Employed worldwide by hundreds of base stations and various types of ships. Frequently heard in North America: KMB, Houston, TX; KRM, Port Arthur, TX; WJK, Jacksonville, FL; WGW, San Juan, PR; KBQ, Miami, FL; KCQ, Galveston, TX; etc.; plus numerous oil rigs
8294.2	USB	....	Atlantic & Pacific Voice Circuits	FC/MS	....	USCG (Includes: NMA, miami, FL; NMC, San Francisco, CA; NMF, Boston, MA; NMG, New Orleans, LA; NMN, Portsmouth, VA; NMO, Honolulu, HI; NMQ, Long Beach, CA; NMR, San Juan, PR; NOJ, Kodiak, AK & NRV, Barrigada, Guam
	USB	....	Coast & Ship Simplex Frequency	FC/MS	....	Inland & inter-coastal waterway traffic. Employed worldwide by hundreds of base stations and various types of ships. Frequently heard in North America: KAH, Seattle, WA; KEJ, New Orleans, LA; KPA, Long Beach, CA; WFE, Houston, TX; WLN, San Juan, PR; WJK, Miami, FL; KZU, Harvey, LA; etc.; plus oil rigs
8360.0 | 8374.4	CW | CW		The band between these two frequencies is divided into 18 channels each 800 Hz wide and assigned numbers from 1 to 18. These are ship transmitting channels. The most commonly used channels are 5 and 6, centered on 8363.6 and 8364.4 kHz, respectively			
8408.0	CW	UJY	Kaliningrad Radio, USSR	FC	5.0	
8436.0	CW	UCO	Yalta Radio, Ukrainian SSR	FC	5.0	
	CW/RTTY	UFH	Petropavlosk Radio, USSR	FC	5.0	
8436.5	CW	ZSD50	Durban Radio, RSA	FC	3.5	
8437.0	CW	7TA6	Algiers Radio, Algeria	FC	10.0	
	CW	OFJ8	Helsinki Radio, Finland	FC	10.0	
	CW	PKN	Balikpapen Radio, Indonesia	FC	0.3	
	CW	JOS	Nagasaki Radio, Japan	FC	3.0	
8437.1	CW	PKC	Palembang Radio, Indonesia	FC	1.0	
8437.2	CW	4XZ	Haifa Naval Radio, Israel	FC	2.0	Israel Navy
8438.0	CW	UNQ	Novorossiysk Radio, USSR	FC	10.0	
8439.0	CW	DFH43	Frankfurt (Usingen) Radio, GFR	FC	10.0	MARPRESS/Reported GG Nx at 0118
	CW	PBC3	Goeree Naval Radio, Netherlands	FC	6.0	Dutch Navy
	CW	UFN	Novorossiysk Radio, USSR	FC	....	
8439.5	CW	C6N	Nassau Radio, Bahamas	FC	0.5	
8439.9	CW	VCS	Halifax CG Radio, NS, Canada	FC	1.0	AMVER
8440.0	CW	YIR	Basrah Control, Iraq	FC	3.0	
8440.6	CW	UAT	Moscow Radio, USSR	FC	10.0	
8441.0	CW	XYR7	Rangoon Radio, Burma	FC	1.0	
	CW	C5G	Banjul Radio, Gambia	FC	1.0	
	CW	9YL	North Post Radio, Trinidad	FC	2.0	
8441.2	CW	70A	Aden Radio, Democratic Yemen	FC	2.0	
8441.3	CW	9HD	Malta Radio, Malta	FC	1.0	
8441.4	CW	5ZF2	Mombasa Radio, Kenya	FC	1.5	
8443.0	CW	VFF	Frobisher CG Radio, NWT, Canada	FC	1.0	AMVER
	CW	Y5M	Ruegen Radio, GDR	FC	5.0	
8444.5	CW	TCR	Tophane Kulesi Naval R., Turkey	FC	2.0	(TCCQ)
	CW	KFS	San Francisco Radio, CA	FC	20.0	NOAA reports wx broadcasts at 0420, 1620, 2200 & 2230
8445.0	CW	PKK	Kupang Radio, Indonesia	FC	1.0	
	CW	PKO	Tarakan Radio, Indonesia	FC	0.3	
	CW	PKR3	Cilacap Radio, Indonesia	FC	1.0	
	CW	PKT	Dili Radio, Indonesia	FC	1.0	
	CW	†XSX	Keelung Radio, Taiwan	FC	1.0	NOAA reports wx broadcasts at 0430, 1030, 1630 & 2230
	CW	YUR4	Rijeka Radio, Yugoslavia	FC	6.0	
	CW	A4M	Muscat Radio, Oman	FC	1.0	
	CW	S7Q	Seychelles Radio, Seychelles	FC	1.0	
	CW	UQK	Riga Radio, Latvian SSR	FC	5.0	
8445.5	CW	WLO	Mobile Radio, AL	FC	20.0	
8446.0	CW	UFM3	Nevelski Radio, USSR	FC	5.0	

kHz	Mode	Call	Location	Service	kW	Remarks
8447.0	CW	....	numerous, waters around Argentina	FS	1.0	Prefectura Naval
			Includes: LSC4, Ushuaia Radio; LSG4, Rio Gallegos Radio; LSM44, Comordo Rivadavia Radio; LSN37, Mar del Plata Radio & LS05, Buenos Aires Radio)			
	CW	EBA	Madrid Radionaval, Spain	FC	2.0	Spanish Navy
8448.0	CW	A9M	Bahrain Radio, Bahrain	FC	2.0	
	CW	4XZ	Haifa Naval Radio, Israel	FC	2.0	Israel Navy
8449.0	CW	TJC7	Douala Radio, Cameroon	FC	0.8	
	CW	ZKR	Rarotonga Radio, Cook Islands	FC	1.0	
	CW	8RB	Demerara Radio, Guyana	FC	1.0	
	CW	ZSC22	Cape Town Radio, RSA	FC	10.0	
8449.4	CW	VRT	Bermuda (Hamilton) Radio, Bermuda	FC	1.0	
8450.2	CW	8PO	Barbados Radio, Barbados	FC	1.0	
8451.0	CW	HAR	Budapest Naval Radio, Hungary	FC	1.0	
	CW	XSG4	Shanghai Radio, PRC	FC	5.0	
	CW	UBF2	Leningrad Radio, USSR	FC	10.0	
8451.5	CW	†CTP	Oeiras Naval Radio, Portugal	FC	....	Port. Navy/NATO
8452.0	CW	VIS35	Sydney Radio, NSW, Australia	FC	2.5	
8452.9	CW	VAI	Vancouver CG Radio, BC, Canada	FC	1.0	AMVER/NOAA reports wx broadcasts at 0130, 0530 & 1730
8453.0	CW	TNA8	Pointe-Noire Radio, Congo	FC	2.0	
	CW	HWN	Paris (Houilles) Naval R., France	FC	2.0	French Navy
	CW	YVM	Puerto Ordaz Radio, Venezuela	FC	....	
8454.0	CW	A9M	Bahrain Radio, Bahrain	FC	2.0	
	CW/RTTY	UJY	Kaliningrad Radio, USSR	FC	5.0	
8454.5	CW	SVG4	Athens Radio, Greece	FC	5.0	
8456.0	CW	ROT	Moscow Naval Radio, USSR	FC	10.0	Russian Navy
8456.5	CW	PKY41	Sorong Radio, Indonesia	FC	1.0	
8457.0	FAX	NOJ	COMMSTA Kodiak, AK	FC	5.0	USCG/Meteo
			[Pics at various times: analysis, prognosis, outlook, surface tem, ice & fisheries aids, etc.]			
	CW	LSA4	Boca Radio, Buenos Aires, Argentina	FC	1.0	
	CW	EDG3	Aranjuez Radio, Spain	FC	10.0	
	CW	XSP	Shantou Radio, PRC	FC	2.0	
	CW	OFJ82	Helsinki Radio, Finland	FC	10.0	
	CW	PKG	Banjarmasin Radio, Indonesia	FC	0.3	Reported NAVAREA broadcasts at 0130 & 0430
	CW	PKP	Dumai Radio, Indonesia	FC	1.0	
	CW	PKY5	Merauke Radio, Indonesia	FC	0.3	
8458.0	CW	†HAR	Budapest Naval Radio, Hungary	FC	0.5	
	CW	YIR	Basrah Control, Iraq	FC	1.0	Reported wx broadcast at 1330
8459.0	CW	YQI5	Constanta Radio, Rumania	FC	2.0	
	CW	GXH	Thurso Naval Radio, Scotland	FC	....	USN
8460.0	CW	OSN48	Oostende Naval Radio, Belgium	FC	1.0	Belgium Navy
	CW	PPJ	Juncao Radio, Brazil	FC	1.0	NOAA reports wx broadcasts at 0130, 0730 & 1900
	CW	LZW42	Varna Radio, Bulgaria	FC	10.0	
	CW	XSM	Xiamen [Amoy] Radio, PRC	FC	1.0	
	CW	PBC	Goeree Naval Radio, Netherlands	FC	6.0	Dutch Navy
	CW	UJY	Kaliningrad Radio, USSR	FC	5.0	
8460.4	CW	PPL	Belem Radio, Brazil	FC	1.0	
8460.5	CW	PKD	Surabaya Radio, Indonesia	FC	1.0	
8461.0	CW	CUG	Sao Miguel Radio, Azores	FC	2.5	
	CW	CBA3	Antofagasto Radiomaritima, Chile	FC	0.5	
	CW	PKR	Semararang Radio, Indonesia	FC	0.3	
	CW	PKY4	Sorong Radio, Indonesia	FC	0.3	
	CW	CUB	Madeira Radio, Madeira Islands	FC	5.0	
	CW	ZSC4	Cape Town Radio, RSA	FC	10.0	TIME SIGNALS NOAA reports wx broadcasts at 0930 & 1730
	CW	FJA8	Mahina Radio, Tahiti	FC	1.0	
8460.7	CW	YVG	La Guiara Radio, Venezuela	FC	0.3	
8461.5	CW	5OZ23	Port Harcourt Radio, Nigeria	FC	1.0	
8461.9	CW	DAL	Norddeich Radio, GFR	FC	5.0	
8463.0	CW	CKN	Vancouver Forces Radio, BC, Canada	FC	10.0	C. Forces/ NAWS

MARITIME MOBILE (con.)

kHz	Mode	Call	Location	Service	kW	Remarks
	CW	Y5M	Ruegen Radio, GDR	FC	10.0	
	CW	JOU	Nagasaki Radio, Japan	FC	10.0	
8465.0	CW	5BA	Nicosia Radio, Cyprus	FC	10.0	
	CW	TUA4	Abidjan Radio, Ivory Coast	FC	4.0	
	CW	5RS8	Tamatave Radio, Madagascar	FC	2.0	
	CW/RTTY	EBA	Madrid Radionaval, Spain	FC	1.0	Spanish Navy
	CW	NMN	COMMSTA Portsmouth, VA	FC	3.0	USCG/AMVER
8465.2	CW	6YI	Kingston Radio, Jamaica	FC	1.0	
8466.0	CW	†HAR3	Budapest Naval Radio, Hungary	FC	2.5	
	CW	HSY62	Sattahip Naval Radio, Thailand	FC	1.0	Royal Thai Navy
	CW	UJY	Kaliningrad Radio, USSR	FC	5.0	
8467.5	FAX	JJC	Tokyo Radio, Japan	FC	10.0	Meteo
8468.8	CW	GKR8	Wick Radio, England	FC	0.3	
8469.0	CW	D4A6	Sao Vicente Radio, Cape Verde	FC	0.5	
	CW	D4D	Praia de Cabo Radio, Cape Verde	FC	1.0	
	CW	EQI	Abbas Radio, Iran	FC	1.0	
	CW	EQK	Khoramshahr Radio, Iran	FC	1.0	
	CW	XXG	Macao Radio, Macao	FC	0.3	
	CW	CUB	Madeira Radio, Madeira Islands	FC	5.0	
	CW	CUL8	Lisbon Radio, Portugal	FC	10.0	
	CW	FFD28	St. Denis Radio, Reunion	FC	2.0	
8470.0	CW	EPC	B. Abbas Naval Radio, Iran	FC	10.0	Iranian Navy
	CW	XFL	Mazatlan Radio, Mexico	FC	0.5	
	CW	URL	Sevastopol Radio, Ukrainian SSR	FC	5.0	
	CW	XVG9	Haiphong Radio, Vietnam	FC	1.0	
8471.0	CW	EQZ	Abadan Radio, Iran	FC	0.5	
	CW	NMR	COMMSTA San Juan, PR	FC	3.0	USCG/AMVER
8471.7	CW	SUP	Port Said Radio, Egypt	FC	2.0	
8472.0	CW/RTTY	3SB	Datong Naval Radio, PRC	FC	10.0	PRC Navy
8472.2	CW	ZRQ4	Cape (Simonstown) Naval Radio, RSA	FC	10.0	RSA Navy
8473.0	CW	PKS	Pontianak Radio, Indonesia	FC	0.3	
	CW	HLG	Seoul Radio, South Korea	FC	10.0	
	CW	EDF3	Aranjuez Radio, Spain	FC	10.0	
	CW	UNM2	Klaipeda Radio, Lithuanian SSR	FC	5.0	
8473.3	CW	PKE	Amboina Radio, Indonesia	FC	1.0	
8474.0	CW	UKJ	Astrakhan Radio, USSR	FC	....	Caspian Sea
8474.5	CW	WLO	Mobile Radio, AL	FC	20.0	NOAA reports wx broadcasts at 1300 & 2300
	CW	UCY2	Astrakhan Radio, USSR	FC	1.0	Caspian Sea
8475.0	CW	4PB	Colombo Radio, Sri Lanka	FC	1.0	TIME SIGNALS NOAA reports wx broadcast at 1330
8475.5	CW	FUX	Le Port Naval Radio, Reunion	FC	10.0	French Navy
8476.0	CW	PPE2	Rio Radio, Brazil	FC	....	
	CW	HCG9	Guayaquil Radio, Ecuador	FC	0.3	AMVER
	CW	OMC	Bratislava Radio, Czechoslovakia	FC	1.0	
	CW	9VG56	Singapore Radio, Singapore	FC	3.0	
	CW	UAH	Tallinn Radio, Estonian SSR	FC	10.0	
8477.0	CW	HWN	Paris (Houilles) Naval R., France	FC	5.0	French Navy
8478.0	CW	VHP4	COMMSTA Canberra, ACT, Australia	FC	20.0	(A13B) RA Navy
	CW	VIS4	Master Station Canberra, Australia	FC	20.0	NOAA reports wx broadcasts at 1730 & 2130/Reported NAVAREA broadcasts at 0100, 0500, 0900, 1300, 1700 & 2100
	CW	OST4	Oostende Radio, Belgium	FC	10.0	
	CW	CBV2	Valparaiso Radiomaritima, Chile	FC	1.0	AMVER
8478.1	CW	TIM	Limon Radio, Costa Rica	FC	3.0	
8478.5	CW/RTTY	FUF	Fort de France Naval R. Martinique	FC	10.0	French Navy
	CW	FUG	La Regine (Castelnaudary), France	FC	10.0	" "
8479.0	CW	JCU	Choshi Radio, Japan	FC	10.0	
8480.0	CW	PWI	Recife Naval Radio, Brazil	FC	2.0	Brazil Navy
	CW	PWN	Natal Naval Radio, Brazil	FC	2.0	" "
	CW	PWP	Florianopolis Naval Radio, Brazil	FC	2.0	" "
	CW	CLQ	Havana (Cojimar) Radio, Cuba	FC	5.0	
	CW	XUK2	Kompong Som-Ville Radio, Cambodia	FC	0.3	
	CW	5AL	Tobruk Radio, Libya	FC	0.5	
	CW	HZY	Ra's Tannurah Radio, Saudi Arabia	FC	3.0	ARAMCO/NOAA reports wx broadcasts at 0430 & 0800

kHz	Mode	Call	Location	Service	kW	Remarks
	CW	UAT	Moscow Radio, USSR	FC	10.0	
8481.0	CW	VIS28	Sydney Radio, NSW, Australia	FC	10.0	Reported Nx broadcast at 1200
8482.0	CW	SPH41	Gydnia Radio, Poland	FC	10.0	[aka SPA4]
8483.5	CW	DAN	Norddeich Radio, GFR	FC	5.0	
8484.0	CW	XSE	Qinhuangdao Radio, PRC	FC	1.0	
	CW	Y5M	Ruegen Radio, GDR	FC	5.0	
8484.2	CW	HLF	Seoul Radio, South Korea	FC	10.0	
8485.0	CW	FUB	Paris (Houilles) Naval R., France	FC	....	French Navy
	CW	4XO	Haifa Radio, Israel	FC	2.5	
	CW	UFN	Novorossiysk Radio, USSR	FC	10.0	
	CW	UFJ	Rostov Radio, USSR	FC	5.0	
8486.0	CW	WOE	Lantana Radio, FL	FC	5.0	
	CW	IDQ4	Rome Naval Radio, Italy	FC	10.0	Italian Navy
			NOAA reports wx broadcasts at 0500, 0830, 1800 & 2100			
8487.0	CW	VID	Darwin Radio, NT, Australia	FC	1.0	
	CW	XSG26	Shanghai Radio, PRC	FC	1.0	Reported NAVAREA broadcast at 0200
8489.0	CW	CLS	Havana (Industria Pesquera) R., Cuba	FC	5.0	
8489.8	CW	CUL7	Lisbon Radio, Portugal	FC	10.0	
8490.0	CW	XSQ4	Guangzhou Radio, PRC	FC	2.0	
	CW	5OZ23	Port Harcourt Radio, Nigeria	FC	1.0	
	CW	AQP3	Karachi Naval Radio, Pakistan	FC	3.5	Pakistan Navy
			Reported wx broadcasts at 0130, 0530, 0930, 1330, 1730 & 2130			
	CW	JJI	Sasebo Naval Radio, Japan	FC	1.0	JSDA
8492.0	CW	PPR	Rio Radio, Brazil	FC	1.0	
	CW	JNA	Tokyo Naval Radio, Japan	FC	5.0	JSDA/Reported
			NAVAREA broadcasts at 0005, 0405, 0805, 1205 & 0220			
	CW	9MG8	Penang Radio, Malaysia	FC	5.0	
8492.6	CW	GYA	Whitehall (London) Naval R., England	FC	5.0	British Navy
8494.8	FAX	GZZ4Ø	Whitehall (Northwood) Naval R., Eng.	FC	5.0	British N./Meteo
8495.0	CW	ARH	Gwadar Naval Radio, Pakistan	FC	6.0	Pakistan Navy
	CW	UBN	Jdanov Radio, Ukrainian SSR	FC	10.0	
8495.5	CW	UFB	Odessa Radio, Ukrainian SSR	FC	10.0	(4KA)
8496.0	CW	CLA2Ø	Havana (Cojimar) Radio, Cuba	FC	5.0	
	CW	5OZ23	Port Harcourt Radio, Nigeria	FC	1.0	
	CW	....	several, USA nationwide	FX/MS	....	USACE
8496.5	CW/TOR	GKS4	Portishead Radio, England	FC	10.0	
8497.0	CW	HLJ	Seoul Radio, South Korea	FC	7.0	
8498.0	CW	SAG4	Goteborg Radio, Sweden	FC	5.0	AMVER
8499.0	CW	GYA	Whitehall (London) Naval R., England	FC	5.0	British Navy
8500.1	FAX	NIK	COMMSTA Boston, MA	FC	10.0	USCG/Transmitts ice pics seasonally at 1600
8502.0	CW	PPL	Belem Radio, Brazil	FC	1.0	NOAA reports wx broadcasts at 0130, 0800 & 1930
	CW	XSG3	Shanghai Radio, PRC	FC	20.0	TIME SIGNALS
			NOAA reports wx broadcasts at 0306 & 0906			
	CW	†IQX	Trieste P.T. Radio, Italy	FC	6.0	
	CW	YJM8	Port-Vila Radio, Vanuatu	FC	0.5	
	CW	ZSC37	Cape Town Radio, RSA	FC	10.0	
	CW	NIK	COMMSTA Boston, MA	FC	10.0	USCG/Transmitts ice information seasonally at 0030 & 1230
	CW	NMF	COMMSTA Boston, MA	FC	10.0	USCG/Notice to Fishermen broadcasts at 1350 & 2150
8504.0	CW	ZLB4	Awarua Radio, New Zealand	FC	5.0	AMVER
8505.0	CW	XFK	La Paz Baja California R., Mexico	FC	0.5	
	CW	TBO3	Izmir Naval Radio, Turkey	FC	....	Turkist Navy
	CW	UCW4	Leningrad Radio, USSR	FC	....	(RJFY)
8505.9	CW	UDH	Riga Radio, Latvian SSR	FC	10.0	
8508.0	CW	RIW	Khiva Naval Radio, Uzbek SSR	FC	....	Russian Navy
8509.0	CW	CCM	Magallenes Radionaval, Chile	FC	2.0	Chilean Navy
8510.0	CW	CLS	Havana (Industria Pesquera) R., Cuba	FC	5.0	
	CW	J2A8	Djibouti Radio, Djibouti	FC	2.5	

kHz	Mode	Call	Location	Service	kW	Remarks
	CW	FFS4	St. Lys Radio, France	FC	10.0	
	CW	UKK3	Nakhodka Radio, USSR	FC	5.0	(4KT)
	CW	UDH	Riga Radio, Latvian SSR	FC	5.0	
	CW/RTTY	UFL	Vladivostok Radio, USSR	FC	80.0	
8511.9	CW	DAL	Norddeich Radio, GFR	FC	5.0	
8513.0	CW	XFY	Guaymas Radio, Mexico	FC	1.0	
8513.8	CW	XSQ4	Guangzhou Radio, PRC	FC	2.0	
8514.0	CW	LPD85	Gen. Pacheco Radio, Argentina	FC	3.0	
	CW	XSG	Shanghai Radio, PRC	FC	10.0	
	CW	VWB	Bombay Radio, India	FC	2.5	NOAA reports wx broadcasts at 0048, 0448, 0848, 1248, 1648 & 2048
	CW	WSL	Amagansett Radio, NY	FC	10.0	NOAA reports wx broadcasts at 0500, 1100, 1700 & 2300
	CW	PBC8	Goeree Naval Radio, Netherlands	FC	10.0	(N13A) Dutch Navy
	CW	XFA	Acapulco Radio, Mexico	FC	0.5	
	CW	XFL	Matzatlan Radio, Mexico	FC	1.0	
8515.0	CW	5AT	Tripoli Radio, Libya	FC	10.0	
	CW	P2M	Port Moresby Radio, Papua New Guinea	FC	1.0	
	CW	P2R	Rabaul Radio, Papua New Guinea	FC	1.0	
	CW/RTTY	UMV	Murmansk Radio, USSR	FC	25.0	
	CW	URL	Sevastopol Radio, Ukrainian SSR	FC	5.0	
	CW	UFL	Vladivostok Radio, USSR	FC	10.0	
8516.0	CW	CLS	Havana (Industria Pesquera) R., Cuba	FC	5.0	
	CW	GKC4	Portishead Radio, England	FC	10.0	
8518.0	CW	ELC	Monrovia Radio, Liberia	FC	2.5	
8518.1	CW	4XZ	Haifa Naval Radio, Israel	FC	2.0	Israel Navy
8520.0	CW	PPO	Olinda Radio, Brazil	FC	1.0	NOAA reports wx broadcasts at 0100, 0600, 1000, 1200, 1500, 1830, 2020 & 2100
	CW	UFB	Odessa Radio, Ukrainian SSR	FC	10.0	
	CW	UKW3	Korsakov Radio, USSR	FC	1.0	
8521.0	CW	VIS26	Sydney Radio, NSW, Australia	FC	2.5	AMVER
8522.0	CW	9WW2Ø	Kuching Radio, Malaysia	FC	0.8	
8522.5	CW	FFL4	St. Lys Radio, France	FC	10.0	
8523.0	CW	RIW	Khiva Naval Radio, Uzbek SSR	FC	....	Russian Navy
8523.4	CW	JOR	Nagasaki Radio, Japan	FC	10.0	
8525.0	CW	9KK6	Kuwait Radio, Kuwait	FC	5.0	NOAA reports wx broadcast on request
	CW	CTW8	Monsanto Naval Radio, Portugal	FC	5.0	Port. Navy NOAA reports wx broadcasts at 0800 & 2000
	CW	WNU33	Slidell Radio, LA	FC	3.0	
8526.0	CW	LPD52	Gen. Pacheco Radio, Argentina	FC	10.0	
	CW	VWC	Calcutta Radio, India	FC	2.5	NOAA reports wx broadcast at 1418
	CW	JMC4	Tokyo Radio, Japan	FC	4.0	NOAA reports wx broadcast at 1518
8527.0	FAX	CTW8	Monsanto Naval Radio, Portugal	FC	5.0	Port. Navy
8528.0	CW	LSO3	Buenos Aires Radio, Argentina	FS	1.0	Prefectura Naval
8528.5	CW	EBA	Madrid Radionaval, Spain	FC	10.0	Spanish Navy Reported NAVAREA broadcasts at 0918, 1248 & 1618
	CW	UFB	Odessa Radio, Ukrainian SSR	FC	10.0	Reported wx broadcasts at 0100 & 0600
8529.3	CW	LFN	Rogaland Radio, Norway	FC	5.0	AMVER
8530.0	CW	PWZ	Rio de Janeiro Naval Radio, Brazil	FC	10.0	Brazil Navy
	CW	SVJ4	Athens Radio, Greece	FC	5.0	
	CW	IAR28	Rome P.T. Radio, Italy	FC	7.0	NOAA reports wx broadcasts at 0700 & 1900
	CW	UQK2	Riga Radio, Latvian SSR	FC	10.0	(4LN)
	CW	9VG35	Singapore Radio, Singapore	FC	3.0	
8532.0	CW	LZW42	Varna Radio, Bulgaria	FC	10.0	
8534.0	CW	GYA	Whitehall (London) Naval R., England	FC	10.0	British Navy
	CW	JCG	Yokohama Radio, Japan	FC	1.0	
	CW	ARL71	Karachi Naval Radio, Pakistan	FC	10.0	Pakistan Navy
	CW	3VP	La Skhirra Radio, Tunisia	FC	0.6	
8535.0	CW	UTA	Tallinn Radio, Latvian SSR	FC	10.0	
	CW	UKA	Vladivostok Radio, USSR	FC	10.0	

MARITIME MOBILE

kHz	Mode	Call	Location	Service	kW	Remarks
	CW	UFW	Vladivostok Radio, USSR	FC	10.0	
8536.5	CW	SVD4	Athens Radio, Greece	FC	5.0	
8538.0	CW	XSA2	Nanjing 1, PRC	FC	10.0	
8539.0	CW	VPS35	Cape d'Aguilar Radio, Hong Kong	FC	3.5	TIME SIGNALS
			NOAA reports wx broadcasts at 0118, 0430, 0718 & 1318			
8540.0	CW	UBN	Jdanov Radio, Ukrainian SSR	FC	....	
	CW	UFN	Novorossiysk Radio, USSR	FC	....	
8540.1	CW	PJK38	Suffisant Dorp Naval Radio, Curacao	FC	10.0	Dutch Navy
8541.0	CW	XSJ	Zhanjiang Radio, PRC	FC	2.0	
	CW	URB2	Klaipeda Radio, Lithuanian SSR	FC	5.0	(4LI) [aka RWWN]
8542.0	CW	PWZ	Rio de Janiero Naval Radio, Brazil	FC	10.0	Brazil Navy
8542.0	CW	9GA	Takoradi Radio, Ghana	FC	10.0	
	CW	PKI	Jakarta Radio, Indonesia	FC	2.0	TIME SIGNALS
8545.8	CW	XST	Qingdao Radio, PRC	FC	5.0	
	CW	DZF	Manila (Bacoor) Radio, Philippines	FC	0.4	Reported wx
				broadcasts at 1600 & 2200		
8545.9	CW	GKA4	Portishead Radio, England	FC	10.0	NOAA reports
			wx broadcasts 0930 & 2130/NAVAREA at 0730 & 1730			
8546.0	CW	ARL	Karachi Naval Radio, Pakistan	FC	10.0	Pakistan Navy
	CW	XSY	Hualien Radio, Taiwan	FC	2.5	
	CW	KLB	Seattle Radio, WA	FC	3.5	
	CW	9PM7	Matadi Radio, Zaire	FC	1.0	
8547.5	CW	JFA	Chuo Gyogyo (Matsudo) Radio, Japan	FC	3.0	
8548.0	CW	PPO	Olinda Radio, Brazil	FC	1.0	
	CW	9PA5	Banana Radio, Zaire	FC	3.0	Reported wx
			broadcasts at 0620, 0820, 1220 & 1620			
8549.0	CW	UXN	Arkhangelsk Radio, USSR	FC	10.0	
8550.0	CW	PWZ	Rio de Janeiro Naval Radio, Brazil	FC	2.0	Brazil Navy
			NOAA reports wx broadcasts at 0030, 0630 & 1730			
	CW	FFT4	St. Lys Radio, France	FC	10.0	
	CW	WPA	Port Arthur Radio, TX	FC	5.0	
8550.4	CW	FUX	Le Port Naval Radio, Reunion	FC	10.0	French Navy
8551.5	CW	CTP95	Oeiras Naval Radio, Portugal	FC	5.0	Port. Navy/NATO
8552.0	CW	GKK4	Portishead Radio, England	FC	10.0	
	CW	UHF3	Yeysk Staro Radio, USSR	FC	....	
8554.0	CW	†GZO4	Hong Kong Naval Radio, Hong Kong	FC	5.0	British Navy
	CW	FUF	Fort de France Naval R., Martinique	FC	10.0	French Navy
	CW	3BA4	Mauritius (Bigara) Radio, Mauritius	FC	5.0	
	CW	UQK2	Riga Radio, Latvian SSR	FC	5.0	
8555.0	CW	DZP	Manila (Novaliches) R. Philippines	FC	3.0	Reported wx
				broadcasts at 0300 & 1100		
	CW	TBB5	unknown, in Turkey	FC	....	Turkist Navy
	CW	UJY	Kaliningrad Radio, USSR	FC	5.0	
	CW	UFH	Petropavlovsk Radio, USSR	FC	5.0	
8557.0	CW	SPE41	Szezecin Radio, Poland	FC	10.0	[aka SPE4]
	CW	UGH2	Juzno-Sakhalinsk Radio, USSR	FC	5.0	
8557.9	CW	GKB4	Portishead Radio, England	FC	10.0	
8558.0	CW	CCV6	Valparaiso Naval Radio, Chile	FC	2.0	Chilean Navy
			TIME SIGNALS/Tfc/NOAA reports wx broadcasts at 0110 1330 & 1430			
8558.4	CW	KFS	San Francisco Radio, CA	FC	20.0	
8560.4	CW	OXZ4	Lyngby Radio, Denmark	FC	10.0	
8562.0	CW	PCH4Ø	Scheveningen Radio, Netherlands	FC	3.0	AMVER
8564.0	CW	LZL4	Bourgas Radio, Bulgaria	FC	5.0	
	CW	DZE	Mandaluyong Radio, Philippines	FC	0.8	Reported wx
				broadcasts at 0000 & 1000		
8565.0	CW	UAT	Moscow Radio, USSR	FC	25.0	
	CW	UKA	Vladivostok Radio, USSR	FC	25.0	
8566.0	CW	VTO4	Vishakhapatnam Naval R., India	FC	....	Indian Navy
	CW	ZSJ4	NAVCOMCEN (Silvermine) Cape R., RSA	FC	5.0	RSA Navy/AMVER
8566.3	CW	D3E2	Luanda Radio, Angola	FC	1.0	
8567.0	CW	XDA	Mexico City Radio, DF, Mexico	FC	10.0	
8567.9	CW	FUV	Djibouti Naval Radio, Djibouti	FC	....	French Navy
8568.0	CW	DZR	Manila Radio, Philippines	FC	10.0	Reported wx
						broadcast at 0120

kHz	Mode	Call	Location	Service	kW	Remarks
8568.3	CW	XFM	Manzanillo Colimas Radio, Mexico	FC	0.5	
8569.0	CW	GKD4	Portishead Radio, England	FC	10.0	
8570.0	CW	NRV	COMMSTA Barrigada, Guam	FC	5.0	USCG
	CW	WNU43	Slidell Radio, LA	FC	3.0	NOAA reports wx broadcasts at 0430 & 1630
	CW	UPB	Providenia Bukhta Radio, USSR	FC	15.0	
	CW	XVT5	Danang Radio, Vietnam	FC	1.0	
8570.5	CW	UQD2	Magadan Radio, USSR	FC	....	
8571.0	CW	UFN	Novorossiysk Radio, USSR	FC	5.0	
8572.0	CW	5TA	Nouadhibou Radio, Mauritania	FC	1.0	
	CW	TBA6	Izmir Naval Radio, Turkey	FC	....	Turkist Navy/NATO
8573.0	CW	CLA21	Havana (Cojimar) Radio, Cuba	FC	5.0	
8573.5	CW	HSA4	Bangkok Radio, Thailand	FC	5.0	
8574.0	CW	XSU	Yantai Radio, PRC	FC	0.5	
	CW	LGB	Rogaland Radio, Norway	FC	5.0	AMVER
	CW	DZD	Manila (Antipolo) R., Philippines	FC	0.4	Reported wx broadcasts at 0300 & 1100
	CW	NMC	COMMSTA San Francisco, CA	FC	10.0	USCG/AMVER
8574.1	CW	HKC3	Buenaventura Radio, Colombia	FC	0.5	
8575.0	CW	URD	Leningrad Radio, USSR	FC	5.0	
	CW/RTTY	UKX	Nakhodka Radio, USSR	FC	5.0	
8575.5	CW	XDA	Mexico City Radio, DF, Mexico	FC	10.0	
8576.0	CW	DZZ	Manila Radio, Philippines	FC	0.1	Reported wx broadcasts at 0905 & 1705
	CW	RCV	Moscow Naval Radio, USSR	FC	....	Russian Navy
8576.6	CW	ZSD45	Durban Radio, RSA	FC	3.5	NOAA reports wx broadcasts at 0905 & 1705
8577.0	CW	HLO	Seoul Radio, South Korea	FC	3.5	Reported wx broadcasts at 0230 & 0930
8578.0	CW	SUH3	Alexandria Radio, Egypt	FC	3.0	
	CW	NGR	Kato Souli Naval Radio, Greece	FC	10.0	USN
	CW	CWM	Montevideo Armada Radio, Uruguay	FC	2.0	Uruguayan Navy
	CW	CWM25	Isla de Flores Armada R., Uruguay	FC	0.3	" "
8578.4	CW	TFA	Reykjavik Radio, Iceland	FC	1.5	
8580.0	CW	DZO	Manila (Bulacan) Radio, Philippines	FC	2.0	NOAA reports wx broadcasts at 1030 & 2330
	CW	RKLM	Arkhangelsk Radio, USSR	FC	5.0	(4LY)
	CW/RTTY	UDK2	Murmansk Radio, USSR	FC	10.0	
	CW/RTTY	UNQ	Novorossiysk Radio, USSR	FC	5.0	
	CW	URL3	Sevastopol Radio, Ukrainian SSR	FC	5.0	(4LA)
8580.5	CW/FAX	GKN4	Portishead Radio, England	FC	10.0	
8581.6	CW	GKM4	Portishead Radio, England	FC	10.0	
8582.0	CW	NRV	COMMSTA Barrigada, Guam	FC	10.0	USCG/AMVER
	CW	XSW	Kaohsiung Radio, Taiwan	FC	1.0	NOAA reports wx broadcasts at 0500, 1100, 1700 & 2300
	CW	KLB	Seattle Radio, WA	FC	3.5	
8582.5	CW	GKO4	Portishead Radio, England	FC	10.0	
8584.0	CW	Y5M	Ruegen Radio, GDR	FC	5.0	
	CW	VPS36	Cape d'Aguilar Radio, Hong Kong	FC	3.5	
	CW	DZN	Manila (Novatas) Radio, Philippines	FC	3.0	
8585.0	CW	KUQ	Pago Pago Radio, American Samoa	FC	2.5	AMVER/NOAA reports wx broadcasts at 0400 & 2000
8586.0	CW	UJ05	Izmail Radio, Ukrainian SSR	FC	....	
	CW	WCC	Chatham Radio, MA	FC	20.0	Reported wx warning broadcast at 1130
8588.0	CW	DZG	Manila (Las Pinas) Radio, Philippines	FC	5.0	Reported wx broadcasts at 0320 & 1520
	CW	4WD3	Hodeidah Port Radio, Yemen A.R.	FC	1.0	
8590.0	CW	UHK	Batumi Radio, USSR	FC	5.0	
	CW	UFN	Novorossiysk Radio, USSR	FC	15.0	
	CW	UOP	Tuapse Radio, USSR	FC	10.0	
	CW	UFL	Vladivostok Radio, USSR	FC	10.0	
	CW	XVS8	Ho Chi Minh-Ville, Vietnam	FC	1.0	NOAA reports wx broadcasts at 0048, 0448 & 1148
8591.0	CW	KOK	Los Angeles Radio, CA	FC	15.0	NOAA reports wx broadcasts at 0450 & 1650

kHz	Mode	Call	Location	Service	kW	Remarks
8591.5	CW	GKG4	Portishead Radio, England	FC	10.0	
8593.0	CW	UFB	Odessa Radio, Ukrainian SSR	FC	10.0	
8595.0	CW	UFL	Vladivostok Radio, USSR	FC	10.0	Reported wx broadcasts at 0900 & 2300/NAVAREA at 0100 & 1500
8595.1	CW	DZK	Manila (Bulacan) Radio, Philippines	FC	2.0	
8595.5	CW	XFP	Chetumal Radio, Mexico	FC	0.5	Reported NAVAREA broadcast at 0000
8596.0	CW	RIT	unknown, probably in USSR	FC	....	
8597.0	CW	VIP3	Perth Radio, WA, Australia	FC	10.0	AMVER (NOAA reports wx broadcasts at 0100, 1200 & 1300
8598.0	CW	XSL	Fuzhou Radio, PRC	FC	1.0	
	CW	OXZ4	Lyngby Radio, Denmark	FC	10.0	
8598.4	CW/RTTY	ZLO4	Irirangi Naval Radio, New Zealand	FC	7.5	RNZ Navy [aka ZLS4]
8599.9	CW	XSV	Tianjin Radio, PRC	FC	4.0	
8600.0	CW	GYA	Whitehall (London) Naval R., England	FC	10.0	British Navy
	CW	DXF	Davao Radio, Philippines	FC	0.4	
	CW	URD	Leningrad Radio, USSR	FC	10.0	
8601.5	CW	HEB	Berne Radio, Switzerland	FC	10.0	
	CW	CWA	Cerrito Radio, Uruguay	FC	10.0	
8604.0	CW	GKH4	Portishead Radio, England	FC	10.0	
8604.2	CW	DZJ	Manila (Bulacan) Radio, Philippines	FC	2.5	Reported wx broadcasts at 0300 & 1000
8605.0	CW/RTTY	ZRH4	Cape (Fisantekraal) Naval R., RSA	FC	10.0	RSA Navy
	CW	KZU	unknown	FC	....	
8606.0	CW	GKI4	Portishead Radio, England	FC	10.0	
	CW	USW2	Rostov Radio, USSR	FC	....	
8607.0	CW	LZW43	Varna Radio, Bulgaria	FC	5.0	
	CW	UBN	Jdanov Radio, Ukrainian SSR	FC	....	
8607.2	CW	RGQ2	unknown	FC	....	
8608.0	CW	HPN6Ø	Canal (Puerto Armuelles) R., Panama	FC	1.0	AMVER
8609.5	CW	9VG73	Singapore Radio, Singapore	FC	5.0	
8610.0	CW	UXN	Arkhangelsk Radio, USSR	FC	25.0	
	CW/RTTY	UBE2	Petropavlovsk Radio, USR	FC	5.0	
	CW	WMH	Baltimore Radio, MD	FC	10.0	
8611.5	CW	TAH	Istanbul Radio, Turkey	FC	3.0	[aka TCCH]
8613.2	CW	GYA	Whitehall (London) Naval R., England	FC	10.0	British Navy
8614.0	CW	CKN	Vancouver Forces Radio, BC, Canada	FC	10.0	C. Forces
	CW	DXU	Catabato Radio, Philippines	FC	....	
	CW	XSJ	Zhanjiang Radio, PRC	FC	2.0	
	CW/RTTY	URL	Sevastopol Radio, Ukrainian SSR	FC	10.0	
8615.0	CW	DXF	Davao Radio, Philippines	FC	0.4	
	CW	UDH2	Riga Radio, Latvian SSR	FC	5.0	
8615.5	CW	WPD	Tampa Radio, FL	FC	5.0	NOAA reports wx broadcasts at 1420 & 2320
8616.0	CW	HAR3	Budapest Naval Radio, Hungary	FC	1.0	
8617.0	CW	XSV	Tianjin Radio, PRC	FC	2.0	
8617.5	CW	URD2	Leningrad Radio, USSR	FC	10.0	
8618.0	CW	ETC	Assab Radio, Ethiopia	FC	2.0	
	CW	VTP3	Vishakhapatnam Naval Radio, India	FC	....	Indian Navy
	CW	EDZ4	Aranjuez Radio, Spain	FC	10.0	
	CW	KPH	San Francisco Radio, CA	FC	20.0	
8619.0	CW	VRN35	Cape d'Aguilar Radio, Hong Kong	FC	3.5	NOAA reports wx broadcasts at 0118 & 1318
8620.0	CW	UBN	Jdanov Radio, Ukrainian SSR	FC	10.0	
	CW	UMV	Murmansk Radio, USSR	FC	10.0	
	CW	UKX	Nakhodka Radio, USSR	FC	5.0	
8622.0	CW	PCH41	Scheveningen Radio, Netherlands	FC	10.0	AMVER
8624.0	CW	XSQ	Guangzhou Radio, PRC	FC	2.0	
8625.0	CW	YKM7	Lattakia Radio, Syria	FC	1.0	
	CW	FUM	Papeete Mahina Naval R., Tahiti	FC	10.0	French Navy
8625.2	CW	GYU4	Gibraltar Naval Radio, Gibraltar	FC	10.0	British Navy
8626.0	CW	OXZ41	Lyngby Radio, Denmark	FC	10.0	
	CW	PKC2	Plaju Radio, Indonesia	FC	1.0	
	CW	PK12	Jakarta Radio, Indonesia	FC	1.0	

kHz	Mode	Call	Location	Service	kW	Remarks
	CW	PKN7	Bontang Radio, Indonesia	FC	1.0	
	CW	PKR6	Cilacap Radio, Indonesia	FC	0.4	
	CW	PKB2Ø	Lhok Seumawe Radio, Indonesia	FC	1.0	
8626.4	CW	9MB4	Penang Naval Radio, Malaysia	FC	2.5	Malaysian Navy
8626.8	CW	KLC	Galveston Radio, TX	FC	3.8	
8627.5	CW	OMC	Bratislava Radio, Czechoslovakia	FC	1.0	
	CW	OMK	Komarno Radio, Czechoslovakia	FC	1.0	
	CW	OMP4	Prague Radio, Czechoslovakia	FC	1.0	
8628.0	CW	VRN	Cape d'Aguilar Radio, Hong Knog	FC	3.5	
	CW	JFA	Chuo Gyogyo (Matsudo) Radio, Japan	FC	3.0	
	CW	DXR	Manila Radio, Philippines	FC	....	Reported wx broadcasts at 0300 & 1000
8628.5	CW	NOJ	COMMSTA Kodiak, AK	FC	5.0	USCG
8630.0	CW	XSV	Tianjin Radio, PRC	FC	6.0	
	CW	VWB	Bombay Radio, India	FC	2.5	NOAA reports wx broadcast at 0650/Reported NAVAREA broadcasts at 0648 & 1548
	CW	UFB	Odessa Radio, Ukrainian SSR	FC	5.0	
	CW	WCC	Chatham Radio, MA	FC	20.0	
8630.5	CW	9MB3	Penang Naval Radio, Malaysia	FC	....	Malaysian Navy
8631.0	CW	XFQ	Salina Cruz Radio, Mexico	FC	0.5	
8634.0	CW	PPR	Rio Radio, Brazil	FC	1.0	TIME SIGNALS/Tfc
	CW	XSH	Basuo Radio, PRC	FC	1.0	
	CW	VTG6	Bombay Naval Radio, India	FC	1.5	Indian Navy
	CW	VTP6	Vishakhapatnam Naval R., India	FC	1.5	" "
	CW	SPH42	Gydnia Radio, Poland	FC	10.0	[aka SPH4]
	CW	CQK	Sao Tome R., Sao Tome & Principe	FC	0.5	
	CW	CQL	Principe R., Sao Tome & Principe	FC	0.5	
8635.0	CW	UJQ2	Kiev Radio, Ukrainian SSR	FC	10.0	(4LA)
8635.3	CW	OXZ4	Lyngby Radio, Denmark	FC	10.0	
8635.8	CW	HLW	Seoul Radio, South Korea	FC	7.0	
8636.0	CW	DZU	Manila (Pasig) Radio, Philippines	FC	1.0	
8638.5	CW	DAM	Norddeich Radio, GFR	FC	10.0	TIME SIGNALS/Tfc (Time signals 1155-1206 & 2355-0006, only)
8640.0	CW	XSK4	Haimen Radio, PRC	FC	0.5	
	CW	UJY	Kaliningrad Radio, USSR	FC	10.0	
	CW	RNR5	unknown	FC	....	
8640.2	CW/RTTY	EBA	Madrid Radionaval Radio, Spain	FC	1.0	Spanish Navy
8641.8	CW	CTV8	Monsanto Naval Radio, Portugal	FC	....	Port. Navy
8642.0	CW	D3L25	Luanda Naval Radio, Angola	FC	6.0	
	CW	C9C6	Maputo Naval Radio, Mozambique	FC	3.0	
	CW	D4A7	Sao Vicente Radio, Cape Verde	FC	3.0	
	CW	HIA2	Domingo Piloto Radio, Dominican Rep.	FC	0.5	Reported wx broadcasts at 0300 & 1700
	CW	HIP2	Puerto Plata R., Dominican Republic	FC	0.5	
	CW	KPH	San Francisco Radio, CA	FC	20.0	
8643.9	CW	VTP6	Vishakhapatnam Naval Radio, India	FC	5.0	Indian Navy Reported NAVAREA & wx broadcasts at 0500, 0900, 1500 & 1915
8644.0	CW	XFS	Tampico Radio, Mexico	FC	0.5	
8644.1	CW/FAX USB	WWD	La Jolla Radio (Scripps), CA	FC	....	Meteo [NOAA reports wx broadcasts 1700 & 2300--June thru October]
8645.0	CW	UAT	Moscow Radio, USSR	FC	25.0	
8646.0	CW	LPD86	Gen. Pacheco Radio, Argentina	FC	3.0	AMVER
	CW	†GYU	Gibraltar Naval Radio, Gibraltar	FC	5.0	British Navy
	CW	SAB4	Goteborg Radio, Sweden	FC	5.0	AMVER
8646.1	CW	FUJ	Noumea Naval Radio, New Caledonia	FC	20.0	French Navy
8647.0	CW	"F"	unknown	RC	....	Beacon
8647.4	CW	VTP4	Vishakhapatnam Naval Radio, India	FC	5.0	Indian Navy Reported wx broadcast at 2320
8647.5	CW	JDC	Choshi Radio, Japan	FC	10.0	
8648.0	CW	DHJ59	Sengwarden Naval Radio, GFR	FC	....	(G23B) German Navy/NATO
	CW	DZH	Manila Radio, Philippines	FC	0.3	Reported wx broadcasts at 1100 & 2320

kHz	Mode	Call	Location	Service	kW	Remarks
8649.5	CW	ICB	Genoa P.T. Radio, Italy	FC	5.0	
8650.0	CW	NMO	COMMSTA Honolulu, HI	FC	10.0	USCG/AMVER
			NOAA reports wx broadcasts at 1745 & 2345			
	CW	OBC3	Callao Radio, Peru	FC	0.5	TIME SIGNALS
	CW	SPE42	Szczecin Radio, Poland	FC	10.0	[aka SPB3]
	CW	UTA	Tallinn Radio, Estonian SSR	FC	5.0	
	CW	....	....	MS	....	USCG (This
			special frequency available to USCG ships for communications with merchant vessels--duplex transmissions)			
8652.0	CW	OST42	Oostende Radio, Belgium	FC	10.0	
	CW	STP	Port Sudan Radio, Sudan	FC	1.5	
8652.5	CW	PZN3	Paramaribo Radio, Suriname	FC	1.0	
8653.0	CW	RNO	Moscow Radio, USSR	FC	....	SAAMC
8653.6	CW	JCS	Choshi Radio, Japan	FC	3.0	
8654.4	CW	PCH42	Scheveningen Radio, Netherlands	FC	10.0	AMVER
8656.0	CW	IAR38	Rome P.T. Radio, Italy	FC	7.0	
	CW	XFU	Veracuz Radio, Mexico	FC	0.5	
8657.0	CW/RTTY	UJQ	Kiev Radio, Ukrainian SSR	FC	10.0	
8658.0	CW	S3E	Khulna Radio, Bangladesh	FC	10.0	
	CW	ASK	Karachi Radio, Pakistan	FC	3.5	
	CW	KLB	Seattle Radio, WA	FC	3.5	
	CW	WSL	Amagansett Radio, NY	FC	10.0	
8659.5	CW	SXA4	Spata Attikis Naval Radio, Greece	FC	....	(K13A/Y)
						Greek Navy/NATO
8660.0	CW	†DZJ	Manila (Bulacan) Radio, Philippines	FC	2.5	
	CW	Y5M	Ruegen Radio, GDR	FC	5.0	
	CW	RPLT	Tuapse Radio, USSR	FC	....	
8662.0	CW	JJH	Kure Radio, Japan	FC	1.0	
	CW	TAH	Istanbul Radio, Turkey	FC	3.0	[aka TCCH]
	CW	TCR	Tophane Kulesi Naval Radio, Turkey	FC	0.2	(TCCQ) Turkish Navy
8663.0	CW	YPO	Giurgiu Radio, Rumania	FC	....	
	CW	UFN	Novorossiysk Radio, USSR	FC	10.0	
8663.5	CW	UJQ7	Kiev Radio, Ukrainian SSR	FC	5.0	(4LA)
8664.0	CW	PWZ	Rio de Janeiro Naval Radio, Brazil	FC	10.0	Brazil Navy
8665.0	CW	XSG3	Shanghai Radio, PRC	FC	4.0	
	CW	6WW	Dakar Naval Radio, Senegal	FC	....	French Navy
8666.0	CW	†HKB	Barranquilla Radio, Colombia	FC	2.5	
	CW	FUG3	La Regine (Castelnaudary), France	FC	....	French Navy
	CW	KLC	Galveston Radio, TX	FC	3.8	NOAA reports
			wx broadcasts at 0530, 1130, 1730 & 2330			
8666.4	CW	SPH43	Gydnia Radio, Poland	FC	10.0	[aka SPI4]
8668.0	CW	YIR	Basrah Control, Iraq	FC	1.0	NOAA reports
						wx broadcasts on request
8669.9	CW	IAR8	Rome P.T. Radio, Italy	FC	7.0	
8671.0	CW	SXA8	Spata Attikis Naval Radio, Greece	FC	5.0	Greek Navy/NATO
8672.0	CW	DZI	Manila (Bacoor) Radio, Philippines	FC	1.0	
8672.5	CW	DAF	Norddeich Radio, GFR	FC	10.0	
8673.3	CW	GYA	Whitehall (London) Naval R., England	FC	....	British Navy
8674.0	CW	FUV	Djibouti Naval Radio, Djibouti	FC	20.0	French Navy
	CW	LGB	Rogaland Radio, Norway	FC	5.0	
	CW	CTP	Oeiras Naval Radio, Portugal	FC	10.0	Port. Navy/NATO
8674.4	CW	VWM	Madras Radio, India	FC	2.5	NOAA reports wx
			warnings broadcast at 0450, 0930, 1350 & 1800			
8675.0	CW	URD	Leningrad Radio, USSR	FC	10.0	
	CW	UQB	Kholmsk Radio, USSR	FC	5.0	
8675.2	CW	FFP3	Fort de France Radio, Martinique	FC	1.0	NOAA reports wx
			warning broadcasts at 0000, 0400, 0830, 1330, 1600 & 2100			
8676.0	CW	DYM	Cebu Radio, Philippines	FC	0.4	
8678.0	CW	XSZ	Darien Radio, PRC	FC	2.0	
	CW	ZLO4	Irirangi Naval Radio, New Zealand	FC	20.0	RNZ Navy
	CW	LFB	Rogaland Radio, Norway	FC	5.0	AMVER
	CW	SPH44	Gydnia Radio, Poland	FC	10.0	[aka SPI4]
8679.0	CW	IQX	Trieste P.T. Radio, Italy	FC	6.0	
	CW	UQB	Kholmsk Radio, USSR	FC	5.0	Reported wx
				broadcasts at 0025 & 0825		

kHz	Mode	Call	Location	Service	kW	Remarks
8680.0	CW	XUK3	Kompong Som-Ville Radio, Cambodia	FC	0.3	
	CW	RMP	Rostov Radio, USSR	FC	....	
8680.1	FAX	NMC	COMMSTA San Francisco, CA	FC	2.0	USCG/Meteo
			USCG reports pics at 0100, 0300, 0500, 0700, 1500, 1700, 2000 & 2300			
8681.0	CW	SVI4	Athens Radio, Greece	FC	5.0	
8682.0	CW	J2A8	Djibouti Radio, Djibouti	FC	2.5	NOAA reports wx broadcasts at 0430, 0900 & 1700
	CW	EAD3	Aranjuez Radio, Spain	FC	10.0	
	CW	NMC	COMMSTA San Francisco, CA	FC	2.0	USCG/Wx broadcasts at 0030, 0630 & 1900
8682.5	CW	CCS	Santiago Naval Radio, Chile	FC	1.0	Chilean Navy
8683.6	CW	LFB2	Rogaland Radio, Norway	FC	5.0	
8684.0	CW	GKJ4	Portishead Radio, England	FC	10.0	
8685.0	CW	IRM6	CIRM, Rome Radio, Italy	FC	10.0	MEDICO/AMVER
8686.0	CW	PKB	Belawan Radio, Indonesia	FC	1.0	
	CW	PKF	Makassar Radio, Indonesia	FC	1.0	
	CW	PKA	Sabang Radio, Indonesia	FC	1.0	
	CW	JCT	Choshi Radio, Japan	FC	3.0	
	CW	CNP	Casablanca Radio, Morocco	FC	2.0	
	CW	C2N	Nauru Radio, Nauru	FC	1.0	
	CW	JWT	Stavanger Naval Radio, Norway	FC	3.0	Norwegian Navy
	CW	HSA2	Bangkok Radio, Thailand	FC	5.0	Reported wx broadcast at 1045
	CW	WNU53	Slidell Radio, LA	FC	3.0	
	CW	WMH	Baltimore Radio, MD	FC	10.0	
8687.0	CW	SVA4	Athens Radio, Greece	FC	0.9	
8687.6	CW	URD	Leningrad Radio, USSR	FC	15.0	
8688.0	CW	9VG36	Singapore Radio, Singapore	FC	3.0	
8688.5	CW	ZSC6	Cape Town Radio, RSA	FC	10.0	TIME SIGNALS/Tfc
8690.0	CW	CLQ	Havana (Cojimar) Radio, Cuba	FC	5.0	
	CW	3DP3	Suva Radio, Fiji	FC	0.5	AMVER
	CW	SVF4	Athens Radio, Greece	FC	5.0	
	CW	TFA3	Reykjavik Radio, Iceland	FC	1.5	
	CW	XFA2	Acapulco Radio, Mexico	FC	1.0	
	CW	XFB2	Campeche Radio, Mexico	FC	1.0	
	CW	XFS2	Ciudad Madero Radio, Mexico	FC	1.0	
	CW	XFQ2	Frontera (Tabasco) Radio, Mexico	FC	1.0	
	CW	XFQ3	Salina Cruz Radio, Mexico	FC	0.5	
	CW	XFY2	Guaymas Radio, Mexico	FC	1.0	
	CW	XFS3	Tampico Radio, Mexico	FC	1.0	
	CW	XFU2	Veracruz Radio, Mexico	FC	1.0	
	CW	FJY4	St. Paul et Amsterdam Radio, French Antarctic	FC	1.0	NOAA reports wx broadcasts at 0245 & 0545
	CW	6VA5	Dakar Radio, Senegal	FC	4.0	NOAA reports wx forecasts at 1000 & 2200
	CW	UMV	Murmansk Radio, USSR	FC	25.0	
8692.5	CW	SVF4	Athens Radio, Greece	FC	5.0	
8694.0	CW	S3D	Chittagong Radio, Bangladesh	FC	1.0	
	CW	CBM2	Magallanes (Punta Arenas) R., Chile	FC	0.7	
	CW	PKM	Bitung Radio, Indonesia	FC	1.0	
	CW	PNK	Jayapura Radio, Indonesia	FC	1.0	
	CW	PJC	Curacao (Willemstad) R., Curacao	FC	2.5	NOAA reports warning broadcast at 2300
	CW	ASK	Karachi Radio, Pakistan	FC	2.0	NOAA reports wx broadcast at 1630
	CW	XSZ	Dairen Radio, PRC	FC	2.0	NOAA reports wx broadcasts at 0050 & 1050
8694.4	CW	4XO	Haifa Radio, Israel	FC	2.5	
8695.1	CW	UHK	Batumi Radio, USSR	FC	5.0	
8696.0	CW	ZAD2	Durres Radio, Albania	FC	4.0	
	CW/RTTY	XSC	Fangshan Radio, PRC	FC	10.0	
	CW/RTTY	Y5M	Ruegen Radio, GDR	FC	5.0	
	CW	9GX	Tema Radio, Ghana	FC	10.0	

kHz	Mode	Call	Location	Service	kW	Remarks
8697.0	CW	CFH	Maritime Command Radio, Halifax, NS, Canada (Canadian Forces)	FC	10.0	(C13L) NOAA reports wx broadcasts at 0200, 0530, 0630, 1400, 1800 & 1930. Seasonal ice warnings at 0230 & 1430
8698.0	CW	JGC	Yokohama Radio, Japan	FC	1.0	
	CW	9MG2	Penang Radio, Malaysia	FC	1.0	
	CW	C9L2	Maputo Radio, Mozambique	FC	0.5	
	CW	C9L4	Beira Radio, Mozambique	FC	1.0	
	CW	C9L6	Mozambique Radio, Mozambique	FC	0.5	
	CW	FJP8	Noumea Radio, New Caledonia	FC	1.0	NOAA reports wx broadcasts at 0000 & 0900
	CW	5OW8	Lagos Radio, Nigeria	FC	1.0	
	CW	UDK2	Murmansk Radio, USSR	FC	10.0	[aka UMV]
8700.0	CW	YUR3	Rijeka Radio, Yugoslavia	FC	6.0	
8700.4	CW	HKB	Barranquilla Radio, Colombia	FC	2.5	
8701.0	CW	UNM2	Klaipeda Radio, Lithuanian SSR	FC	5.0	
8701.3	CW	CTU28	Monsanto Naval Radio, Portugal	FC	1.0	Port. Navy
8701.9	CW	CLA22	Havana (Cojimar) Radio, Cuba	FC	5.0	
8702.0	CW	OSN	Oostende Naval Radio, Belgium	FC	....	Belgium Navy
	CW	XSR	Haikou Radio, PRC	FC	1.0	
	CW	ODR3	Beirut Radio, Lebanon	FC	1.0	
8702.9	CW	CTU28	Monsanto Naval Radio, Portugal	FC	10.0	(P13A) Port. Navy
8703.0	CW	CTV73	Monsanto Naval Radio, Portugal	FC	1.0	Port. Navy
	CW	UXN	Arkhangelsk Radio, USSR	FC	10.0	
8703.2	CW	"E"	unknown	RC	....	Beacon
8704.0	CW	SVB4	Athens Radio, Greece	FC	20.0	
8705.0	CW/TOR	ZUD79	Olifantsfontein Radio, RSA	FC	5.0	
8705.5	CW/TOR	GKE4	Portishead Radio, England	FC	10.0	
8705.9	CW	†4XZ	Haifa Naval Radio, Israel	FC	....	Israel Navy
8707.0	CW/TOR	LGB2	Rogaland Radio, Norway	FC	5.0	
	CW/TOR	WLO	Mobile Radio, AL	FC	15.0	
8707.4	CW/TOR	VIP33	Perth Radio, WA, Australia	FC	10.0	
8709.0	CW/TOR	9VG78	Singapore Radio, Singapore	FC	10.0	
	CW/TOR	HEC18	Berne Radio, Switzerland	FC	10.0	
8710.0	CW	9LL	Freetown Radio, Sierra Leone	FC	1.5	
8711.6	CW/TOR	VIS65	Sydney Radio, NSW, Australia	FC	5.0	
8711.9	CW/TOR	WLO	Mobile Radio, AL	FC	10.0	
8712.6	CW	SVU5	Athens Radio, Greece	FC	5.0	
8713.0	CW	UDK2	Murmansk Radio, USSR	FC	....	
8714.0	CW	9PA5	Banana Radio, Zaire	FC	3.0	
8717.0	CW/TOR	WLO	Mobile Radio, AL	FC	15.0	
8718.0	CW	TJC8	Douala Radio, Cameroon	FC	0.8	

8718.9 - 8811.9 kHz — MARITIME COAST USB DUPLEX (Worldwide)

kHz	Mode	Call	Location	Service	kW	Remarks
8718.9	USB	....	Goteborg Radio, Sweden	FC	5.0	CH#801 (8195.0)
	USB		This frequency used by USCG Coast Stations: NMA, Miami, FL; NMC, San Francisco, CA; NMF, Boston, MA; NMG, New Orleans, LA; NMN, Portsmouth, VA; NMR, San Juan, PR; NOJ, Kodiak, AK & NRV, Barrigada, Guam			
8721.0	CW	PPE	Rio de Janeiro Radio, Brazil	FC	2.0	TIME SIGNALS
8722.0	USB	VIS	Sydney Radio, NSW, Australia	FC	5.0	CH#802 (8198.1)
	USB	Y5P	Ruegen Radio, GDR	FC	10.0	" "
	USB	CUL	Lisbon Radio, Portugal	FC	5.0	" "
	USB	WOM	Miami Radio, FL	FC	10.0	" " [NOAA reports wx broadcast at 1230]
8725.1	USB	....	Goteborg Radio, Sweden	FC	5.0	CH#803 (8201.2)
	USB	WFN	Jeffersonville, IN	FC/MS	0.8	Inland Waterway
	USB	WJG	Memphis, TN	FC/MS	0.8	Inland Waterway
8728.2	USB	3AC8	Monaco Radio, Monaco	FC	10.0	CH#804 (8204.3)
	USB	9VG63	Singapore Radio, Singapore	FC	10.0	" "
	USB	KMI	San Francisco (Dixon) Radio, CA	FC	10.0	" " [NOAA reports wx broadcasts at 0000, 0600 & 1500]
	USB	EHY	Pozuelo del Rey Radio, Spain	FC	10.0	CH#804 (8204.3)
8731.3	USB	PCG42	Scheveningen Radio, Netherlands	FC	10.0	CH#805 (8207.4)

kHz	Mode	Call	Location	Service	kW	Remarks
	USB	P2M	Port Moresby Radio, Papua New Guinea	FC	1.0	" "
	USB	P2R	Rabaul Radio, Papua New Guinea	FC	1.0	" "
	USB	ZSC26	Cape Town Radio, RSA	FC	5.0	" "
	USB	WOM	Miami Radio, FL	FC	10.0	" "
			[NOAA reports wx broadcast at 1330]			
8734.4	USB	SVN4	Athens Radio, Greece	FC	5.0	CH#806 (8210.5)
	USB	ICB	Genoa Radio 8, Italy	FC	7.0	" "
	USB	PCG43	Scheveningen Radio, Netherlands	FC	10.0	" "
8737.5	USB	5BA42	Nicosia Radio, Cyprus	FC	10.0	CH#807 (8213.6)
	USB	ZLW	Wellington Radio, New Zealand	FC	10.0	" "
	USB	WCM	Pittsburgh, PA	FC/MS	0.8	Inland Waterway
	USB	WGK	St. Louis, MO	FC/MS	0.8	Inland Waterway
8740.6	USB	....	Lyngby Radio, Denmark	FC	10.0	CH#803 (8216.7)
	USB	KMV	Agana Radio, Guam	FC	1.0	" "
	USB	KQM	Kahuku (Honolulu) Radio, HI	FC	0.8	" "
	A3H	ZSD38	Durban Radio, RSA	FC	5.0	CH#803 (8216.7)
			[NOAA reports wx broadcast at 0918]			
	USB	WOO	New York (Ocean Gate, NJ) Radio, NY	FC	10.0	" "
			[NOAA reports wx broadcasts at 0100, 1300 & 1900]			
8743.7	USB	3AC9	Monaco Radio, Monaco	FC	10.0	CH#809 (8219.8)
	USB	LFL5	Rogaland Radio, Norway	FC	10.0	" "
	USB	KMI	San Francisco (Dixon) Radio, CA	FC	10.0	" "
8746.8	USB	3DP	Suva Radio, Fiji	FC	1.0	CH#810 (8222.9)
	USB	JBO	Tokyo Radio, Japan	FC	10.0	" "
	USB	LFL6	Rogaland Radio, Norway	FC	10.0	" "
	USB	EHY	Pozuelo del Rey Radio, Spain	FC	10.0	" "
	USB	WOM	Miami Radio, FL	FC	10.0	" "
			[NOAA reports wx broadcast at 2230]			
	USB	....	Rijeka Radio, Yugoslavia	FC	7.5	CH#810 (8222.9)
8749.9	USB	....	Hong Kong (Cape d'Aguilar) Radio	FC	6.0	CH#811 (8226.0)
	USB	LFL7	Rogaland Radio, Norway	FC	10.0	" "
	USB	WOO	New York (Ocean Gate, NJ) Radio, NY	FC	10.0	" "
			[NOAA reports wx broadcasts at 0100, 1300 &1900]			
8752.0	CW	"K"	unknown	RC	....	Beacon
8753.0	USB	JBO	Tokyo Radio, Japan	FC	10.0	CH#812 (8229.1)
	USB	SPC41	Gydnia Radio, Poland	FC	10.0	" "
8756.1	USB	OSU45	Oostende Radio, Belgium	FC	10.0	CH#813 (8232.2)
	USB	LFL8	Rogaland Radio, Norway	FC	10.0	" "
8759.2	USB	LPL3	Gen. Pacheco Radio, Argentina	FC	10.0	CH#814 (8235.3)
	USB	SVN4	Athens Radio, Greece	FC	5.0	" "
			[NOAA reports wx broadcasts at 1215 & 2015]			
8762.3	USB	VID	Darwin Radio, NT, Australia	FC	1.0	CH#815 (8238.4)
	USB	OSU41	Oostende Radio, Belgium	FC	10.0	" "
	USB	9VG64	Singapore Radio, Singapore	FC	10.0	" "
	USB	WOO	New York (Ocean Gate, NJ) Radio, NY	FC	10.0	" "
8765.4	USB	....	Barbados Radio, Barbados	FC	2.0	CH#816 (8241.5)
	USB	GKU46	Portishead Radio, England	FC	10.0	" "
	USB	NMN	COMMSTA Portsmouth, VA	FC	10.0	USCG/AMVER CH#816 (8241.5)
			[USCG reports wx broadcasts at 0400, 0530, 1000, 1130, 1600, 1730, 2200 & 2300]			
	USB	NMO	COMMSTA Honolulu, HI	FC	10.0	USCG CH#816 (8241.5)
			[USCG reports wx broadcasts at 0545, 1145, 1745 & 2345]			
	USB	NMC	COMMSTA San Francisco, CA	FC	10.0	USCG CH#816 (8241.5)
			[NOAA wx broadcasts at 0430 & 1030]			
	USB		This frequency is used by USCG Coastal Stations: NMA, Miami, FL; NMF, Boston, MA; NMG, New Orleans, LA; NOJ, Kodiak, AK & NRV, Barrigada, Guam			
8768.5	USB	VIT	Townsville Radio, Qld., Australia	FC	1.0	CH#817 (8244.6)
	USB	VRT	Bermuda (Hamilton) Radio, Bermuda	FC	1.0	" "
	USB	FFL43	St. Lys Radio, France	FC	10.0	" "
	USB	DAJ	Norddeich Radio, GFR	FC	10.0	" "

kHz	Mode	Call	Location	Service	kW	Remarks
	USB		This frequency is used by USCG Coastal Stations: NMA, Miami, FL; NMG, New Orleans, LA & NOJ, Kodiak, AK [May be used for special law enforcement activities when required]			
8771.6	USB	5BA44	Nicosia Radio, Cyprus	FC	10.0	CH#818 (8247.7)
	USB	....	Lyngby Radio, Denmark	FC	10.0	" "
	USB	EHY	Pozuelo del Rey Radio, Spain	FC	10.0	" "
8774.7	USB	GKU49	Portishead Radio, England	FC	10.0	CH#819 (8250.8)
	USB	....	Rome Radio 8, Italy	FC	7.0	" "
			This frequency is also used by USCG Coast Stations: NMC, San Francisco, CA; NMF, Boston, MA & NMO, Honolulu, HI			
8777.8	USB	DAI	Norddeich Radio, GFR	FC	10.0	CH#820 (8253.9)
	USB	JBO	Tokyo Radio, Japan	FC	2.5	" "
	USB	....	Rome Radio 8, Italy	FC	7.0	" "
8778.0	USB	....	numerous, around USA coastline	FC/MS	....	USN
8780.9	USB/AM	....	INTERNATIONAL COAST STATION CALLING FREQUENCY			CH#821 (8257.0)
	USB	9VG89	Singapore Radio, Singapore	FC	10.0	CH#821 (8357.0)
	USB	....	Rogaland Radio, Norway	FC	10.0	" "
8784.0	USB	PPL	Belem Radio, Brazil	FC	1.0	CH#822 (8260.1)
	USB	HEB18	Berne Radio, Switzerland	FC	10.0	" "
	USB	KMI	San Francisco (Dixon) Radio, CA [NOAA reports wx broadcasts at 0000, 0600 & 1500]	FC	10.0	" "
8787.1	USB	VCS	Halifax CG Radio, NS, Canada [NOAA reports wx broadcasts at 0215, 0815, 1615 & 2215]	FC	5.0	CH#823 (8263.2)
8787.1	USB	ICB	Genoa Radio 8, Italy	FC	5.0	CH#823 (8263.2)
8790.2	USB	PPO	Olinda Radio, Brazil	FC	1.0	CH#824 (8266.3)
	USB	PJC	Curacao (Willemstad) R., Curacao [NOAA reports wx broadcasts on request]	FC	1.0	" "
	USB	SPC42	Gdynia Radio, Poland	FC	10.0	" "
	USB	9VG65	Singapore Radio, Singapore	FC	10.0	" "
	USB	HEB28	Berne Radio, Switzerland	FC	10.0	" "
	USB	WLO	Mobile Radio, AL	FC	10.0	" "
8793.3	USB	KWJ91	Anchorage Radio, AK	FC	1.0	CH#825 (8269.4)
	USB	ZKR	Rarotonga Radio, Cook Islands	FC	1.5	" "
	USB	FFL42	St. Lys Radio, France	FC	10.0	" "
	USB	....	Goteborg Radio, Sweden	FC	5.0	" "
	USB	WOM	Miami Radio, FL [NOAA reports wx broadcast at 2330]	FC	10.0	" "
8794.0	AM	3DP	Suva Radio, Fiji	FC	1.0	NOAA reports wx broadcasts at 0003, 0403, 0803 & 2003
8796.4	USB	KMV	Agana Radio, Guam	FC	1.0	CH#826 (8272.5)
	USB	PCG41	Scheveningen Radio, Netherlands	FC	10.0	AMVER CH#826 (8272.5)
	USB	....	Rome Radio 8, Italy	FC	7.0	CH#826 (8272.5)
	USB	WOO	New York (Ocean Gate, NJ) Radio, NY	FC	10.0	" "
	USB	WBL	Buffalo Radio, NY	FC	1.0	" "
	USB	WLC	Rogers City Radio, MI	FC	1.0	" "
	USB	WMI	Lorain Radio, OH	FC	1.0	" "
8799.5	USB	4XO	Haifa Radio, Israel	FC	10.0	CH#827 (8275.6)
	USB	....	Goteborg Radio, Sweden	FC	5.0	" "
8802.6	USB	PPJ	Juncao Radio, Brazil	FC	1.0	CH#828 (8278.7
	USB	FFL44	St. Lys Radio, France	FC	10.0	" "
	USB	DAI	Norddeich Radio, GFR	FC	10.0	" "
8805.7	USB	VIS	Sydney Radio, NSW, Australia	FC	5.0	CH#829 (8281.8)
	USB	OHG2	Helsinki Radio, Finland	FC	10.0	" "
	USB	....	Banjul Radio, Gambia	FC	1.0	" "
	USB	....	Mahina (Papeete) Radio, Tahiti	FC	1.0	AMVER (CH#829) (8281.8)
	USB	WLO	Mobile Radio, AL	FC	10.0	" "
	USB	....	Rijeka Radio, Yugoslavia	FC	7.5	" "
8807.1	CW	"E"	unknown	RC	....	Beacon
8808.8	USB	PPR	Rio Radio, Brazil	FC	5.0	CH#830 (8284.9)
	USB	FFL41	St. Lys Radio, France	FC	10.0	" "
	USB	H4H7	Honiara Radio, Solomon Islands	FC	0.5	CH#830 (8284.9)

kHz	Mode	Call	Location	Service	kW	Remarks
	USB	WLO	Mobile Radio, AL	FC	10.0	" "
			[NOAA reports wx broadcasts at 0000, 1300 & 1800]			
8811.9	USB	†....	Falkland Islands Radio	FC	1.0	CH#831 (8288.0)
	USB	....	Rome Radio 8, Italy	FC	7.0	" "
	USB	KRV	Ponce Radio, PR	FC	1.0	" "
	USB	HEB38	Berne Radio, Switzerland	FC	10.0	" "
	USB	WOM	Miami Radio, FL	FC	10.0	" "

8815.0 - 9040.0 kHz -- AERONAUTICAL MOBILE (Worldwide)

kHz	Mode	Call	Location	Service	kW	Remarks
8818.0	FAX	LOK	Orcadas Radio, South Orkney Is.	FX	FX 0.2	NOAA reports
			seasonal [summer] broadcasts at 0300, 1700 & 1915			
	CW	ZCB4	unknown	FX	....	
8819.0	AM	....	Teheran Aeradio, Iran	FA	....	VOLMET (05+35)
	AM	....	Baghdad Aeradio, Iraq	FA	....	" (00+30)
	AM	....	Beirut Aeradio, Lebanon	FA	....	" (15+45)
	AM	SUC5Ø	Cairo Aeradio, Egypt	FA	1.3	" (20+50)
	AM	....	Istanbul Aeradio, Turkey	FA	....	" (25+55)
			§VOLMET broadcasts move to 8945.0 in 1982			
8826.0	AM/A3H	....	numerous (ICAO SAM & AFI)	FA/MA	....	Airports & in-
			flight covering all of the west coast of South America, plus the north to south central part of Africa from the Mediterranean Sea to the Cape of Good Hope, RSA			
			§SAM moves to 10,024.0 & AFI moves to 8903.0 in 1982			
8828.0	A3H/USB	....	§VOLMET broadcasts on 8903.0 move here in 1982			
8828.5	AM	A9G	Bahrain Aeradio, Bahrain	FA	0.3	VOLMET
8833.0	A3H	EIP	Shannon Aeradio (Rineanna), Ireland	FA	2.0	VOLMET (cont.)
8840.0	AM/A3H	....	numerous (ICAO SEA & CAR)	FA/MA	....	Airports & in-
			flight covering Southeast Asia, Central America & western Caribbean.			
			§SEA moves to 10,066.0 & CAR moves to 8846.0 in 1982			
8842.0	CW	COL	Havana Aeradio, Cuba	FA/MA	....	Aeroflot
	CW	RFNV	Moscow Aeradio, USSR	FA/MA	....	Aeroflot (Mostly
			working in-flight to/from Havana)			
8843.0	A3H/USB	....	§CEP moves here from 8875.0 in 1983			
8846.0	A3H/USB	....	§VOLMET broadcasts on 8819.0 move here in 1982			
	A3H/USB	...	§CAR moves here in 1982 from 8840.0			
8847.0	AM/A3H	....	numerous (ICAO SP, SAM & MID)	FA/MA	....	Airports & in-
			flight covering all of the South Pacific Ocean, all of the east coast of South America, plus all of the Middle East.			
			§MID moves to 8918.0, SAM moves to 8855.0 & SP moves to 8867.0 in 1982			
8852.0	A3H/USB	....	§VOLMET moves here from 11,279.0 in 1983			
8854.0	AM/A3H	....	numerous (ICAO CWP, SEA & NAT)	FA/MA	....	Airports & in-
			flight covering all of the western Pacific--Honolulu to Ulan Bator; all of Indonesia and western Australia; all the North Atlantic region			
			§CWP moves to 8903.0; SEA moves to 11,396.0 & NAT moves to 8879.0 in 1982			
8855.0	A3H/USB	....	§SAM moves here from 8847.0 in 1983			
8858.0	A3H/USB	....	§VOLMET moves here from 8896.0 in 1983			
8861.0	A3H/USB	....	§SAT moves from 8882.0 in 1983			
8864.0	A3H/USB	....	§NAT moves here from 8889.0 in 1983			
8867.0	A3H/USB	....	§SP moves here from 8847.0 in 1983			
8868.0	AM/A3H	....	numerous (ICAO SEA)	FA/MA	....	Airports & in-
			flight covering all of Southeast Asia, Indonesia and western Australia			
			§SEA moves to 11,396.0 in 1982			
	USB	VFG	Gander Aeradio, Nfld., Canada	FA	1.0	VOLMET (20+50)
	USB	WSY7Ø	New York Aeradio, NY	FA	1.5	VOLMET (05+35)
			§Gander & New York VOLMET move to 8870.0 in 1983			
8875.0	AM/A3H	....	numerous (ICAO CEP & EUR)	FA/MA	....	Airport & in-
			flight covering Honolulu to San Francisco, plus all of central Europe, Mediterranean, Middle East and north to Leningrad & Moscow			

AERONAUTICAL MOBILE (con.)

kHz	Mode	Call	Location	Service	kW	Remarks
			§CEP moves to 8843.0 in 1982, EUR moves to 10,084 in 1982			
8879.0	A3H/USB	....	§NAT moves here from 8854.0 in 1982			
8882.0	AM/A3H	....	numerous (ICAO SEA, AFI & SAT)	FA/MA	....	Airports & in-flight covering India southeast to Singapore, western bulge of Africa, north to Portugal and routes between Dakar and Rio de Janeiro
			§SEA moves to 8942.0 & SAT moves to 8861.0 in 1983			
8889.0	AM/A3H	....	numerous (ICAO NAT)	FA/MA	....	Airports & in-flight covering North Atlantic
8891.0	A3H/USB	....	§NAT moves here from 8910.0 in 1982			
8896.0	AM	....	Algiers Aeradio, Algeria	FA	0.4	VOLMET (00+30)
	AM	....	Kano Aeradio, Nigeria	FA	....	" (05+35)
	AM	....	Dakar Aeradio, Senegal	FA	1.5	" (15+45)
	AM	....	Khartoum Aeradio, Sudan	FA	....	" (25+55)
			§VOLMET moves to 8858.0 in 1983			
8903.0	A3H	KIS7Ø	Anchorage Aeradio, AK	FA	1.5	VOLMET (25+55)
	A3H	KVM7Ø	Honolulu Aeradio, HI	FA	1.5	" (00+30)
	AM	....	Kai Tak Aeradio, Hong Kong	FA	1.5	TIME SIGNALS & VOLMET (15+45)
	A3H	....	Narita Aeradio, Tokyo, Japan	FA	1.5	" (10+40)
	A3H	....	Auckland Aeradio, New Zealand	FA	5.0	" (20+50)
	A3H	KSF7Ø	Oakland Aeradio, CA	FA	1.5	" (05+35)
			§VOLMET moves to 8828.0 in 1983			
	A3H/USB	....	§CWP moves here from 8854.0 in 1983			
	AM	....	Kiev Aeradio, Ukrainian SSR	FA	....	VOLMET/RR (20+50)
	AM	....	Leningrad Aeradio, USSR	FA	....	" " (05+35)
	AM	....	Moscow Aeradio, USSR	FA	....	" " (15+45)
	A3H	....	Riga Aeradio, Latvian SSR	FA	....	" " (00+30)
	A3H	....	Rostov Aeradio, USSR	FA	....	" " (10+40)
8910.0	AM/A3H	....	numerous (ICAO NAT)	FA/MA	....	Airports & in-flight covering North Atlantic Arctic region
			§NAT moves to 8891.0 in 1982			
	AM/A3H	....	numerous, worldwide	FA/MA	....	Airports & in-flight Paid Services for International Carriers: particularly in Australia & Canada
8913.5	CW	C7A	Ocean Station Vessel 62°N & 32°W	FA	....	[aka "ALFA"]
	CW	C7I	Ocean Station Vessel 59°N & 19°W	FA	....	[aka "INDIA]
	CW	C7J	Ocean Station Vessel 52°N & 20°W	FA	....	[aka "JULIETT"]
			(Ocean Station Vessels will discontinue use of CW in 1982)			
8917.0	USB	....	several, in the USA	FA/MA	....	In-flight Testin
	AM	....	Aktyubinsk Aeradio, Kazakh SSR	FA	....	VOLMET/RR (05+35)
	A3H	....	Alma Ata Aeradio, USSR	FA	....	" " (15+45)
	A3H	....	Baku Aeradio, Azerbaijan SSR	FA	....	" " (25+55)
	AM	....	Krasnodar Aeradio, USSR	FA	....	" " (10+40)
	AM	....	Tbilisi Aeradio, Georgian SSR	FA	....	" " (00+30)
8918.0	A3H/USB	....	§CAR moves here from 8847.0 in 1982/MID moves here from 8847.0 in 1983			
8922.0	A3H	....	Leningrad (Pulkovo), USSR	FA	....	VOLMET/RR
	A3H	....	Moscow (Sheremetyevo), USSR	FA	....	VOLMET/FF/RR
8924.0	AM	....	numerous (ICAO AFI)	FA/MA	....	Airports & in-flight covering bulge of West Africa including Cameroon, Chad, Congo, Gabon, Equatorial Guinea, etc.
			§AFI moves to 11,330.0 in 1982			
8925.0	CW	J3R	unknown	FX	....	
	USB	KPA2	unknown	FX	....	
8930.0	USB	9VA27	Singapore Aeradio, Singapore	FA	6.0	
	USB	HEE4Ø	Berne Aeradio, Switzerland	FA/MA	6.0	
8931.0	AM/A3H	....	numerous (ICAO CEP)	FA/MA	....	Airports & in-flight covering to Honolulu to San Francisco
			§CEP moves to 11,282.0 in 1982			
8938.0	AM/A3H	....	numerous (ICAO NP)	FA/MA	,,,,	Airports & in-flight covering northern Pacific Ocean & Arctic regions
			§NP moves to 10,048.0 in 1983			
	CW	C7P	Ocean Station Vessel 50°N & 145°W	FA	....	(aka "PAPA")
8942.0	A3H/USB	....	§SEA moves here from 8882.0 in 1983			

kHz	Mode	Call	Location	Service	kW	Remarks
8945.0	AM/A3H	....	numerous (ICAO NAT)	FA/MA	....	Airports & in-flight covering all of the North Atlantic region §NAT moves to 8825.0 in 1983
	A3H/USB	....	§VOLMET moves here from 8819.0 in 1983			
8952.0	USB	PBW	Amsterdam (Schiphol), Netherlands	FA/MA	1.7	Tfc/Oil Rigs
8955.0	CW	CLP6	unknown	FX	....	Cuban EMBASSY
8959.0	AM/A3H	....	numerous (ICAO AFI & CAR)	FA/MA	....	Airports & in-flight covering northeastern Africa, Arabia and Indian Ocean as far east as Bombay, plus all of the USA east coast, eastern Caribbean and northern South America §AFI moves to 8894.0 in 1983 & CAR moves to 8918.0 in 1982
8960.0	AM	†....	Kuybyshev (Kurumoch) Aeradio, USSR	FA	....	VOLMET/RR
8963.0	CW	CAK6H	Santiago (Los Cerrillos AB), Chile	FA/MA	1.0	Chilean AF
8964.0	USB	....	numerous, worldwide	FA/MA	....	USAF [Used mostly for SKY KING broadcasts]
8967.0	USB	....	numerous, worldwide	FA/MA	....	USAF (Usually AFA3, Andrews AFB, MD; AIE2, Andersen AFB, Guam; AIF2, Yokota AB, Japan; AKA5, Elemendorf AFB, AK; CUW, Lajes Field, Azores; AGA2. Hickam AFB, HI, etc.)
8970.0	USB	PBV4	Valkenburg AB, Netherlands	FA/MA	1.0	Dutch Naval Air
8972.0	USB	SHA38	Stockholm Aeradio, Sweden	FA	5.0	
	USB	....	numerous, worldwide	FA	....	USN [Sometimes referred to as the "KILO" frequency]
8975.0	USB	....	numerous, in Australasia	FA/MA	....	RAAF/RNZAF (Includes: AXD, "Air Force Laverton", Vic.; AXF, "Air Force Sydney", NSW; AXH, "Air Force Townsville", Qld.; AXI, "Air Force Darwin", NT; AXJ, "Air Force Perth", WA; AXK, "Air Force Sale", NSW; AXS, "Air Force Learmouth", WA; AXT, "Air Force Edinburgh", SA; ZKW, RNZAF, Christchurch; ZKX, RNZAF, Auckland; ZKY, RNZAF, Palmerston North & MRX, Buttersworth, Malaysia)
		....	numerous, worldwide	FA/MA	....	USN
8976.0	USB	....	several, throughout Australia	FA/MA	....	RAAF
8980.0	USB	....	numerous, mostly North America	FA/MA	....	USN/USCG (Primary frequency for heliocopters)
8984.0	USB	....	numerous, mostly North America	MA/MS	....	USCG (Primary frequency for aircraft to ships) [Available to USAF/USN]
	USB	....	Kingston Coast Radio, Jamaica	FA/FC	....	
8989.0	CW	ONY77	Brussels National Airport, Belgium	FA/MA	2.0	Belgium AF
	USB	....	numerous, worldwide	FA/MA	....	USAF (Usually AFE8, MacDill AFB, FL; AFI2, McClellan AFB, CA; AIE2, Andersen AFB, Guam; AKA5, Elemendorf AFB, AK; etc.)
8992.9	CW	6WW	Dakar Naval Radio, Senegal	FA	....	French Naval Air
8993.0	USB	....	numerous, worldwide	FA/MA	....	USAF (Usually AFE8, MacDill AFB, FL; AFH3, Albrook AFS, Panama; AGA8, Clark AB, Philippines; AFD14, Ascension Island; etc.) [May be used for MAINSAIL & SKY KING broadcasts]
8997.0	USB	....	numerous, New Zealand & Antarctica	FA/MA	....	RNZAF/USAF/USN (Includes: NBY, Bird Station; NGD, McMurdo Station; NPX, South Pole Station; NQU, Siple Station; etc., all Antarctica; plus, Williams Field & ZLK, Christchurch, New Zealand)
9000.0	CW	FUM	Papeete (Mahina) Naval Radio, Tahiti	FX	....	French Navy
9006.0	USB	AFE71	"Cape Radio", Patrick AFB, FL	FX	45.0	USAF/NASA Shuttle Support [Alt: 6708.0]
9009.0	ISB	....	numerous, worldwide	FA/MA	....	USAF
	AM	....	Kuybyshev (Kurumoch) Aeradio, USSR	FA	....	VOLMET/RR (15+45)
	USB	....	Salekhard Aeradio, USSR	FA	....	" " (25+55)
	A3H	....	Sverdlovsk Aeradio, USSR	FA	....	" " (05+35)
	A3H	....	Syktyvka Aeradio, USSR	FA	....	" " (00+30)
	A3H	....	Yeniseysk Aeradio, USSR	FA	....	" " (10+40)
9010.0	FAX/USB	....	several, throughout Canada	FA	....	RCAF
9011.0	USB	AJE	Croughton (Barford), England	FA/MA	....	USAF
9014.0	CW	LFP	Rogaland Radio, Norway	FC	2.0	
	USB	AFL2	Loring AFB, ME	FA/MA	....	USAF
	USB	AFG9	Scott AFB, IL	FA/MA	....	USAF (May also be used by the USN)

kHz	Mode	Call	Location	Service	kW	Remarks
9018.0	USB	AFG9	Scott AFB, IL	FA/MA	....	USAF
	USB	....	"AIR FORCE TWO"	MA	....	USAF
9020.0	USB	AFE8	MacDill AFB, FL	FA/MA	....	USAF Tfc (Also heard of this frequency during the hurricane season are in-flight aircraft and KJY74, "Miami Monitor") [Occasional use by AHF3, Albrook AB, Panama; AFL, Loring AFB, ME & AFI2, McClellan AFB, CA, etc.]
9023.0	USB	....	several, USA nationwide	FA/MA	....	USAF/SAC
9025.0	CW	†MQY	Pitreavie Castle, Scotland	FA	1.0	RAF
9027.0	USB	....	numerous	FA/MA	....	USAF (Used for SKY KING broadcasts--some code names for Air Bases--has been called "ROMEO" frequency)
9032.0	USB	....	numerous, mostly Pacific Ocean area	FA/MA	....	USN
9036.0	USB	....	numerous, mostly within England	FA/MA	....	RAF
	USB	....	numerous, worldwide	FA/MA	....	USN

9040.0 - 9500.0 kHz -- FIXED SERVICE (Worldwide)

kHz	Mode	Call	Location	Service	kW	Remarks
9040.0	AM	....	unknown	FX	....	NUMBERS/EE/SS/YL
9040.5	CW	†KNY25	Washington, DC	FX	1.0	Rumanian EMBASSY
9041.5	CW/RTTY	†KNY26	Washington, DC	FX	1.0	Hungarian EMBASSY
9042.9	CW	"K"	unknown	RC	....	Beacon [// 11,155.5]
9043.0	CW	5YE	Nairobi, Kenya	FX	10.0	NOAA reports wx broadcast at 1200
	FAX	5YE	Nairobi, Kenya	FX	30.0	Meteo
	ISB	AFE71	"Cape Radio", Patrick AFB, FL	FX	45.0	USAF/NASA
9044.0	USB	TJF9Ø	Douala, Cameroon	FX	2.0	Telcom
9046.0	USB	VNV52	Sydney (Doonside), NSW, Australia	FX	3.0	Tfc/Casey
	CW/RTTY	ONN39	Brussels, Belgium	FX	1.0	Hungarian EMBASSY
9048.0	CW	....	Ankara, Turkey	FX	1.0	EMBASSY
9050.0	CW	NMO	COMMSTA Honolulu, HI	FX	10.0	USCG/NERK Wx broadcasts at 0100, 0400, 0700, 1300 & 2000
9051.5	USB	†FZO9Ø	Papeete, Tahiti	FX	20.0	Telcom
9057.0	USB	....	numerous, worldwide	FX	....	USAF (Includes: AFC24, Westover AFB, MA; AFF5Ø, Altus AFB, OK; AFH39, March AFB, CA; AFX, Barksdale AFB, LA; AGA, Hickam AFB, HI; etc.)
9057.5	CW	"U"	unknown	RC	....	Beacon
9060.0	CW	AQP9	Karachi Naval Radio, Pakistan	FX	10.0	Pakistan Navy
	FAX	RTA21	Novosibirsk, USSR	FX	15.0	Meteo
9062.0	USB	CME9	Havana, Cuba	FX	2.0	GDR EMBASSY
9062.0	CW/RTTY	OMZ29	Bratislava, Czechoslovakia	FX	1.0	GDR EMBASSY
9064.0	CW	Y7A36	Berlin, GDR	FX	5.0	EMBASSY
9065.0	USB	....	unknown, probably in Australia	FX	....	Net
	CW	IDR2Ø	Rome Naval Radio, Italy	FX	15.0	Italian Navy [Reported Nx in II at 0300, 1800 & 2230]
9073.0	USB	....	Aleutian LORAN Network	FX	....	USCG (Includes: NMJ, Attu; NRW2, St. Paul & NRW3, Pt. Clarence--all AK) [Alt: 12,150.0]
9074.0	AM	....	unknown	FX	....	NUMBERS/OM/SS
9077.0	USB	....	several, throughout Australia	FX	....	Commercial Tfc
9078.0	CW/RTTY	†OMZ28	Prague, Czechoslovakia	FX	1.0	GDR EMBASSY [Alt: 9086.0]
	CW	Y7A37	Berlin, GDR	FX	....	EMBASSY
9083.5	LSB	†CNR25	Rabat, Morocco	FX	20.0	Telcom
9085.9	CW	FDC4	Reims Air, France	FX	....	French AF
	CW	FDC7	Strasbourg Air, France	FX	....	" "
	CW	FDY4	Orleans Air, France	FX	....	" "
9086.0	CW/RTTY	OMZ29	Bratislava, Czechoslovakia	FX	1.0	GDR EMBASSY
	USB	5YE	Nairobi, Kenya	FX	....	Meteo (NOAA reports wx forecast at 1240)
9090.0	ISB	....	Holzkirchen, GFR	FX	10.0	RFE Feeder
	CW	IDR4	Rome Naval Radio, Italy	FX	15.0	Italian Navy
9093.0	ISB	†YAF	Kabul, Afghanistan	FX	3.0	Telcom/Tfc
9095.0	USB	ZKS	Rarotonga, Cook Islands	FX	1.0	Net Control [Alt: 9368.0]

FIXED SERVICE (con.)

kHz	Mode	Call	Location	Service	kW	Remarks
9097.0	USB	....	Daventry, England	FX	30.0	BBC Feeder
9100.0	ISB	....	numerous, throughout Angola	FX	1.0	Police
9101.0	USB	CNR	Rabat, Morocco	FX	20.0	Telcom/Tfc
9105.0	CW	XJD48	Ottawa (Almonte), Ont., Canada	FX	1.0	INTERPOL
	USB	....	numerous, throughout Canada	FX	....	RCMP Backup Net
9106.0	CW/USB	VSD	Halley Base, Antarctica	FX	0.5	[All stations in
	CW/USB	†ZBH	Grytviken, So. Georgia Island	FX	0.3	this list are
	CW/USB	ZHF33	Signy Island, S. Orkney Islands	FX	0.3	part of the
	CW	ZHF44	Faraday Base, Antarctica	FX	0.3	British Antarctic
	CW/USB	ZHF45	Rothera Base, Adelaide Island	FX	0.3	Survey]
9111.0	CW	J5G	Bissau, Guinea-Bisseau	AX	3.0	Aero Tfc
	CW	D4B44	Sal, Cape Verde	AX	3.0	Aero Tfc
9112.5	ISB	†JBE39	Tokyo, Japan	FX	15.0	Telcom/Taipei
9115.0	LSB	LRB91	Buenos Aires (G. Pacheco), Argentina	FX	10.0	Feeder/Telcom
9116.0	USB	....	Papeete, Tahiti	AX	0.2	Net Control
9117.5	USB	FYJ3	Paris (St. Assise), France	FX	20.0	Telcom
9125.6	USB	....	Atlantic Domestic Emergency Net	FX	....	USCG (Includes: NMA, Miami, FL; NMF, Boston, MA; NMG, New Orleans, LA; NMN, Portsmouth, VA; NMR, San Juan, PR & NOZ, Elizabeth City, NC--plus transportables) [Alt: 7528.6 & 10,136.6]
	USB	....	Pacific Area Domestic Emergency Net	FX	....	USCG (Includes: NMC, San Francisco, CA; NMO, Honolulu, HI; NMQ, Long Beach, CA; NOJ, Kodiak, AK & NRV, Barrigada, Guam--plus transportables) [Alt: 7528.6 & 10,166.6]
9129.0	USB	AGA	Hickam AFB, HI	FX	3.0	USAF/Tfc Osan
9130.0	USB	....	numerous, throughout Australia	FX	....	Commercial Tfc
9135.0	ISB	CME396	Havana, Cuba	FX	0.5	British EMBASSY
	FAX	JKB4	Tokyo (Kemigawa), Japan	FX	5.0	
9138.5	USB	XYQ65	Mandalay, Burma	FX	0.5	Net Control
9140.0	LSB	ROW25	Moscow, USSR	FX	15.0	R. Moscow Feeder
9145.0	ISB	....	Holzkirchen, GFR	FX	10.0	RFE Feeder
	CW	RIW	Khiva Naval Radio, Uzbek SSR	FX	....	Russian Navy
9150.0	FAX	....	Alma Ata, USSR	FX	15.0	Meteo/Nx
9160.0	CW	BFP99	Xian, PRC	FX	10.0	Meteo
9165.0	CW	ONN38	Brussels, Belgium	FX	0.1	Swiss EMBASSY
9166.9	ISB	†LDN2	Oslo (Jeloey), Norway	FX	10.0	Telcom/Tfc
9170.0	ISB	....	Holzkirchen, GFR	FX	10.0	RFE Feeder
	USB	†WMA29	San Francisco (Dixon), CA	FX	30.0	Telcom
9181.0	CW	FUM	Papeete Mahina Naval R. Tahiti	FX	10.0	French Navy
9196.0	CW	RYZ	Komsomolskamur, USSR	FX	...	Tfc
9200.0	CW	LJP29	Oslo, Norway	FX	0.5	INTERPOL
	CW	JPA6Ø	Nagoya (Komaki), Japan	FX	10.0	INTERPOL (Also occasionally using this frequency are: AYA, Buenos Aires; GMP, London, HEP92, Zurich; PDB2, Utrecht; SHX, Stockholm; 5BP5, Nicosia & 5OP25, Lagos)
	CW	ZBP	Pitcairn Island	FX	0.7	Tfc
	LSB	....	Khabarovsk, USSR	FX	15.0	R. Moscow Feeder
9200.1	CW	....	unknown, probably in Mexico	FX	....	
9203.0	FAX	GFE22	Bracknell, England	FX	5.0	Meteo/Reported seasonal ice broadcast at 1413
9205.0	LSB	FJD291	Papeete, Tahiti	FX	0.1	Net/Meteo
9205.5	USB	TYK92	Cotonou, Benin	FX	6.0	Telcom/Paris
9210.0	USB	....	Lyndhurst, Vic., Australia	FX	....	R. Australia Feeder
	ISB	RME2Ø	Moscow, USSR	FX	15.0	R. Moscow Feeder
9215.0	CW	†RUZU	Molodezhnaya Base, Antarctica	FX	....	SAAM
	CW	†UGE2	Bellingshausen Base, Antarctica	FX	....	SAAM
9217.0	CW	RLX	Dublin, Ireland	FX	0.5	USSR EMBASSY [Alt: 6842.0 & 9842.0]
9220.0	FAX	....	Novosibirsk, USSR	FX	15.0	Meteo
	USB	....	numerous, USA nationwide	FX	....	USAF (Includes: AFF7, Davis-Monthan AFB, AZ; AFG8, Malstrom AFB, MT; AFG29, Whiteman AFB, MO; AFH28, Beale AFB, CA; etc.)
9222.5	USB	VJP	Perth (Gnangara), WA, Australia	FX	10.0	Telcom/VLU
9223.6	USB	....	Northwest Pacific LORAN Net	FX	....	USCG (Includes: NRT, "Yokota Monitor", Japan; NRT2, Gesashi, Japan; NRT3

kHz	Mode	Call	Location	Service	kW	Remarks
			Iwo Jima; NRT 9, Hokkaido, Japan; NRV, Barrigada, Guam; NRV6, Marcus Is.,; NRV7, Yap Island) [Alt: 8063.6 & 10,523.6]			
9224.0	CW	KWL9Ø	Tokyo, Japan	FX	....	USA EMBASSY
9230.0	FAX	RHB/RHO	Khabarovsk, USSR	FX	....	Meteo
9235.0	CW	NBL	New London Naval Radio, CT	FX	....	USN
9236.2	LSB	MKT	London (Stanbridge), England	FX	10.0	RAF Telcom/Tfc
9240.0	CW/RTTY	....	several, USA nationwide	FX	....	FCC Backup
	USB	RVO7Ø	Moscow, USSR	FX	20.0	R. Moscow Feeder
9241.9	ISB	TNI92	Brazzaville, Congo	FX	20.0	Telcom/Tfc
9242.4	LSB	AJE	Croughton (Barford), England	FX	4.0	AFRTS Feeder
9245.0	ISB	FZE93	Fort de France, Martinique	FX	20.0	Telcom/Tfc
9250.0	USB	VLN2Ø	Sydney (Doonside), NSW, Australia	FX	3.5	Telcom/Ld. Howe [Alt: 9330.0]
	ISB	....	Holzkirchen, GFR	FX	10.0	RFE Feeder
	ISB	60A24	Mogadiscio, Somalia	FX	5.0	Telcom
	CW	....	numerous, mostly east coast USA	FX	....	USN
9260.0	FAX	JKA3	Tokyo (Kemigawa), Japan	FX	5.0	
9273.0	CW/FAX	JJC2Ø	Tokyo, Japan	FX	1.0	Meteo
9277.5	CW	NPG	San Francisco Naval Radio, CA	FX/FC	1.0	USN/TIME SIGNALS
9278.0	ISB	†HSD93	Bangkok Aeradio, Thailand	AX	7.5	Tfc/Singapore
9278.5	USB	†....	North Atlantic (East Coast) LORAN Net	FX	....	USCG (Includes:
			NMA7, Jupiter, FL; NMP32, Nantucket, MA; NMF33, Caribou, ME; NMN73, Carolina Beach [Cape Fear], NC & NOL, "Bermuda Monitor") [Alt: 11,513.5]			
9280.0	FAX	†RUZU	Molodezhnaya USSR Base, Antarctica	FX	....	SAAM/Meteo/Ice
	CW	†UGE2	Bellingshausen USSR Base, Antarctica	FX	....	SAAM
9385.0	CW	8UF75	New Delhi, India	FX	5.0	INTERPOL
	CW/RTTY	JPA33	Tokyo, Japan	FX	10.0	INTERPOL (Far
			East distribution point working DUN356, Manila, Philippines; HMA22, Seoul, S. Korea; HSQ, Bangkok, Thailand & VRD, Hong Kong)			
9290.0	CW	FUJ	Noumea Naval Radio, New Caledonia	FX	1.0	French Navy
	FAX	WFA29	New York, NY	FX	50.0	Meteo (0712-1212)
9295.0	ISB	JBE59	Tokyo, Japan	FX	10.0	Telcom/Tfc
9297.0	CW	DFR28	Frankfurt (Usingen), GFR	FX	....	Marpress
9300.0	USB	....	numerous, throughout Australia	FX	....	Commercial Tfc
9303.0	USB	....	Hawaii & Central Pacific LORAN Net	FX	....	USCG (Includes:
			NMO, Lualualei, HI; NRO, Johnston Is.; NRO5, Upolu Point, HI & NRO7, Kure Island [Ocean Is.], HI) [Alt: 12,205.0]			
9305.0	LSB	TLZ93	Bangui, Central African Republic	FX	6.0	Telcom
9310.5	CW/ISB	LCB	Oslo (Jeloey), Norway	FX	30.0	Tfc/Rome
9320.0	ISB	GWC	Daventry, England	FX	30.0	BBC Feeder
9325.0	CW	RUZU	Molodezhnaya USSR Base, Antarctica	FX	....	SAAM
9330.0	USB	VLN2Ø	Sydney (Doonside), NSW, Australia	FX	30.0	Telcom/Rawalpindi
9331.0	ISB	ZEN42	Cape d'Aguilar, Hong Kong	FX	5.0	Telcom/Brunei
9338.0	CW	NPM	Lualualei Naval Radio, HI	FX	10.0	USN
9340.0	FAX	....	Irkutsk, USSR	FX	....	Nx/Meteo
9342.0	CW	FDY	Orleans Air, France	FX	....	French AF
9345.0	AM	....	unknown	FX	....	NUMBERS/YL/SS
9350.0	ISB	....	off coast of Durban, RSA	FX	....	Oil rigs
	ISB	....	Delano, CA	FX	50.0	VOA Feeder
9351.0	CW	BPV	Shanghai, PRC	FX	15.0	TIME SIGNALS
9360.0	FAX	OXI	Skamlebaek Radio, Denmark	FX	20.0	Danish Marine
			Meteo (0005-0025; 1010-1215; 1245-1305 & 1830-1850)			
9374.6	CW	D4D	Praia, Cape Verde	FX	....	Tfc
9375.0	CW	CQF42	Bissau, Guinea-Bissau	FX	1.0	Tfc/London
	CW/ISB	CSF37	Funchal, Madeira Islands	FX	1.0	Port. AF
	CW	HMK21	Pyongyang, North Korea	FX	....	
	CW/ISB	CSF25	Lisbon, Portugal	FX	1.0	Port. AF
9375.1	CW/ISB	CSF36	Ponta Delgada, Azores	FX	1.0	Port. AF
9380.0	USB	NMR	COMMSTA San Juan, PR	FX	10.0	USCG
			[Hurricane Warning Net Control]			
9384.0	USB	....	Papeete, Tahiti	AX	0.2	Net Control
9386.0	USB	ZKS	Rarotonga, Cook Islands	FX	1.0	Net Control [Alt: 9095.0]

kHz	Mode	Call	Location	Service	kW	Remarks
9389.5	FAX	WFH29	New York, NY	FX	60.0	Meteo (0712-1212)
9391.0	CW	SOJ249	Warsaw (Radom), Poland	FX	20.0	Tfc/Reported PAP nx broadcast at 1800
9392.0	CW	RUQ	Krasnoyarsk USSR	FX	1.0	
9392.5	USB	FTJ39	Paris (St. Assise), France	FX	20.0	Telcom
9396.0	USB	....	Mediterranean LORAN Net	FX	....	USCG (Includes: NCI, Sellia Marina, Italy & NCI3, Lampedusa, Italy)
9405.0	USB	....	numerous, throughout NW Australia	FX	....	Commerical Tfc
9410.5	FAX	JKE6	Tokyo (Usui), Japan	FX	5.0	
9415.0	CW	UJY2	Kaliningrad , USSR	FX	10.0	Tfc/Havana
9419.0	CW	CUA48	Lisbon (Alfragide), Portugal	FX	10.0	Tfc
9434.0	CW/RTTY	†KAD26Ø	New Orleans, LA	FX	0.5	Border Patrol (Also: KAD68Ø, Buffalo, NY & KAK7ØØ, Detroit, MI)
9438.0	FAX	JMJ3	Tokyo (Usui), Japan	FX	5.0	Meteo
9440.0	CW/USB	†....	several, in Australia	FX	1.0	Monitoring Network (Includes: VNA3, Melbourne; VNA4, Brisbane; VNA5, Adelaide; VNA6, Perth; VNA7, Hobart & VNA21, Townsville) [Alt: 6810.0 & 18,880.0]
9440.0	FAX	NPM	Lualualei Naval Radio, HI	FX	20.0	USN/Meteo (cont.)
9442.5	ISB	†ZEN44	Cape d'Aguilar, Hong Kong	FX	10.0	Tfc/Nx
9450.0	ISB	†....	Lyndhurst, NSW, Australia	FX	....	R. Australia Feeder
9463.0	AM	....	unknown	FX	....	NUMBERS/YL/SS
9473.0	USB	PUZ4	Brasilia, Brazil	FX	1.0	Governmental [Telcom/Asuncion]
	ISB	AFH3	Albrook AFS, Panama	FX	3.0	USAF [Alt: 7320.0 & 15,675.]
9474.0	USB	....	Mediterranean LORAN Net	FX	....	USCG (Includes: AOB5Ø, Estartit, Spain; NCI, Sellia Marina, Italy; NCI3, Lampedusa, Italy; NCI4, Kargabarun, Turkey & NCI1Ø, Rhodes) [Alt: 9396.0 & 10,333.6]
9475.5	USB	VNV	Sydney (Doonside), NSW, Australia	FX	10.0	Telcom
9480.0	CW	†HMH56	Pyongyang, North Korea	FX	....	Tfc
9481.0	CW	CME301	Havana, Cuba	FX	1.0	Polish EMBASSY

9500.0 - 9775.0 kHz -- INTERNATIONAL BROADCASTING (Worldwide)

9500.0 - 9900.0 kHz -- INTERNATIONAL BROADCASTING (Worldwide--post WARC)

kHz	Mode	Call	Location	Service	kW	Remarks
9630.0	USB	....	Hawaii & Central Pacific LORAN Net	FX	....	USCG (Includes: NMO, Lualualei, HI; NRO, Johnson Is.; NRO5, Upolu Pt., HI; & NRO7, Kure Island [Ocean Is.], HI) [Alt: 12,205.0]

9775.0 - 9995.0 kHz -- FIXED SERVICE (Worldwide)

kHz	Mode	Call	Location	Service	kW	Remarks
9784.0	CW/RTTY	†OMZ28	Prague, Czechoslovakia	FX	1.0	GDR EMBASSY
	CW/RTTY	OMZ29	Bratislava, Czechoslovakia	FX	1.0	GDR EMBASSY
9785.0	CW	LWB	Buenos Aires (Moron), Argentina	AX	2.4	Aero Tfc
9796.0	CW	EEQ	Madrid, Spain	FX	1.0	Police Net
9815.0	USB	VZLH	Lord Howe Island, Australia	AX	1.0	Tfc/Sydney
	ISB	†9VE55	Singapore, Singapore	AX	2.5	Tfc
9817.5	USB	....	offshore, southeast Asia	FX	0.5	Oil Rigs
9840.3	LSB	FTJ84	Paris (St. Assise), France	FX	20.0	Telcom
9842.0	CW/RTTY	RLX	Dublin, Ireland	FX	0.5	Russian EMBASSY [Alt: 9217.0 & 11,148.0]
9846.0	CW	OMC2	Bratislava, Czechoslovakia	FX	1.0	Tfc/Izmail
9854.0	LSB	†....	Daventry, England	FX	30.0	BBC Feeder
9855.0	FAX	JKC2	Tokyo (Kemigawa), Japan	FX	5.0	
9865.0	CW	VTK	Tuticorin Naval Radio, India	FX	....	Indian Navy
9880.0	CW	DDK9	Pinneberg, GFR	FX	1.2	Meteo
9885.0	FAX	JKD3	Tokyo, Japan	FX	5.0	
9890.0	CW/RTTY	CFH	Maritime Command Radio, Halifax, NS, Canada	FX	5.0	NAWS (Canadian Forces)
	FAX	CFH	Maritime Command Radio, Halifax, NS,	FX	10.0	Meteo (NOAA reports seasonal ice broadcasts at 0000, 1300 & 2000: also test charts, analysis & prognosis throughout each day)

kHz	Mode	Call	Location	Service	kW	Remarks
9894.8	LSB	5LF9	Monrovia, Liberia	FX	6.0	Telcom

9900.0 - 9995.0 kHz -- FIXED SERVICE (Worldwide--post WARC)

kHz	Mode	Call	Location	Service	kW	Remarks
9905.0	USB	RVL24	Khabarovsk, USSR	FX	20.0	R. Moscow Feeder
9909.5	ISB	ZFD44	Hamilton, Bermuda	FX	10.0	Telcom
9910.0	CW/RTTY	ONN3Ø	Brussels, Belgium	FX	0.5	GDR EMBASSY [Alt: 9920.0]
9918.5	LSB	6VK49	Dakar, Senegal	FX	20.0	Telcom
9921.5	USB	AJE	Croughton (Barford), England	FX	4.0	AFRTS Feeder
9930.9	CW	....	several, in central Africa	FX	....	Police (Infrequently used net between Gabon, Kenya, Malawi, Tanzania & Zambia)
9932.5	USB	AGA	Hickam AFB, HI	FX	1.0	USAF/SAC
	USB	AIE	Andersen AFB, Guam	FX	1.0	USAF/SAC
9935.0	CW	FUF	Fort de Fort Naval Radio, Martinique	FX	....	French Navy
9942.0	CW	ONN36	Brussels, Belgium	FX	1.0	Angola EMBASSY
9948.5	USB	....	numerous, throughout France	FX	0.1	Net
9950.0	USB	ZLBC4	Campbell Is., New Zealand	FX	1.0	Telcom/Wellington
	USB	ZME3	Raoul (Sunday) Is., Kermadec Islands	FX	1.0	Telcom/Wellington
9960.0	CW/ISB	LJN23	Ny Aalesund Naval Radio, Spitzbergen	FX	10.0	Norwegian Navy
9965.0	USB	....	numerous, throughout Australia	FX	0.1	Oil rigs/Mining
9970.0	FAX	JMH3	Tokyo, Japan	FX	5.0	Meteo
	USB	....	unknown	FX	....	Net
9972.0	USB	AJO	Adana AFS, Turkey	FX	4.0	USAF
9974.0	USB	AIE	Andersen AFB, Guam	FX	1.0	USAF
9980.0	CW	LCJ	Oslo (Jeloey), Norway	FX	40.0	Tfc/Nx
9982.0	USB	....	several, throughout Australia	FX	0.1	Commercial Tfc
9982.5	FAX	KVM7Ø	Honolulu (International), HI	AX	10.0	Meteo (1140-1330)
9983.0	CW/FAX	LOK	Orcadas Radio, S. Orkney Is.	FX	0.2	NOAA reports wx broadcast at 1625--September to March
9990.0	CW	LTY99	Buenos Aires, Argentina	FX	5.0	Police
9993.0	CW	YQBF	unknown	FX	....	clg XUQC

9995.0 - 10,003.0 kHz -- STANDARD FREQUENCY & TIME SIGNALS

kHz	Mode	Call	Location	Service	kW	Remarks
9996.0	CW	RWM	Moscow, USSR	SS	10.0	TIME SIGNALS
10,000.0	A2	LOL1	Buenos Aires, Argentina	SS	2.0	" "
	CW	BPM	Xian, PRC	SS	50.0	" "
	A2	MSF	Teddington, England	SS	0.5	" "
	AM	WWVH	Kauai, HI	SS	10.0	" "
	A2	ATA	New Delhi, India	SS	8.0	" "
	AM	JJY	Tokyo (Sanwa, Ibaraki), Japan	SS	2.0	" "
	CW	RTA	Novosibirsk, USSR	SS	5.0	" "
	CW	RCH	Tashkent, USSR	SS	1.0	" "
	AM	WWV	Fort Collins, CO	SS	10.0	" "

10,003.0 - 10,0005.0 kHz -- STANDARD FREQUENCY/TIME/SPACE RESEARCH

kHz	Mode	Call	Location	Service	kW	Remarks
10,004.0	CW	RID	Irkutsk, USSR	SS	1.0	TIME SIGNAL

10,005.0 - 10,100.0 kHz -- AERONAUTICAL MOBILE (Worldwide)

kHz	Mode	Call	Location	Service	kW	Remarks
10 009.0	AM/A3H	....	numerous (ICAO MID)	FA/MA	....	Airports & inflight throughout the entire Middle East
			§MID moves to 10,018.0 in 1982			
	USB	....	Moscow (Vnukovo) Aeradio, USSR	FA	....	VOLMET/EE (25+55)
	USB	....	Novosibirsk Aeradio, USSR	FA	....	" " (10+40)
	USB	....	Tashkent Aeradio, Uzbek SSR	FA	....	" " (05+35)
10,017.0	AM/A3H	....	numerous (ICAO CAR)	FA/MA	....	Airports & inflight throughout Central America, Caribbean & FL
			§CAR moves to 11,396.0 in 1982			
			§VOLMET had moved to 11,387.0 at press time			
10,023.5	AM	....	Brazzaville Aeradio, Congo	FA	....	VOLMET/FF
10,024.0	A3H/USB	....	§SAM moves here from 8826.0 in 1982			

kHz	Mode	Call	Location	Service	kW	Remarks
10,025.0	AM/A3H	....	numerous (ICAO AFI)	FA/MA	....	Airports & in-flight over the Arabian Sea from Bombay, India to Johannesburg, RSA
			§AFI moves to 11,300,0 in 1982			
	CW	COL	Havana Aeradio, Cuba	FA/MA	....	USSR Aeroflot
	CW	RFNV	Moscow Aeradio, Cuba	FA/MA	....	USSR Aeroflot
						Mostly in-flight to/from Havana/Moscow
10,048.0	A3H/USB	....	§NP moves here from 8938.0 in 1982			
10,049.0	AM/USB	....	numerous (ICAO SAT)	FA/MA	....	Airports & in-flight over the South Atlantic Ocean--mostly between Dakar to Belem, Recife & Rio de Janeiro
			§SAT moves to 11,291.0 in 1982			
10,051.0	A3H/USB	....	§New VOLMET frequency for North Atlantic area in 1982			
10,057.0	USB	KW03	Anchorage Aeradio, AK	FA	1.0	
			§VOLMET moves here from 10,073.0 in 1982.0			
10,073.0	AM	....	Brazzaville Aeradio, Congo	FA	1.5	VOLMET/FF (15+45)
	AM	....	Nairobi Aeradio, Kenya	FA	....	VOLMET/EE (05+35)
	AM	....	Antananarivo Aeradio, Madagascar	FA	1.5	VOLMET/FF (25+55)
	AM	....	Johannesburg Aeradio, RSA	FA	....	VOLMET/EE (00+30)
			§VOLMET moves to 10,057.0 in 1982			
10,078.0	USB	9VA28	Singapore Aeradio, Singapore	FA	6.0	
			§New VOLMET frequency for South America area in 1982			
10,084.0	A3H/USB	....	§EUR moves here from 8875.0 in 1982			
10,089.0	USB	....	Berne Aeradio, Switzerland	FA/MA	6.0	
10,090.0	AM/A3H	....	Novosibirsk Aeradio, USSR	FA	....	VOLMET/EE/RR
	AM/A3H	....	Tashkent Aeradio, Uzbek SSR	FA	....	VOLMET/RR
	AM/A3H	....	Vnukovo-Sheremetievo Aeradio, USSR	FA	....	VOLMET/EE
			§VOLMET moves here from 11,279.0 in 1982			
10,093.0	A3H/USB	....	numerous worldwide	FA/MA	....	Airports & in-flight--particularly between New York and Honolulu [planes have in-flight telephones]
	USB	VIS56	Sydney Aeradio, Australia	FA	1.5	
	USB	9VA28	Singapore Aeradio, Singapore	FA	6.0	
10,096.0	A3H/USB	....	§SAM moves here from 11,327.0 in 1982			
10,097.0	CW	OBT	Talara, Peru	AX	....	Aero Tfc

10,100.0 - 11,175.0 kHz -- FIXED STATIONS (Worldwide)

10,100.0 - 10,150.0 kHz -- AMATEUR RADIO (Shared--post WARC)

kHz	Mode	Call	Location	Service	kW	Remarks
10,100.5	CW	†KNY28	Washington, DC	FX	1.0	Algerian EMBASSY
10,105.0	USB	....	unknown	FX	....	Telcom/SS
	USB	....	unknown, probably in Australia	FX	....	Net
	CW/ISB	....	numerous, among Caribbean islands	FX	0.1	Hurricane Emergency Net--Antigua, Barbados, Dominica, Grenada, Montserrat, St. Kitts, St. Vincent, etc.
10,115.0	CW	8PX45	Bridgetown, Barbados	FX	0.2	Tfc/HIA36
	FAX	BAF4	Beijing, PRC	FX	35.0	Meteo
10,120.0	USB	RGI24	Moscow, USSR	FX	15.0	R. Moscow Feeder
10,123.0	FAX	SUU2	Cairo, Egypt	FX	2.0	Meteo
10,125.5	CW	YQBF	unknown	FX	....	clg XUQC
10,128.0	ISB	JBE3Ø	Tokyo, Japan	FX	10.0	Tfc
10,130.0	FSK	NAA	Cutler Naval Radio, ME	FX	40.0	USN
10,130.0	FAX	RBW48	Murmansk, USSR	FX	20.0	Meteo
10,135.0	ISB	IRH31	Rome (Torrenova), Italy	FX	10.0	Telcom
10,136.6	USB	....	Atlantic Domestic Emergency Net	FX	....	USCG (Includes: NMA, Miami, FL; NMF, Boston, MA; NMG, New Orleans, LA, NMN, Portsmouth, VA; NMR; San Juan, PR & NOZ, Elizabeth City, NC--plus transportables) [Alt: 9125.6 & 11,434.6]
10,140.0	USB	....	several, throughout Australia	FX	0.1	Commercial Tfc
	CW	RUZU	Molodezhnaya USSR Base, Antarctica	FX	....	SAAM
	CW	UGE2	Bellingshausen USSR Base, Antarctica	FX	....	SAAM
10,144.5	LSB	....	several, Ghana, Ivory Coast & Niger	FX	1.0	W.H.O.
10,144.9	ISB	HBO2Ø	Geneva, Switzerland	FX	10.0	Tfc
10,147.0	USB	TUP	Abidjan, Ivory Coast	FX	2.0	Telcom

FIXED SERVICE (con.)

kHz	Mode	Call	Location	Service	kW	Remarks
10,148.0	LSB	....	unknown	FX	....	Telcom/SS
10,148.4	CW	....	unknown	FX	....	Coded Tfc
10,150.0	USB	SAM	Stockholm, Sweden	FX	1.2	EMBASSY (Includes: SAM2Ø, Athens, Greece; SAM25, Lisbon, Portugal; SAM3Ø, Madrid, Spain; SAM35, Belgrade, Yugoslavia; SAM36, Budapest, Hungary & SAM39, Prague, Czechoslovakia)
10,155.0	ISB	FPK5	Paris (St. Assise), France	FX	10.0	Telcom/Tfc
	CW	RMD53	Moscow, USSR	FX	20.0	Tfc
	USB	....	unknown, probably in Central America	FX	....	Net/SS
10,164.0	USB	SAM	Stockholm, Sweden	FX	1.2	EMBASSY
	USB	SAM38	Moscow, USSR	FX	1.0	Swedish EMBASSY
10,165.0	CW/FAX	RPT31	Tashkent, Uzbek USSR	FX	50.0	Tfc
10,166.6	USB	....	Pacific Domestic Emergency Net	FX	....	USCG (Includes: NMC, San Francisco, CA; NMO, Honolulu, HI; NMQ, Long Beach, CA; NOJ, Kodiak, AK & NRV, Barrigada, Guam--plus transportables) [Alt: 9125.6 & 11,434.6]
10,170.0	USB	....	several, throughout Australia	FX	....	Commercial Tfc
10,175.0	ISB	CUA53	Lisbon (Alfragide), Portugal	FX	5.0	Telcom/Tfc
10,185.0	FAX	JTD27	Ulan Bator, Mongolia	FX	10.0	Meteo
10,185.0	FAX	AFA	Washington (Andrews AFB, MD), DC	FX	3.0	USAF/Meteo
10,186.0	USB	....	unknown	FX	....	Net/SS
10,190.0	ISB	....	Holzkirchen, GFR	FX	10.0	RFE Feeder
10,190.5	USB	KFK92	Tucson, AZ	FX	1.0	Tfc/Chile
10,193.5	USB	FPK9	Paris (St. Assise), France	FX	20.0	Telcom
10,197.4	CW/ISB	LCN2	Oslo (Jeloey), Norway	FX	10.0	Tfc
10,200.3	ISB	†MUA2	London (Stanbridge), England	FX	10.0	RAF Telcom/Tfc
10,206.4	LSB	ART	Rawalpindi, Pakistan	FX	1.0	Tfc/Ankara
10,208.0	USB	....	Belem Aeradio, Brazil	FX	1.0	Tfc/Paramaribo
10,210.0	USB	MKA	London (Stanbridge), England	FX	1.0	RAF Telcom/Tfc
	CW	....	numerous, worldwide	FX	....	US Army
10,211.0	CW	"U"	unknown	RC	....	Beacon
10,216.2	USB	....	Guantanamo, Cuba	FX	....	Telcom
10,217.0	USB	....	several, throughout RSA	FX	1.0	Governmental
10,220.1	FAX	RDW76	Khabarovsk, USSR	FX	20.0	Meteo
10,222.6	USB	WEM2Ø	New York, NY	FX	20.0	Telcom
10,230.0	FAX	RKA78	Moscow, USSR	FX	15.0	Meteo
10,233.4	USB	....	unknown, possibly in Cuba	FX	....	Telcom/SS
10,233.5	CW	....	unknown	FX	....	TIME Pips/Tfc
10,235.0	LSB	....	Greenville, NC	FX	50.0	VOA Feeder
10,241.9	USB	....	unknown	FX	....	Telcom/PP
10,246.0	USB	....	several, along USA west coast	FX	....	USN
10,250.0	FAX	....	Madrid (Vicalv), Spain	AX	5.0	Meteo (NOAA reports transmissions at 0410 & 1555)
	CW	NDT6	Totsuka Naval Radio, Japan	FC	5.0	USN
10,254.0	LSB	....	Greenville, NC	FX	....	VOA Feeder/PP
10,255.0	CW	KWS78	Athens, Greece	FX	....	USA EMBASSY
	FAX	NPN	Barrigada (Agana NAS), Guam	FX	15.0	USN/Meteo
10,256.4	USB	AQI286	Karachi, Pakistan	FX	1.0	Net Control
10,264.0	CW	ZBP	Pitcairn Island	FX	0.7	Tfc/Suva
10,269.0	AM	....	Monaco, Monaco	FX	....	Feeder
10,275.0	ISB	9VF251	Singapore (Jurong), Singapore	FX	10.0	Telcom/Tfc
10,280.0	USB	LTZ	Buenos Aires, Argentina	FX	5.0	Government
10,285.0	CW	KWS78	Athens, Greece	FX	....	USA EMBASSY
10,289.0	USB	CSY53	Santa Maria Aeradio, Azores	AX	10.0	Tfc
10,289.5	CW	RGG42	Kuybychev, USSR	FX	20.0	Tfc
10,290.0	USB	....	unknown	FX	....	Scrambled
10,295.0	CW	FSB71	Paris (St. Martin Abbat), France	FX	2.5	INTERPOL (Also includes: OEQ44, Vienna, Austria; ONA2Ø, Brussels, Belgium; SHX, Stockholm, Sweden & HEP96, Zurich, Switzerland)
10,296.0	USB	OEC44	Tel Aviv, Israel	FX	2.0	Austrian EMBASSY [Alt: 7890.0 & 13,617.0]
10,301.0	USB	....	numerous, throughout Australia	FX	....	Commercial Tfc
10,305.0	ISB	AFE71	"Cape Radio", Patrick AFB, FL	FX	45.0	USAF/NASA [Alt: 9043.0 & 11,104.0]
10,307.5	CW	OLG4	Prague (Podebrad), Czechoslovakia	FX	3.0	Tfc/Shanghai
10,310.0	CW	ELE1Ø	Harbel, Liberia	FX	15.0	Tfc

kHz	Mode	Call	Location	Service	kW	Remarks
10,315.0	ISB	....	Holzkirchen, GFR	FX	10.0	RFE Feeder
10,323.0	CW	CWQ	San Gregorio Tac, Uruguay	FX	....	
10,333.6	USB	....	Mediterranean LORAN Net NCI, Sellia Marina, Italy; NCI4, Kargabarun, Turkey and NCI1Ø, Rhodes) [Alt: 9474.0]	FX	....	USCG (Includes:
10,335.0	CW	RPC21	Kiev, Ukranian SSR	FX	20.0	Tfc
10,337.5	USB	†....	Norwegian Sea & N. Atlantic Ocean LORAN Net DML, Sylt, GFR; JXL, BØ, Norway; JXP, Jan Mayen Is.; NMS, Shetland Is.; OUN, Ejde, Denmark; OVY, Angissog, Greenland; TFR, Sandur, Iceland & TFR2, Reykjavik, Iceland) [Alt: 7512.5]	..		USCG (Includes:
10,338.0	USB	†RME21	Moscow, USSR	FX	50.0	R. Moscow Feeder
10,340 0	FAX/FSK	WFK5Ø	New York, NY	FX	50.0	Tfc
10,344.0	USB	AGB4	Wake Island	FX	1.0	USAF
	USB	AIE	Andersen AFB, Guam	FX	1.0	USAF
10,354.0	LSB	....	Greenville, NC	FX	50.0	VOA Feeder/PP
10,368.5	USB	....	Aleutian LORAN NET NMJ22, Attu; NOJ, Kodiak; NRW2, St. Paul Is. & NRW3, Pt. Clarence--all AK)	FX	....	USCG (Includes:
10,382.5	LSB	....	Greenville, NC	FX	50.0	VOA Feeder
	CW	RLU8	Tiksi Bukhta, USSR	FX	....	Tfc
10,388.0	USB	....	numerous, throughout Australasia	FX	....	War Games 12/81
10,388.5	USB	FSB57	Paris (St. Martin Abbat), France	FX	2.5	INTERPOL
10,390.0	CW	7RA2Ø	Algiers, Algeria	FX	0.4	"
	CW	AYA47	Buenos Aires, Argentina	FX	1.0	"
	CW	ONA2Ø	Brussels, Belgium	FX	3.0	"
	CW	XJE57	Ottawa (Almonte), Ont., Canada	FX	1.5	"
	CW	OWS4	Copenhagen, Denmark	FX	1.0	"
	CW	5BP6	Nicosia, Cyprus	FX	3.0	"
	CW	GMP	London (W. Wickham), England	FX	0.8	"
	CW	OGX	Helsinki, Finland	FX	0.5	"
	CW	FSB57	Paris (St. Martin Abbat), France	FX	2.5	"
	CW	DEB	Wiesbaden, GFR	FX	1.5	"
	CW	8UF75	New Delhi, India	FX	5.0	"
	CW	4XP41	Tel Aviv, Israel	FX	1.0	"
	CW	IU481	Rome, Italy	FX	0.1	"
	CW	JPA56	Nagoya (Komaki), Japan	FX	10.0	"
	CW	ODW22	Beirut, Lebanon	FX	1.0	"
	CW	LXF5Ø	Luxembourg	FX	1.0	"
	CW	CNP	Rabat, Morocco	FX	0.5	"
	CW	PDB2	Utrecht (Bilthoven), Netherlands	FX	1.0	"
	CW	LJP2Ø	Oslo, Norway	FX	0.1	"
	CW	CSJ26	Lisbon, Portugal	FX	1.0	"
	CW	YO99	Bucharest, Rumania	FX	3.0	"
	CW	EEQ	Madrid, Spain	FX	1.0	"
	CW	SHX	Stockholm, Sweden	FX	1.5	"
	CW	HEP39	Zurich (Waltikon), Switzerland	FX	3.0	"
	CW	TCC2	Ankara, Turkey	FX	0.8	"
	CW	4NX7	Belgrade, Yogoslvia	FX	0.4	"
	AM	9TK21	Kinshasa, Zaire	FX	1.0	"
10,391.5	LSB	TTR1Ø3	N'Djamena, Chad	FX	0.3	
10,396.0	ISB	JBF4Ø	Tokyo, Japan	FX	30.0	Telcom/Tfc
10,400.0	CW/RTTY	HGX31	Warsaw, Poland	FX	1.0	Hungarian EMBASSY
10,402.0	USB	....	Riyadh, Saudi Arabia	FX	....	Feeder (reported)
10,410.0	CW	....	unknown	FX	....	Tfc
10,415.0	CW	JJD	Tokyo, Japan	FX	10.0	Solar & Geomagnetic Reports
	ISB	ZPG4	Asuncion, Paraguay	FX	10.0	Telcom/Tfc
10,420.0	USB	....	Holzkirchen, GFR	FX	10.0	RFE Feeder
10.422.5	CW/LSB	OEC57	Pretoria, RSA	FX	1.0	Austrian EMBASSY [Alt: 13,617.0]
	CW/LSB	OEC72	Lisbon, Portugal	FX	1.0	Austrian EMBASSY
10,425.1	ISB	HLQ62	Seoul, South Korea	FX	5.0	Telcom/Tfc
10,425.5	CW/LSB	OEC	Vienna, Austria	FX	10.0	EMBASSY
	CW/LSB	OEC61	Rome, Italy	FX	1.0	Austrian EMBASSY [Alt: 7894.0]

kHz	Mode	Call	Location	Service	kW	Remarks
10,427.0	USB	....	unknown	FX	....	Telcom/EE
10,430.0	USB	AGA	Hickam AFB, HI	FX	3.0	USAF/SAC
	USB	AIE	Andersen AFB, Guam	FX	3.0	USAF/SAC
	USB	5LF1Ø	Monrovia, Liberia	FX	6.0	Telcom
10,433.5	CW	CMU967	Santiago Naval Radio, Cuba	FX	....	Russian Navy
10,434.0	CW	RIW	Khiva Naval Radio, Uzbek SSR	FX	....	Russian Navy
10,440.5	CW	NPO	San Miguel (Capas), Philippines	FX	20.0	USN/TIME SIGNALS/Tfc/NOAA reports wx broadcasts at 0400 & 1300
	CW	NRV	COMMSTA Barrigada, Guam	FX	3.0	USCG
10,445.0	CW	....	unknown, possibly southeast Asia	FX	....	Tfc/Net/EE
10,446.1	CW	"E"	unknown	RC	....	Beacon
10,450.0	LSB	....	unknown	FX	....	NUMBERS/KK/YL
10,452.0	USB	AGA	Hickam AFB, HI	FX/MA	3.0	USAF/SAC (Used for SKY KING broadcasts)
10,454.0	ISB	....	Greenville, NC	FX	50.0	VOA Feeder
10,457.2	USB	....	Kabul, Afghanistan	FX	....	R. Afghanistan Feeder
10,460.0	CW	†WKB2Ø	Slidell, LA	FX	3.0	Tfc/Havana
10,462.5	ISB	FTK46	Paris (St. Assise), France	FX	20.0	Telcom
10,463.4	CW	KWL9Ø	Tokyo, Japan	FX	....	USA EMBASSY
10,477.5	CW/ISB	JBE7Ø	Tokyo (Oyamo), Japan	FX	5.0	Tfc
10,484.0	CW	CME304	Havana, Cuba	FX	1.0	Polish EMBASSY
10,493.0	USB	....	numerous, USA nationwide	FX	....	FEMA (Recently reported have included: KPA64, Battle Creek, MI: KPA66, Denver, CO; WGY9Ø9, Santa Rosa, CA; etc.)
10,495.0	ISB	MKD	Akrotiri, Cyprus	FX	30.0	Telcom
10,500.0	AM	....	unknown	FX	....	NUMBERS/GG/YL
10,505.0	USB	....	numerous, in Australia	FX	1.0	Police (Includes: VKA, Adelaide; VKC, Melbourne; VKM, Darwin; VKR, Brisbane; VKT, Hobart; VKX, Canberra & VKI, Perth) [Alt: 7660.0 & 13,700.0]
	CW	....	unknown	FX	....	Coded Tfc/Net
10,509.7	CW	RIW	Khiva Naval Radio, Uzbek SSR	FX	....	Russian Navy
10,510.0	ISB	AGA	Hickam AFB, HI	FX	10.0	USAF
10,515.0	ISB	5RY5	Antananarivo, Madagascar	FX	20.0	Telcom/Tfc
10,523.6	USB	....	Northwest Pacific LORAN Net	FX	....	USCG (Includes: NRT, "Yokota Monitor", Japan; NRT2, Gesashi, Japan; NRT3, Iwo Jima; NRT9, Hokkaido, Japan; NRV, Barrigada, Guam; NRV6, Marcus Is.; NRV7, Yap Is. Also reported on this frequency have been two LORAN stations in South Korea)
10,525.2	CW	....	several, in Central America	AX	....	Aero Tfc
10,526.3	ISB	5VH3Ø5	Lome, Togo	FX	6.0	Telcom/Tfc
10,530.0	ISB	....	unknown	FX	....	Telcom/EE
10,531.5	LSB	....	unknown	FX	....	Telcom/EE
10,538.0	ISB	AJE	Croughton (Barford), England	FX	4.0	AFTRS Feeder
10,554.0	LSB	....	unknown	FX	....	Telcom/Net/SS
10,553.7	CW	....	unknown	FX	....	
10,554.5	CW	Y7A46	Berlin, GDR	FX	....	EMBASSY
10,555.0	CW	†RUZU	Molodezhnaya USSR Base, Antarctica	FX	....	SAAM
	FAX	AXI34	Darwin, NT, Australia	FX	5.0	Meteo (cont.)
	ISB	CUA54	Lisbon (Alfragide), Portugal	FX	3.0	Telcom/Tfc
10,563.0	CW	FTK56A	Paris (St. Assise), France	FX	20.0	Tfc
	ISB	XDD229	Mexico City, DF, Mexico	FX	10.0	Telcom/Tfc
10,570.3	CW	"K"	unknown	RC	....	Beacon
10,575.0	USB	....	Stockholm Aeradio, Sweden	FA	....	Aero Tfc
	CW	....	unknown	FX	....	Net
10,580.0	ISB	MKE	Akrotiri, Cyprus	FX	30.0	RAF Telcom/Tfc
10,582.5	ISB	FPK58	Paris (St. Assise), France	FX	20.0	Telcom/Tfc
10,595.0	USB	†RKD48	Moscow, USSR	FX	20.0	R. Moscow Feeder
10,605.1	CW	....	unknown	FX	....	Coded Tfc
10,606.0	USB	....	several naval bases in Japan	FX	3.0	USN
10,610.0	USB	....	unknown	FX	....	Net/SS
10,614.0	CW	"F"	unknown	RC	....	Beacon
10,620.0	ISB	†....	Moscow, USSR	FX	....	R. Moscow Feeder
10,635.9	ISB	ZFD49	Hamilton, Bermuda	FX	30.0	Telcom/La Paz

kHz	Mode	Call	Location	Service	kW	Remarks
10,636.9	CW	KKN5Ø	Washington, DC	FX	....	EMBASSY
10,638.0	CW	"K"	unknown	RC	....	Beacon
10,639.0	CW	KDR3	unknown	FX	....	
10,640.0	USB	PVX	Rio de Janeiro (Sepetiba), Brazil	FX	2.5	Telcom
	ISB	JBF6Ø	Tokyo, Japan	FX	10.0	Tfc
10,642.5	CW	KNY26	Washington, DC	FX	1.0	Hungarian EMBASSY
10,643.5	CW	"D"	unknown	RC	....	Beacon
10,645.0	ISB	ZEO47	Cape d'Aguilar, Hong Kong	FX	3.0	Telcom/Tfc
	CW	"F"	unknown	RC	....	Beacon
10,645.9	CW	"R"	unknown	RC	....	Beacon
10,646.0	CW	"K"	unknown	RC	....	Beacon
10,646.8	CW	†CSF37	Funchal, Madeira Islands	FX	0.5	Port. AF
10,650.4	ISB	ODI65	Beirut, Lebanon	FX	2.0	Tfc
10,660.0	USB	†RZA27	Moscow, USSR	FX	15.0	R. Moscow Feeder
10,670.0	ISB	FEB21	Chichi Jima, Bonin Is., Japan	FX	2.0	Tfc
10,672.0	ISB	....	Kathmandu, Nepal	FX	5.0	
10,676.0	CW/USB	....	unknown	FX	....	Coded Tfc
10,678.0	ISB	†MKG	London (Stanbridge), England	FX	30.0	RAF Telcom/Tfc
10,680.5	CW	†KRH5Ø	London, England	FX	....	USA EMBASSY
10,690.0	USB	....	Yuzhno-Sakhalinsk, USSR	FX	....	R. Moscow Feeder
10,695.0	ISB	FTK69	Paris (St. Assise), France	FX	20.0	Telcom/Tfc
10,700.5	ISB	PCL2Ø	Amsterdam (Kootwijk), Netherlands	FX	30.0	R. Nederlands Backup Feeder
10,708.5	LSB	KUQ2ØA	Pago Pago, American Samoa	FX	10.0	Telcom/Tfc
10,715.0	ISB	IRH37	Rome (Torrenova), Italy	FX	10.0	Telcom
10,720.0	FAX	LRB72	Buenos Aires (G. Pacheco), Argentina	FX	....	Meteo
			(0300, 1600, 1900 & 2000)			
10,725.0	CW	CMU967	Santiago Naval Radio, Cuba	FX	....	Russian Navy
10,732.0	CW	P6Z	Paris, France	FX	....	EMBASSY
10,736.0	USB	OHU2Ø	Helsinki, Finland	FX	1.0	EMBASSY
	USB	OHU21	Paris, France	FX	1.0	Finnish EMBASSY [Alt: 10,885.0]
	USB	OHU26	Belgrade, Yugoslavia	FX	1.0	" "
10,740.0	LSB	RKA72	Moscow, USSR	FX	20.0	R. Moscow Feeder
10,743.5	USB	BEG33	Guangzhou, PRC	FX	7.0	Tfc/Shanghai
10,745.0	CW	....	several, along Argentine coastline	FX	....	Prefectura Naval
10,746.5	USB	OHU2Ø	Helsinki, Finland	FX	1.0	EMBASSY
	USB	OHU23	Tel Aviv, Israel	FX	1.0	Finnish EMBASSY
	USB	OHU22	Bucharest, Rumania	FX	1.0	" "
10,763.5	DATA	....	unknown	FX	....	
10,765.0	CW	OVC	Groennedal, Greenland	FX	3.0	Danish Marine
			Reported wx broadcast at 2100			
	ISB	FTK76	Paris (St. Asssise), France	FX	20.0	Telcom/Tfc
10,770.0	ISB	†CXL24	Montevideo, Uruguay	FX	2.5	Telcom
10,775.0	ISB	FTK77	Paris (St. Assise), France	FX	20.0	TIME SIGNALS/ Telcom
10,779.0	CW	NAM	Norfolk Naval Radio, VA	FX	....	USN
10,780.0	CW	VCS838	Ottawa, Ont., Canada	FX	5.0	Yugoslav EMBASSY [Alt: 12,295.0]
	USB	AFE71	"Cape Radio", Patrick AFB, FL	FX/MA	45.0	USAF/NASA
			(Also active is "Cape Leader" also Patrick AFB, FL-- Both Shuttle Support--May be called "HOTEL" channel) [Alt: 10,305.0 & 11,104.0]			
10,785.0	USB	....	unknown	FX	....	Telcom/II
10,787.0	ISB	BCO24	Shanghai, PRC	FX	25.0	Tfc/Berne
10,788.0	ISB	CLN3ØØ	Havana (Bauta), Cuba	FX	10.0	Telcom/Mx Mkr
10,792.5	CW	P6Z	Paris, France	FX	2.0	EMBASSY
10,799.5	USB	5VH3Ø8	Lome, Togo	FX	20.0	Telcom
10,803.0	USB	....	numerous, throughout Australia	FX	....	Commercial Tfc
10,805.0	ISB	....	Vandenberg AFB, CA	FX	....	USAF/SAMTO
			[Frequently uses callsigns starting with "Abnormal"]			
10,820.7	ISB	FTK82	Paris (St. Assise), France	FX	20.0	Telcom/Tfc
10,833.0	LSB	....	numerous, throughout Argentina	FX	1.0	Governmental
10,836.0	LSB	TUP8	Abidjan, Ivory Coast	FX	2.0	Telcom

kHz	Mode	Call	Location	Service	kW	Remarks
10,844.5	USB	5UR8	Niamey, Niger	FX	20.0	Telcom
10,845.0	USB	†A9M43	Hamala, Bahrain	FX	10.0	Telcom/Aden
10,846.0	ISB	†IRH28	Rome (Torrenova), Italy	FX	10.0	Telcom
10,847.5	CW	....	several, throughout Caribbean	AX	....	Aero Tfc
10,853.0	ISB	CLN295	Havana (Bauta), Cuba	FX	20.0	Telcom
10,855.0	LSB	†....	Sverdlovsk, USSR	FX	15.0	R. Moscow Feeder
10,857.5	ISB	JBF5Ø	Tokyo, Japan	FX	30.0	Telcom/Tfc
10,860.0	LSB	....	Holzkirchen, GFR	FX	10.0	RFE Feeder
10,861.2	CW	RSF23	Tbilisi, Georgian SSR	FX	....	Coded Tfc
	FAX	NAM	Norfolk Naval Radio, VA	FX	50.0	USN Meteo
	CW	....	numerous, eastern USA	FX	....	USN
10,865.0	USB	....	Decimonannu AS, Sardinia	FX	....	RAF
10,869.0	LSB	....	Bethany, OH	FX	40.0	VOA Feeder
10,870.0	FAX	RBA2	Magadan, USSR	FX	20.0	Meteo/Nx
10,873.0	USB	....	unknown	FX	....	Telcom/SS
10,885.0	USB	OHU21	Paris, France	FX	1.0	Finnish EMBASSY [Alt: 10,736.0]
	USB	OHU2Ø	Helsinki, Finland	FX	1.0	EMBASSY [Alt: 14,954.0]
10,892.0	USB	4WA49	Ghuraff, Yemen, Arab Republic of	FX	....	Telcom/Bahrain
10,894.5	LSB	†TYK8	Cotonou, Benin	FX	....	Telcom/Tfc
10,895.0	CW	LRB39	Buenos Aires (G. Pacheco), Argentina	FX	1.0	Governmental
10,900.0	CW	....	several, Japan, Philippines & Taiwan	FX	2.5	USAF
10,902.1	CW/RTTY	....	numerous, USA nationwide	FX	....	FCC Backup Net
			(Includes: KAA6Ø, Grand Island, NE; KCA35, Belfast, ME; KCA38, Canandaigua, NY; KGA91, Laurel, MD; KGA93, Washington, DC; KIA84, Powder Springs, GA; KIA85, Ft. Lauderdale, FL & KQA62, Allegan, MI) [Most traffic on this channel is via bit-inversion RTTY, however CW has been monitored]			
10,905.0	ISB	CUA55	Lisbon (Alfragide), Portugal	FX	10.0	Telcom/Tfc
10,911.0	USB	....	unknown	FX	....	Telcom/II
10,912.0	CW	RIW	Khiva Naval Radio, Uzbek SSR	FX	....	Russian Navy
10,913.4	USB	KGE33	Washington, DC	FX	3.0	FBI Net Control
10,917.0	USB	5TN27	Nouakchott, Mauritania	FX	20.0	Telcom
10,922.0	USB	DFK92	Frankfurt (Bonames), GFR	FX	20.0	Telcom
			[Sometimes used as Deutsche Welle feeder]			
10,929.0	CW/FAX	GLQ3Ø	London (Rugby), England	FX	30.0	AP Nx
10,932.0	USB	....	Riyadh, Saudi Arabia	FX	....	Feeder
10,935.0	....	†5KO225	Bogota, Colombia	FX	....	USMILGP
	USB	HIP491	Santo Domingo, Dominican Republic	FX	....	USMILGP [Alt: 13,950.0]
10,938.5	USB	....	Mediterranean LORAN Net	FX	....	USCG (Includes:
			NCI, Sellia Marina, Italy; NCI3, Lampedusa, Italy and NCI4, Kargabarun, Turkey) [Alt: 8021.5 & 10,958.5]			
10,945.0	CW	CFH	Maritime Command Radio, Halifax, NS, Canada	FX	5.0	(C13L) NAWS (Canadian Forces)
10,953.0	ISB	VJU	Norfolk Island	FX	15.0	Telcom/Sydney
10,958.5	USB	....	Mediterranean LORAN Net	FX	....	USCG (For List of stations see 10,938.5)
10,961.0	ISB	9VF284	Singapore, Singapore	FX	10.0	Telcom
10,962.0	CW	AQJ	Islamabad Naval Radio, Pakistan	FX	10.0	Pakistan Navy
10,966.0	FAX	NPO	San Miguel (Capas), Philippines	FX	15.0	USN/Meteo
10,967.0	CW	ASK	unknown	FX	....	Mkr
10,980.0	FAX	RDD79	Moscow, USSR	FX	....	Meteo (cont.)
10,982.2	CW	CTH	Horta Naval Radio, Azores	FX	3.0	Port. Navy
			NOAA reports wx broadcast at 0930			
10,995.0	CW	7RP7Ø	Rome, Italy	FX	0.2	Algerian EMBASSY
10,996.0	CW	7RV7Ø	Warsaw, Poland	FX	0.8	Algerian EMBASSY
11,000.0	CW	ONN3Ø	Brussels, Blegium	FX	0.5	GDR EMBASSY
11,004.0	USB	CNR39	Rabat, Morocco	FX	20.0	Telcom
11,005.0	CW	7RV5Ø	Islamabad, Pakistan	FX	1.0	Algerian EMBASSY
11,007.0	CW	JWT/JXU	Stavanger Naval Radio, Norway	FX	....	Norwegian Navy
11,009.0	FAX	....	unknown, probably in North America	FX	....	
11,010.0	CW	CNO98	Casablanca, Morocco	AX	0.5	NOAA reports wx broadcast at 0835

kHz	Mode	Call	Location	Service	kW	Remarks
11,012.5	FAX	JAL21	Tokyo, Japan	FX	5.0	Nx
11,015.0	CW	CTV81	Monsanto Naval Radio, Portugal	FX	3.0	Port. Navy
	USB	....	numerous, throughout Brazil	FX	....	Net
11,018.5	USB	CTN81	Apulia Naval Radio, Portugal	FX	5.0	Port. Navy/ Tfc London
11,021.3	USB	TLZ2	Bangui, Central African Republic	FX	20.0	Telcom/Tfc
11,023.5	USB	CTH71	Horta Naval Radio, Azores	FX	5.0	Port. Navy/ Tfc Apulia
11,028.0	USB	....	Northwest Pacific LORAN Net NRT, "Yokota Monitor", Japan; NRT3, Iwo Jima; NRV6, Marcus Is.; NRV, Barrigada, Guam & NRV7, Yap Island) [Alt: 5320.0 & 7580.0]	FX	....	USCG (Includes:
11,030.0	FAX	AXM34	Canberra, ACT, Australia	FX	20.0	Meteo
	ISB	CME397	Havana, Cuba	FX	0.5	British EMBASSY
11,033.9	USB	FTL3	Paris (St. Assise), France	FX	20.0	Telcom
11,035.0	FAX	WFL51	New York, NY	FX	50.0	Meteo (0712-1212)
11,039.0	CW	DDH9	Pinneberg, GFR	FX	1.0	Meteo/North Sea
11,040.0	ISB	A9M5Ø	Hamala, Bahrain	FX	3.5	Telcom/Aden
11,046.0	CW	RIW	Khiva Naval Radio, Uzbek SSR	FC	....	Russian Navy
11,065.0	CW	PWI	Recife Naval Radio, Brazil	FX	2.0	Brazil Navy
	ISB	FTL6	Paris (St. Assise), France	FX	20.0	Telcom/Tfc
	ISB	9VF68	Singapore, Singapore	FX	7.0	Telcom/Tfc
11,067.0	ISB	....	Daventry, England	FX	30.0	BBC Feeder
11,076.0	USB	....	numerous, southwestern USA [Alt: 14,686.0 & 18,666.0]	FX/MA	....	DEA
11,086.0	FAX	GFA24	Bracknell, England	FX	2.5	Meteo (cont.)
11,090.0	FAX	KVM7Ø	Honolulu (International), HI	AX	10.0	Meteo (1900-2045)
	CW/RTTY	†KNY25	Washington, DC	FX	1.0	Rumanian EMBASSY
11,095.0	CW	KKN5Ø	Washington, DC	FX	10.0	EMBASSY
11,095.3	USB	†XTA9	Ouagadougou, Upper Volta	FX	6.0	Telcom
11,097.0	FAX	LMO	Oslo, Norway	FX	0.5	Meteo
11,100.0	ISB	ZKS28	Rarotonga, Cook Islands	FX	10.0	Telcom/Wellington
11,104.0	USB	AFE71	"Cape Radio", Patrick, AFB, FL Shuttle Support [Alt: 10,780.0]	FX	45.0	USAF/NASA
11,106.0	CW	†KNY29	Washington, DC	FX	1.0	British EMBASSY
11,108.0	USB	CLN321	Havana (Bauta), Cuba	FX	....	Telcom
11,112.0	LSB	AJO	Adana AS, Turkey	FA	....	USAF
11,114.0	CW	CMU967	Santiago Naval Radio, Cuba	FX	....	Russian Navy
11,116.2	LSB	6VK311	Dakar, Senegal	FX	4.0	Telcom
11,116.7	USB	....	Goteborg Radio, Sweden	FC	20.0	
11,121.0	ISB	AGA1WP	Wright-Patterson AFB, OH	FX	2.0	USAF
	USB	AIR	Andrews AFB, MD	FX	2.0	USAF
11,121.5	LSB	6VK311	Dakar, Senegal	FX	4.0	Telcom
11,125.0	USB	7OB452	Aden (Khormaksar), Democratic Yemen	FX	1.0	Libyan EMBASSY
11,130.0	USB	NAW34	Guantanamo, Cuba	FX	....	Telcom
	CW	ZLX22	Wellington (Himatangi), New Zealand wx broadcasts at 0200, 0500, 0520, 0815, 0840, 0910, 1800, 2040 & 2100	FX	5.0	NOAA reports
11,132.0	USB	NAW	Guantanamo Base, Cuba	FX	....	Telcom
	CW	UMS	Moscow Naval Radio, USSR	FX	5.0	Russian Navy
11,135.0	CW/RTTY	PBC911	Goeree Naval Radio, Netherlands	FX	3.0	(N13A) Dutch Navy
11,140.0	CW	CTV71	Monsanto Naval Radio, Portugal	FX	3.0	Port. Navy
11,142.0	CW	KRH5Ø	London, England	FX	....	USA EMBASSY
11,147.0	CW/FAX	LOK	Orcadas, South Orkney Islands wx broadcast at 1625--September to March	FX	0.2	NOAA reports
11,148.0	CW/RTTY	RLX	Dublin, Ireland	FX	0.5	RUSSIAN EMBASSY [Alt: 9842.0 & 12,211.0]
11,150.0	ISB	JEB2Ø	Choshi, Japan	FX	2.0	Telcom
	ISB	....	Greenville, NC	FX	45.0	VOA Feeder
11,152.0	CW/RTTY	RXO	Warsaw, Poland	FX	0.5	Russian EMBASSY
11,155.5	CW	"K"	unknown	RC	....	//9043.0
11,156.0	CW	XSG	Shanghai Radio, PRC	FX	....	Coded Tfc
11,161.2	CW	....	unknown	FX	....	Nx/EE/1000
11,165.2	CW	LMB	Bergen, Norway wx broadcasts at 1305 & 2240	FX	0.5	NOAA reports
	ISB	5TA17	Nouadhibou, Mauritania	FX	6.0	Telcom/Tfc

kHz	Mode	Call	Location	Service	kW	Remarks
11,169.0	USB	AIE	Andersen AFB, Guam	FX	3.0	USAF

11,175.0 - 11,400.0 kHz -- AERONAUTICAL MOBILE (Worldwide)

kHz	Mode	Call	Location	Service	kW	Remarks
11,176.0	USB	....	numerous, worldwide	FA/MA	....	USAF (Usually AIE2, Andersen AFB, Guam; AJE, Croughton, England; NKW, Diego Garcia; AKA5, Elemendorf AFB, AK; AIF2, Yokota AB, Japan; AGA2, Hickam AB, HI, AJG9, Incirlik AB, Turkey & AIC2, Clark AB, Philippines)
	USB	....	several, West Germany	FA/MA	....	RAF (Includes: Bruggen AB, Wildenrath AB, etc.)
11,179.0	USB	....	numerous, worldwide	FA/MA	....	USAF (Usually AFA3, Andrews AFB, MD; AFE8, MacDill AFB, FL; AGA2, Hickam AFB, HI; AIE2, Andersen AFB, Guam; AIF2, Yokota AB, Japan)
11,180.0	USB	AFA3	Andrews AFB, MD	FA/MA	....	USAF [Has been contacted by "AIR FORCE ONE" when overseas]
11,180.5	USB	ZKW	Christchurch, New Zealand	FA/MA	1.0	RNZAF
11,182.0	USB	....	numerous, in the USA	FA/MA	....	USAF/USN (In addition to use by AFL, Loring AFB, ME; AFE38, Maxwell, AFB, AL; AGF37, Scott AFB, IL; etc. this channel used by foreign Allied in-flight within USA air space) [Occasional use for SKY KING and MAINSAIL broadcasts]
11,186.0	CW	....	numerous, worldwide	FA	....	USN
11,191.0	USB	....	numerous, worldwide	FA/MA	....	USN
11,192.0	USB	....	Novosibirsk Aeradio, USSR	FA	....	VOLMET/RR (cont.)
11,193.5	USB	GFO	Lossiemouth (Milltown), Scotland	FA	1.0	British Naval Air
11,194.0	LSB	....	several, USA nationwide	FA	....	USN
	USB	....	several, western USA	FA/MA	....	Test Flights
11,195.0	USB	....	numerous, in USA	MA	....	USN/USCG
11,198.0	A3H	....	Kiev Aeradio, Ukrainian SSR	FA	....	VOLMET/RR (20+50)
	A3H	....	Leningrad Aeradio, USSR	FA	....	" " (05+35)
	A3H	....	Moscow (Sheremetyevo), USSR	FA	....	" " (15+45)
	A3H	....	Riga Aeradio, Latvian SSR	FA	....	" " (00+30)
	A3H	....	Rostov Aeradio, USSR	FA	....	" " (10+40)
	USB	....	numerous, in the USA	FA/MA	....	USCG Helicopters [shared with USN]
11,200.0	USB/CW	MVU	West Drayton (Upavon), England	FA	2.5	RAF/Frequent wx broadcasts
11,201.0	USB	....	numerous, USA nationwide	MA	....	USCG/USN (Major air/surface frequency)
11,208.0	CW	FKO	Djibouti, Djibouti	FA	....	French AF
11,209.0	USB	....	several, in Canada & Europe	FA/MA	....	RCAF (Usually CHR, "Trenton Military", Ont; CJX, "St. Johns Military", Nfld.; DHM95, "Lahr Military", GFR; VAF, "Alert Military", NWT & VXA, "Edmonton Military". Alta.)
	ISB	CUW3	Lajes Field, Azores	FA/MA	....	USAF
11,214.0	USB	....	several, throughout Australia	FA/MA	....	RAAF
11,215.0	USB	....	several, USA nationwide	MA	....	AWACS
11,222.0	USB	....	Auckland AB, New Zealand	FA	....	RNZAF
11,222.0	USB	....	Stockholm Aeradio, Sweden	FA	5.0	Meteo/Tfc
11,223.0	USB	CHR	"Trenton Military", Ont., Canada	FA	....	RCAF
11,226.0	LSB	....	numerous, worldwide	FA/MA	....	USAF (For list of possible stations see 11,176.0)
11,228.0	USB	AFE8	MacDill AFB, FL	FA/MA	....	USAF
11,233.0	USB	....	several, in Canada & Europe	FA/MA	....	RCAF (For list of stations see 11,209.0)
	USB	DHM95	"Lahr Military", GFR	FA/MA	....	RCAF/Wx (+20)
11,234.0	USB	....	several, around the world	FA/MA	....	RAF/RCAF (Includes: "Belize Flight Watch"; "Cyprus Flight Watch"; "Gibraltar Flight Watch"; etc.)
	USB	NKW	"Diego Tower", Diego Garcia	FA/MA	....	USN/USAF
11,235.0	USB	....	several, throughout Australia	FA/MA	....	RAAF (Usually AXD, "Air Force Laverton", Vic.; AXF, "Air Force Sydney", NSW; AXH, "Air Force Townsville"; Qld.; AXI, "Air Force Darwin", NT; AXK, "Air Force Sale", NSW; AXS, "Air Force Learmouth", WA & AXT, "Air Force Edinburgh", SA)

kHz	Mode	Call	Location	Service	kW	Remarks
11,235.6	FAX	....	unknown	FA	....	
11,239.0	USB	....	several, throughout Australia	FA/MA	....	RAAF [Alternate to 11,235.0]
11,243.0	USB	MJV	Falmouth (Culdrose), England	FA	1.0	RAF
	USB	....	numerous, worldwide	FA/MA	....	USAF (Used for SKY KING broadcasts--some code names used for various Air Bases--was called the "ALFA ONE" frequency--now part of "GIANT TALK")
11,246.0	USB	....	numerous, worldwide	FA/MA	....	USAF/USCG/USN [May be used by "AIR FORCE TWO"]
11,247.0	USB	....	numerous, worldwide	FA/MA	....	USAF
11,249.0	USB	....	several, USA nationwide	FA/MA	....	USAF
	USB	CZW	"Halifax Military", NS, Canada	FA	0.4	RCAF
11,255.0	USB	....	numerous, worldwide & Antarctica	FA/MA	....	RNZAF/USN
11,258.0	USB	....	numerous, mostly Pacific area	FA/MA	....	USN
11,258.5	FAX	....	unknown	FA	....	
11,265.0	USB	....	several, in Canada & Europe	FA/MA	....	RCAF (For list of stations see 11,209.0)
11,267.0	USB	....	numerous, worldwide	FA/MA	....	USAF/USN (USN uses this frequency for coordination/USAF uses this frequency for SKY KING broadcasts)

The frequencies between 11,175.0 and 11,275.0 kHz are used by all Allied Air and Naval Forces in 1.0 kHz increments. There are hundreds of code names for various bases that are changed or rotated on a random basis to confuse and circumvent positive identification.

kHz	Mode	Call	Location	Service	kW	Remarks
11,278.0	CW	CMU967	Santiago Naval Radio, Cuba	FX	....	Russian Navy
11,279.0	AM	....	Algiers Aeradio, Algeria	FA	....	VOLMET/FF (00+30)
	AM	....	Kano Aeradio, Nigeria	FA	....	" " (05+35)
	AM	....	Dakar Aeradio, Senegal	FA	1.3	" " (15+45)
	AM	....	Khartoum Aeradio, Sudan	FA	....	" " (25+55)
			§AFI VOLMET move to 8852.0 in 1982			
	AM	....	Khabarovsk Aeradio, USSR	FA	....	VOLMET/RR (00+30)
	AM	....	Novosibirsk Aeradio, USSR	FA	....	" " (10+40)
	AM	....	Tashkent Aeradio, Uzbek SSR	FA	....	" " (05+35)
11,287.0	USB	....	several, western USA	FA/MA	....	In-flight Tests
11,303.0	AM/A3H	....	numerous (ICAO CWP & EUR)	FA/MA	....	Airports & in-flight covering all of the northern Pacific Ocean from Honolulu, Manila and Tokyo to Ulan Bator. Also all of central Europe from Cairo north to Leningrad and east to Madrid
			§CWP moves to 11,384.0 in 1983			
	AM/USB	....	This frequency is internationally used for Paid In-Flight Services by commercial carriers			
11,303.5	CW/RTTY	KNY21	Washington, DC	FX	1.0	Yugoslav EMBASSY
11,312.0	CW	COL	Havana Aeradio, Cuba	FA/MA	....	Aeroflot
	CW	RFNV	Moscow, USSR	FA/MA	....	Aeroflot (Mostly working in-flight to/from Moscow/Havana)
11,319.0	USB	....	numerous, throughout Alaska	FA/MA	....	Private Carriers
	USB	VYW61	Montreal Aeradio, PQ, Canada	FA	1.0	
	USB	ZKRG	Rarotonga, Cook Islands	FA	1.0	
	A3H	....	several, deep interior of USSR	FA	....	VOLMET/RR (Daytime use--see 9009.0 for list of possible stations)
11,327.0	AM/A3H	....	numerous (ICAO SAM)	FA/MA	....	Airports & in-flight covering entire eastern half of South America
			§SAM moves to 10,096.0 in 1982			
11,343.0	AM/A3H	....	numerous (ICAO SAM)	FA/MA	....	Airports & in-flight covering entire western half of South America
			§CAR moved to 11,387.0 & SAM moves to 11,360.0 in 1982 11,355.0			
11,355.0	CW	FDA35	Lyons Air, France	FA	....	French AF
11,357.0	CW	YRA	Bucharest Aeradio, Rumania	FA/MA	....	
	A3H/USB		§SAM moves here from 11,343.0 in 1982			
11,359.0	A3H	....	several, in interior of the USSR	FA	....	VOLMET/RR (Daytime use--see 8917.0 for list of possible stations)
11,367.0	....	....	§CAR moved to 11,387.0 in 1982			

AERONAUTICAL MOBILE (con.)

kHz	Mode	Call	Location	Service	kW	Remarks
11,382.0	A3H/USB		§VOLMET broadcasts move here from 11,391.0 in 1982			
	A3H/USB		§CAR moves here from 11,343.0 & 11,367.0 in 1982			
11,386.0	USB	....	§VOLMET moves here from 11,391.0 in 1982			
11,387.0	A3H/USB	....	Bangkok Aeradio, Thailand	FA	1.6	VOLMET (10+40)
	A3H/USB	....	Bombay Aeradio, India	FA	0.8	" (25+55)
	USB	...	Calcutta Aeradio, India	FA	....	" (05+35)
	USB	....	Karachi Aeradio, Pakistan	FA	....	" (15+45)
	USB	9VA43	Singapore Aeradio, Singapore	FA	4.0	" (20+50)
	USB	VLS	Sydney Aeradio, NSW, Australia	FA	3.0	" (00+30)
			[VOLMET originally on 10,017.0 has completed its move to this new frequency--observed at press time]			
	A3H/USB	....	numerous (ICAO CAR)	FA/MA	....	Airports &
			in-flight covering the eastern Caribeean north to New York and south to Bogota and Paramaribo			)
11,390.0	CW	UGAB	Moscow (Tuchino) Aeradio, USSR	FA	....	)
11,391.0	AM	OKL	Prague Aeradio, Czechoslovakia	FA	1.6	VOLMET/EE (15+45
	AM	4XL	Ben Gurion Aeradio, Israel	FA	2.0	VOLMET/EE (05+35
			§VOLMET broadcasts move to 11,378.0 in 1982			
11,395.0	CW	CCS	Santiago Naval Radio, Chile	FX	....	
11,396.0	USB	KJY74	"Miami Monitor", FL	FA/MA	....	NOAA
11,396.0	A3H/USB	....	§CAR moves here from 11,343.0 in 1982			
	A3H/USB	....	§SEA moves here from 8868.0 in 1982			

11,400.0 - 11,700.0 kHz — FIXED SERVICE (Worldwide)

11,400.0 - 11,650.0 kHz -- FIXED SERVICE (Worldwide--post WARC)

kHz	Mode	Call	Location	Service	kW	Remarks
11,401.0	USB	KUP67	Truk Aeradio, Caroline Islands	AX	1.0	Aero Tfc
11,402.0	USB	KUP65	Majuro Aeradio, Marshall Islands	AX	1.0	Aero Tfc
11,402.5	USB	BAZ21	Beijing, PRC	FX	10.0	Telcom/Berlin
	USB	KGA57	Agana Aeradio, Guam	AX	0.1	Aero Tfc
	USB	KWT73	Honolulu Aeradio, HI	AX	1.0	Aero Tfc
11,403.5	USB	....	several, throughout Zaire	FX	....	UNO
11,406.0	CW	COY895	Havana Naval Radio, Cuba	FC	....	Russian Navy
	FAX	CLN324	Havana (Bauta), Cuba	FX	20.0	Nx
11,407.0	USB	AFH39	March AFB, CA	FX	2.0	USAF/SAC
	USB	AFS	Offutt AFB, NE	FX	0.5	USAF/SAC
	USB	AGA	Hickam AFB, HI	FX	1.0	USAF/SAC
	USB	AIE	Andersen AFB, Guam	FX	1.0	USAF/SAC
11,412.0	CW	CSF37	Funchal, Madeira Islands	FX	1.0	Port. AF
	CW	CSF25	Lisbon, Portugal	FX	1.0	Port. AF
11,414.0	ISB	AFE71	"Cape Radio", Patrick AFB, FL	FX	45.0	USAF/NASA
11,425.0	LSB	†CUG3Ø	Lisbon (Alfragide), Portugal	FX	10.0	Telcom
	USB	....	unknown	FX	....	Net/SS
11,427.0	ISB	ZPG14	Asuncion, Paraguay	FX	10.0	Telcom/Tfc
11,430.0	ISB	†MKG	London (Stanbridge), England	FX	30.0	RAF Telcom/Tfc
	CW	HMN51	Pyongyang, North Korea	FX	....	Tfc/Nx
11,430.0	CW	UMS	Moscow Naval Radio, USSR	FX	....	Russian Navy
11,434.6	USB	....	Atlantic Domestic Emergency Net	FX	....	USCG (Includes:
			NMA, Miami, FL; NMF, Boston, MA; NMG, New Orleans, LA; NMN, Portsmouth, VA; NMR, San Juan, PR & NOZ, Elizabeth City, NC; plus transportables) [Alt: 10,136.8 & 11.513.6]			
	USB	....	Pacific Domestic Emergency Net	FXX	....	USCG (Includes:
			NMC, San Francisco, CA; NMO, Honolulu, HI; NMQ, Long Beach, CA; NRV, Barrigada, Guam; plus transportables) [Alt: 10,166.6 & 11,513.6]			
	CW/USB	OVC	Groennedal, Greenland	FX	5.0	Danish Marine
11,435.2	USB	VNV13Ø	Sydney (Doonside), NSW, Australia	FX	....	Telcom/Mawson
11,440.0	CW	PLC	Jakarta, Indonesia	FX	3.0	TIME SIGNALS
11,440.0	CW	SOL244	Warsaw, Poland	FX	....	Tfc/Reported
				PAP nx broadcast at 1400		
11,448.0	CW/USB	CME31Ø	Havana, Cuba	FX	2.0	GDR EMBASSY
	CW	Y7A49	Berlin, GDR	FX	....	EMBASSY
	CW	KNY37	Washington, DC	FX	1.0	GDR EMBASSY
	CW	Y7L36	unknown	FX	....	GDR EMBASSY
11,453.0	USB	....	Daventry, England	FX	30.0	BBC Feeder

kHz	Mode	Call	Location	Service	kW	Remarks
11,454.0	CW	LBA9	Stavanger Naval Radio, Norway	FX	....	Norwegian Navy (aka JWT & LBA2)
11,461.0	CW/FAX	WFK21	New York, NY	FX	50.0	Tfc/AP Nx
11,465.0	ISB	VNV49	Sydney (Doonside), NSW, Australia	FX	30.0	Telcom/Norfolk
11,465.8	ISB	HBO51	Geneva, Switzerland	FX	20.0	Tfc
11,474.0	CW	KKN44	Monrovia, Liberia	FX	....	USA EMBASSY
11,483.0	USB	....	several, USA nationwide	FX	....	USAF (includes:
			AFC24, Westover AFB, MA; AFF7, Davis-Monthan, AZ; AFX, Barksdale, AZ; etc.)			
11,487.3	ISB	ASH	Rawalpindi, Pakistan	FX	....	Tfc
11,488.0	CW	RIW	Khiva Naval Radio, Uzbek SSR	FX	....	Russian Navy
	USB	....	numerous, USA nationwide	FX	....	FBI Backup Net
			(Most recently reported have included: KAG69, Denver, CO; KAG78, Kansas City, MO; KAG98, Omaha, NE; etc.)			
11,494.0	CW/RTTY	SOL249	Warsaw, Poland	FX	20.0	
	USB	AFS	Offutt AFB, NE	FX	0.5	USAF/SAC [Alt: 11,607.0]
11,495.0	ISB	†5YF52	Nairobi, Kenya	FX	8.0	Telcom
	LSB	†....	Moscow, USSR	FX	....	R. Moscow Feeder
11,498.8	CW	OMZ	Prague Czechoslovakia	FX	....	EMBASSY
11,500.0	CW	POB35	Jakarta, Indonesia	FX	3.0	NOAA reports wx broadcast at 0830
	CW	C9E22Ø	Maputo, Mozambique	AX	3.0	Tfc/Luanda
11,501.4	LSB	ART	Rawalpindi, Pakistan	FX	10.0	Telcom/Teheran
11,508.0	CW/USB	CME31Ø	Havana, Cuba	FX	1.0	Libyan EMBASSY
11,513.5	USB	†....	North Atlantic (East Coast) LORAN Net	FX	....	USCG (Includes:
			NMA7, Jupiter, FL; NMF32, Nantucket, MA; NMF33, Caribou, ME; NMN73, Carolina Beach [Cape Fear], NC & NOL, "Bermuda Monitor" [Alt: 9278.5]			
11,513.6	USB	....	Atlantic Area Emergency Net	FX	....	USCG (Includes:
			NMA, Miami, FL; NMF, Boston, MA; NMG, New Orleans, LA; NMN, Portsmouth, VA; NMR, San Juan, PR & NOZ, Elizabeth City, NC--plus transportables) [Alt: 11,434.6 & 12,148.6]			
	CW/USB	....	Pacific Domestic Emergency Net	FX	....	USCG (Includes:
			NMC, San Francisco, CA; NMO, Honolulu, HI; NMQ, Long Beach, CA & NRV, Barrigada, Guam--plus transportables) [Alt: 11,434.6 & 12,173.6]			
11,520.0	CW	KKN44	Monrovia, Liberia	FX	....	USA EMBASSY
11,522.5	FAX	NPN	Barrigada (Agana NAS), Guam	FX	15.0	USN/Meteo
11,525.0	USB	CLN362	Havana (Bauta), Cuba	FX	....	Telcom
11,532.0	AM	....	unknown	FX	....	NUMBERS/SS/YL
11,538.0	CW	HEP58	Zurick (Waltikon), Switzerland	FX	3.0	INTERPOL
	CW	OEQ45	Vienna, Austria	FX	1.0	"
	CW	ONA2Ø	Brussels, Belgium	FX	3.0	"
	CW	SHX	Stockholm, Sweden	FX	1.5	"
11,545.0	USB	....	unknown	FX	....	NUMBERS/GG/YL
11,550.0	USB	ZLM32	Wellington, New Zealand	FX	1.0	Telcom
	ISB	MKG	London (Stanbridge), England	FX	5.0	RAF Telcom/Tfc
11,552.0	CW/USB	NGD	McMurdo Station, Antarctica	FX	5.0	USN
	CW/USB	NQU	Siple Station, Antarctica	FX	1.0	USN
	USB	NPX	South Pole Station, Antarctica	FX	....	USN
	USB	ZLBC	Campbell Island, New Zealand	FX	0.3	
11,555.0	CW	CMU967	Santiago Naval Radio, Cuba	FX	....	Russian Navy
	USB	....	several, USA nationwide	FX	....	DOE
11,557.0	USB	ZHF45	Rothera Base, Antarctica	FX	....	Tfc
11,560.0	USB	TUA211	Abidjan, Ivory Coast	FX	....	Telcom
11,562.0	ISB	CSS414	Ponta Delgada, Azores	FX	....	Telcom
11,565.0	ISB	HMU94	unknown, undoubtedly Korean	FX	....	
11,570.0	CW	HMN53	Pyongyang, North Korea	FX	....	Tfc/Nx
11,575.0	ISB	†....	Holzkirchen, GFR	FX	10.0	RFE Feeder
	ISB	....	Moscow, USSR	FX	....	R. Moscow Feeder
11,600.0	USB	....	†Tashkent, Uzbek SSR	FX	....	R. Moscow Feeder
11,606.0	USB	....	Northwest Pacific LORAN Net	FX	....	USCG (Includes:
			NRT, "Yokota Monitor", Japan; NRT2, Gesashi, Japan; NRT3, Iwo Jima; NRT9, Hokkaido, Japan; NRV, Barrigada, Guam; NRV6, Marcus Island & NRV7, Yap Island) [Alt: 10,523.6 & 13,608.6]			

kHz	Mode	Call	Location	Service	kW	Remarks
11,607.0	USB/RTTY	AFK	Peterson AFB, CO	FX	1.0	USAF/SAC
	USB	AFS	Offutt AFB, NE	FX	0.5	USAF/SAC
				[Alt: 11,494.0 & 15,962.0]		
11,615.5	FAX	VFE	Edmonton, Alta., Canada	FX	5.0	Meteo
11,617.5	LSB	....	unknown	FX	....	NUMBERS/GG/YL
11,620.0	CW	HLL	Seoul Radio, South Korea	FC	0.5	NOAA reports
				wx broadcasts at 0000 & 1200		
11,623.0	CW	UFB	Odessa Radio, Ukrainian SSR	FC	....	
11,625.0	CW	GZU	Portsmouth (Petersfield), England	FX	1.0	British Naval Air
11,630.0	ISB	....	Beijing, PRC	FX	....	R. Beijing Feeder
11,635.0	CW	†KKN44	Monrovia, Liberia	FX	....	USA EMBASSY
	AM	....	unknown	FX	....	NUMBERS/YL/SS
11,640.0	USB	....	several in PQ, Canada	FX	....	Net

11,650.0 - 12,050.0 kHz -- INTERNATIONAL BROADCASTING (Worldwide--post WARC)

kHz	Mode	Call	Location	Service	kW	Remarks
11,650.0	ISB	CML32	Havana (Bauta), Cuba	FX	10.0	Telcom/Mexico
11,660.0	ISB	....	Beijing, PRC	FX	....	R. Beijing Feeder
11,662.5	CW/ISB	C8N27	Maputo, Mozambique	FX	10.0	Tfc/Telcom
	CW	CAN6D	Punta Arenas (Montalva), Chile	FX	0.5	Meteo (NOAA reports wx broadcasts at 1430 & 2200)
11,674.5	ISB	JEB21	Chichi Jima, Bonin Is., Japan	FX	2.0	Tfc
11,675.0	ISB	CLN328	Havana (Bauta), Cuba	FX	10.0	Telcom/Caracas
11,675.0	ISB	....	†Holzkirchen, GFR	FX	10.0	RFE Feeder
11,687.5	CW	NAM	Norfolk Naval Radio, VA	FC	100.00	USN
11,688.0	CW/RTTY	CYS22	Ottawa, Ont., Canada	FX	1.0	Polish EMBASSY [Alt: 14,692.0]
11,692.0	CW	....	unknown	FX	....	Coded Tfc

11,700.0 - 11,975.0 kHz -- INTERNATIONAL BROADCASTING (Worldwide)

11,975.0 - 12,330.0 kHz -- FIXED SERVICE (Worldwide)

kHz	Mode	Call	Location	Service	kW	Remarks
11,985.0	USB	7OB452	Aden (Khormaksar), Democratic Yemen	FX	1.0	Libyan EMBASSY
11,994.0	CW/RTTY	GFH	Hong Kong Naval Radio, Hong Kong	FX	3.5	British Navy
12,000.0	AM	VNG	Lyndhurst, Vic., Australia	SS	10.0	TIME SIGNALS
	CW	JJD	Tokyo, Japan	FX	10.0	Solar & Geomagnetic Reports
12,003.5	CW/USB	....	several, west coast of USA	FX	....	USN (Usually NPC, Seattle, WA; NPG, San Francisco, CA & NPL, San Diego, CA)
12,017.0	ISB	8CJ	unknown	FX	....	
12,022.5	CW	KKN5Ø	Washington, DC	FX	10.0	EMBASSY
12,025.0	FAX	PWZ	Rio de Janeiro Naval Radio, Brazil	FX	1.5	Brazil Navy/Meteo
12,035.0	CW	EEQ	Madrid, Spain	FX	1.0	Police/Net [Alt: 9796.0]

12,050.0 - 12,230.0 kHz -- FIXED SERVICE (Worldwide--post WARC)

kHz	Mode	Call	Location	Service	kW	Remarks
12,056.0	CW	†RIW	Khiva Naval Radio, Uzbek SSR	FX	....	Russian Navy
12,058.5	USB	†NAM	Norfolk Naval Radio, VA	FC	40.0	Telcom
12,068.5	USB	CTH29	Horta Naval Radio, Azores	FX	5.0	Port. Navy
12,087.5	USB	....	several, among South Pacific Islands	AX	....	Aero Tfc
12,095.0	CW	FUV	Djibouti Naval Radio, Djibouti	FX	....	French Navy
12,100.0	USB	RTU4Ø	Moscow, USSR	FX	15.0	R. Moscow Feeder
12,101.0	USB	SAM	Stockholm, Sweden	FX	1.2	EMBASSY
12,110.0	FAX	BAF33	Beijing, PRC	FX	....	Meteo
	CW	ZBP	Pitcairn Island	FX	0.7	Tfc
12,111.5	CW	KKN5Ø	Washington, DC	FX	10.0	EMBASSY
12,118.6	USB	ZLK43	Christchurch (Weedons), New Zealand	FX	10.0	Telcom/McMurdo
12,125.0	CW	CKN	Vancouver Forces Radio, BC, Canada	FX	10.0	(C13E) Canadian Forces
			NOAA reports wx broadcasts at 0030, 0430 & 1330			
	FAX	CKN	Vancouver Forces Radio, BC, Canada	FX	10.0	Canadian Forces/ Meteo/Prognosis & Surface Analysis--0230-2200

FIXED SERVICE (con.)

kHz	Mode	Call	Location	Service	kW	Remarks
	USB	TNH21	Brazzaville, Congo	FX	20.0	Telcom
12,127.0	USB	....	Daventry, England	FX	30.0	BBC Feeder
12,129.0	USB	AIE	Andersen AFB, Guam	FX	1.0	USAF
12,135.0	LSB	RNE39	Moscow, USSR	FX	20.0	R. Moscow Feeder
	CW	NAM	Norfolk Naval Radio, VA	FX	50.0	USN/NOAA reports wx broadcasts at 0030, 0630, 1230 & 1900--seasonal ice warning broadcasts at 1000 & 2300
	CW/RTTY	....	several, in the USA	FX	....	USN (Usually NAA, Cutler, ME; NAR, Key West, FL; NPG, San Francisco, CA & NSS, Washington, DC
12,138.5	CW/USB	KUR2Ø	Honolulu, HI	FX	1.0	Tfc/Agana
	CW/USB	KUR5Ø	Agana, Guam	FX	1.0	Tfc/Honolulu
12,140.5	USB	....	several, probably in Caribbean	FX	....	Net/EE/SS
12,148.6	USB	....	Atlantic Domestic Emergency Net	FX	....	USCG (Includes: NMA, Miami, FL; NMF, Boston, MA; NMG, New Orleans, LA; NMN, Portsmouth, VA; NMR, San Juan, PR & NOZ, Elizabeth City, NC --plus transportables) [Alt: 11,513.6 & 15,548.6]
12,150.0	USB	....	Aleutian LORAN Network	FX	....	USCG (Includes: NMJ22, Attu; NRW2, St. Paul Is. & NRW3 Pt. Clarence--all AK) [Alt: 9073.0]
	USB	....	West Indies LORAN Net	FX	....	USCG (Includes: NMA, Miami, FL; NMA4, San Salvador [Watling Is.], Bahamas; NMA5, South Caicos, Bahamas; NMA7, Jupiter, FL & NMR, San Juan, PR) [Alt: 7530.0 & 8083.5]
12,151.0	CW	"K"	unknown	RC	....	Beacon
12,152.0	LSB	MKT	London (Stanbridge), England	FX	10.0	RAF Telcom/Tfc
12,152.5	USB	ZLBC6	Campbell Is., New Zealand	FX	0.3	Telcom/Wellington
	USB	ZME6	Raoul (Sunday) Is., Kermandec Islands	FX	0.3	Telcom/Wellington
12,154.0	CW	....	unknown, probably Cuba	FX	....	Tfc/RR
12,155.1	ISB	LDR	Oslo (Jeloey), Norway	FX	10.0	Telcom/Tfc
12,165.0	USB	†RKB78	Moscow, USSR	FX	20.0	R. Moscow Feeder
12,173.6	USB	....	Pacific Domestic Emergency Net	FX	....	USCG (Includes: NMC, San Francisco, CA; NMO, Honolulu, HI; NMQ, Long Beach, CA; NOJ, Kodiak, AK & NRV, Barrigada, Guam--plus transportables) [Alt: 11,513.6 & 15,473.6]
12,175.0	CW	HMR23	Pyongyang, North Korea	FX	....	Tfc
	LSB	†RCD34	Moscow, USSR	FX	20.0	R. Moscow Feeder
12,179.0	USB	GWC	Daventry, England	FX	30.0	BBC Feeder
12,180.1	CW/ISB	CSR26	Lisbon, Portugal	FX	3.0	Tfc/Telcom
12,185.0	CW	"U"	unknown	FX	....	Beacon
12,190.0	ISB	†....	Lyndhurst, Vic., Australia	FX	....	R. Australia Feeder
12,200.0	CW	NPO	San Miguel (Capas), Philippines	FX	5.0	USN/NOAA reports wx broadcasts at 0400 & 1300
12,201.0	FAX	AFA	Washington (Andrews AFB, MD), DC	FX	3.0	USAF/Meteo
12,205.0	USB	....	Hawaii & Central Pacific LORAN Net	FX	....	USCG (Includes: NMO, Lualualei, HI; NRO, Johnston Is.; NRO5, Upolu Pt., HI & NRO7, Kure [Ocean Is.] Island, HI) [Alt: 9630.0]
	CW	NDT4	Yokosuka Naval Radio, Japan	FX	5.0	USN Tfc
	LSB	RDT72	Alma Ata, USSR	FX	15.0	R. Moscow Feeder
12,207.5	ISB	†CUG41	Ponta Delgada, Azores	FX	10.0	Telcom/Tfc
12,210.0	CW	KWL9Ø	Tokyo, Japan	FX	....	USA EMBASSY
12,211.0	CW	RLX	Dublin, Ireland	FX	0.5	USSR EMBASSY [Alt: 11,148.0 & 14,426.0]
12,216.0	USB	....	numerous, USA nationwide	FX	....	FEMA (This is one of several "working" frequencies assigned to the US Federal Emergency Management Agency [FEMA]. Stations reported have included: KPA65, Denton, TX; KPA71, Maynard, MA; WGY9Ø3, Olney, MD; WGY9Ø9, Santa Rosa, CA; WAR42, Mt. Weather, VA., etc.)
12,224.5	CW/RTTY	....	several, throughout Europe	FX	....	INTERPOL (Operating on this frequency are: DEB, Weisbaden, GFR; HEP25, Zurich, Switzerland; OEQ46, Vienna, Austria; SHX, Stockholm, Sweden & ONA2Ø, Brussels)
12,225.0	USB	SAM	Stockholm, Sweden	FX	1.2	EMBASSY [Alt: 12,101.0 & 12,306.0]

kHz	Mode	Call	Location	Service	kW	Remarks

12,230.0 - 12,330.0 kHz — FIXED & MARITIME SERVICES (Shared Worldwide--post WARC)

kHz	Mode	Call	Location	Service	kW	Remarks
12,232.0	CW	GFT42	Bracknell, England	FX	2.5	Meteo
12,235.9	CW	CBV	Valparaiso , Chile	FX	....	Coded Tfc
12,238.5	USB	†ONN34	Brussels, Belgium	FX	1.0	Iranian EMBASSY
12,240.0	ISB	RHA41	Moscow, USSR	FX	....	R. Moscow Feeder
12,246.5	USB	....	Holzkirchen, GFR	FX	10.0	RFE Feeder
12,247.0	CW	....	several (ICAO AFI) in northeastern corner of the African continent, Arabia and as far east as Bombay, India]	AX	....	Aero Tfc [Net
12,250.0	ISB	FTM25	Paris (St. Assise), France	FX	2.0	Telcom/Rabat
12,255.0	CW/USB	VJM	Macquarie Island, Antarctica	FX	1.5	ANARE
	CW/USB	VLV	Mawson Base, Antarctica	FX	5.0	"
	CW/USB	VLZ	Davis Base, Antarctica	FX	1.0	"
	CW/USB	VNJ	Casey Base, Antarctica	FX	10.0	"
	CW/USB	VNH	Hobart, Tasmania, Australia (Australian National Antarctic Research Expedition--ANARE)	FX	10.0	"
	ISB	TZA512	Bamako, Mali	FX	....	Telcom/Tfc
12,257.0	CW/RTTY	ONN39	Brussels, Belgium	FX	1.0	Hungarian EMBASSY
12,260.5	LSB	....	Dakar, Senegal	FX	....	Telcom
12,270.0	ISB	BAZ53	Beijing, PRC	FX	10.0	Telcom/Tfc
	CW	....	several, in southern Pacific Ocean the major islands including: Fiji, Solomon, Tonga, Tuvalu, Nauru, etc.)	AX	....	Aero (Net among
12,282.0	CW	X3Q	unknown	FX	....	VVV's
12,290.0	USB	....	Lyndhurst, Vic., Australia	FX	....	R. Australia Feeder
12,294.0	USB	3AD1	Monaco, Monaco	FX	30.0	R. Monaco Feeder
12,294.7	ISB	FZX322	Miquelon, St. Pierre & Miquelon Is.	FX	6.0	Telcom/Tfc
12,295.0	CW	VCS838	Ottawa, Ont., Canada	FX	1.0	Yugoslav EMBASSY [Alt: 10,780.0 & 14,515.0]
12,304.5	USB	SAM	Stockholm, Sweden heard on this frequency: SAM24, Copenhagen, Denmark; SAM25, Lisbon, Portugal; SAM3Ø, Madrid, Spain; SAM48, Damascus, Syria & SAM52, Tel Aviv, Israel)	FX	1.2	EMBASSY (Also
12,305.0	FAX	FTM3Ø	Paris (St. Assise), France	FX	20.0	Meteo
	USB	3AD1	Monaco, Monaco	FX	30.0	R. Monaco Feeder
12,306.0	USB	SAM	Stockholm, Sweden	FX	1.2	EMBASSY
12,307.0	CW	OBC3	Callao, Peru	FX	....	TIME SIGNALS/Tfc
12,310.0	CW	NPM	Lualualei Naval Radio, HI	FX	2.0	USN
12,322.0	FAX	JKC3	Tokyo (Kemigawa), Japan	FX	5.0	Meteo/Nx
12,329.0	CW	OVC	Groennedal, Greenland	FX	5.0	Danish Marine
	CW	OVG12	Frederikshaven Naval Radio, Denmark	FX	5.0	Danish Marine
	CW	"U"	unknown	RC	....	Beacon

12,330.0 — 13,200.0 kHz — MARITIME MOBILE (Worldwide)

kHz	Mode	Call	Location	Service	kW	Remarks
12,330.0 – 12,426.1	USB		The band between these frequencies is divided into 32 channels spaced 3.1 kHz apart and assigned numbers from 1201 to 1232. These are ship transmitting channels and are paired with Coastal USB stations operating between 13,100.8 and 13,196.9 kHz			
12,333.1	USB	WCM	Pittsburgh, PA	FC/MS	0.8	Inland Waterway
	USB	WJG	Memphis, TN	FC/MS	0.8	Inland Waterway
12,429.2	USB	....	Atlantic & Pacific Voice Circuits NMA, Miami, FL; NMC, San Francisco, CA; NMF, Boston, MA; NMG, New Orleans, LA; NMN, Portsmouth, VA; NMO, Honolulu, HI; NMQ, Long Beach, CA; NMR, San Juan, PR; NOJ, Kodiak, AK & NRV, Barrigada, Guam)	FC/MS	....	USCG (Includes:
	USB	....	Coast & Ship Simplex Voice Frequency coastal waterway traffic. Employed worldwide by hundreds of base stations and various types of ships. Frequently heard in North America: KHF, Amelia, LA; WRF, Elizabeth, NJ; KUX, San Diego, CA: WBD, Houston, TX; etc.	FC/MS	..	Inland & inter-

kHz	Mode	Call	Location	Service	kW	Remarks
12,432.3	USB	NMC	COMMSTA San Francisco, CA	FC/MS	10.0	USCG
	USB	....	Coast & Ship Simplex Voice Frequency	FC/MS	..	Inland and inter-coastal waterway traffic. Employed worldwide by hundreds of base stations and various types of ships. Frequently heard in North America: KMC955, Seattle, WA; WXY998, Miami, FL; KHT, Cedar Rapids, IA; etc.)
12,435.4	USB	....	Atlantic & Pacific Voice Circuits	FC/MS	....	USCG (Includes same stations as listed on 12,429.2)
	USB	....	Coast & Ship Simplex Voice Frequency	FC/MS	..	Inland and inter-coastal waterway traffic. Employed worldwide by hundreds of base stations and various types of ships. Frequently heard in North America: KYX593, Seattle, WA; WMP, West Palm Beach, FL; etc.)

kHz	Mode	
12,540.0	CW	The band between these two frequencies is divided into 18 channels spaced 1.2 kHz apart and assigned numbers from 1 to 18. These are ship transmitting channels. The most commonly used channels are 5 and 6, 12,544.8 and 12,546.0 kHz, respectively.
| |		
12,561.6	CW	

kHz	Mode	Call	Location	Service	kW	Remarks
12,653.5	CW	HKC	Beunaventura Radio, Colombia	FC	0.5	
	CW	6VA6	Dakar Radio, Senegal	FC	4.0	
12,655.2	CW	PJK213	Suffisant Dorp Naval Radio, Curacao	FC	1.0	Dutch Navy
12,655.5	CW	FFT6	St. Lys Radio, France	FC	10.0	
	FAX	....	unknown	FC	....	Meteo
12,656.5	CW	UJQ	Kiev Radio, Ukrainian SSR	FC	20.0	
12,658.0	CW	5OW12	Lagos Radio, Nigeria	FC	1.0	
12,659.0	CW	LZW5	Varna Radio, Bulgaria	FC	15.0	
12,659.4	CW	9VG37	Singapore Radio, Singapore	FC	3.0	
12,659.9	CW	YIR	Basrah Control, Iraq	FC	1.0	
12,660.0	CW	JNE	Hiroshima Radio, Japan	FC	0.5	
	CW	JGC	Yokohama Radio, Japan	FC	1.0	
	CW	6VA6	Dakar Radio, Senegal	FC	4.0	
	CW	S7Q	Seychelles Radio, Seychelles	FC	1.0	
	CW	CWM	Montevideo Armada Radio, Uruguay	FC	2.0	Uruguayan Navy
	CW	WSL	Amagansett Radio, NY	FC	10.0	
12,661.5	CW	XSG	Shanghai Radio, PRC	FC	2.0	
	CW	UMV	Murmansk Radio, USSR	FC	15.0	
12,662.0	CW	7TA8	Algiers Radio, Algeria	FC	1.0	
12,663.0	CW	XSG	Shanghai Radio, PRC	FC	2.0	
	CW	ZSD51	Durban Radio, RSA	FC	3.5	
12,663.5	CW	UBN	Jdanov Radio, Ukrainian SSR	FC	15.0	
12,664.5	CW	FUO	Toulon Naval Radio, France	FC	2.0	French Navy
	CW	FUM	Papeete Naval Radio, Tahiti	FC	10.0	French Navy
	CW/RTTY	UKA	Vladivostok Radio, USSR	FC	10.0	
12,666.5	CW	5BA	Cyprus (Nicosia) Radio, Cyprus	FC	15.0	
12,667.0	FAX	VFF	Frobisher CG Radio, NWT, Canada	FC	5.0	Meteo
12,667.5	CW	JCX	Naha Radio, Okinawa, Japan	FC	2.0	
12,669.0	CW	9GA	Takoradi Radio, Ghana	FC	10.0	
12,669.1	CW	OFJ3	Helsinki Radio, Finland	FC	10.0	
12,671.0	CW	VFF	Frobisher CG Radio, NWT, Canada	FC	1.0	AMVER
12,671.5	CW	UDH	Riga Radio, Latvian SSR	FC	5.0	
12,673.5	CW	CLA33	Havana (Cojimar) Radio, Cuba	FC	5.0	
	CW	ETC	Assab Radio, Ethiopia	FC	2.0	
12,673.6	CW	JOU	Nagasaki Radio, Japan	FC	15.0	
12,674.0	CW	UXN	Arkhangelsk Radio, USSR	FC	25.0	
12,675.0	CW	XFK	La Paz Baja California Radio, Mexico	FC	0.5	
12,675.5	CW	A4M	Muscat Radio, Oman	FC	1.0	
12,677.9	CW	9MG2	Penang Naval Radio, Malaysia	FC	1.0	Malaysian Navy
12,678.0	CW	FFS6	St. Lys Radio, France	FC	10.0	
	CW	YJM3	Port-Vila Radio, Vanuatu	FC	0.5	
	CW	UQB	Kholmsk Radio, USSR	FC	15.0	
12,681.5	CW	Y5M	Ruegen Radio, GDR	FC	15.0	
12,682.5	CW	TNA12	Pointe-Noire Radio, Congo	FC	2.0	
	CW	PKE	Ambina Radio, Indonesia	FC	1.0	
	CW	PKF	Makassar Radio, Indonesia	FC	1.0	

kHz	Mode	Call	Location	Service	kW	Remarks
	CW	PKP	Dumai Radio, Indonesia	FC	1.0	
	CW	PNK	Jayapura Radio, Indonesia	FC	1.0	
	CW	ODR4	Beirut Radio, Lebanon	FC	1.0	
	CW	LFC	Rogaland Radio, Norway	FC	5.0	
12,683.0	CW	UFB	Odessa Radio, Ukrainian SSR	FC	10.0	
12,684.5	CW	UFM3	Nevelsk Radio, USSR	FC	....	
	CW/RTTY	UGH2	Juzno-Sakhalinsk Radio, USSR	FC	15.0	
12,687.0	CW	PPR	Rio Radio, Brazil	FC	1.0	TIME SIGNALS
	CW	OFJ4	Helsinki Radio, Finland	FC	10.0	
	CW	JCT	Choshi Radio, Japan	FC	6.0	
	CW	XFU	Veracruz Radio, Mexico	FC	0.5	
12,688.5	CW	UQK2	Riga Radio, Estonian SSR	FC	5.0	
12,689.0	CW	LZL5	Bourgas Radio, Bulgaria	FC	1.0	
	CW	PPJ	Juncao Radio, Brazil	FC	1.0	NOAA reports wx broadcasts at 0130, 0730 & 1900
	CW	SXA3	Spata Attikis Radio, Greece	FC	....	Greek Navy/NATO
12,690.0	CW	ZAD2	Durres Radio, Albania	FC	1.0	
	CW	PBC9	Goeree Naval Radio, Netherlands	FC	....	Dutch Navy
	CW	UJY	Kaliningrad Radio, USSR	FC	5.0	
12,691.0	CW	GXH	Thurso Naval Radio, Scotland	FC	10.0	USN/NOAA reports wx broadcast at 1230
	CW	EDF44	Aranjuez Radio, Spain	FC	15.0	
	CW	NAM	Norfolk Naval Radio, VA	FC	....	USN
12,691.4	CW	FUX	Le Port Naval Radio, Reunion	FC	5.0	French Navy
12,692.2	CW	ZRQ5	Cape (Simonstown) Naval Radio, RSA	FC	15.0	RSA Navy
12,693.0	CW	URD	Leningrad Radio, USSR	FC	15.0	
12,695.2	CW	XSX	Keelung Radio, Taiwan	FC	2.5	NOAA reports wx broadcasts at 0430, 1030, 1630 & 2230
12,695.5	CW	CNP	Casablanca Radio, Morocco	FC	2.0	
	CW	KFS	San Francisco Radio, CA	FC	10.0	NOAA reports wx broadcasts at 0430, 1620, 2200 & 2300
12,696.5	CW	5OZ23	Port Harcourt Radio, Nigeria	FC	1.0	
12,697.0	CW	UBN	Jdanov Radio, Ukrainian SSR	FC	15.0	
12,697.8	CW	5RS	Tamatave Radio, Madagascar	FC	2.0	
	CW	FUJ	Noumea Naval Radio, New Caledonia	FC	10.0	French Navy
12,698.0	CW	A9M	Bahrain Radio, Bahrain	FC	2.0	
	CW	PPL	Belem Radio, Brazil	FC	1.0	
	CW	ZSC9	Cape Town Radio, RSA	FC	10.0	TIME SIGNALS/Tfc
	CW	YKI	Tartous Radio, Syria	FC	1.0	
12,698.9	CW	XSQ4	Guangzhou Radio, PRC	FC	2.0	
12,700.0	CW	D4A7	Sao Vicente Radio, Cape Verde	FC	3.0	
	CW	D4D	Praia de Cabo Radio, Cape Verde	FC	1.0	
	CW	3DP	Suva Radio, Fiji	FC	0.5	AMVER
	CW	NMN	COMMSTA Portsmouth, VA	FC	10.0	USCG
	CW	H4H29	Honiara Radio, Solomon Islands	FC	0.6	
	CW	NMR	COMMSTA San Juan, PR	FC	10.0	USCG/AMVER
	CW	UXN	Arkhangelsk Radio, USSR	FC	15.0	SAAMC
12,701.0	CW/RTTY	FUB	Paris (Houilles) Naval R., France	FC	....	French Navy
12,702.0	CW	CKN	Vancouver Forces Radio, BC, Canada	FC	10.0	C. Forces
	CW	Y5M	Ruegen Radio, GDR	FC	15.0	
12,703.0	CW	XFL	Mazatlan Radio, Mexico	FC	0.5	
	CW	UNM2	Klaipeda Radio, Lithuanian SSR	FC	15.0	
12,703.6	CW	PKD	Surabaya Radio, Indonesia	FC	1.0	
12,704.5	CW	PKM	Bitung Radio, Indonesia	FC	1.0	
	CW	WLO	Mobile Radio, AL	FC	30.0	NOAA reports wx broadcasts at 1300 & 2300
	CW	JFQ	Makuraki Gyogyo Radio, Japan	FC	0.5	
12,705.0	CW	DHJ59	Sengwarden Naval Radio, GFR	FC	....	GFR Navy/NATO
12,706.0	CW	UQK2	Riga Radio, Latvian SSR	FC	5.0	
12,706.5	CW	LSU20	Bahai Blanca Radio, Argentina	FC	1.0	
12,707.0	CW	LST9	Buenos Aires Radio, Argentina	FC	1.0	
	CW	9VG34	Singapore Radio, Singapore	FC	3.0	
12,708.0	CW	FJP23	Noumea Radio, New Caledonia	FC	1.0	
12,709.0	CW	LSA5	Boca Radio, Argentina	FC	1.0	
	CW	A9M	Bahrain Radio, Bahrain	FC	2.0	
	CW	8PO	Barbados Radio, Barbados	FC	1.0	

kHz	Mode	Call	Location	Service	kW	Remarks
	CW	LZL5	Bourgas Radio, Bulgaria	FC	1.0	
	CW	XSP	Shantou Radio, PRC	FC	2.0	
	CW	GKR3	Wick Radio, England	FC	0.3	
	CW	8RB	Demerara Radio, Guyana	FC	1.0	
	CW	ELC	Monrovia Radio, Liberia	FC	2.5	
	CW	ZSC23	Cape Town Radio, RSA	FC	10.0	
12,709.2	CW	VRT	Bermuda (Hamilton) Radio, Bermuda	FC	1.0	
	CW	9HD	Malta Radio, Malta	FC	1.0	
12,710.0	CW	VWB	Bombay Radio, India	FC	2.5	AMVER/Reported wx broadcasts at 0848 & 1648/NAVAREA 0648 & 1548
	CW	UFL6	Vladivostok Radio, USSR	FC	15.0	
12,711.0	CW	HCG	Guayaquil Radio, Ecuador	FC	0.3	AMVER
	CW	UBN	Jdanov Radio, Ukrainian SSR	FC	15.0	
12,712.0	CW	HLW3	Seoul Radio, South Korea	FC	15.0	
12,712.9	CW/FAX	GKN5	Portishead Radio, England	FC	15.0	
12,714.0	CW	CBV3	Valparaiso Radiomaritima, Chile	FC	0.7	AMVER
	CW	GKM5	Portishead Radio, England	FC	15.0	
	CW	UXN	Arkhangelsk Radio, USSR	FC	5.0	
12,714.9	CW	GK05	Portishead Radio, England	FC	15.0	
12,715.0	CW	OSN42	Oostende Naval Radio, Belgium	FC	1.0	Belgium Navy
	CW	UJQ7	Kiev Radio, Ukranian SSR	FC	....	
	CW	UFH	Petropavlovsk Radio, USSR		20.0	
12,716.9	CW/RTTY	ZL05	Irirangi Naval Radio, New Zealand	FC	7.5	RNZ Navy
12,717.0	CW	EBA	Madrid Radionaval, Spain	FC	1.0	Spanish Navy
12,718.5	CW	VWM	Madras Radio, India	FC	2.5	
12,718.5	CW	NMN	COMMSTA Portsmouth, VA	FC	10.0	USCG/AMVER
12,720.0	CW	SVG5	Athens Radio, Greece	FC	5.0	
	CW	UAT	Moscow Radio, USSR	FC	15.0	
	CW	UPB	Providenia Bukhta Radio, USSR	FC	15.0	
	CW	RMP	Rostov Radio, USSR	FC	....	
12,721.0	CW	SPH61	Gdynia Radio, Poland	FC	10.0	[aka SPA6]
12,722.0	CW	FJY4	St. Paul et Amsterdam Radio, French Antarctic	FC	0.3	
12,723.0	CW	5OZ23	Port Harcourt Radio, Nigeria	FC	1.0	
	CW	UAH	Tallinn Radio, Estonian SSR	FC	5.0	
	CW/RTTY	....	numerous, in Pacific area	FC	....	USN
	CW	†RCV	Moscow Naval Radio, USSR	FC	....	Russian Navy
12,724.0	CW	9VG57	Singapore Radio, Singapore	FC	3.0	
	CW	ZSC43	Cape Town Radio, RSA	FC	10.0	TIME SIGNALS/NOAA reports wx broadcasts at 0930 & 1730
12,725.0	CW	OSN	Oostende Naval Radio, Belgium	FC	....	Belgium Navy
	CW	URB2	Klaipeda Radio, Lithuanian SSR	FC	....	
12,726.0	CW/RTTY	CFH	Maritime Command Radio, Halifax, NS Canada (Canadian Forces)	FC	10.0	(C13L) NOAA reports wx broadcasts at 0200, 0630, 1400 & 1800/seasonal ice warnings at 0230 & 1430/Reported NAVAREA broadcast at 1700
12,727.0	CW	HLJ	Seoul Radio, South Korea	FC	7.0	
	CW	RKDF	Parnu Radio, Estonian SSR	FC	....	
12,727.5	CW	SVG5	Athens Radio, Greece	FC	5.0	
	CW	XSW	Kaohsiung Radio, Taiwan	FC	2.5	
12,727.6	CW	LGJ	Rogaland Radio, Norway	FC	5.0	AMVER
12,728.0	CW	LS05	Buenos Aires Radio, Argentina	FC	1.0	Prefectura Naval Reported wx broadcasts at 0210, 1210 & 1810
	CW	S3V	Cox's Bazaar Radio, Bangladesh	FC	1.0	
	CW	J2A9	Djibouti Radio, Djibouti	FC	1.0	
	CW	YKM7	Lattakuia Radio, Syria	FC	1.0	
12,729.0	CW	UFL	Vladivostok Radio, USSR	FC	15.0	Reported wx & NAVAREA broadcasts at 0100, 0900 & 1500
12,730.0	CW	NMC	COMMSTA San Francisco, CA	FC	10.0	USCG/NOAA reports wx broadcasts 0000 (Mondays), 0300, 0630 & 1900
	FAX	NMC	COMMSTA San Francisco, CA	FC	5.0	USCG/Wx pics at 0100, 0300, 0500, 0700, 1500, 1700, 2000 & 2300
	CW	UMV	Murmansk Radio, USSR	FC	15.0	
12,732.0	CW	DUN	Manila Radio, Philippines	FC	0.5	
	CW	UBN	Jdanov Radio, Ukrainian SSR	FC	15.0	

kHz	Mode	Call	Location	Service	kW	Remarks
12,734.0	CW	UGK2	Kaliningrad Radio, USSR	FC	15.0	
12,735.0	CW	URL	Sevastopol Radio, Ukrainian SSR	FC	5.0	
12,735.8	CW	TAH	Istanbul Radio, Turkey	FC	3.0	
12,738.0	CW	PPR	Rio Radio, Brazil	FC	1.0	
	CW	UXN	Arkhangelsk Radio, USSR	FC	5.0	
	CW	UAT	Moscow Radio, USSR	FC	15.0	
12,740.0	CW	ZLB5	Awarua Radio, New Zealand	FC	5.0	AMVER
	CW	UNQ2	Novorossiysk Radio, USSR	FC	5.0	
12,740.2	CW	GYA	Whitehall (London) Naval R., England	FC	10.0	British Navy
12,741.0	CW	CBV	Valparaiso Radiomartima, Chile	FC	6.0	
12,741.8	FAX	GZZ44	Whitehall (Northwood) Naval R., Eng.	FC	5.0	British Navy/ Meteo
12,743.0	CW	NRV	COMMSTA Barrigada, Guam	FC	10.0	USCG/AMVER
	CW	NMC	COMMSTA San Francisco, CA	FC	10.0	USCG
12,744.0	CW	RCV	Moscow Naval Radio, USSR	FC	....	Russian Navy
	CW	URB	Leningrad Radio, USSR	FC	....	
12,745.0	CW	Y5M	Ruegen Radio, GDR	FC	10.0	
	CW	UNM2	Klaipeda Radio, Lithuanian SSR	FC	15.0	
12,745.5	CW	VWC	Calcutta Radio, India	FC	2.5	TIME SIGNALS/ Tfc/NOAA reports wx broadcast at 0630
12,747.0	USB	"KPA2"	unknown	FX	....	YL rpts (1980)
	USB	"MIW2"	unknown	FX	....	YL rpts (1981)
	FAX	JJC	Tokyo, Japan	FC	15.0	NTM
12,747.5	CW	CBV	Valparaiso Radiomaritima, Chile	FC	1.0	
12,748.0	CW	CLQ	Havana (Cojimar) Radio, Cuba	FC	5.0	
	CW	CLA3Ø	Havana (Cojimar) Radio, Cuba	FC	5.0	
	CW	IRM8	CIRM, Rome Radio, Italy	FC	15.0	MEDICO/AMVER
12,748.1	FAX	NIK	COMMSTA Boston, MA	FC	5.0	USCG [Seasonal ice pics at 1600]
12,750.0	CW	PWB	Belem Radio, Brazil	FC	2.0	Brazil Navy
	CW	PWF	Salvador Radio, Brazil	FC	2.0	" "
	CW	PWP	Florianopolis Radio, Brazil	FC	2.0	" "
	CW	DHJ59	Sengwarden Naval Radio, GFR	FC	....	GFR Navy/NATO
	CW	CWA	Cerrito Radio, Uruguay	FC	10.0	
	CW	NIK	COMMSTA Boston, MA	FC	5.0	USCG [Seasonal ice warnings at 1218]
12,752.0	CW	C6N	Nassau Radio, Bahamas	FC	0.5	
	CW	XFA	Acapulco Radio, Mexico	FC	0.5	
	CW	RIW	Khiva Naval Radio, Uzbek SSR	FC	....	Russian Navy
12,753.5	CW	OXZ62	Lyngby Radio, Denmark	FC	5.0	
12,754.0	CW	PWZ	Rio de Janeiro Naval Radio, Brazil	FC	10.0	Brazil Navy
12,754.5	CW	UJO5	Izmail Radio, Ukrainian SSR	FC	....	
12,755.0	CW	UTA	Tallinn Radio, Estonian SSR	FC	5.0	
12,755.5	CW	SAB6	Goteborg Radio, Sweden	FC	5.0	AMVER
12,756.0	CW	RKDF	Parnu Radio, Estonian SSR	FC	....	
12,757.0	CW	YQI2	Constanta Radio, Rumania	FC	1.0	
12,759.0	CW	JCE82	Naha Radio, Okinawa, Japan	FC	5.0	
12,760.0	CW	IDQ26	Rome Naval Radio, Italy	FC	1.0	Italian Navy [Reportedly broadcasts II Nx at 0300, 1800 & 2230]
	CW	IRM7	CIRM, Rome Radio, Italy	FC	....	MEDICO/AMVER
	CW	UFB	Odessa Radio, Ukrainian SSR	FC	15.0	
12,761.0	CW	VFC	Cambridge Bay CG Radio, NWT, Canada	FC	0.3	AMVER
	CW	XFS	Tampico Radio, Mexico	FC	0.5	
12,763.5	CW	LPD76	Gen. Pacheco Radio, Argentina	FC	3.0	NOAA reports wx broadcasts at 0118 & 1600
870907 1339	CW	DAM	Norddeich Radio, GFR	FC	10.0	TIME SIGNALS (2355-0006 spring & summer months only)
12,764.5	CW	EAD	Aranjuez Radio, Spain	FC	15.0	
12,765.0	CW	9GX	Tema Radio, Ghana	FC	10.0	
	CW	UCW4	Leningrad Radio, USSR	FC	....	
	CW	UKW3	Korsakov Sakalinsk Radio, USSR	FC	5.0	
12,768.0	CW	PCH5Ø	Scheveningen Radio, Netherlands	FC	10.0	AMVER
12,772.5	CW	ZSC38	Cape Town Radio, RSA	FC	10.0	
12,775.0	CW	XFU	Veracruz Radio, Mexico	FC	2.5	
12,775.5	CW	SAB6	Goteborg Radio, Sweden	FC	5.0	

kHz	Mode	Call	Location	Service	kW	Remarks
12,770.0	CW	HAR	Budapest Naval Radio, Hungary	FC	1.0	
12,779.5	CW	AQP4	Karachi Naval Radio, Pakistan	FC	....	Pakistan Navy/Reported wx broadcasts at 0130, 0530, 0930, 1330 1730 & 2130
12,780.0	CW	UUK	unknown, probably in USSR	FC	....	
12,780.5	CW	YUR5	Rijeka Radio, Yugoslavia	FC	6.0	
12,781.5	CW	OST5	Oostende Radio, Belgium	FC	15.0	
	CW	HKB	Barranquilla Radio, Colombia	FC	2.5	
	CW	CWA	Cerrito Radio, Uruguay	FC	10.0	
12,783.0	CW	†NMF	COMMSTA Boston, MA	FC	10.0	USCG
12,785.0	CW	UPW2	Liepaja Radio, Latvian SSR	FC	5.0	
	CW	UJY	Kaliningrad Radio, USSR	FC	5.0	
12,788.0	CW	JFA	Chuo Gyogyo (Matsudo) Radio, Japan	FC	3.0	
12,788.5	CW	GKD5	Portishead Radio, England	FC	15.0	
12,790.0	CW	GKG5	Portishead Radio, England	FC	15.0	
	CW	XFP	Campeche Radio, Mexico	FC	1.0	
	CW	XFB3	Ciudad del Carmen Radio, Mexico	FC	1.0	
	CW	XFS2	Ciudad Madero Radio, Mexico	FC	1.0	
	CW	XFF2	Coatzacoalcos Radio, Mexico	FC	0.5	
	CW	XFQ2	Frontera (Tabasco) Radio, Mexico	FC	0.5	
	CW	XFY2	Guaymas Radio, Mexico	FC	1.0	
	CW	XKF2	La Paz Baja California Radio, Mexico	FC	1.0	
	CW	XFM2	Manzanillo Radio, Mexico	FC	0.5	
	CW	XFL2	Mazatlan Radio, Mexico	FC	1.0	
	CW	XFU2	Veracruz Radio, Mexico	FC	1.0	
12,791.5	CW	GKH5	Portishead Radio, England	FC	15.0	
12,792.0	CW	CLA31	Havana (Cojimar) Radio, Cuba	FC	5.0	
12,793.0	CW	RNO	Moscow Radio, USSR	FC	5.0	SAAMC
12,795.0	CW	PWZ	Rio de Janeiro Naval Radio, Brazil	FC	10.0	Brazil Navy NOAA reports wx broadcasts at 0030, 0630 & 1730
	CW	UXN	Arkhangelsk Radio, USSR	FC	15.0	
12,797.0	CW	UDK2	Murmansk Radio, USSR	FC	15.0	
12,799.5	CW	XSZ	Darien Radio, PRC	FC	4.0	
	CW	XST	Qingdao Radio, PRC	FC	30.0	
	CW	PCH51	Scheveningen Radio, Netherlands	FC	10.0	AMVER
12,800.0	CW	HSA3	Bangkok Radio, Thailand	FC	10.0	
12,801.0	CW	EAD44	Aranjuez Radio, Spain	FC	15.0	
12,803.0	CW	UDK2	Murmansk Radio, USSR	FC	15.0	
12,804.0	CW	PWB	Belem Naval Radio, Brazil	FC	2.0	Brazil Navy
	CW	NPO	San Miguel (Capas) Naval Radio, Philippines	FC	15.0	USN/TIME SIGNALS
12,805.0	CW	UJQ7	Kiev Radio, Ukrainian SSR	FC	10.0	
12,806.0	CW	3VP8	La Skhirra Radio, Tunisia	FC	0.6	
12,807.5	CW	CTU3	Monsanto Naval R., Portugal	FC	5.0	Port. Navy
12,807.7	CW	GYA	Whitehall (London) Naval R., England	FC	10.0	British Navy
12,808.0	CW	3SB	Datong Naval Radio, PRC [Also operating on this frequency are 3SM, Qingdao; 3SQ, Ningbo & 3SV, Zhanjiang--all PRC Navy]	FC	15.0	PRC Navy
12,808.5	CW	VTG7	Bombay Naval Radio, India	FC	....	Indian Navy
	CW	KPH	San Francisco Radio, CA	FC	30.0	
12,810.0	CW	CTV28	Monsanto Naval Radio, Portugal	FC	....	Port. Navy
	CW	UJE	Moscow Radio, USSR	FC	....	
12,811.3	CW	HZY	Ra's Tannurah Radio, Saudi Arabia	FC	3.0	NOAA reports wx broadcast at 1630
12,811.5	CW	JFH	Hamajima Gyogyo Radio, Japan	FC	1.0	
12,815.0	CW	GKF2	Portishead Radio, England	FC	15.0	
	CW	URB2	Klaipeda Radio, Lithuanian SSR	FC	15.0	(4LI) RWWN
12,817.5	CW	XSU	Yantai Radio, PRC	FC	0.5	
	CW	9VG26	Singapore Radio, Singapore	FC	3.0	
	CW	YVG	La Guaira Radio, Venezuela	FC	0.3	
12,818.0	CW	OMP51	Prague Radio, Czechoslovakia	FC	1.0	
	CW	OMK	Komarno Radio, Czechoslovakia	FC	1.0	
12,822.0	CW	XSV	Tianjin Radio, PRC	FC	2.0	
	CW	GKA5	Portishead Radio, England	FC	5.0	NOAA reports wx broadcasts at 0930 & 2130/NAVAREA 0730 & 1730
12,822.2	CW	DZF	Manila (Bacoor) Radio, Philippines	FC	0.4	Reported wx broadcasts at 1600 & 2200

MARITIME MOBILE (con.)

kHz	Mode	Call	Location	Service	kW	Remarks
12,823.5	CW	CTP96	Oeiras Naval Radio, Portugal	FC	5.0	Port. Navy/NATO
12,824.0	CW	RIW	Khiva Naval Radio, Uzbek SSR	FC	....	Russian Navy
12,824.2	CW	GYU	Gibraltar Naval Radio, Gibraltar	FC	5.0	British Navy
12,825.0	CW	UFH	Petropavlovsk Radio, USSR	FC	20.0	
12,826.5	CW	CBM2	Magallanes (Punta Arenas) R., Chile	FC	4.0	
	CW	JCS	Choshi Radio, Japan	FC	3.0	
	CW	SPH62	Gydnia Radio, Poland	FC	10.0	[SPH6]
	CW	WNU34	Slidell Radio, LA	FC	3.0	NOAA reports wx broadcasts at 0430 & 1630
12,828.0	CW	DZF	Manila (Bacoor), Philippines	FC	0.4	
12,829.5	CW	XFM	Manzanillo Radio, Mexico	FC	0.5	
12,830.0	CW	RKLM	Arkhangelsk Radio, USSR	FC	15.0	
12,831.0	CW	VHP	COMMSTA Canberra, ACT, Australia	FC	5.0	RA Navy
	CW	VHM5	COMMSTA Darwin, NT, Australia	FC	5.0	RA Navy
	CW	3BA5	Mauritius (Bigara) Radio, Mauritius	FC	5.0	
	CW	FFP7	Fort de France Radio, Martinique	FC	1.0	NOAA reports warning broadcasts at 1145, 1500, 1730 & 2000
12,832.5	CW	DAF	Norddeich Radio, GFR	FC	10.0	
12,833.0	CW	SVF5	Athens Radio, Greece	FC	5.0	
12,834.5	CW	DZP	Manila (Novaliches) R., Philippines	FC	1.0	Reported wx broadcasts at 0300 & 1100
12,835.5	CW	GKB5	Portishead Radio, England	FC	5.0	
	CW	UBE2	Petropavlovsk Radio, USSR	FC	30.0	
12,839.3	CW	VTP6	Vishakhapatnam Naval Radio, India	FC	5.0	Indian Navy
12,839.6	CW	WPA	Port Arthur Radio, TX	FC	5.0	
12,840.0	CW	PPJ	Juncao Radio, Brazil	FC	1.0	
	CW	PPO	Olinda Radio, Brazil	FC	1.0	NOAA reports wx broadcasts at 0100, 0600, 1000, 1200, 1500, 1830, 2020 & 2100
	CW	LZW5	Varna Radio, Bulgaria	FC	5.0	
	CW	JMC5	Tokyo Radio, Japan	FC	4.0	NOAA reports wx broadcasts at 0318, 0648, 0918, 1148, 1518, 1748, 2118 & 2348
12,840.5	CW	PBC3	Goeree Naval Radio, Netherlands	FC	10.0	Dutch Navy
12,843.0	CW	HLO	Seoul Radio, South Korea	FC	7.0	
	CW	XFP	Chetumal Radio, Mexico	FC	0.5	
12,844.5	FAX	GZZ	Whitehall (Northwood) Naval R., Eng.	FC	6.0	British Navy/ Meteo
	CW	UON	Baku Radio, Azerbaijan SSR	FC	4.0	
	CW	KFS	San Francisco Radio, CA	FC	30.0	
12,846.0	CW	DZE	Mandaluyong Radio, Philippines	FC	0.4	Reported wx broadcasts at 0000 & 1000
12,847.5	CW	JFA	Chuo Gyogyo (Matsudo) Radio, Japan	FC	3.0	
12,849.0	CW	GZO5	Hong Kong Naval Radio, Hong Kong	FC	5.0	British Navy
	CW	ZSJ5	NAVCOMCEN (Silvermine) Cape R., RSA	FC	5.0	RSA Navy/AMVER
12,852.0	CW	DZR	Manila Radio, Philippines	FC	5.0	Reported wx broadcast at 0120
	CW	UKA	Vladivostok Radio, USSR	FC	25.0	Reported wx broadcast at 1000
12,853.3	CW	HKC	Buenaventura Radio, Colombia	FC	0.5	
12,853.5	CW	PCH52	Scheveningen Radio, Netherlands	FC	3.0	AMVER
12,854.0	CW	UFW	Vladivostok Radio, USSR	FC	25.0	
12,855.0	CW	UBF2	Odessa Radio, Ukrainian SSR	FC	5.0	
	CW	UKA	Vladivostok Radio, USSR	FC	25.0	
12,855.8	CW	4WD3	Hodeidah Port Radio, Yemen A.R.	FC	2.0	
12,856.0	CW	XSG7	Shanghai Radio, PRC	FC	10.0	
12,858.0	CW	FUV	Djibouti Naval Radio, Djibouti	FC	2.0	French Navy
	CW	JFS	Hachinohe Gyogyo Radio, Japan	FC	0.5	
	CW	JFU	Ishinomaki Gyogyo Radio, Japan	FC	1.0	
	CW	JFW	Iwaki Gyogyo Radio, Japan	FC	0.5	
	CW	JFF	Yaizu Gyogyo Radio, Japan	FC	....	
	CW	FUJ2	Noumea Naval Radio, New Caledonia	FC	20.0	French Navy
	CW	DZD	Manila (Antipolo) R., Philippines	FC	0.4	Reported wx broadcast at 0300
12,859.0	CW	OMC	Bratislava Radio, Czechoslovakia	FC	1.0	

kHz	Mode	Call	Location	Service	kW	Remarks
	CW	SVF5	Athens Radio, Greece	FC	10.0	
12,860.0	CW	4XO	Haifa Radio, Israel	FC	2.5	
	CW	UGG9	Belomorsk Radio, USSR	FC	....	
12,860.5	CW	Y5M	Ruegen Radio, GDR	FC	10.0	
12,860.8	CW	5RS	Tamatave Radio, Madagascar	FC	2.0	
12,862.0	CW	....	unknown	FC	....	Reported wx broadcast at 1150
12,862.5	CW	....	numerous, worldwide	FC	....	USN
12,864.7	CW	DZZ	Manila Radio, Philippines	FC	0.5	Reported wx broadcasts at 0230 & 0930
12,865.0	CW/RTTY	UJY	Kaliningrad Radio, USSR	FC	15.0	
	CW	UQK2	Riga Radio, Latvian SSR	FC	20.0	
12,865.5	CW	NRV	COMMSTA Barrigada, Guam	FC	10.0	USCG
12,866.0	CW	NMO	COMMSTA Honolulu, HI	FC	10.0	USCG/NERK/NUKO
12,867.0	CW	XYR8	Rangoon Radio, Burma	FC	5.0	
	CW	....	numerous, worldwide	FC	....	USN
12,868.8	CW	LZL5	Bourgas Radio, Bulgaria	FC	1.0	
12,869.0	CW	WNU54	Slidell Radio, LA	FC	3.0	
12,870.0	CW	LZS32	Sofia Naval Radio, Bulgaria	FC	....	
	CW	DZO	Manila (Bulacan) Radio, Philippines	FC	2.0	Reported wx broadcast at 1030
	CW	UKA	Vladivostok Radio, USSR	FC	25.0	
12,871.5	CW	XSG	Shanghai Radio, PRC	FC	10.0	TIME SIGNALS/Tfc NOAA reports wx broadcasts at 0306 & 0906
	CW	GKJ5	Portishead Radio, England	FC	15.0	
	CW	KUQ	Pago Pago, American Samoa	FC	2.5	
12,872.0	CW	SPE62	Szczecin Radio, Poland	FC	10.0	[aka SPB6]
	CW/RTTY	UKX	Nakhodka Radio, USSR	FC	5.0	
12,874.0	CW	VCS	Halifax CG Radio, NS, Canada	FC	1.0	AMVER
12,874.5	CW	HPN6Ø	Canal (Puerto Armuelles) R., Panama	FC	1.0	AMVER
12,875.0	CW	FUG3	La Regine (Castelnaudary), France	FC	....	French Navy
	CW	FUO	Toulon Naval Radio, France	FC	2.0	French Navy
12,876.0	CW	VAI	Vancouver CG Radio, BC, Canada	FC	5.0	AMVER/NOAA reports wx broadcasts at 0130, 0530 & 1730
	CW	LFJ	Rogaland Radio, Norway	FC	5.0	AMVER
	CW/RTTY	URL	Sevastopol Radio, Ukrainian SSR	FC	15.0	
12,877.0	CW	XSQ	Guangzhou Radio, PRC	FC	10.0	
12,877.5	CW	UJY	Kaliningrad Radio, USSR	FC	5.0	(4LC)
12,878.0	CW	JCU	Choshi Radio, Japan	FC	15.0	
	CW	CLQ	Havana (Cojimar) Radio, Cuba	FC	5.0	
12,880.0	CW	LZW	Varna Radio, Bulgaria	FC	15.0	
	CW	JFE	Naha Gyogyo Radio, Okinawa, Japan	FC	1.0	
12,880.5	CW	SAG6	Goteborg Radio, Sweden	FC	5.0	AMVER
12,882.0	CW	DZG	Manila (Las Pinas) Radio, Philippines	FC	5.0	NOAA reports wx broadcasts at 0320 & 1520
18,883.0	CW	UKA	Vladivostok Radio, USSR	FC	25.0	
12,885.0	CW	FUM	Papeete Naval Radio, Tahiti	FC	....	French Navy
	CW	9YL	North Post Radio, Trinidad	FC	2.0	NOAA reports warning broadcasts at 1200 & 1530
	CW	WLO	Mobile Radio, AL	FC	30.0	
12,886.5	CW/TOR	WLO	Mobile Radio, AL	FC	30.0	
12,887.5	CW	EAD44	Aranjuez Radio, Spain	FC	15.0	
12,889.0	CW	UJY	Kaliningrad Radio, USSR	FC	5.0	(4LC)
12,889.5	CW/RTTY	....	numerous	FC/MA/MS	...	USCG Command Net Frequency
	CW	NMO	COMMSTA Honolulu, HI	FC	10.0	USCG/AMVER
12,890.0	CW	XSJ	Zhanjiang Radio, PRC	FC	1.0	
	CW	IDQ6	Rome Naval Radio, Italy	FC	10.0	Italian Navy
	CW	UOP	Tuapse Radio, USSR	FC	5.0	
12,891.0	CW	DZK	Manila (Bulacan) Radio, Philippines	FC	2.5	
	CW	UFN	Novorossiysk Radio, USSR	FC	5.0	
12,894.0	CW	FUF	Fort de France Naval R., Martinique	FC	4.0	French Navy
	CW	6WW	Dakar Naval Radio, Senegal	FC	10.0	" "
12,895.0	CW	9KK8	Kuwait Radio, Kuwait	FC	5.0	NOAA reports wx broadcasts on request

kHz	Mode	Call	Location	Service	kW	Remarks
	CW	UDH	Riga Radio, Latvian SSR	FC	....	
	CW	UNM2	Klaipeda Radio, Lithuanian Radio	FC	....	
12,898.0	CW	XSG	Shanghai Radio, PRC	FC	2.0	
12,898.5	CW	DAN	Norddeich Radio, GFR	FC	5.0	
12,900.0	CW	PWZ	Rio de Janeiro Naval Radio, Brazil	FC	10.0	Brazil Navy
	CW	DXF	Davao Radio, Philippines	FC	0.4	
12,901.5	CW	URD	Leningrad Radio, USSR	FC	15.0	
12,902.0	CW	HLJ	Seoul Radio, South Korea	FC	7.0	
12,903.0	FAX	NGR	Kato Souli Naval Radio, Greece	FC	15.0	USN/Meteo (0800-2000)
12,905.0	CW	UMV	Murmansk Radio, USSR	FC	15.0	
12,905.5	FAX	XSC	Fangshan 1 Radio, PRC	FC	10.0	Meteo
12,906.0	CW	DZJ	Manila (Bulacan) Radio, Philippines	FC	2.5	NOAA reports wx broadcasts at 0300 & 1000
12,907.5	CW	VHP5	COMMSTA Canberra, ACT, Australia	FC	15.0	RA Navy NOAA reports wx broadcast at 1730
	CW	VIX	Master Station Canberra, Australia	FC	15.0	
	CW	KLB	Seattle Radio, WA	FC	3.5	
	CW	YUR6	Rijeka Radio, Yugoslavia	FC	6.0	
12,910.0	CW	UAT	Moscow Radio, USSR	FC	15.0	
12,911.0	CW	CLS	Havana (Industria Pesquera) R., Cuba	FC	5.0	
12,912.0	CW	XSH	Basuo Radio, PRC	FC	1.0	
	CW	UHK	Batumi Radio, USSR	FC	5.0	
12,912.6	CW	FFL6	St. Lys Radio, France	FC	10.0	
12,915.0	CW	URD2	Leningrad Radio, USSR	FC	....	
12,916.5	CW	OXZ6	Lyngby Radio, Denmark	FC	10.0	
	CW	HLF	Seoul Radio, South Korea	FC	10.0	
	CW	KLB	Seattle Radio, WA	FC	3.5	
12,919.0	CW	UDN	Jdanov Radio, Ukrainian SSR	FC	5.0	
12,920.0	CW	UFN	Novorossiysk Radio, USSR	FC	....	
12.921.0	CW	GYA	Whitehall (London) Naval R., England	FC	10.0	British Navy
12,923.1	CW	HLW2	Seoul Radio, South Korea	FC	15.0	
12,925.0	CW	9KK20	Kuwait Radio, Kuwait	FC	5.0	
	CW	UDK2	Murmansk Radio, USSR	FC	15.0	
12,925.5	CW	WCC	Chatham Radio, MA	FC	30.0	
12,927.0	CW	4PB	Colombo Radio, Sri Lanka	FC	1.0	
12,928.0	CW	SPH63	Gydnia Radio, Poland	FC	10.0	[aka SPI6]
12,932.5	CW/RTTY	EBA	Madrid Radionaval, Spain	FC	1.0	Spanish Navy
12,933.0	CW/RTTY	UKA	Vladivostok Radio, USSR	FC	15.0	
	CW	KOK	Los Angeles Radio, CA	FC	15.0	NOAA reports wx broadcasts at 0450 & 1650
12,934.5	CW	EDZ5	Aranjuez Radio, Spain	FC	3.5	
12,935.0	CW	HLG	Seoul Radio, South Korea	FC	10.0	
12,935.5	CW	5OZ23	Port Harcourt Radio, Nigeria	FC	1.0	
12,936.0	CW	DZN	Manila (Navotas) Radio, Philippines	FC	3.0	NOAA reports wx broadcasts at 0200 & 1400
12,939.0	CW	SPE61	Szczecin Radio, Poland	FC	10.0	
	CW/RTTY	....	numerous, worldwide	FC	....	USN
12,940.0	CW	LZW5	Varna Radio, Bulgaria	FC	15.0	
12,942.0	CW	SVD5	Athens Radio, Greece	FC	5.0	
	CW	JNA	Tokyo Naval Radio, Japan	FC	....	JSDA/Reported NAVAREA at 1200
	CW	DXR	Manila Radio, Philippines	FC	....	Reported wx broadcasts at 0300 & 1000
	CW	UFD9	Arkhangelsk Radio, USSR	FC	5.0	(RKLM)
12,943.5	CW	CUG27	Sao Miguel Radio, Azores	FC	1.0	
	CW	CUB	Madeira Radio, Madeira Islands	FC	1.0	
	CW	CUL2Ø	Lisbon Radio, Portugal	FC	15.0	
	CW	ZLP5	Irirangi Naval Radio, New Zealand	FC	10.0	RNZ Navy
12,943.6	CW	CUL7	Monsanto Naval Radio, Portugal	FC	10.0	Port. Navy
12,947.0	CW	UFB	Odessa Radio, Ukrainian SSR	FC	15.0	
12,947.3	CW/RTTY	ZRH5	Cape (Fisantekraal) Naval R., RSA	FC	15.0	RSA Navy
12,948.0	CW	JJF	Tokyo Naval Radio, Japan	FC	4.0	JSDA
12,948.5	CW	UFB	Odessa Radio, Ukrainian SSR	FC	5.0	(4KA)
12,949.0	CW	9VT5	Singapore Radio, Singapore	FC	1.0	
12,952.5	CW	VIS5	Sydney Radio, NSW, Australia	FC	2.0	AMVER

kHz	Mode	Call	Location	Service	kW	Remarks
12,952.9	CW	WMH	Baltimore Radio, MD	FC	10.0	NOAA reports wx broadcasts at 0130, 1330, 1600 & 1930
12,953.0	CW	DFM95	Frankfurt (Usingen) Radio, GFR	FC	15.0	Reported MAR-PRESS GG nx broadcasts at 0118 & 1618
12,953.5	CW	XVG28	Shanghai Radio, PRC	FC	10.0	Reported wx broadcasts at 0000 & every 3rd hour
12,954.9	CW	CLS	Havana (Industria Pesquera) R., Cuba	FC	5.0	
12,955.0	CW	UFL	Vladivostok Radio, USSR	FC	15.0	
12,958.5	CW	PPO	Olinda Radio, Brazil	FC	1.0	
12,960.0	CW	CCV6	Valparaiso Naval Radio, Chile	FC	....	Chilean Navy NOAA reports wx broadcasts at 0110, 1330 & 1430
	CW	ULY4	Aleksandrovsk Radio, USSR	FC	....	
12,961.5	CW	LFI	Rogaland Radio, Norway	FC	5.0	AMVER
	CW	WCC	Chatham Radio, MA	FC	30.0	
12,962.9	CW	HAR	Budapest Naval Radio, Hungary	FC	2.5	
12,963.0	CW	UJO5	Izmail Radio, Ukrainian SSR	FC	....	
12,963.5	CW	ZHH	St. Helena Radio, St. Helena Is.	FC	1.0	
12,966.0	CW	PWZ	Rio de Janeiro Naval Radio, Brazil	FC	10.0	Brazil Navy
	CW	VWB	Bombay Radio, India	FC	2.5	
	CW	PCH8	Scheveningen Radio, Netherlands	FC	15.0	
	CW	NPG	San Francisco Naval Radio, CA	FC	30.0	USN/TIME SIGNALS
	CW	UFB	Odessa Radio, Ukrainian SSR	FC	15.0	
12,967.0	CW	UJE	Moscow Naval Radio, USSR	FC	....	Russian Navy
12,969.0	CW	XSV	Tianjin Radio, PRC	FC	10.0	
12,970.0	CW	UJY	Kaliningrad Radio, USSR	FC	15.0	
12,970.0	CW	UAH	Tallinn Radio, Estonian SSR	FC	5.0	
	CW/RTTY	URL	Sevastopol Radio, Ukrainian SSR	FC	15.0	
12,970.5	CW	SUH4	Alexandria Radio, Egypt	FC	3.0	
	CW	PKB	Belwana Radio, Indonesia	FC	1.0	
	CW	PKI	Jakarta Radio, Indonesia	FC	1.0	
12,970.7	CW	WOE	Lantana Radio, FL	FC	5.0	
12,972.0	CW	HAR	Budapest Naval Radio, Hungary	FC	2.5	
	CW	DZH	Manila Radio, Philippines	FC	0.3	Reported wx broadcasts at 1100 & 2320
12,973.0	CW	UJY	Kaliningrad Radio, USSR	FC	15.0	
12,975.0	CW	IQX	Trieste P.T. Radio, Italy	FC	3.0	
	CW/RTTY	....	numerous, worldwide	FC	....	USN
12,977.0	CW/RTTY	URL	Sevastopol Radio, Ukrainian SSR	FC	5.0	
12,978.0	CW	ICB	Genoa P.T. Radio, Italy	FC	5.0	
12,979.5	CW	PPL	Belem Radio, Brazil	FC	1.0	NOAA reports wx broadcasts at 0130, 0800 & 1930
	CW	VIS49	Sydney Radio, NSW, Australia	FC	5.0	
12,980.0	CW	PKB2Ø	Lhok Seumawe Radio, Indonesia	FC	1.0	
	CW	PKI2	Jakarta Radio, Indonesia	FC	1.0	
	CW	PKN7	Bontang Radio, Indonesial	FC	1.0	
	CW	PKR6	Cilacap Radio, Indonesia	FC	0.4	
	CW	UAT	Moscow Radio, USSR	FC	15.0	
	CW	UPB	Providenia Bukhta Radio, USSR	FC	5.0	
12,983.0	CW	USW2	Rostov Radio, USSR	FC	5.0	
12,984.0	CW	DZM	Manila (Bulacan) Radio, Philippines	FC	10.0	
12,984.2	CW	4XZ	Haifa Naval Radio, Israel	FC	2.0	Israel Navy
12,985.0	CW	OSN12	Oostende Naval Radio, Belgium	FC	....	Belgium Navy
12,987.2	CW	GYU	Gibraltar Naval Radio, Gibraltar	FC	....	British Navy
12,988.5	CW	3BM5	Mauritius (Bigara) Radio, Mauritius	FC	4.0	NOAA reports wx broadcasts at 0130, 0430, 0900 & 1600/cyclones 1330
	CW	LPD88	Gen. Pacheco Radio, Argentina	FC	15.0	AMVER
12,990.0	CW	DZJ	Manila (Bulacan) Radio, Philippines	FC	2.5	
	CW	UJQ	Kiev Radio, Ukrainian SSR	FC	5.0	
	CW	UKK3	Nakhodka Radio, USSR	FC	15.0	
	CW	UOP	Tuapse Radio, USSR	FC	15.0	
12,993.0	CW	D4A7	S. Vicente Radio, Cape Verde	FC	3.0	
	CW	UKG	Kerch Radio, Ukrainian SSR	FC	5.0	
	CW	URL	Sevastopol Radio, Ukrainian SSR	FC	5.0	
	CW	KOK	Los Angeles Radio, CA	FC	10.0	NOAA reports wx broadcasts at 0450 & 1650

MARITIME MOBILE (con.)

kHz	Mode	Call	Location	Service	kW	Remarks
12,994.0	CW	VIP4	Perth Radio, WA, Australia	FC	10.0	AMVER/NOAA reports wx broadcasts at 0100 & 1300
	CW	CTH55	Horta Naval Radio, Azores	FC	5.0	(P13C) Port. Navy/Reported wx broadcasts at 0930 & 1030
12,995.0	CW	ROT	Moscow Naval Radio, USSR	FC	25.0	Russian Navy
12,996.0	CW	IAR33	Rome P.T. Radio, Italy	FC	7.0	Reported nx broadcast at 0600
12,997.0	CW	UFB	Odessa Radio, Ukrainian SSR	FC	15.0	
12,997.5	CW	WSL	Amagansett Radio, NY	FC	10.0	NOAA reports wx broadcasts at 0500, 1100, 1700 & 2300
13,000.0	CW	LZW5	Varna Radio, Bulgaria	FC	15.0	
	CW	UBE2	Petropavlovsk Radio, USSR	FC	15.0	NOAA reports wx broadcast at 0500
	CW	UAH	Tallinn Radio, Estonian SSR	FC	5.0	
13,001.9	CW	CTU2	Monsanto Naval Radio, Portugal	FC	5.0	Port. Navy NOAA reports wx broadcasts at 0800 & 2000
13,002.0	CW	LPD32	Gen. Pacheco Radio, Argentina	FC	3.0	
	CW	KPH	San Franisco Radio, CA	FC	30.0	
13,003.0	CW	CTV2	Monsanto Naval Radio, Portugal	FC	10.0	Port. Navy
13,004.0	CW	UDK2	Murmansk Radio, USSR	FC	....	
13,005.5	CW	HLW	Seoul Radio, South Korea	FC	3.5	
13,006.5	CW	GKK5	Portishead Radio, England	FC	15.0	
13,007.0	CW	OMP5	Prague Radio, Czechoslovakia	FC	1.0	
13,008.0	CW	JOR	Nagasaki Radio, Japan	FC	15.0	
	CW	DZI	Manila (Bacoor) Radio, Philippines	FC	1.0	
13,010.0	CW	UQA4	Murmansk Radio, USSR	FC	....	
	CW	UFJ	Rostov Radio, USSR	FC	5.0	
13,010.8	CW	AQP4	Karachi Naval Radio, Pakistan	FC	7.0	Pakistan Navy
13,011.0	CW	IAR23	Rome P.T. Radio, Italy	FC	7.0	NOAA reports wx broadcasts at 0700 & 1900
	CW	WNU44	Slidell Radio, LA	FC	3.0	
13,015.5	CW	IAR3	Rome P.T. Radio, Italy	FC	7.0	
	CW	JGC	Yokohama Radio, Japan	FC	1.0	
13,019.9	CW	GKC5	Portishead Radio, England	FC	15.0	
13,020.0	CW	UBE	Petropavlovsk Radio, USSR	FC	....	
13,020.4	CW	VPS6Ø	Cape d'Aguilar Radio, Hong Kong	FC	3.0	TIME SIGNALS/Tfc NOAA reports wx broadcasts at 0118, 0430, 0718 & 1318
13,022.0	CW	SPE63	Szczecin Radio, Poland	FC	10.0	[aka SPL6]
13,023.0	CW	D3E3	Luanda Radio, Angola	FC	3.0	
13,023.7	CW	HEB	Berne Radio, Switzerland	FC	10.0	
13,024.0	CW	S3E	Khulna Radio, Bangladesh	FC	10.0	
13,024.5	CW	ASK	Karachi Radio, Pakistan	FC	3.5	NOAA reports wx broadcast at 0830
13,024.9	CW	WSL	Amagansett Radio, NY	FC	10.0	NOAA reports wx broadcasts at 1100, 1700 & 2300
13,025.0	CW	HLX	Seoul Radio, South Korea	FC	10.0	
13,027.5	CW	DAL	Norddeich Radio, GFR	FC	15.0	
13,028.0	CW	ZSD46	Durban Radio, RSA	FC	3.5	
13,029.0	CW	SVB5	Athens Radio, Greece	FC	7.0	
	CW	UQB	Kholmsk Radio, USSR	FC	10.0	
13,030.0	CW	URD	Leningrad Radio, USSR	FC	15.0	
13,031.0	CW	VRN6Ø	Cape d'Aguilar Radio, Hong Kong	FC	3.0	NOAA reports wx broadcasts at 0430 & 0718
13,031.2	CW	FUF	Fort de France Naval R., Martinique	FC	10.0	French Navy
13,032.0	CW	DUN	Manila Naval Radio, Philippines	FC	0.5	Philippine Navy
13,033.5	CW	WCC	Chatham Radio, MA	FC	10.0	
13,036.0	CW	OMC	Bratislava Radio, Czechoslovakia	FC	1.0	
13,038.0	CW	OXZ61	Lyngby Radio, Denmark	FC	10.0	
	CW	KLC	Galveston Radio, TX	FC	3.8	NOAA reports wx broadcast at 1730
	CW	UFN	Novorossiysk Radio, USSR	FC	15.0	
13,041.0	CW	UFB	Odessa Radio, Ukrainian SSR	FC	5.0	
13,042.0	CW	C5G	Banjul Radio, Gambia	FC	1.0	

kHz	Mode	Call	Location	Service	kW	Remarks
	CW	C9C7	Maputo Radionaval, Mozambique	FC	3.0	TIME SIGNALS
	CW	P2M	Port Moresby Radio, Papua New Guinea	FC	1.0	Reported wx broadcasts at 0100 & 0900
	CW	P2R	Rabaul Radio, Papua New Guinea	FC	1.0	
	CW	CTV3	Monsanto Naval Radio, Portugal	FC	2.0	Port. Navy
	CW	9LL	Freetown Radio, Sierra Leone	FC	1.5	
	CW	XVS9	Ho Chi Minh-Ville Radio, Vietnam	FC	1.0	
13,042.5	CW	PJC	Curacao (Willemstad) R., Curacao	FC	2.5	
13,042.6	CW	FUV	Djibouti Naval Radio, Djibouti	FC	20.0	French Navy
13,044.0	CW	VPS61	Cape d'Aguilar Radio, Hong Kong	FC	3.5	
13,044.9	CW	PZN4	Paramaribo Radio, Suriname	FC	1.0	
13,045.0	CW	ROT	Moscow Naval Radio, USSR	FC	15.0	Russian Navy
13,046.0	CW	RIT	unknown, probably in the USSR	FC	....	
13,046.9	CW	SVA5	Athens Radio, Greece	FC	7.0	
13,047.0	CW	D4AB	Sao Vicente Radio, Cape Verde	FC	3.0	
	CW	UXN	Arkhangelsk Radio, USSR	FC	15.0	
	CW	UKX	Nakhodka Radio, USSR	FC	5.0	
	CW	UKA	Vladivostok Radio, USSR	FC	30.0	
13,050.0	CW	UDK2	Murmansk Radio, USSR	FC	5.0	
13,051.5	CW	4XO	Haifa Radio, Israel	FC	2.5	
	CW	WPD	Tampa Radio, FL	FC	5.0	NOAA reports wx broadcasts at 1420 & 2320
13,052.4	CW	EBA	Madrid Radionaval, Spain	FC	1.0	(EBCQ) Spanish Navy/Reported NAVAREA 0918, 1218 & 1618
13,054.0	CW	XSQ	Guangzhou Radio, PRC	FC	1.0	
	CW	SVZ	Athens Radio, Greece	FC	....	
	CW	JDC	Choshi Radio, Japan	FC	15.0	
	CW/RTTY	UJY7	Kaliningrad Radio, USSR	FC	5.0	
13,055.0	CW	UJQ7	Kiev Radio, Ukrainian SSR	FC	15.0	
13,056.0	CW	S3D	Chittagong Radio, Bangladesh	FC	1.0	
	CW	C9L6	Mozambique Radio, Mozambique	FC	0.5	
	CW	EDG4	Aranjuez Radio, Spain	FC	15.0	
13,057.0	CW	OMC	Bratislava Radio, Czechoslovakia	FC	1.0	
	CW	OMK	Komarno Radio, Czechoslovakia	FC	1.0	
13,059.0	CW	EBA	Madrid Radionaval Spain	FC	20.0	(EBCQ) Spanish Navy
13,060.0	CW	UFJ	Rostov Radio, USSR	FC	....	
13,060.1	CW	....	unknown	FC	....	METAR/1340
13,060.5	CW	7OA	Aden Radio, Democratic Yemen	FC	2.0	
	CW	TUA5	Abidjan Radio, Ivory Coast	FC	4.0	
13,062.0	CW	CLA32	Havana (Cojimar) Radio, Cuba	FC	5.0	
13,062.5	CW	Y5M	Ruegen Radio, GDR	FC	15.0	
13,063.0	CW	JDB	Nagasaki Radio, Japan	FC	15.0	
13,064.0	CW	RIW	Khiva Naval Radio, Uzbek SSR	FC	....	Russian Navy
13,065.0	CW	VIY	Christmas Island, Australia	FC	0.5	
13,065.0	CW	5ZF3	Mombasa Radio, Kenya	FC	1.5	
	CW	6YI	Kingston Radio, Jamaica	FC	1.0	
	CW	EAD4	Aranjuez Radio, Spain	FC	15.0	
	CW	RKDF	Parnu Radio, Estonian SSR	FC	....	
13,065.5	CW	5OW13	Lagos Radio, Nigeria	FC	1.0	
13.065.9	CW	UDH	Riga Radio, Latvian SSR	FC	5.0	
13,067.0	CW	OST52	Oostende Radio, Belgium	FC	15.0	
	CW	UKX	Nakhodka Radio, USSR	FC	5.0	
13,069.5	CW	TJC9	Douala Radio, Cameroon	FC	0.8	
	CW	EQI	Abbas Radio, Iran	FC	1.0	
	CW	EQZ	Abadan Radio, Iran	FC	0.5	
	CW	†EQK	Khoramshahr Radio, Iran	FC	1.0	
	CW	TFA13	Reykjavik Radio, Iceland	FC	1.5	
	CW	JOS	Nagasaki Radio, Japan	FC	3.0	
13,070.0	CW	TRP8	Banjul Shell Radio, Gambia	FC	....	
	CW	UDH	Riga Radio, Latvian SSR	FC	5.0	
13,071.5	CW/TOR	9VG8Ø	Singapore Radio, Singapore	FC	10.0	
13,072.0	CW	GKE5	Portishead Radio, England	FC	5.0	
13,073.2	CW/TOR	ZUD81	Olifantsfonten Radio, RSA	FC	5.0	

kHz	Mode	Call	Location	Service	kW	Remarks
13,073.5	CW/TOR	WLO	Mobile Radio, AL	FC	15.0	
13,074.0	CW/TOR	VIP24	Perth Radio, WA, Australia	FC	1.0	
	CW	9GA	Takoradi Ragio, Ghana	FC	10.0	
13,076.0	CW	VIP4Ø	Perth Radio, WA, Australia	FX	1.0	
13,078.1	CW/TOR	VIS67	Sydney Radio, NSW, Australia	FC	5.0	
13,078.5	CW/TOR	WLO	Mobile Radio, AL	FC	15.0	
13,080.0	CW	HEC13	Berne Radio, Switzerland	FC	10.0	
13,082.5	CW/TOR	SVT5	Athens Radio, Greece	FC	5.0	
13,083.5	CW/TOR	WLO	Mobile Radio, AL	FC	15.0	
13,085.0	CW	GKP5	Portishead Radio, England	FC	15.0	
13,087.2	CW	HKB	Barranquilla Radio, Colombia	FC	2.5	
		HIA	Domingo Piloto Radio, Dominican Rep.	FC	0.5	
13,097.0	CW/TOR	LJG4	Rogaland Radio, Norway	FC	5.0	
13,097.9	CW/TOR	WLO	Mobile Radio, AL	FC	15.0	
13,100.0	CW	TIM	Limon Radio, Costa Rica	FC	3.0	
13,101.0	CW	ODR4	Beirut Radio, Lebanon	FC	1.0	

13,100.8 - 13,196.9 kHz -- MARITIME COAST USB DUPLEX (Worldwide)

kHz	Mode	Call	Location	Service	kW	Remarks
13,100.8	USB	GTK51	Portishead Radio, England	FC	10.0	CH#1201 (12,330.0)
	USB	....	Moscow Radio, USSR	FC	10.0	" "
	USB	KMI	San Francisco (Dixon) Radio, CA NOAA reports wx broadcasts at 0000, 0600 & 1500	FC	10.0	" "
13,103.9	USB	GTK52	Portishead Radio, England	FC	10.0	CH#1202 (12,333.1)
	USB	HEB13	Berne Radio, Switzerland	FC	10.0	" "
	USB	WFN	Jeffersonville, IN	FC/MS	0.8	Inland Waterway
	USB	WGK	St. Louis, MO	FC/MS	0.8	Inland Waterway
	USB	WCM	Pittsburgh, PA	FC/MS	0.8	Inland Waterway
13,107.0	USB	VIS	Sydney Radio, NSW, Australia	FC	5.0	CH#1203 (12,336.2)
	USB	....	Goteborg Radio, Sweden	FC	5.0	" "
	USB	WOO	New York (Ocean Gate, NJ) Radio, NY [NOAA reports wx broadcasts at 0100, 1300 & 1900]	FC	10.0	" "
13,110.1	USB	LFL31	Rogaland Radio, Norway	FC	10.0	CH#1204 (12,339.3)
	A3H	....	Leningrad Radio, USSR	FC	10.0	" "
	USB	....	numerous, worldwide	FC	....	USN
13,113.2	USB	....	Rome Radio 3	FC	7.0	CH#1205 (12,342.4)
	USB	....	Riga Radio, Latvian SSR	FC	10.0	" "
	USB	NRV	COMMSTA Barrigada, Guam [Wx broadcasts at 0130 & 0730]	FC	10.0	USCG CH#1205 AMVER (12,342.4)
	USB	NMO	COMMSTA Honolulu, HI [Wx broadcasts at 1745 & 2345]	FC	10.0	USCG CH#1205 (12,342.4)
	USB	NMC	COMMSTA San Francisco, CA [Wx broadcasts at 0430, 1030, 1630 & 2230]	FC	10.0	USCG CH#1205 (12,342.4)
	USB	NMG	COMMSTA New Orleans, LA	FC	10.0	USCG CH#1205 AMVER (12,342.4)
	USB	NMN	COMMSTA Portsmouth, VA [Wx broadcasts at 1130, 1600, 1730, 2200 & 2230]	FC	10.0	USCG CH#1205 (12,342.4)
	USB		This frequency also used by USCG Coast Stations: NMA, Miami, FL; NMF, Boston, MA; NMR, San Juan, PR & NOJ, Kodiak, AK; plus ships and aircraft			
13,116.3	USB	WOM	Miami Radio, FL [NOAA reports wx broadcast at 1230]	FC	10.0	CH#1206 (12,345.5)
13,119.4	USB	CUG	Sao Miguel Radio, Azores	FC	5.0	CH#1207 (12,348.6)
	USB	OSU51	Oostende Radio, Belgium	FC	10.0	CH#1207
	USB	4XO	Haifa Radio, Israel	FC	10.0	" "
	USB	JBO	Tokyo Radio, Japan	FC	10.0	" "
	USB	PCG52	Scheveningen Radio, Netherlands	FC	10.0	" "

MARITIME COAST USB DUPLEX (con.)

kHz	Mode	Call	Location	Service	kW	Remarks
	USB	....	Curacao (Willemstad) R., Curacao	FC	1.0	" "
	A3H	....	Murmansk Radio, USSR	FC	10.0	" "
13,122.5	USB	5BA54	Nicosia Radio, Cyprus	FC	10.0	CH#1208 (12,351.7)
	USB	EHY	Pozuelo del Rey Radio, Spain	FC	10.0	" "
	USB	WOM	Miami Radio, FL	FC	10.0	" "
			[NOAA reports wx broadcasts at 1330 & 2330]			
13,125.6	AM/USB	....	Rome Radio 3, Italy	FC	7.0	CH#1209 (12,354.8)
	USB	XFU	Veracruz Radio, Mexico	FC	1.0	" "
	USB	ZLW	Wellington Radio, New Zealand	FC	10.0	" "
	USB	WOM	Miami Radio, FL	FC	10.0	" "
			[NOAA reports wx broadcast at 2230]			
13,128.7	USB	....	Lyngby Radio, Denmark	FC	10.0	CH#1210 (12,357.9)
	USB	EHY	Pozuelo del Rey Radio, Spain	FC	10.0	" "
	USB	WOO	New York (Ocean Gate, NJ) Radio, NY	FC	10.0	" "
			[NOAA reports wx broadcasts at 0100, 1300 & 1900]			
13,131.8	USB	LFL34	Rogaland Radio, Norway	FC	10.0	CH#1211 (12,361.0)
	USB	....	Goteborg Radio, Sweden	FC	5.0	" "
	USB	WOO	New York (Ocean Gate, NJ) Radio, NY	FC	10.0	" "
			[NOAA reports wx broadcasts at 0200, 1400 & 2000]			
13,134.9	USB	SVN5	Athens Radio, Greece	FC	5.0	CH#1212 (12,364.1)
	USB	KQM	Kahuku Radio, HI	FC	1.0	" "
	USB	JBO	Tokyo Radio, Japan	FC	10.0	CH#1212 (12,364.1)
	USB	WLO	Mobile Radio, AL	FC	10.0	" "
13,138.0	USB	OSU53	Oostende Radio, Belgium	FC	10.0	CH#1213 (12,367.2)
	USB	VCS	Halifax CG Radio, NS, Canada	FC	5.0	" "
			[NOAA reports wx broadcasts at 1625 & 2225]			
	USB	PCG51	Scheveningen Radio, Netherlands	FC	10.0	CH#1213/AMVER (12,367.2)
13,141.1	USB	PPR	Rio de Janeiro Radio, Brazil	FC	1.0	CH#1214 (12,370.3)
	USB	KVH	Atlantic Marine Center, Norfolk, VA	FC	1.0	NOAA/CH#1214 (12,370.3)
13,144.2	USB	4XO	Haifa Radio, Israel	FC	10.0	CH#1215 (12,373.4)
	USB	....	Goteborg Radio, Sweden	FC	5.0	" "
	USB	WOM	Miami Radio, FL	FC	10.0	" "
13,147.3	USB	OHG2	Helsinki Radio, Finland	FC	10.0	CH#1216 (12,376.5)
	USB	SPC61	Gdynia Radio, Poland	FC	10.0	" "
	USB	9VG66	Singapore Radio, Singapore	FC	10.0	" "
	USB	WWD	La Jolla (Scripps) Radio, CA	FC	1.0	
			[NOAA reports wx forecasts on request]			
13,150.0	USB	"KPA2"	unknown	FX	....	YL rpts (1981)
13,150.4	USB	LFL37	Rogaland Radio, Norway	FC	10.0	CH#1217 (12,379.6)
	USB		This frequency also used by USCG Coast Stations: NMA, Miami, FL; NMC, San Francisco, CA; NMF, Boston, MA & NRV, Barrigada, Guam) [NMA may use this frequency for special law enforcement activities when required]			
13,153.5	USB	....	Rome Radio 3, Italy	FC	7.0	CH#1218 (12,382.7)
	USB	JBO	Tokyo Radio, Japan	FC	2.5	" "
13,156.6	USB	OSU57	Oostende Radio, Belgium	FC	10.0	CH#1219 (12,385.8)
	USB	PCG53	Scheveningen Radio, Netherlands	FC	10.0	" "
	USB	....	Goteborg Radio, Sweden	FC	5.0	" "
	USB		This frequency also used by USCG Coast Stations: NMA, Miami, FL; NMG, New Orleans, LA & NOJ, Kodiak, AK			

MARITIME COAST USB DUPLEX (con.)

kHz	Mode	Call	Location	Service	kW	Remarks
13,159.7	USB	LPL4	Gen. Pacheco Radio, Argentina	FC	3.0	CH#1220 (12,388.9)
	USB	....	Bermuda (Hamilton) Radio, Bermuda	FC	1.0	" "
	USB	SVN5	Athens Radio, Greece	FC	5.0	" "
13,162.8	AM/USB	....	INTERNATIONAL COAST STATION CALLING FREQUENCY			CH#1221 (12,392.0)
	USB	LPL	Gen. Pacheco Radio, Argentina	FC	3.0	" "
	USB	....	Rogaland Radio, Norway	FC	10.0	" "
13,164.0	CW	MI3T	unknown	FX	....	Coded Tfc
13,165.9	USB	FFL64	St. Lys Radio, France	FC	10.0	CH#1222 (12,395.1)
	USB	LFL4Ø	Rogaland Radio, Norway	FC	10.0	" "
13,169.0	USB	....	Lyngby Radio, Denmark	FC	10.0	CH#1223 (12,398.2)
	USB	....	Goteborg Radio, Sweden	FC	5.0	" "
13,172.1	USB	DAJ	Norddeich Radio, GFR	FC	10.0	CH#1224 (12,401.3)
	USB	GKV54	Portishead Radio, England	FC	10.0	" "
	USB	ZSD39	Durban Radio, RSA	FC	10.0	" "
	USB	....	numerous, worldwide	FC	....	USN
	USB	....	Rijeka Radio, Yugoslavia	FC	7.5	CH#1225 (12.401.3)
13,175.2	USB	LFL42	Rogaland Radio, Norway	FC	10.0	CH#1225 (12,404.4)
	USB	EHY	Pozuelo del Rey Radio, Spain	FC	10.0	" "
	USB	WLO	Mobile Radio, AL	FC	10.0	" "
13,178.3	USB	VIP	Perth Radio, WA, Australia	FC	5.0	CH#1226 (12,407.5)
	USB	....	Lyngby Radio, Denmark	FC	10.0	" "
	USB	FFL61	St. Lys Radio, France	FC	10.0	" "
	USB	WLO	Mobile Radio, AL [NOAA reports wx broadcasts at 1300, 1800 & 0000]	FC	10.0	" "
13,181.0	USB	....	numerous, off USA coastline "OVERWORK" broadcasts)	FC/MS	....	USN (Used for [Alt: 6720.0]
13,181.4	USB	VID	Darwin Radio, NT, Australia	FC	1.0	CH#1227 (12,410.6)
	USB	OHG2	Helsinki Radio, Finland	FC	10.0	" "
	USB	SPO61	Szczecin Radio, Poland	FC	10.0	" "
	AM/USB	EHY	Pozuelo del Rey Radio, Spain	FC	10.0	" "
	USB	HEB23	Berne Radio, Switzerland	FC	10.0	CH#1227 (12,410.6)
13,182.8	USB	....	numerous, worldwide	FC	....	USN
13,184.5	USB	LFL44	Rogaland Radio, Norway	FC	10.0	CH#1228 (12,413.7)
	USB	WOO	New York (Ocean Gate, NJ) Radio, NY	FC	10.0	" "
13,187.6	USB	VIB	Brisbane Radio, Qld., Australia	FC	1.0	CH#1229 (12,416.8)
	USB	FFL62	St. Lys Radio, France	FC	10.0	" "
	USB	KMI	San Francisco (Dixon) Radio, CA [NOAA reports wx broadcasts at 0000, 0600 & 1500]	FC	10.0	" "
13,189.0	CW	NAM	Norfolk Naval Radio, VA [Reported wx broadcast at 1230]	FC	....	USN/NUKO
13,190.7	USB	HEB33	Berne Radio, Switzerland	FC	10.0	CH#1230 (12,419.9)
	USB	WOO	New York (Ocean Gate, NJ) Radio, NY	FC	10.0	" "
13,193.8	USB	FFL63	St. Lys Radio, France	FC	10.0	CH#1231 (12,423.0)
	USB	....	Hong Kong (Cape d'Aguilar) Radio	FC	6.0	" "
	USB	LFL45	Rogaland Radio, Norway	FC	10.0	" "
13,196.9	USB	SVN5	Athens Radio, Greece	FC	5.0	CH#1232 (12,426.1)
	USB	KUQ	Pago Pago, American Samoa	FC	1.0	" "

This frequency is also used by USCG Coastal stations: NMC, San Francisco, CA; NMF, Boston, MA; NMN, Portsmouth, VA and NMO, Honolulu, HI

kHz	Mode	Call	Location	Service	kW	Remarks
			13,200.0 - 13,360.0 kHz — AERONAUTICAL MOBILE (Worldwide)			
13,201.0	USB	....	numerous, worldwide	FA/MA	....	USAF (Usually AFE8, MacDill AFB, FL; AFI2, McClellan AFB, CA; AFL, Loring AFB, ME; AGA2, Hickam AFB, HI; AIC2, Clark AB, Philippines; AIE2, Andersen AFB, Guam; AIF, Yokota AB, Japan; AJE, Croughton, England; AKA5, Elemendorf AFB, AK; XPH, Thule, Greenland; AJG9, Incirlik AB, Turkey; AFA3, Andrews AFB, MD & "AIR FORCE ONE")
	USB	....	numerous, throughout Australia	FA/MA	....	RAAF
13,204.0	USB	....	several, throughout North America	FA/MA	....	USAF (Appears to be frequently used by "AIR FORCE ONE & TWO")
13,205.0	USB	....	numerous, in Australasia	FA/MA	....	RAAF/RNZAF (Usually AXD, "Air Force Laverton", Vic., AXF, "Air Force Sydney", NSW; AXH, "Air Force Townsville", Qld.; AXI, "Air Force Darwin", NT; AXS, "Air Force Learmouth", WA; AXT, "Air Force Edinburgh", SA; ZKW, RNZAF, Christchurch; ZKX, RNZAF, Auckland; ZKY, RNZAF, Wanganui & MRX, Buttersworth, Malaysia
13,210.0	USB	AFE8	MacDill AFB, FL	FA/MA	....	USAF
13,215.0	ISB	....	numerous, worldwide	FA/MA	....	USAF (Usually AFA3, Andrews AFB, MD; AGA2, Hickam AFB, HI; AIF2, Yokota AB, Japan; AKA5, Elemendorf AFB, AK; CUW2, Lajes Field, Azores; AJO, Adana AB, Turkey & "AIR FORCE ONE & TWO")
13,215.5	CW	FUX	St. Denis Naval Radio, Reunion	FA	....	French Navy
13,218.0	USB	....	Vandenburg AFB, CA	FA/MA	....	USAF/AFSC [Frequently uses callsign starting with "ABNORMAL"]
13,220.0	USB	....	numerous, within Soviet Union	FA/MA	....	Aeroflot
13,221.0	USB	....	several, in Canada	FA/MA	....	RCAF (Usually CHR, "Trenton Military", Ont.; CJX, "St. Johns Military", Nfld. & VXA, "Edmonton Military", Alta.)
13,224.0	USB	....	numerous, throughout Australia	FA/MA	....	RAAF
13,227.0	USB	....	numerous, throughout Australia	FA/MA	....	RAAF
13,231.0	USB	....	numerous, worldwide	FA/MA	....	USN
13,235.5	CW	HWN	Paris (Houilles) Naval R., France	FA	....	French Navy
13,241.0	USB/RTTY	....	numerous, worldwide	FA/MA	....	USAF (Used for SKY KING broadcasts--some alternative code names for Air Bases--sometimes called SAC "SIERRA" frequency)
13,244.0	USB	....	numerous, worldwide	FA/MA	....	USAF (Usually AFE8, MacDill AFB, FL; AFL, Loring AFB, ME; CUW2, Lages Field, Azores; AFD14, Ascension Island; etc.) [This frequency also used by the USN]
13,247.0	USB	AFA3	Andrews AFB, MD	FA/MA	1.5	USAF
	LSB	AFG37	Scott AFB, IL	FA/MA	....	USAF
	CW	COL	Havana Aeradio, Cuba	FA/MA	....	Aeroflot
		RFNV	Moscow Aeradio, USSR	FA/MA	....	Aeroflot (Mostly working in-flight to/from Havana/Moscow)
	CW	BPA	Urumqi, PRC	FA	....	Tfc
13,251.0	USB	....	numerous, in Pacific Ocean area	FA/MA	....	USN (Usually NGD, McMurdo Station, Antarctica; "Liberty Lady" [Moffett NAS], CA; NPX, Siple Station, Antarctica; NCN, Awase, Okinawa, Japan
13,254.0	USB	....	several, in Canada	FA/MA	4.0	RCAF (Usually CHR, "Trenton Military", Ont.; CJX, "St. Johns Military", Nfld. & VXA, "Edmonton Military", Alta.)
13,255.0	CW	....	several, in Mexico	FA	4.0	Tfc (Includes: XGF24(P), Cozumel; XGF35(X), La Paz, etc.)
13,257.0	USB	....	Reported Bampton AB, England	FA	3.5	RAF Wx Bcbcst

The frequencies between 13,200.0 and 13,260.0 kHz are used by all Allied Air and Naval Forces in 1.0 kHz increments. There are hundreds of code names for various bases that are changed or rotated on a random basis to confuse and circumvent positive identification.

kHz	Mode	Call	Location	Service	kW	Remarks
13,264.0	AM/A3H	....	numerous (ICAO NP)	FA/MA		Airports & in-flight covering the North Pacific Ocean into the Arctic §NP moves to 13,300.0 in 1982

kHz	Mode	Call	Location	Service	kW	Remarks
	CW/RTTY	C7P	Ocean Station Vessel 50°N & 145°W	FA	....	[aka "PAPA"]
	AM/A3H/USB		INTERNATIONAL AIR-TO-GROUND FREQUENCY			
13,270.0	USB	VFG	Gander Aeradio, Nfld., Canada	FA	1.5	VOLMET (20+50)
	USB	WSY7Ø	New York Aeradio, NY	FA	3.0	" (5+35)
		....	§VOLMET moved here from 13,272.0 in 1982			
13,272.0	A3H/USB		§VOLMET broadcasts moved to 13,270.0 in 1982			
	A3H	....	Irkutsk Aeradio, USSR	FA	....	VOLMET/RR (05+35)
	A3H	....	Magadan Aeradio, USSR	FA	....	" " (10+40)
13,279.0	USB	....	Moscow (Unukovo) Aeradio, USSR	FA	....	" " (25+55)
	USB	....	Novosibirsk Aeradio, USSR	FA	....	" " (10+40)
13,282.0	USB	....	§VOLMET broadcasts move here from 13,344.0 in 1983			
13,288.0	AM/A3H	....	numerous (ICAO SEA & NAT)	FA/MA	....	Airports & in-flight covering from New Delhi southeast to Perth and Darwin--east to Biak. Also all routes across the North Atlantic Ocean and into the Arctic regions.
			§SEA moves to 13,318.0 in 1982			
			§AFI moves here from 13,336.0 in 1982			
13,294.0	A3H/USB	....	§AFI moves here from 13,304.0 in 1983			
13,296.0	AM/A3H	....	numerous (ICAO CWP)	FA/MA	....	Airports & in-flight covering all of the western Pacific ocean area from Honolulu and as far to the west as Ulan Bator
			§CWP moves to 13,300.0 in 1982			
13,300.0	A3H/USB	....	§CEP moves here from 13,336.0 in 1982; CWP moves here from 13,296.0 in 1982 & SP moves here from 13,304.0 in 1982			
13,304.0	AM/A3H	....	numerous (ICAO AFI & SP)	FA/MA	....	Airports & in-flight covering from the Mediterranean Sea south to RSA. Also all of the South Pacific Ocean area from Honolulu to Easter Island and Sydney
			§AFI moves to 13,294.0 in 1983 & SP moves to 13,300.0 in 1982			
13,306.0	A3H/USB	....	§AFI moves here from 13,336.0 in 1982			
13,312.0	AM/A3H	....	numerous (ICAO CEP & SEA)	FA/MA	....	Airports & in-flight covering the San Francisco to Honolulu route, plus all of the South China Sea between Hong Kong, Manila, Bangkok, Jakarta and Singapore
13,312.0	USB	EIP	Shannon Aeradio (Rineanna), Ireland	FA	2.0	VOLMET (cont.)
13,320.0	AM/A3H	....	numerous (ICAO CAR & SAM)	FA/MA	....	Airports &in-flight covering the western Caribbean and Central America, plus all of the eastern part of the South American continent from north to south
13,324.0	USB	9VA25	Singapore Aeradio, Singapore	FA/MA	6.0	
	USB	HEE61	Berne Aeradio, Switzerland	FA/MA	6.0	
13,324.5	CW/RTTY	....	Ocean Station Vessels C7A, C7I & C7J	FA	....	(Ocean Station Vessels will discontinue use of CW in 1982) [Alt: 8913.5]
13,326.0	CW	†CAK	Santiago (Los Cerrillos AB), Chile	FA/MA	....	Chilean AF
13,328.0	AM/A3H	....	numerous (ICAO NAT)	FA/MA	....	Airports & in-flight covering all of the North Atlantic Ocean in the Arctic and northern Canada--occasionally referred to the "ALFA" frequency
			§NAT moves to 13,291.0 in 1982			
	CW	"U"	unknown	RC	....	Beacon
13,336.0	AM/A3H	....	numerous (ICAO AFI, CEP & MID)	FA/MA	....	Airports & in-flight covering all of eastern Africa, Arabian Sea to Bombay and west to Malta. All of the Middle East and Indian sub-continent to Calcutta. Also in Pacific on routes between San Francisco and Honolulu
			§AFI moves to 13,288.0 in 1982, CEP moves to 13,330.0 in 1982			
			§MID moves to 13,312.0 in 1982			
13,344.0	AM/A3H	....	numerous (ICAO AFI & SAT)	FA/MA	....	Airports & in-flight from Portugal to the west coast of Africa and across the South Atlantic Ocean to Rio de Janeiro
			§AFI & SAT move to 13,357.0 in 1982			
	A3H	KIS7Ø	Anchorage Aeradio, AK	FA	....	VOLMET (25+55)

kHz	Mode	Call	Location	Service	kW	Remarks
	AM/CW	....	Kai Tak Aeradio, Hong Kong	FA	2.0	VOLMET (15+45) TIME SIGNALS
	A3H	KVM7Ø	Honolulu (International) Aeradio, HI	FA	....	VOLMET (00+30)
	A3H	JMA	Narita Aeradio, Japan	FA	3.0	" (10+40)
	A3H	....	Auckland, New Zealand	FA	5.0	" (20+50)
	A3H	KSF7Ø	Oakland Aeradio, CA	FA	....	" (05+35)
			§VOLMET broadcasts move to 13,282.0 in 1983			
13,356.0	USB	....	numerous, worldwide, operational	FA/MA	....	Airports &
			in-flight--particularly ARINC New York to Honolulu (phone patches are available)			
13,357.0	A3H/USB	....	§AFI moves here from 13,344.0 in 1982			

13,360.0 - 14,000.0 kHz -- FIXED SERVICE (Worldwide)

13,360.0 - 13,600.0 kHz -- FIXED SERVICE (Worldwide--post WARC)

kHz	Mode	Call	Location	Service	kW	Remarks
13,364.9	ISB	CSY46	Santa Maria Aeradio, Azores	AX	10.0	
13,365.0	CW	GFT29	Bracknell, England	FX	2.5	Meteo
13,371.2	CW	†CWO4	Cerrito, Uruguay	FX	5.0	Tfc
13,372.5	CW	NGR	Kato Souli Naval Radio, Greece	FX	10.0	USN/NOAA re-
			ports wx broadcasts at 0030, 0630 & 1230			
13,375.0	USB	....	Berlin, GDR	FX	....	Telcom
13,377.5	CW/RTTY	†KNY21	Washington, DC	FX	1.0	Yugoslav EMBASSY
	CW/RTTY	†KNY23	Washington, DC	FX	1.0	Czechoslovakian EMBASSY
13,377.6	USB	XBD492	Mexico City, DF, Mexico	FX	0.1	
13,379.0	CW/RTTY	†KNY26	Washington, DC	FX	1.0	Hungarian EMBASSY
13,380.0	CW	NRV	COMMSTA Barrigada, Guam	FX	10.0	USCG/NUKO
			NOAA reports wx broadcasts at 0400, 1300, 1700 & 2300			
	LSB	....	Moscow, USSR	FX	....	R. Moscow Feeder
13,382.0	CW	GFT	Bracknell, England	FX	....	Meteo
	CW	MLU	Gibraltar, Gibraltar	FX	1.0	RAF Tfc
	CW	†MPU	Luca, Malta	FX	1.0	RAF Tfc
13,385.0	CW	RMN3	Omsk, USSR	FX	....	SAAM
	CW	RULE	Vostok USSR Base, Antarctica	FX	....	"
	CW	RUZU	Molodezhnaya USSR Base, Antarctica	FX	....	"
	CW	UDY	Novolazarevskaya USSR Base, Antarctica	FX	...	"
	CW	UFE	Mirnyy USSR Base, Antarctica	FX	....	"
	CW	UGE2	Bellingshausen USSR Base, Antarctica	FX	....	"
			[Also heard on this frequency are supply ships]			
13,386.0	USB/RTTY	LJA21	Tel Aviv, Israel	FX	0.4	Norwegian EMBASSY
13,390.0	CW	COY851	Havana Naval Radio, Cuba	FX	....	Russian Navy
13,394.0	ISB	IRK43	Rome (Torrenova), Italy	FX	....	Telcom
13,395.0	CW	....	unknown	FX	....	Coded Tfc
13,404.5	USB	....	numerous, throughout Japan	FX	1.0	
13,405.9	CW	....	unknown, probably in Cuba	FX	....	Tfc/SS
13,410.0	ISB	YIE35	Baghdad (Abu Ghuraib), Iraq	FX	....	Telcom
	CW	6WW	Dakar Naval Radio, Senegal	FX	2.0	French Navy
13,420.1	CW	CUA69	Lisbon (Alfragide), Portugal	FX	30.0	Tfc
13,423.0	USB	....	Norwegian Sea & North Atlantic	FX	....	USCG (Includes:
			Ocean LORAN Net DML, Sylt, GFR: JXL, BØ, Norway JXP, Jan Mayen Is.; NMS, Shetland Is.; OUN, Ejde, Denmark; OVY, Angissoq, Greenland; TFR, Sandur, Iceland & TFR2, Reykjavik, Iceland)			
13,425.0	CW	RIW	Khiva Naval Radio, Uzbek SSR	FX	....	Russian Navy
13,427.0	USB	†TTZ34	N'Djamena, Chad	FX	2.0	Telcom
13,427.5	USB	WEK53	New York, NY	FX	20.0	Telcom
13,431.2	USB	"XS"	unknown	FX	....	Net Control
13,433.0	USB	OHU2Ø	Helsinki, Finland	FX	1.0	EMBASSY
	USB	OHU25	Beijing, PRC	FX	1.0	Finnish EMBASSY
			[Alt: 10,885.0 & 13,913.0]			
13,434.0	LSB	WEK53	New York, NY	FX	20.0	Telcom
13,440.0	CW	EEQ	Madrid, Spain	FX	1.0	Police
13,455.0	CW	....	unknown	FX	....	Coded Tfc
13,460.0	USB	....	Riyadh, Saudi Arabia	FX	....	Feeder
13,466.0	USB	....	unknown	FX	....	SEDCO Oil Rigs

FIXED SERVICE (con.)

kHz	Mode	Call	Location	Service	kW	Remarks
13,474.0	USB	CLN404	Havana (Bauta), Cuba	FX	100.0	Telcom/Moscow
13,482.0	ISB	†VPC	Port Stanley, Falkland Islands	FX	7.5	Telcom/Tfc
13,483.5	USB	....	numerous, worldwide	FX	....	USN MARS
13,485.1	CW	KWL9Ø	Tokyo, Japan	FX	....	USA EMBASSY
13,490.0	USB	OZU25	Copenhagen, Denmark	FX	1.5	EMBASSY (Also used by OZU3Ø, Cairo, Egypt; OZU33, Tel Aviv, Israel; OZU34, Jeddah, Saudi Arabia; OZU35, Beijing, PRC; OZU38 Lagos, Nigeria & OZU39, Nairobi, Kenya) [Alt: 16,400.0 & 18,587.0]
13,493.0	LSB	†....	Greenville, NC	FX	40.0	VOA Feeder
13,499.0	CW/RTTY	SON249	Warsaw, Poland	FX	10.0	Tfc/Nx
13,500.0	CW	D3I44	Luanda Aeradio, Angola	AX	3.0	Aero Tfc
13,503.0	A3H	VLN32	Sydney (Doonside), NSW, Australia	FX	1.0	Telcom/Papeete
13,503.5	USB	ZFD62	Hamilton, Bermuda	FX	30.0	Emergency
13,510.0	CW/FAX	CFH	Maritime Command Radio, Halifax, NS, Canada (Canadian Forces)	FX	10.0	Meteo/Seasonal ice at 0000, 1300 & 2200 + test charts, analysis & prognosis
13,511.8	CW	....	unknown, possibly in Indonesia	FX	....	Nx/EE
13,520.0	CW	OWS3	Copenhagen, Denmark	FX	0.4	INTERPOL
	CW	OGX	Helsinki, Finland	FX	....	"
	CW	8UF75	New Delhi, India	FX	5.0	"
	CW	JPA61	Nagoya (Komaki), Japan	FX	10.0	" [Alt: 13,820.0]
13,525.0	CW/FAX	CCS	Santiago Naval Radio, Chile	FX	....	Chilean Navy
	USB	....	Riyadh, Saudi Arabia	FX	....	Feeder
13,527.0	USB	XDA358	Mexico City, DF, Mexico	FX	10.0	Telcom/Havana
13,529.0	USB	NBW	Guantanamo Bay Naval Radio, Cuba	FX	1.0	USN
	USB	NQX	Key West Naval Radio, FL	FX	1.0	USN [Emergency only]
13,533.4	LSB	....	unknown, possibly in South America	FX	....	Telcom/SS
13,535.0	ISB	FTN53	Paris (St. Assise), France	FX	20.0	Telcom/Tfc
	ISB	†HLB28	Seoul, South Korea	FX	30.0	Telcom/Tfc
13,537.0	CW	BXM51	unknown	FX	....	PRC EMBASSY
13,547.0	USB	AFG4Ø	Colorado Springs. CO	FX	1.0	USAF
13,552.0	CW	....	Ankara, Turkey	FX	1.0	EMBASSY
13,555.0	CW	CFH	Maritime Command Radio, Halifax, NS, Canada	FX	10.0	Canadian Forces
13,560.0	CW	BMB	Taipei Naval Radio, Taiwan	FX	....	Rep. China Navy NOAA reports wx broadcasts at 0400, 100, 1600 & 2000
13,565.0	USB	FTN56	Paris (St. Assise), France	FX	20.0	Telcom
13,570.0	USB	ZLX82	Wellington (Himatangi), New Zealand	FX	4.0	Telcom/Sydney
13,576.0	USB	....	Stockholm Aeradio, Sweden	FA	7.0	
13,580.0	CW/FSK	HMS19	Pyongyang, North Korea	FX	....	Tfc/Nx
13,582.0	CW	EC3Y	unknown	FX	....	"VVV DE EC3Y"
13,585.0	USB	GAB33	London (Rugby), England	FX	30.0	Telcom/Accra
13,590.0	USB	RGD23	Moscow, USSR	FX	20.0	R. Moscow Feeder
13,593.0	ISB	†HBE53	Berne, Switzerland	FX	40.0	Tfc
13,597.0	FAX	JMH4	Tokyo, Japan	FX	5.0	Meteo

13,600.0 - 13,800.0 kHz -- INTERNATIONAL BROADCASTING (shared post WARC)

kHz	Mode	Call	Location	Service	kW	Remarks
13,600.0	CW	CAK	Santiago (Los Cerillos AB), Chile	FA	0.4	Chilean AF NOAA reports wx broadcasts at 0130, 1330 & 2000
	FAX	IMB56	Rome, Italy	AX	1.0	Italian AF/Meteo (0600-2030)
	CW	....	In the USA this frequency assigned to radio wave welding			
13,603.0	LSB	FTN6Ø	Paris (St. Assise), France	FX	8.0	Telcom
	ISB	6VK413	Dakar, Senegal	FX	6.0	Telcom
13,605.0	AM	ZPG36	Asuncion, Paraguay	FX	0.5	Telcom (Has been used for relay of official programs)
	CW/RTTY	†KNY27	Washington, DC	FX	1.0	Swiss EMBASSY
13,608.6	USB	....	Northwest Pacific LORAN Net	FX	....	USCG (Includes: NRT, "Yokota Monitor", Japan; NRT2, Gesashi, Japan; NRT3, Iwo Jima; NRT9, Hokkaido, Japan; NRV, Barrigada, Guam; NRV6, Marcus Is.; NRV7, Yap Island) [Alt: 11,606.0 & 15,875.0]

FIXED SERVICE (con.)

kHz	Mode	Call	Location	Service	kW	Remarks
13,612.0	ISB	†WEL73	New York, NY	FX	20.0	Telcom
13,616.0	CW	SON261	Warsaw, Poland	FX	20.0	Reported PAP nx broadcast at 1800
13,617.0	CW/USB	OEC44	Tel Aviv, Israel	FX	....	Austrian EMBASSY [Alt: 10,298.0 & 14,480.0]
	CW	OEC57	Pretoria, RSA	FX	1.0	Austrian EMBASSY [Alt: 10,422.5]
13,618.8	USB	....	unknown	FX	....	Net/FF
13,627.1	FAX	DGN62	Quickborn, GFR	FX	....	Meteo
13,627.5	FAX	KVM7Ø	Honolulu (International), HI	AX	10.0	Meteo (1900-2045)
13,630.0	USB	KEM8Ø	Washington, DC	FX	....	FAA HQ
	USB	KDM5Ø	Atlanta (Hampton), GA	FX	....	FAA
	USB	....	several, USA nationwide such as "Dragnet" etc.)	FX	....	FAA (Code names [Alt: 8125.0 & 16,348.0]
13,637.0	CW	"F"	unknown	RC	....	Beacon
13,640.0	USB	VLN63	Sydney (Doonside), NSW, Australia	FX	30.0	Telcom
13,651.5	LSB	AJE	Croughton (Barford), England	FX	4.0	AFRTS Feeder
13,655.0	CW	NMO	COMMSTA Honolulu, HI	FX	10.0	USCG/Wx broadcasts at 0100, 0400, 0700, 1300 & 2000
13,657.0	FAX	DDH8	Pinneberg, GFR	FX	20.0	Meteo
	CW/RTTY	DDJ8	Frankfurt (Usingen), GFR	FX	20.0	Meteo
	CW	IDR	Rome Naval Radio, Italy	FX	15.0	Italian Navy
13,660.0	CW	6XH57	Antananarivo, Madagascar	FX	1.2	Tfc/Dakar
13,665.5	USB	....	Mediterranean LORAN Net	FX	....	USCG (Includes: AOB5Ø, Estartit, Spain; NCI, Sellia Marina, Italy; NCI3, Lampedusa, Italy; NCI4, Kargabarun, Turkey and NCI1Ø, Rhodes) [Alt: 10,958.5 & 15,723.0]
13,667.5	FAX	6VU73	Dakar, Senegal	FX	5.0	Meteo (cont.)
13,678.5	USB	NGD	McMurdo Station, Antarctica	FX	1.0	USN
13,685.0	ISB	FZC36	Cayenne, Guiana	FX	20.0	Telcom/Tfc
	CW	2Z7	unknown	FX	....	French EMBASSY
13,688.0	ISB	ZEN63	Cape d'Aguilar, Hong Kong	FX	8.0	Telcom
13,690.0	ISB	....	Holzkirchen, GFR	FX	10.0	RFE Feeder
13,695.0	CW	ONN38	Brussels, Belgium	FX	0.1	Swiss EMBASSY
	ISB	MKG	London (Stanbridge), England	FX	30.0	Telcom/Tfc
13,700.0	USB	PUZ4	Brasilia, Brazil	AX	1.0	Govermental
13,707.0	FAX	RXA77	Alma Ata, USSR	FX	20.0	Meteo
13,710.0	USB	†RGD22	Moscow, USSR	FX	80.0	R. Moscow Feeder
13,712.0	ISB	FTN71	Paris (St. Assise), France	FX	20.0	Telcom
13,720.0	ISB	†RIC77	Moscow, USSR	FX	....	Telcom
13,730.0	USB	....	numerous, in Australia	FX	1.0	Police (Includes: VKA, Adelaide; VKC, Melbourne; VKI, Perth; VKM, Darwin; VKR, Brisbane; VKT, Hobart & VKX, Canberra) [Alt: 10,505.0]
13,742.0	ISB	AFE83	Ascencion Island	FX	20.0	USAF
13,750.0	ISB	NAW31	Guantanamo, Cuba	FX	....	Telcom
13,759.0	USB	NAW32	Guantanamo, Cuba	FX	....	Telcom
13,760.0	ISB	FZF37	Fort de France, Martinique	FX	20.0	Telcom
13,765.2	USB	NAW36	Guantanamo, Cuba	FX	....	Telcom
13,768.5	USB	PCW2Ø	Warsaw, Poland	FX	1.0	Dutch EMBASSY
	USB	PCW11	Bangkok, Thailand	FX	1.0	Dutch EMBASSY
13,773.0	FAX	ZRO3	Pretoria, RSA	FX	5.0	Meteo (0300-1730)
13,780.0	CW/FSK	HME28	Pyongyang, North Korea	FX	....	Tfc/Nx
13,793.0	CW	SON279	Warsaw, Poland	FX	....	Tfc/Reported PAP nx broadcast at 1400
13,797.0	ISB	C5B64	Banjul, Gambia	FX	3.5	Telcom
	USB	....	Reykjavik, Iceland	FX	....	Feeder
13,799.0	ISB	ODL79	Beirut, Lebanon	FX	5.0	Telcom/Tfc
13,800.0	ISB	VLN33	Sydney (Doonside), NSW, Australia	FX	30.0	Telcom/Rawalpindi

13,800.0 - 14,000.0 kHz — FIXED SERVICE (Worldwide—post WARC)

kHz	Mode	Call	Location	Service	kW	Remarks
13,807.5	FAX	NPN	Barrigada (Agana NAS), Guam	FX	15.0	USN/Meteo
	ISB	†PCK93	Hilversum (Kootwijk), Netherlands	FX	30.0	Telcom/Feeder
13,812.0	USB	....	several, in Caroline Islands	AX	....	Aero Tfc

kHz	Mode	Call	Location	Service	kW	Remarks
	CW	....	unknown	FX	....	Net
13,815.0	CW	KRH5Ø	London, England	FX	....	USA EMBASSY
13,817.0	ISB	†HB073	Geneva, Switzerland	FX	30.0	Tfc
13,820.0	CW	OWS3	Copenhagen, Denmark	FX	0.4	INTERPOL
	CW	8UF75	New Delhi, India	FX	5.0	"
	CW	JPA58	Nagoya (Komaki), Japan	FX	10.0	"
	USB	RVW55	Moscow, USSR	FX	20.0	R. Moscow Feeder
13,829.5	ISB	MKG	London (Stanbridge), England	FX	30.0	RAF Telcom/Tfc
	CW/RTTY	....	several, USA nationwide	FX	....	FCC Backup Net
			[For list of stations see 10,902.1]			
13,843.0	USB	†FZF38	Fort de France, Martinique	FX	20.0	Telcom
13,850.0	CW	ETM85	Addis Ababa, Ethiopia	FX	1.0	Vietnam EMBASSY
13,855.0	CW	OXI	Godthaab, Greenland	FX	....	Danish Marine
	FAX	OXT	Skamlebaek Radio, Denmark	FX	20.0	Danish Marine
			Meteo (1220-1240; 1310-1330 & 1805-1825)			
13,860.0	ISB	....	Dixon, CA	FX	50.0	VOA Feeder
13,862.5	CW/FAX	NPM	Lualualei Naval Radio, HI	FX	15.0	USN Meteo (cont.)
	ISB	TUP38	Abidjan, Ivory Coast	FX	20.0	Telcom
13,865.0	USB	....	numerous, throughout Australia	FX	....	Commercial Tfc
13,873.0	CW	FTN87	Paris (St. Assise), France	FX	20.0	TIME SIGNALS/ Tfc
13,878.0	ISB	FYN87	Paris (St. Assise), France	FX	20.0	Telcom/RSA
13,882.5	CW	DDK2	Quickborn, GFR	FX	5.0	Meteo
13,891.0	A3H	CLN413	Havana (Bauta), Cuba	FX	10.0	Mx Mirror
13,895.0	FAX	Y2V47	Berlin (Nauen), GDR	FX	20.0	Nx
13,903.0	USB	....	unknown, probably in Brazil	FX	....	Net/PP
13,907.0	USB	AFH39	March AFB, CA	FX	0.5	USAF/SAC
13,909.0	LSB	....	unknown, probably in South America	FX	....	Telcom/SS
13,913.0	USB	OHU2Ø	Helsinki, Finland	FX	1.0	EMBASSY
	USB	OHU25	Beijing, PRC	FX	1.0	Finnish EMBASSY
			[Alt: 13,433.0 & 14,693.0]			
13,915.0	USB	SAR88	Stockholm, Sweden	FX	1.0	INTERNATIONAL
			RED CROSS--Special Emergency [Alt: 13,965.0, 13,973.0 & 13,997.0]			
13,920.0	FAX	AXM35	Canberra, ACT, Australia	FX	20.0	Meteo
13,927.0	USB	CUW2Ø	Lajes Field, Azores	FX	1.0	USAF
			[Alt: 13,977.0]			
	DATA	....	unknown, possibly CUW2Ø	FX	....	
13,931.0	CW/RTTY	YMS	Warsaw, Poland	FX	0.4	Turkist EMBASSY
13,937.5	CW	ONY24	Rouveroy, Belgium	FX	....	Belgium Army/ NATO
13,937.7	CW	CLP1	Havana, Cuba	FX	....	EMBASSY
13,941.5	USB	†ZE066	Cape d'Aguilar, Hong Kong	FX	3.0	Telcom
13,945.0	CW/RTTY	ELE13	Harbel, Liberia	FX	15.0	Tfc
	CW	3VA3Ø	Tunis, Tunesia	FX	1.0	Tfc
	CW	WQB23	Akron, OH	FX	15.0	Tfc
13,950.0	CW	Y7A55	Berlin, GDR	FX	....	EMBASSY
	LSB	AHF4	Albrook AFS, Panama	FX	1.0	USMAAG (At
			press time also included: AHF1C, Lojac, Ecuador; OAE21 Lima, Peru; YWA6 Caracas, Venezuela; YS1UKE, San Salvador; AFF4, Bergstrom AFB, TX; AFF3, Brooks AFB, TX; HR1MM, Tegucigalpa, Honduras; TI2USA, San Jose, Costa Rica; TDMG3, Guatemala City, Guatemala; ACH54, Haiti & HCUS1, Quito, Ecuador) [Alt: 15,675.0]			
13,952.5	CW	CLP1	Havana, Cuba	FX	....	EMBASSY
13,960.8	LSB	ROW27	Moscow, USSR	FX	15.0	R. Moscow Feeder
13,965.0	ISB	CNR28	Rabat, Morocco	FX	20.0	Telcom/Tfc
13,970.0	CW	CLP1	Havana, Cuba	FX	....	EMBASSY
	ISB	CUA7Ø	Lisbon (Alfragide), Portugal	FX	3.0	Telcom
13,971.0	USB	....	several, in Canada & Middle East	FX	....	Canadian Forces
			(Reported: VXN9, Nicosia, Cyprus; VXV9, Golan Heights, Syria; VET9, Damascus, Syria; plus various stations in in Canada--phone patches for peace keeping force--may be referred to as the "ALFA" frequency) [Alt: 14,445.0]			
	USB	VDH9	Alert, NWT, Canada	FX	....	Canadian Forces

kHz	Mode	Call	Location	Service	kW	Remarks
13,974.0	USB	....	numerous, worldwide	FX	....	USN MARS (Frequently used from Antarctica by NNNØICE, McMurdo Station and NNNØKMR, Palmer Station, plus NNNØNRS, Diego Garcia)
13,977.0	USB	CUW2Ø	Lajes Field, Azores	FX	1.0	USAF [Alt: 13,927.0 & 13,993.0]
13,977.3	USB	RAN	Moscow, USSR	FX	15.0	Telcom
13,980.0	CW	LCO	Oslo (Jeloey), Norway	FX	20.0	Tfc/Nx/Wx
13,993.0	USB	....	numerous, worldwide	FX	....	USAF MARS
13,995.0	ISB	....	†Monrovia (Careysburg), Liberia	FX	50.0	VOA Feeder
13,996.0	USB	....	numerous, worldwide	FX	....	USAF MARS
13,997.0	USB	ETM99	Addis Ababa, Ethiopia	FX	0.6	Yemen (DR) EMBASSY
	USB	PGA88	The Hague, Netherlands	FX	0.3	INTERNATIONAL RED CROSS
	USB	SAR88	Stockholm, Sweden	FX	1.0	" " "
	USB	HBC88	Geneva, Switzerland	FX	0.8	" " "

14,000.0 - 14,350.0 kHz -- AMATEUR RADIO (Worldwide)

14,000.0 - 14,250.0 kHz -- AMATEUR RADIO (Worldwide--post WARC)

14,250.0 - 14,350.0 kHz -- AMATEUR RADIO (Shared with Iran, PRC & USSR--post WARC)

14,350.0 - 14,990.0 kHz -- FIXED SERVICE (Worldwide)

kHz	Mode	Call	Location	Service	kW	Remarks
14,353.5	CW/RTTY	†KNY34	Washington, DC	FX	1.0	Swedish EMBASSY
14,357.0	USB	FTO35	Paris (St. Assise), France	FX	20.0	Telcom
14,360.0	CW	KWS78	Athens, Greece	FX	....	USA EMBASSY
14,362.0	CW/RTTY	SOO236	Warsaw, Poland	FX	20.0	
14,365.0	FAX	BAF8	Beijing, PRC	FX	35.0	Meteo
	CW	....	unknown	FX	....	Coded Tfc
14,366.0	ISB	5TN29	Nouakchott, Mauritania	FX	20.0	Telcom/Tfc
14,370.0	LSB	LSE	Buenos Aires, Argentina	FX	0.5	Net Control
14,378.6	USB	ICG	Milan, Italy	FX	1.0	Telcom/Kaduna
14,385.0	CW	....	numerous, USA nationwide	FX	....	USN [Reserves]
14,387.0	USB	OMZ3Ø	Prague, Czechoslovakia	FX	4.0	Algerian EMBASSY
14,388.6	USB	ICG	Milan, Italy	FX	1.0	Telcom/Shiroru
14,389.0	USB	....	numerous, in the USA/South Korea	FX	....	USAF MARS
14,398.0	ISB	....	Delano, CA	FX	50.0	VOA Feeder
14,400.0	CW/RTTY	ETN4Ø	Addis Ababa, Ethiopia	FX	1.0	Czech EMBASSY
14,402.0	USB	....	numerous, worldwide	FX	....	US Army MARS
14,403.0	CW	WAR	Washington, DC	FX	....	US Army
14,405.0	CW	RIW	Khiva Naval Radio, Uzbek SSR	FX	....	Russian Navy
	CW/RTTY	....	numerous, USA nationwide	FX	....	US Army
14,408.0	LSB	....	numerous, USA nationwide	FX	....	US Army MARS
14,410.0	CW	Y7F54	Algiers, Algeria	FX	....	GDR EMBASSY
	CW	Y7L36	Havana, Cuba	FX	....	GDR EMBASSY
14,411.0	CW	Y7F68	Beirut, Lebanon	FX	....	GDR EMBASSY
14,415.0	USB	VNM	Melbourne, Vic., Australia	FX	1.0	ANARE
	USB	VJM	Macquarie Island, Antarctica	FX	1.5	"
	USB	VLV	Mawson Base, Antarctica	FX	5.0	"
	USB	VLZ	Davis Base, Antarctica	FX	1.0	"
	USB/RTTY	VNJ	Casey Base, Antarctica	FX	10.0	"
	USB	CLN422	Havana (Bauta), Cuba	FX	100.0	Telcom/Luanda
14,423.0	ISB	†FTO42	Paris (St. Assise), France	FX	20.0	Telcom
	USB	AOE5Ø	Torrejon AB, Spain	FX	0.5	USAF [Alt: 14,448.0]
14,424.9	ISB	†9UB69	Bujumbura, Burundi	FX	5.0	Telcom
14,426.0	CW/RTTY	RLX	Dublin, Ireland	FX	0.5	USSR EMBASSY [Alt: 12,211.0 & 17,528.0]
14,432.0	ISB	†FTO43	Paris (St. Assise), France	FX	30.0	Telcom
14,436.0	FAX	GFE23	Bracknell, England	FX	7.0	Meteo/Reported seasonal ice warnings pics at 1413
14,436.2	LSB	....	unknown, numerous, Central America	FX	....	Net/SS
14,441.5	USB	....	numerous, USA nationwide	FX	....	USN MARS
14,442.5	CW	ART	Rawalpindi, Pakistan	FX	1.0	Tfc/Ankara

kHz	Mode	Call	Location	Service	kW	Remarks
14,445.0	USB	....	several, in Canada & Middle East	FX	....	Canadian Forces
			(Reported: VXV9, Golan Heights, Syria & CIW2Ø2, Vancouver, BC/phone patches for peace keeping forces) [Alt: 13,971.0]			
	USB	ZKS	Rarotonga, Cook Islands	FX	1.0	Net (Including:
			ZKI, Tauhunu, Manihiki Atoll; ZKP5, Nassau; ZKL, Palmerston Atoll; ZKJ, Omoka, Tongareva Atoll; etc.) [Alt: 14,900.0]			
14,448.0	USB	AOE5Ø	Torrejon AB, Spain	FX	1.0	USAF
				[Alt: 14,423.0]		
14,450.0	USB	CUA71	Lisbon (Alfragide), Portugal	FX	10.0	Telcom
14,451.0	CW	UAQ	unknown	FX	....	"RFI DE UAQ"
14,455.2	USB	....	unknown	FX	....	Net/EE
14,464.9	ISB	D4E25	Praia, Cape Verde	FX	10.0	Telcom/Tfc
14,467.0	USB	....	numerous, worldwide	MS	....	USN
14,470.0	CW	NPG	San Francisco Naval Radio, CA	FX	....	USN
	USB	....	numerous, worldwide	FX	....	USN MARS
14,476.0	USB	†5BC67	Nicosia, Cyprus	FX	3.0	Telcom
14,477.2	USB	....	numerous, worldwide	FX	....	MARS
14,477.6	CW	"K"	unknown	FX	....	Beacon
14,478.5	CW/LSB	OEC44	Tel Aviv, Israel	FX	1.0	Austrian EMBASSY
						[Alt: 20,753.0]
14,483.4	CW/LSB	OEC	Vienna, Austria	FX	10.0	EMBASSY
	CW/LSB	OEC64	Lagos, Nigeria	FX	1.0	Austrian EMBASSY
	CW/LSB	OEC72	Lisbon, Portugal	FX	1.0	" "
14,489.2	CW	....	unknown	FX	....	Net ("NEON")
14,492.0	CW	UAA	unknown	FX	....	"VVV DE UAA"
14,495.0	USB	FTO49	Paris (St. Assise), France	FX	20.0	Telcom
14,500.0	USB	PPR48	Rio de Janeiro (Sepetiba), Brazil	FX	3.0	Telcom/La Paz
	USB	ZKRG	Rarotonga, Cook Islands	AX	5.0	Aero/Auckland
	ISB	HBE35	Berne, Switzerland	FX	30.0	Tfc/Telcom
	ISB	RRRF	Moscow, USSR	FX	15.0	Telcom
14,505.0	CW	HXZ24	Paris (St. Assise), France	FX	10.0	Meteo
14,510.0	CW	RIW	Khiva Naval Radio, Uzbek SSR	FX	....	Russian Navy
14,515.0	CW/RTTY	VCS838	Ottawa, Ont., Canada	FX	5.0	Yugoslav EMBASSY
					[Alt: 12,295.0 & 18,247.0]	
14,520.0	ISB	†GIN34	London (Rugby), England	FX	30.0	Telcom
14,525.0	CW	ETN52	Addis Ababa, Ethiopia	FX	3.0	Algerian EMBASSY
14,526.0	ISB	†....	Dixon, CA	FX	50.0	VOA Feeder
14,534.0	USB	WEK34	New York, NY	FX	50.0	Telcom
14,535.0	ISB	VJP	Perth (Gnangara), WA, Australia	FX	10.0	Telcom/Cocos Is.
14,541.5	LSB	AUY67	Murud, India	FX	2.5	Telcom/Narnaul
	LSB	AUY8Ø	Narnaul, India	FX	2.5	Telcom/Murud
14,544.5	CW	RIW	Khiva Naval Radio, Uzbek SSR	FX	....	Russian Navy
14,545.8	LSB	PYP	Rio de Janeiro, Brazil	FX	....	Telcom/Tfc
14,550.0	CW/ISB	LCP	Oslo (Jeloey), Norway	FX	10.0	Tfc
14,556.0	CW	RIW	Khiva Naval Radio, Uzbek SSR	FX	....	Russian Navy
14,580.0	USB	ZPG45	Asuncion, Paraguay	FX	5.0	Telcom
14,582.5	FAX	GFA25	Bracknell, England	FX	2.5	Meteo (0600-1800)
14,587.0	CW	CCS	Santiago Naval Radio, Chile	FX	....	Chilean Navy
	CW	"K"	unknown	RC	....	Beacon
14,596.0	FAX	CSZ69	Lisbon, Portugal	AX	2.0	Meteo/Aero
14,600.0	USB	WEK54	New York, NY	FX	80.0	Telcom
14,605.0	LSB	PPQ7	Rio de Janeiro (Santa Cruz), Brazil	FX	2.5	Telcom
	CW	JMB3	Tokyo, Japan	FX	2.0	Meteo
	CW/RTTY	†ELE14	Harbel, Liberia	FX	3.0	Tfc
14,606.0	USB	....	several, in West Germany & USA	FX	....	US Army MARS
14,607.5	CW	8UF75	New Delhi, India	FX	5.0	INTERPOL
	CW/RTTY	JPA23	Tokyo, Japan	FX	5.0	INTERPOL
			(Distribution of INTERPOL Bulletins to Bangkok, Hong Kong, Jakarta, Kuala Lumpur, Melbourne & Singapore) [Alt: 14,623.5 & 14,707.0]			
	CW	HMA22	Seoul, South Korea	FX	1.0	INTERPOL
14,609.0	USB	WEK64	New York, NY	FX	30.0	Telcom
14,613.5	ISB	†5VH346	Lome, Togo	FX	20.0	Telcom
14,616.0	CW	KWL9Ø	Tokyo, Japan	FX	....	USA EMBASSY
14,620.0	USB	XJE57	Ottawa (Almonte), Ont., Canada	FX	....	RCMP Net Control

kHz	Mode	Call	Location	Service	kW	Remarks
14,623.5	CW	JPA34	Tokyo, Japan	FX	5.0	INTERPOL
			[Alt: 14,607.5 & 14,707.0]			
14,625.0	ISB	RCG79	Moscow, USSR	FX	15.0	Telcom
14,630.0	ISB	WFK34	New York, NY	FX	40.0	Telcom
			[Has been heard with voice scrambler]			
14,632.0	CW	YZC2	Belgrade (Makis), Yugoslavia	FX	30.0	Tfc/Buenos Aires
14,640.9	CW	EBA	Madrid Radionaval Radio, Spain	FX	....	Spanish Navy
14,646.0	ISB	GIW34	London (Rugby), England	FX	10.0	Telcom
14,648.0	USB	....	unknown	FX	....	Net/FF
14,649.0	CW	KNY23	Washington, DC	FX	1.0	Czech EMBASSY
14,650.0	CW	....	unknown	FX	....	Coded Tfc
14,660.0	FAX	....	Taipei, Taiwan	FX	....	Meteo
14,661.0	USB	OHU2Ø	Helsinki, Finland	FX	1.0	EMBASSY (Includes:
			ETN66, Addis Ababa, Ethiopia; OHU21, Paris, France; OHU22, Bucharest, Rumania; OHU23, Tel Aviv, Israel; OHU26, Belgrade, Yugoslavia & OHU28, Warsaw, Poland) [Alt: 14,693.0]			
14,664.0	USB	....	numerous, worldwide	FX	....	USAF MARS
14,655.0	USB	ZLQ8	Scott Base, Antarctica	FX	0.7	Telcom/Wellington
	USB	....	numerous, worldwide	FX	....	US Army MARS
14,670.0	A3H	CHU	Ottawa, Ont. Canada	SS	3.0	TIME SIGNALS
	ISB	CLN447	Havana (Bauta), Cuba	FX	100.0	Tfc/Luanda
	ISB	MKD	Akrotiri, Cyprus	FX	10.0	RAF Telcom/Tfc
14,672.0	FAX	AFA	Washington (Andrews AFB, MD), DC	FX	3.0	USAF/Meteo
14,676.0	ISB	9NB24	presumed Kathmandu, Nepal	FX	....	Telcom
14,678.5	USB	....	Tel Aviv, Israel	FX	1.0	Philippine EMBASSY
14,680.0	CW/RTTY	UJY2	Kaliningrad, USSR	FX	5.0	Tfc/Havana
14,684.0	CW	CYS22	Ottawa, Ont., Canada	FX	....	Polish EMBASSY
			[Alt: 11,688.0 & 15,582.0]			
14,686.0	USB	....	numerous, USA nationwide	FX	....	DEA (Many
			code names preclude positive ID. "ATLAS" appears to be Net Control. Others heard include "CONDOR", "FLINT", "SHARK", "SUNDANCE", etc.) [Alt: 11,076.0 & 18,666.0]			
14,687.0	USB	....	Hanoi, Vietnam	FX	....	Telcom
14,690.0	CW	GYU	Gibraltar Naval Radio, Gibraltar	FX	....	British Navy
14,692.5	FAX	JMJ4	Tokyo (Usui), Japan	FX	5.0	Meteo
14,693.0	USB	†OHU2Ø	Helsinki, Finland	FX	1.0	EMBASSY (In-
			cludes: OHU22, Bucharest, Rumania; OHU25, Beijing, PRC; OHU26, Belgrade, Yugoslavia & OHU28, Warsaw, Poland) [Alt: 14,661.0]			
14,696.5	CW	COY895	Havana Naval Radio, Cuba	FX	....	Russian Navy
14,707.0	CW/RTTY	JPA35	Tokyo, Japan	FX	5.0	INTERPOL
			[Alt: 14,623.5]			
14,712.5	ISB	....	Holzkirchen, GFR	FX	10.0	RFE Feeder
14,730.0	USB	†FTO73	Paris (St. Assise), France	FX	20.0	Telcom
	USB	†JBL54	Tokyo (Oyama), Japan	FX	30.0	Telcom
14,736.0	USB	IRL27	Rome (Torrenova), Italy	FX	10.0	Telcom
14,737.0	FAX	RHB/RHO	Khabarovsk, USSR	FX	....	Meteo [aka UMN5]
14,741.0	CW/RTTY	YMS	Warsaw, Poland	FX	0.4	Turkist EMBASSY [Alt: 13,942.0]
14,744.0	USB	AGA2	Hickam AFB, HI	FX	1.0	USAF/SAC
	USB	AIE	Andersen AFB, Guam	FX	1.0	USAF/SAC
	USB	....	several, USA nationwide	FX	....	USAF
			[May be called "ALFA TANGO ONE" frequency]			
14,747.5	USB	YKY32	Damascus, Syria	FX	....	Telcom
14,748.5	USB	VCR976	Ottawa, Ont., Canada	FX	1.0	Saudi Arabian
			[Alt: 5806.0 & 15,595.5]			EMBASSY
14,766.0	USB	†NGD	McMurdo Station, Antarctica	FX	1.0	USN/Telcom
14,767.0	USB	ZBI67	Georgetown, Ascension Island	FX	....	Telcom
14,771.0	LSB	....	numerous, worldwide	FX	....	USN MARS
14,776.2	USB	FTO77	Paris (St. Assise), France	FX	20.0	Telcom
14,777.0	ISB	5VH347	Lome, Togo	FX	20.0	Telcom
14,778.0	USB	....	several, in Pacific Ocean area	FA	....	USAF [Has been
			used for SKY KING broadcasts]			
14,782.0	CW	KWL9Ø	Tokyo, Japan	FX	....	USA EMBASSY

FIXED SERVICE (con.)

kHz	Mode	Call	Location	Service	kW	Remarks
14,786.0	CW	....	unknown	FX	....	
14,792.0	CW	CMU967	Santiago Naval Radio, Cuba	FX	....	Russian Navy
	CW	RIW	Khiva Naval Radio, Uzbek SSR	FX	....	Russian Navy
14,792.5	USB	WMC94	San Francisco (Oakland), CA	FX	20.0	Telcom/Tfc
14,794.8	ISB	XY084	Rangoon, Burma	FX	4.0	Telcom
14,817.5	CW	....	numerous, worldwide	FX	....	INTERPOL [For list of stations see 10,390.0--there will be some call-sign variations--most recently heard have been AYA48, CNT, DEB, EPX5, FSB, Y099, LJP34, 4XP27 & 8UF75]
14,818.0	CW	Y7A6Ø	Berlin, GDR	FX	....	EMBASSY
14,818.5	USB	....	unknown	FX	....	EE/Telcom
14,819.0	USB	....	possibly in RSA	FX	....	Tfc/Telcom
14,825.0	FAX	Y2V25	Berlin (Nauen), GDR	FX	20.0	Nx
14,827.0	CW	TUW22Ø	Abidjan, Ivory Coast	FX	1.0	INTERPOL/Police (Bulletins to Central African Republic, Gabon, Mali, Niamey, Niger, Nigeria, Senegal, Togo & Zaire) [Alt: 19,487.0]
14,834.7	CW	YBFQ	unknown	FX	....	clg "XUQC"
14,842.0	FAX	ATV65	New Delhi, India	FX	18.0	Meteo (0230-1400)
14,850.0	ISB	ROW26	Moscow, USSR	FX	20.0	R. Moscow Feeder
	CW	ZLX37	Wellington (Himatangi), New Zealand	FX	5.0	NOAA reports wx broadcasts at 0200, 0500, 0520, 0815, 0850, 0910, 1800, 2040 & 2110
14,855.0	USB	PSR	Rio de Janeiro (Sepetiba), Brazil	FX	2.5	Telcom
	USB	FT085	Paris (St. Assise), France	FX	15.0	Telcom
14,862.0	CW	NAM	Norfolk Naval Radio, VA	FX	....	USN
14,870.0	CW	CTN37	Apulia Naval Radio, Portugal	FX	....	Port. Navy
14,875.0	ISB	WFE34	New York, NY	FX	....	Telcom
	CW/RTTY	†KNY21	Washington, DC	FX	1.0	Yugoslav EMBASSY
14,880.0	CW	†KKN5Ø	Washington, DC	FX	10.0	EMBASSY
14,891.0	LSB	FT088	Paris (St. Assise), France	FX	15.0	Telcom
14,894.1	CW	A4I	unknown, not in Middle East	FX	....	"VVV's"
14,900.0	USB	ZKS	Rarotonga, Cook Islands	FX	1.0	[Alt: 14,445.0]
	USB	FT09Ø	Paris (St. Assise), France	FX	15.0	Telcom
	USB	....	Hilversum (Kootwijk), Netherlands	FX	....	Radio Netherlands Feeder
14,902.0	USB	AFF21	Tyler, TX	FX	1.6	USAF
	USB	....	numerous, USA nationwide	FX	....	CAP
14,921.0	ISB	†XTA49	Ouagadougou, Upper Volta	FX	20.0	Telcom
14,925.0	USB	OHU2Ø	Helsinki, Finland	FX	1.0	EMBASSY
	USB	OHU25	Beijing, PRC	FX	1.0	Finnish EMBASSY [Alt: 13,913.0 & 15,870.0]
	CW/RTTY	FUB	Paris (Houilles) Naval R., France	FX	15.0	French Navy
14,927.0	CW	8BY	unknown	FX	....	Coded Tfc
14,930.0	ISB	VNV113	Sydney (Doonside), NSW, Australia	FX	30.0	Telcom/PNG
14,945.0	ISB	VNV71	Sydney (Doonside), NSW, Australia	FX	30.0	Telcom/Mawson
14,947.0	USB	....	unknown	FX	....	NUMBERS/YL/GG
14,954.0	USB	OHU2Ø	Helsinki, Finland	FX	1.0	EMBASSY
14,955.0	USB	AFG4	Offutt AFB, NE	FX	0.5	USAF/SAC [Alt: 20,890.0]
14,959.0	USB	....	Sao Jose Campos, Brazil	FX	1.0	USAF
14,960.0	ISB	....	Moscow, USSR	FX	....	Telcom
14,961.0	CW	BPM	Xian, PRC	FX	50.0	TIME SIGNALS
14,967.0	CW	CMU967	Santiago Naval Radio, Cuba	FX	....	Russian Navy
14,967.0	CW	"K"	unknown	RC	....	Beacon
14,968.5	USB	SAM	Stockholm, Sweden	FX	1.2	EMBASSY (This one of the principal channels for Swedish Embassies. On this frequency are: SAM21, Berlin, GFR; SAM25, Lisbon, Portugal; SAM3Ø, Madrid, Spain; SAM4Ø, Warsaw, Poland; SAM48, Damascus, Syria; SAM54, Vientiane, Laos; SAM58, Jakarta, Indonesia; SAM61, New Delhi, India; SAM64, Pyong-yang, N. Korea; SAM79, Lusaka, Zambia; SAM8Ø, Garborone, Botswana; SAM81, Bissau, Guinea-Bissau & SAM82, Maputo, Mozambique) [Alt: 12,304.5 & 18,808.0]

kHz	Mode	Call	Location	Service	kW	Remarks
14,971.5	LSB	SAM	Stockholm, Sweden	FX	1.2	EMBASSY (This is the other principal channel for Embassies and includes: SAM21, Berlin, GFR; SAM26, London, England; SAM35, Belgrade, Yugoslavia; SAM36, Budapest, Hungary; SAM37, Bucharest, Rumania; SAM39, Prague, Czechoslovakia; SAM45, Ankara, Turkey; SAM47, Beirut, Lebanon; SAM49, Jeddah, Saudi Arabia; SAM51, Teheran, Iran; SAM52, Tel Aviv, Israel; SAM55, Bangkok, Thailand; SAM57, Dacca, Bangladesh; SAM59, Hanoi, Vietnam; SAM6Ø, Islamabad, Pakistan; SAM62, Beijing, PRC; SAM71, Dar es Salaam, Tanzania; SAM72, Kinshasa, Zaire; SAM73, Lagos, Nigeria; SAM74, Monrovia, Liberia; SAM75, Nairobi, Kenya; SAM8Ø, Gaborone, Botswana & SAM82, Maputo, Mozambique [Alt: 17,430.0]
14,975.0	CW	XVS	Ho Chi Minh-Ville, Vietnam	FX	4.0	NOAA reports wx broadcasts at 0420, 1020 & 1620
14,976.0	USB	....	Berlin, GDR	FX	....	Telcom
14,982.0	FAX	RBV76	Tashkent, Uzbek SSR	FX	20.0	Meteo
14,985.0	AM	EHY26	Madrid (Pozuelo del Rey), Spain	FX	10.0	Telcom
14,987.0	CW	ZEN69	Cape d'Aguilar, Hong Kong	FX	3.5	Tfc/Vientiane

14,990.0 - 15,005.0 kHz -- STANDARD FREQUENCY & TIME SIGNALS

kHz	Mode	Call	Location	Service	kW	Remarks
14,996.0	CW	RWM	Moscow, USSR	SS	8.0	TIME SIGNALS
15,000.0	A2	LOL1	Buenos Aires, Argentina	SS	2.0	" "
	AM/CW	BPM	Xian, PRC	SS	50.0	" "
	AM	WWVH	Kauai, HI	SS	10.0	" "
	A2	ATA	New Delhi, India	SS	8.0	" "
	A2	JJY	Tokyo (Sanwa, Ibaraki), Japan	SS	2.0	" "
	A2	BSF	Chung-li, Taiwan	SS	5.0	" "
	CW	RTA	Novosibirsk, USSR	SS	5.0	" "
	AM	WWV	Fort Collins, CO	SS	10.0	" "
15,004.0	CW	RID	Irkutsk, USSR	SS	1.0	" "

15,005.0 - 15,010.0 kHz -- STANDARD FREQUENCY/TIME & SPACE RESEARCH

15,010.0 - 15,100.0 kHz -- AERONAUTICAL MOBILE (Worldwide)

kHz	Mode	Call	Location	Service	kW	Remarks
15,015.0	USB	....	numerous, worldwide	FA/MA	....	USAF (Usually AFA3, Andrews AFB, MD; AFL2, Loring AFB, ME; AHF3, Albrook AFS, Panama; AJG9, Incirlik AB, Turkey; AFG37, Scott AFB, IL; AFD14, Ascension Is.; AJO, Adana AB, Turkey; etc.) [May be used for MAINSAIL broadcasts]
15,021.0	USB	....	Stockholm Aeradio, Sweden	FA	5.0	
15,022.5	USB	....	numerous, worldwide	FA/MA	....	USN
15,024.0	CW	COL	Havana Aeradio, Cuba	FA/MA	....	Aeroflot
	CW	RFNV	Moscow Aeradio, USSR	FA/MA	....	Aeroflot (Mostly in-flight to/from Havana/Moscow)
15,027.0	USB	....	numerous, worldwide	FA/MA	....	USN
15,031.0	USB	†....	several, in Canada & Europe	FA	4.0	RCAF (See 15,035.0 for list of stations
	USB	AIF2	Yokota AB, Japan	FA/MA	1.3	USAF
15,034.5	USB	....	numerous, worldwide	FA/MA	....	French AF
15,035.0	USB	CHR	"Trenton Military", Ont., Canada	FA	4.0	RCAF Wx (+30)
		CJX	"St. Johns Military", Nfld., Canada	FA	4.0	" Wx (+40)
		DHM95	"Lahr Military", GFR	FA	4.0	" Wx (+10)
		VXA	"Edmonton Military", Alta., Canada	FA	4.Ø	" Wx (+20)
	USB	AIE2	Andersen AFB, Guam	FA/MA	1.3	USAF
15,038.0	USB	AIF2Ø	Yokota AB, Japan	FA/MA	1.3	USAF
15,041.0	USB	....	numerous	FA	....	USAF (Used for SKY KING broadcasts--some code names for Air Bases--sometimes called SAC "MIKE" frequency)
15,046.0	USB	....	"Cyprus Flight Watch", Cyprus	FA	....	RAF/Wx (+30)
15,048.0	ISB	....	numerous, mostly in the USA	FA/MA	....	USAF (Usually AFA3, Andrews AFB, MD; AFE8, MacDill AFB, FL and VIP a/c "AIR FORCE ONE" & "AIR FORCE TWO")

kHz	Mode	Call	Location	Service	kW	Remarks
15,056.0	USB	ZRB2	Pretoria (Waterkloof AFB), RSA	FA/MA	....	RSA Air Force
15,061.0	USB	....	several, in Japan	FA/MA	....	USN
15,066.0	ISB	CUW3	Lajes Field, Azores	FA/MA	5.0	USAF
15,081.0	USB	....	numerous, worldwide	FA/MA	....	USCG/USN
15,084.0	USB	....	numerous, mostly in the USA	FA/MA	....	USCG/USN [Particularly used by heliocopters]
15,087.0	USB	....	numerous, around the USA coastline	FA/MA	....	USCG/USN
	USB	....	several, in Australia	FA/MA	....	RAAF

15,100.0 - 15,450.0 kHz -- INTERNATIONAL BROADCASTING (Worldwide)

15,100.0 - 15,600.0 kHz -- INTERNATIONAL BROADCASTING (Worldwide--post WARC)

15,450.0 - 16,460.0 kHz -- FIXED SERVICE (Worldwide)

15,600.0 - 16,360.0 kHz -- FIXED SERVICE (Worldwide--post WARC)

kHz	Mode	Call	Location	Service	kW	Remarks
15,462.5	USB	6YF21	Kingston (Coopers HIll), Jamaica	FX	1.0	
15,465.0	CW	RCV	Moscow Naval Radio, USSR	FX	....	Russian Navy
15,470.0	CW	CAN6D	Punta Arenas (Montalva) R., Chile	FX	0.5	Meteo/Reported wx broadcast at 2200
15,473.6	USB	....	Pacific Domestic Emergency Net	FX	....	USCG (Includes: NMC, San Francisco, CA; NMO, Honolulu, HI; NMQ, Long Beach, CA; NOJ, Kodiak, AK & NRV, Barrigada, Guam, plus transportables) [Alt: 12,173.6 & 18,196.1]
15,490.0	LSB	RUC27	Kiev, Ukrainian SSR	FX	20.0	R. Moscow Feeder
15,492.5	CW	KKN5Ø	Washington, DC	FX	10.0	EMBASSY
15,497.0	CW	CMU967	Santiago Naval Radio, Cuba	FX	....	Russian Navy
15,500.0	ISB	†....	Beijing, PRC	FX	120.0	R. Beijing Feeder
15,502.5	CW	AYA5Ø1	Buenos Aires, Argentina	FX	1.0	INTERPOL [Distribution of INTERPOL Bulletins to Brasilia, Caracas, La Paz, Lima, Santiago & Sao Paulo]
15,510.0	CW	SOP251	Warsaw, Poland	FX	20.0	Tfc/Reported PAP nx broadcast at 1800
15,524.0	CW	NPM	Lualualei Naval Radio, HI	FX	15.0	USN/NOTAM
15,527.0	USB	KFK92	Kitts Peak (Tucson), AZ	FX	1.0	Tfc/Chile
15,535.0	USB	....	Tula, USSR	FX	50.0	R. Moscow Feeder
15,540.0	CW	KKN5Ø	Washington, DC	FX	10.0	EMBASSY
15,545.0	USB	ROW23	Moscow, USSR	FX	15.0	R. Moscow Feeder
15,548.6	USB	....	Atlantic Domestic Emergency Net	FX	....	USCG (Includes: NMA, Miami, FL; NMF, Boston, MA; NMG, New Orleans, LA; NMN. Portsmouth, VA; NMR, San Juan, PR & NOZ, Elizabeth City, NC) [Alt: 12,124.6 & 18,196.1]
15,556.0	USB	†CLN483	Havana (Bauta), Cuba	FX	10.0	Tfc
15,564.0	CW/RTTY	SOP256	Warsaw, Poland	FX	20.0	Tfc/Reported PAP nx broadcast at 1400
15,570.0	LSB	S7Q73	Victoria, Seychelles	FX	7.5	Telcom
15,575.0	A3H	4MJ4	Caracas, Venezuela	FX	5.0	Telcom
	USB	....	Riyadh, Saudi Arabia	FX	....	Feeder
15,582.0	CW/RTTY	CYS22	Ottawa, Ont., Canada	FX	1.0	Polish EMBASSY [Alt: 14,684.0 & 16,024.0]
15,582.1	ISB	9HC72	Malta, Malta	FX	6.0	Telcom/Tfc
15,589.0	ISB	....	Daventry, England	FX	30.0	BBC Feeder
15,595.0	USB	V3F	Berlin (Nauen), GDR	FX	20.0	Telcom
15,595.5	USB	VCR976	Ottawa, Ont., Canada	FX	1.0	Saudi Arabian EMBASSY [Alt: 14,748.5 & 17,448.5]
15,600.0	LSB	†RBM76	Moscow, USSR	FX	15.0	R. Moscow Feeder
15,605.0	ISB	FTP6Ø	Paris (St. Assise), France	FX	15.0	Telcom/Tfc
15,610.0	ISB	VLN	Sydney (Doonside), NSW, Australia	FX	10.0	Telcom/Nauru
15,615.0	FAX	AXI35	Darwin, WA, Australia	FX	5.0	Meteo (2100-0800)
15,620.5	FAX	AFA	Washington (Andrews AFB, MD), DC	FX	4.0	USAF/Meteo
15,623.5	USB	†KUQ2ØC	Pago Pago, American Samoa	FX	2.0	Telcom/Tfc
15,628.5	CW/RTTY	CTW32	Monsanto Naval Radio, Portugal	FX	10.0	Port. Navy
15,630.0	ISB	HSP27	Bangkok (Laki), Thailand	FX	10.0	Telcom/Tokyo
15,635.0	ISB	5TN277	Nouakchott, Mauritania	FX	....	Telcom

FIXED SERVICE (con.)

kHz	Mode	Call	Location	Service	kW	Remarks
15,640.0	USB	....	Beijing, PRC	FX	....	PRC Feeder
15,644.0	FAX	VFF	Frobisher CG Radio, NWT, Canada	FX	....	Meteo (July-Oct)
15,652.0	LSB	....	Greenville, NC	FX	50.0	VOA Feeder
15,655.6	CW	"U"	unknown	RC	....	Beacon
15,657.0	CW	ONN32	Brussels, Belgium	FX	1.0	Turkish EMBASSY
15,660.0	USB	5TN2Ø5	Nouakchott, Mauritania	FX	....	Telcom
15,665.0	USB	HZQ566	Riyadh, Saudi Arabia	FX	20.0	Feeder
	USB	....	Stockholm Aeradio, Sweden	FA	7.0	
15,667.0	CW	FDY	Orleans Air, France	FX	....	French AF
15,670.0	USB	....	Daventry, England	FX	30.0	BBC Feeder
15,675.0	LSB	AFH3	Albrook AFS. Panama	FX	3.0	USAF
				[Alt: 13,950.0]		
15,682.0	ISB	FTP68	Paris (St. Assise), France	FX	20.0	Telcom
15,684.0	CW	8UF75	New Delhi, India	FX	5.0	INTERPOL
	CW/RTTY	JPA62	Nagoya (Komaki), Japan	FX	10.0	INTERPOL
15,690.0	USB	ZKY	Palmerston North, New Zealand	FX	5.0	RNZAF/Telcom
					Melbourne/Emergency	
15,700.0	CW	"U"	unknown	RC	....	Beacon
15,704.0	CW	KNY21	Washington, DC	FX	1.0	Yugoslav EMBASSY
	CW	†KNY23	Washington, DC	FX	1.0	Czech EMBASSY
15,705.0	CW	VCS838	Ottawa, Ont., Canada	FX	5.0	Yugoslav EMBASSY
						[Alt: 7445.0 & 18,247]
	CW	"U"	unknown	RC	....	Beacon
15,715.0	ISB	†....	Greenville, NC	FX	50.0	VOA Feeder
15,718.0	CW	ZBP	Pitcairn Island	FX	0.7	Tfc/Suva
15,720.0	LSB	RWN74	Moscow, USSR	FX	15.0	R. Moscow Feeder
15,723.0	USB	....	Mediterranean LORAN Net	FX	....	USCG (Includes:
			AOB5Ø, Estartit, Spain; NCI, Sellia Marina, Italy; NCI3, Lampedusa, Italy & NCI4, Kargabarun, Turkey) [Alt: [13,665.5 & 16,044.5]			
15,724.0	CW/FAX	ISX57	Rome (Santa Rosa), Italy	FX	3.0	ANSA Nx
15,726.0	USB	OHU2Ø	Helsinki, Finland	FX	1.0	EMBASSY
	USB	OHU25	Beijing, PRC	FX	1.0	Finnish EMBASSY
						[Alt: 15,870.0]
15,730.0	ISB	†WFL25	New York, NY	FX	....	Telcom
			[Has been used as an RFE feeder]			
15,735.0	CW	AWB	Bombay Aeradio, India	AX	....	Tfc/70C
	ISB	6VK815	Dakar, Senegal	FX	6.0	Telcom
15,738.2	CW	AYA26	Buenos Aires, Argentina	FX	1.0	INTERPOL [Bul-
			letins to South American capitol cities--some answerback from: PPC, Brasilia, Brazil; ZPZ, Asuncion, Paraguay; OAV86, Lima, Peru & YVZ3, Caracas, Venezuela]			
15,741.2	LSB	FZF57	Fort de France, Martinique	FX	4.0	Telcom/Cayenne
15,752.0	USB	....	Greenville, NC	FX	50.0	VOA Feeder
15,755.0	ISB	MKE	Akrotiri, Cyprus	FX	30.0	RAF Telcom/Tfc
15,760.0	LSB	GYU	Gibraltar Naval Radio, Gibraltar	FX	10.0	British Navy
15,770.0	ISB	....	Greenville, NC	FX	50.0	VOA Feeder
15,770.5	FAX	VFE	Edmonton, Alta., Canada	FX	5.0	Meteo
15,771.4	CW	CCS	Santiago Naval Radio, Chile	FX	....	Chilean Navy
15,775.0	USB	....	Holzkirchen, GFR	FX	10.0	RFE Feeder
15,778.0	CW	WFM55	New York, NY	FX	50.0	Tfc
15,779.8	USB	†RWM71	Moscow, USSR	FX	20.0	R. Moscow Feeder
15,780.0	CW	ZPG57	Asuncion, Paraguay	FX	1.0	Tfc/Buenos Aires
	CW	WQB35	Akron, OH	FX	15.0	Tfc/Harbel
15,785.0	FAX	WFM55	New York, NY	FX	50.0	Nx
15,792.0	LSB	....	Vandenberg AFB, CA	FX	....	USAF/SAMTO
			[Frequently uses callsign with the word "Abnormal"]			
15,800.0	ISB	VNV56	Sydney (Doonside), NSW, Australia	FX	30.0	Telcom/Colombo
15,804.0	CW	†KNY2Ø	Washington, DC	FX	1.0	Polish EMBASSY
	CW	†KNY23	Washington, DC	FX	1.0	Czech EMBASSY
15,805.0	LSB	†FTP8Ø	Paris (St. Assise), France	FX	20.0	Telcom
15,815.0	ISB	MKG	London (Stanbridge), England	FX	30.0	RAF Telcom/Tfc
15,820.0	LSB	NAW38	Guantanamo, Cuba	FX	....	Telcom/Tfc
15,823.5	CW	WFL35	New York, NY	FX	60.0	UPI Nx
15,827.4	ISB	HB075	Geneva, Switzerland	FX	10.0	Telcom/Tfc

kHz	Mode	Call	Location	Service	kW	Remarks
15,832.0	ISB	†TJF58	Douala, Cameroon	FX	20.0	Telcom
	CW	....	unknown, possibly in Sri Lanka	FX	....	
15,835.5	ISB	JBL55	Tokyo, Japan	FX	20.0	Telcom/Tfc
15,839.9	ISB	3XF26	Conakry, Guinea	FX	6.0	Telcom/Tfc
15,845.0	USB	VLV	Mawson Base, Antarctica	FX	5.0	ANARE
	CW/USB	VLZ	Davis Base, Antarctica	FX	1.0	"
	USB	VNJ	Casey Base, Antarctica	FX	10.0	"
	USB	VNM	Melbourne, Vic., Australia	FX	1.0	"
15,849.0	ISB	....	Daventry, England	FX	30.0	BBC Feeder
15,870.0	USB	....	Lyndhurst, Australia	FX	....	R. Australia Feeder
	USB	OHU2Ø	Helsinki, Finland	FX	1.0	EMBASSY
	USB	OHU25	Beijing, PRC	FX	1.0	Finnish EMBASSY [Alt: 15,726.0 & 17,443.0]
15,875.0	USB	....	Northwest Pacific LORAN Net	FX	....	USCG (Includes NRT, "Yokota Monitor", Japan; NRT2, Gesashi, Japan; NRT3, Iwo Jima; NRT9, Hokkaido, Japan; NRV, Barrigada, Guam; NRV6, Marcus Is.; NRV7, Yap Island) [Alt: 13,608.6 & 15,922.0]
15,904.1	CW	RUZU	Molodezhnaya USSR Base, Antarctica	FX	....	SAAM
15,910.0	ISB	....	Daventry, England	FX	30.0	BBC Feeder
15,917.0	CW	KKN44	Monrovia, Liberia	FX	....	USA EMBASSY
15,920.0	CW	CFH	Maritime Command Radio, Halifax, NS, Canada	FX	10.0	NAWS (Canadian Forces)
	ISB	DEP92	Frankfurt, GFR	FX	20.0	Telcom [Has been used as a Deutsche Welle feeder]
15,922.0	USB	....	Northwest Pacific LORAN Net	FX	....	USCG (For list of stations see 15,875.0)
15,935.0	ISB	FPP93	Paris (St. Assise), France	FX	20.0	Telcom/Tfc
15,940.0	CW/RTTY	ELE25	Harbel, Liberia	FX	15.0	Tfc
15,941.5	FAX	CNL	Kenitra, Morocco	FX	15.0	USN/Meteo
15,945.2	FAX	JAM55	Tokyo, Japan	FX	5.0	
15,950.0	CW	JJD2	Tokyo, Japan	FX	10.0	Reported to transmitt solar & geomagnetic reports at 0800
	FAX	RBI77	Moscow, USSR	FX	15.0	Meteo (0220-1745)
15,955.0	CW	3BT4	Bigara, Mauritius	FX	3.0	Meteo
15,957.0	ISB	†CUA77	Lisbon (Alfragide), Portugal	FX	10.0	Telcom/Tfc
15,966.1	USB	PCW1Ø	Islamabad, Pakistan	FX	1.0	Dutch EMBASSY
15,962.0	USB	AFS	Offutt AFB, NE	FX	0.5	USAF/SAC [Alt: 11,607.0 & 17,617.0]
	USB	AFC24	Westover AFB, MA	FX	0.5	USAF/SAC
	USB	AFG14	Grand Forks AFB, ND	FX	0.5	" "
	USB	AFH24	March AFB, CA	FX	0.5	" "
15,982.0	CW	CKN	Vancouver Forces Radio, BC, Canada	FX	....	NOAA reports wx broadcasts at 0030, 0430 & 1330
15,983.0	CW/RTTY	SOP298	Warsaw, Poland	FX	20.0	Tfc
15,990.0	FAX	RAD21	Tashkent, Uzbek SSR	FX	20.0	
15,996.0	USB	ONN27	Brussels, Belgium	FX	1.0	French EMBASSY
	CW	SOP299	Warsaw, Poland	AX	1.0	Tfc/Cairo
16,004.0	CW	IDR5	Rome Naval Radio, Italy	FX	....	Italian Navy
16,016.0	CW	RMP	Rostov, USSR	FX	....	
16,022.0	CW	....	unknown, probably in North America	FX	....	Code Tfc
16,024.0	CW	CYS22	Ottawa, Ont., Canada	FX	....	Polish EMBASSY [Alt: 15,582.0 & 17,532.0]
16,030.0	CW	PBC5	Goeree Naval Radio, Netherlands	FX	5.0	Dutch Navy
	USB	RBI75	Moscow, USSR	FX	15.0	R. Moscow Feeder
16,038.7	CW	....	unknown	FX	....	Nx/FF
16,041.4	USB	5TP294	Nouakchott, Mauritania	FX	0.2	GFR EMBASSY
16,041.5	USB	CTH38	Horta Naval Radio, Azores	FX	5.0	Port. Navy
	LSB	AJE	Croughton (Barford), England	FX	4.0	AFRTS Feeder
16,044.5	USB	....	Mediterranean LORAN Net	FX	....	USCG (Includes AOB5Ø, Estartit. Spain & NCI, Sellia Marina, Italy--days)
16,045.0	CW	FUB	Paris (Houilles) Naval R., France	FX	2.0	French Navy
16,053.2	ISB	HBP26	Geneva, Switzerland	FX	20.0	Telcom
16,056.5	ISB	EHY96	Madrid (Pozuelo del Rey), Spain	FX	4.0	Telcom/Nouadhibou

kHz	Mode	Call	Location	Service	kW	Remarks
16,061.0	CW/RTTY	NAU	San Juan Naval Radio, PR	FX	15.0	USN
16,065.0	ISB	....	Holzkirchen, GFR	FX	10.0	RFE Feeder
	CW/RTTY	†KNY25	Washington, DC	FX	1.0	Rumanian EMBASSY
	CW	RBNF	Moscow Aeradio, USSR	AX	....	
	CW	YZJ7	Belgrade (Makis), Yugoslavia	FX	30.0	Tfc/Tokyo [Alt: 16,343.]
16,090.0	CW	D3M93	Luanda Naval Radio, Angola	FX	....	Russian Navy
16,095.0	USB	GZU	Portsmouth (Petersfield), England	FX	1.0	British Naval Air
	USB	FZS6Ø	St. Denis (Bel Air), Reunion	FX	....	Telcom
16,100.0	CW	†OMZ	Prague, Czechoslovakia	FX	1.0	EMBASSY
16,108.5	USB	ONN34	Brussels, Belgium	FX	1.0	Iranian EMBASSY
16,117.0	USB	6VK317	Dakar, Senegal	FX	6.0	Telcom
16,122.0	USB	....	unknown	FX	....	RR Feeder
16,125.0	CW/RTTY	ZEF81	Salisbury, Zimbabwe	FX	1.0	PRC EMBASSY [Alt: 17,385.0]
16,131.5	USB	OEC36	Beijing, PRC	FX	1.0	Austrian EMBASSY [Alt: 20,972.0]
16,135.0	CW	6WW	Dakar Naval Radio, Senegal	FX	....	French Navy
	FAX	KVM7Ø	Honolulu (International), HI	AX	....	Meteo (1900-2045)
16,140.0	LSB	RGW28	Moscow, USSR	FX	15.0	R. Moscow Feeder
16,152.0	ISB	ZLK36	Christchurch (Weedons), New Zealand	FX	5.0	Telcom/McMurdo
16,162.5	CW/USB	OEC	Vienna, Austria	FX	10.0	EMBASSY
16,165.0	USB	7UP16	Algiers, Algeria	FX	7.0	Telcom
16,170.0	ISB	D2J32	Luanda, Angola	FX	10.0	Telcom
16,175.0	FAX	CML49	Havana (Bauta), Cuba	FX	15.0	Nx
16,180.0	CW	†OMZ	Prague, Czechoslovakia	FX	1.0	EMBASSY
	CW	NAM	Norfolk Naval Radio, VA	FX	10.0	USN/NOAA re- reports wx broadcasts at 0030, 0630, 1230 & 1900/seasonal ice warnings at 1000 & 2300
16,193.0	LSB	RIF35	Moscow, USSR	FX	20.0	R. Moscow Feeder
16,200.0	CW	8BB39	Jakarta, Indonesia	FX	3.0	NOAA reports wx broadcast at 0830
	USB	SOQ22Ø	Warsaw, Poland	FX	20.0	Telcom
16,215.0	ISB	RCF47	Moscow, USSR	FX	80.0	Telcom
16,218.5	CW	NDT6	Totsuka Naval Radio, Japan	FC	5.0	USN Tfc
16,222.0	ISB	....	Bethany, OH	FX	50.0	VOA Feeder
16,228.5	CW/USB	JBQ36	Tokyo, Japan	FX	30.0	Tfc
16,234.0	ISB/RTTY	CSY46	Santa Maria Aeradio, Azores	AX	10.0	Tfc
16,238.0	USB	YAB8	Kabul, Afghanistan	FX	....	Telcom/Moscow
16,238.6	CW	CNL	Kenitra, Morocco	FX	3.0	
16,240.0	ISB	....	Holzkirchen, GFR	FX	10.0	RFE Feeder
16,243.1	CW/RTTY	Y7A64	Berlin (Nauen), GDR	FX	20.0	EMBASSY
16,245.0	CW	....	numerous (ICAO AFI & MID)	AX	....	Aero Tfc (Net in Middle East and northeastern Africa--including: TNL, Brazzaville, Congo; 4XI, Lod, Israel; 5YD, Nairobi, Kenya; 5ST81, Antananarivo, Madagascar, etc.)
16,250.0	USB	†....	Moscow, USSR	FX	....	R. Moscow Feeder
16,251.0	USB	LJA21	Tel Aviv, Israel	FX	0.4	Norwegian EMBASSY [Alt: 16,380.5]
16,265.0	CW	NAM	Norfolk Naval Radio, VA	FC	1.0	USN/NUKO
16,265.1	CW	DHJ59	Sengwarden Naval Radio, GFR	FX	....	(G23B) GFR Navy/ NATO
16,268.0	CW/RTTY	Y7A65	Berlin, GDR	FX	....	EMBASSY
16,270.5	CW/USB	Z2H27	Salisbury, Zimbabwe	FX	0.5	British EMBASSY
16,274.5	ISB	†ZEN77	Cape d'Aguilar, Hong Kong	FX	10.0	Telcom/Tfc
16,276.0	ISB	†D4E29	Praia, Cape Verde	FX	3.0	Telcom/Tfc
16,280.0	ISB	MKD	Akrotiri, Cyprus	FX	10.0	RAF Telcom/Tfc
	USB	†....	numerous, USA nationwide	FX	....	FAA Backup Net
16,281.0	USB	RKD41	Moscow, USSR	FX	15.0	Telcom
16,293.4	CW	F71	unknown, possibly in India	FX	....	Tfc/F72 & F74
16,298.5	USB	OMZ26	Prague, Czechoslovakia	FX	1.0	Egyptian EMBASSY
	USB	....	numerous, worldwide	FX	....	USN MARS
16,310.0	CW	SSL	Alexandria Naval Radio, Egypt	FX	....	Egyptian Navy
16,315.0	FAX	†5YE	Nairobi, Kenya	FX	30.0	Meteo
16,320.0	ISB	AFA	Andrews AFB, MD	FX	....	USAF

kHz	Mode	Call	Location	Service	kW	Remarks
16,323.0	CW	OVG16	Frederikshaven Naval Radio, Denmark	FX	2.0	Danish Marine
16,330.0	ISB	VJU	Kingston Radio, Norfolk Island	FX	15.0	Tfc
	USB	RKA75	Moscow, USSR	FX	20.0	R. Moscow Feeder
16,330.5	LSB	EHY24	Madrid (Pozuelo del Rey), Spain	FX	10.0	Telcom/Las Paz
16,335.0	FAX	FZS63	St. Denis (Bel Air), Reunion	FX	10.0	Meteo (1145)
16,338.0	CW	RIW	Khiva Naval Radio, Uzbek SSR	FX	....	Russian Navy
16,343.0	CW/RTTY	YZI4	Belgrade (Makis), Yugoslavia	FX	30.0	Tfc/Tokyo [Alt: 16,065.0]
16,343.5	USB	CTN84	Apulia Naval Radio, Portugal	FX	5.0	Port. Navy
16,345.0	USB	GZU	Portsmouth (Petersfield), England	FX	1.0	British Naval Air
16,348.0	USB	....	several, USA nationwide	FX	....	FAA (Recently reported included: KDM5Ø, Hampton, GA & "Dragnet") [Alt: 13,630.0]
16,354.0	CW	Y7B94	unknown, probably in Far East	FX	....	GDR EMBASSY
16,355.5	ISB	FZY363	Miquelon, St. Pierre & Miquelon Is.	FX	16.0	Tfc
16,356.0	CW	Y7F48	unknown, probably in Middle East	FX	5.0	GDR EMBASSY

16,360.0 - 17,410.0 kHz -- FIXED SERVICE & MARITIME MOBILE (Worldwide--shared post WARC)

kHz	Mode	Call	Location	Service	kW	Remarks
16,364.0	CW	SNN213	Warsaw, Poland	FX	....	EMBASSY
16,365.0	USB	†FTQ36	Paris (St. Assise), France	FX	20.0	Telcom
16,380.0	ISB	GAP36	London (Rugby), England	FX	30.0	Tfc/Cairo
16,384.0	USB	†FZP64	Papeete (Papenoo), Tahiti	FX	20.0	Telcom
16,386.0	ISB	†TLZ63	Bangui, Central African Republic	FX	20.0	Telcom/Tfc
	ISB	†XTA63	Ouagadougou, Upper Volta	FX	6.0	Telcom
16,392.0	CW	RIW	Khiva Naval Radio, Uzbek SSR	FX	....	Russian Navy
	CW/RTTY	†KNY26	Washington, DC	FX	1.0	Hungarian EMBASSY
16,394.9	CW	OMZ	Prague, Czechoslovakia	FX	1.0	EMBASSY
16,397.0	USB	YKW1Ø7	Damascus, Syria	FX	5.0	Telcom
16,400.0	USB	OZU25	Copenhagen, Denmark	FX	1.5	EMBASSY
	USB	OZU35	Beijing, PRC	FX	0.4	Dannish EMBASSY [Alt: 13,490.0 & 18,587.0]
	FAX	NPM	Lualualei Naval Radio, HI	FX	15.0	USN/Meteo (1600-0630)
16,407.1	LSB	....	unknown, probably in the USA	FX	....	Telcom
16,408.6	USB	FTQ41	Paris (St. Assise), France	FX	20.0	Telcom
16,410.0	USB	LSM4ØØ	Buenos Aires, Argentina	FX	0.5	Prefectura Naval [Net Control]
	FAX	NAM	Norfolk Naval Radio, VA	FX	2.5	USN/Meteo
16,421.5	CW/LSB	....	unknown	FX	....	EMBASSY Tfc
16,425.0	USB	TZB216	Bamako, Mali	FX	20.0	Telcom
16,433.0	LSB	....	Bairiki, Tarawa, Kiribati	FX	....	Feeder
16,438.4	CW	OD044	Beirut, Lebanon	FX	2.0	Tfc
16,440.0	CW	VTG8	Bombay Naval Radio, India	FC	3.5	Indian Navy
16,440.0	USB	....	Moscow, USSR	FX	....	Telcom
16,440.8	CW/USB	ODN22	Damascus, Syria	FX	....	Telcom/Tfc
16,447.0	LSB	GLJ368	London (Ongar), England	FX	20.0	Telcom/Tfc/Nx
16,450.0	CW	ETR45	Addis Ababa, Ethiopia	FX	0.4	Bulgarian EMBASSY
16,452.5	USB	†8RB78	Georgetown, Guyana	FX	3.5	Telcom
16,454.4	LSB	AJE	Croughton (Barford), England	FX	4.0	AFRTS Feeder
16,455.0	CW	YMN6	Ankara, Turkey	FX	1.0	EMBASSY (Also rarely reported are YMN, New Delhi, India; YMN5, Karachi, Pakistan & YMN12, Madrid, Spain)
16,457.0	CW/RTTY	†....	several, in Africa & Middle East	FX	....	Hungarian EMBASSY [Most use call prefix "HGX"]
16,457.5	CW	NMO	COMMSTA Honolulu, HI	FX	10.0	USCG/NERK/NUKO Wx broadcasts at 0100, 0400, 0700, 1300 & 2000
16,458.0	CW	KRH5Ø	London, England	FX	....	USA EMBASSY
16,459.0	CW/RTTY	VCS838	Ottawa, Ont., Canada	FX	5.0	Yugoslav EMBASSY [Alt: 14,515.0]

16,460.0 - 17,360.0 kHz -- MARITIME MOBILE (Worldwide)

MARITIME MOBILE

kHz	Mode	Call	Location	Service	kW	Remarks
16,460.0	USB		The band between these two frequencies is divided into 41 channels spaced 3.1 kHz apart and assigned numbers from 1601 to 1641. These are ship transmitting channels and are paired with Coastal USB Stations operating between 17,232.9 and 17,356.9 kHz.			
16,584.0	USB					
16,463.1	USB	....	off northeast Australian coast	FX	....	SEDCO Oil Rigs
16,518.9	USB	WCM	Pittsburgh, PA	FC	0.8	Inland Waterway
	USB	WJG	Memphis, TN	FC	0.8	Inland Waterway
16,523.4	AM/USB	....	INTERNATIONAL SHIP CALLING FREQUENCY			
16,587.1	USB	....	Atlantic & Pacific Voice Circuits	FC/MS	....	USCG (Includes: NMA, Miami, FL; NMC, San Francisco, CA; NMF, Boston, Ma; NMG, New Orleans, LA; NMN, Portsmouth, VA; NMO, Honolulu, HI; NMQ, Long Beach, CA; NMR, San Juan, PR; NOJ, Kodiak, AK & NRV, Barrigada, Guam)
	USB	....	Coast & Ship Simplex Voice Frequency	FC/MS	..	Inland & inter-coastal waterway traffic. Employed worldwide by hundreds of base stations and various types of ships. Frequently heard in North America: KHT, Cedar Rapids, IA; WJK, Jacksonville, FL; KHU, San Diego, CA; KFC, Seattle, WA; WMP, West Palm Beach, FL; etc.
16,590.2	USB	....	Coast & Ship Simplex Voice Frequency	FC/MS	..	Inland & inter-coastal waterway traffic. Similar stations heard here as on 16,587.1
16,593.3	USB	....	Coast & Ship Simplex Voice Frequency	FC/MS	..	Inland & inter-coastal waterway traffic. Similar stations heard here as on 16,587.1
16,599.5	CW	FUF	Fort de France Naval R., Martinique	FC	10.0	French Navy
16,720.0	CW		The band between these two frequencies is divided into 18 channels spaced 1.6 kHz apart and assigned numbers from 1 to 18. These are ship calling/transmitting channels.			
16,748.8	CW					
16,861.7	CW	PKB	Belawan Radio, Indonesia	FC	1.0	
	CW	PKI	Jakarta Radio, Indonesia	FC	1.0	
	CW	PKD	Surabaya Radio, Indonesia	FC	1.0	
	CW	SPE82	Szczecin Radio, Poland	FC	10.0	[aka SPB8]
	CW	WNU35	Slidell Radio, LA	FC	3.0	
16,861.8	CW	5OW16	Lagos Radio, Nigeria	FC	1.0	
16,862.5	CW	XSV	Tianjin Radio, PRC	FC	5.0	
	CW/RTTY	UKW3	Korsakov Sakhalinsk Radio, USSR	FC	15.0	
16,863.3	CW	HEB	Berne Radio, Switzerland	FC	10.0	
16,865.0	CW	UJY	Kaliningrad Radio, USSR	FC	15.0	
	CW	URB2	Klaipeda Radio, Lithuanian SSR	FC	....	
16,867.5	CW	UXN	Arkhangelsk Radio, USSR	FC	15.0	
16,868.0	CW	A4M	Muscat Radio, Oman	FC	1.0	
16,868.5	CW	9VG53	Singapore Radio, Singapore	FC	5.0	
16,870.0	CW	RIT	unknown, probably in the USSR	FC	....	
16,870.4	CW	DZJ	Manila (Bulacan) Radio, Philippines	FC	1.5	Reported wx broadcasts at 0300 & 1000
16,871.3	CW	XSG	Shanghai Radio, PRC	FC	30.0	
	CW	CWA	Cerrito Radio, Uruguay	FC	2.0	Reported wx broadcasts at 0000, 1400 & 1900
16,871.4	CW	KLC	Galveston Radio, TX	FC	3.8	NOAA reports wx broadcasts at 1130 & 1730
16,872.5	CW	UXN	Arkhangelsk Radio, USSR	FC	15.0	
	CW	UAT	Moscow Radio, USSR	FC	5.0	
16,873.5	USB	ZLS6	Irirangi Naval Radio, New Zealand	FC	10.0	RNZ Navy
16,874.1	CW	TIM	Limon Radio, Costa Rica	FC	10.0	
16,874.2	CW	ZSD52	Durban Radio, RSA	FC	3.5	
16,874.5	CW	ZLO6	Irirangi Naval Radio, New Zealand	FC	....	RNZ Navy

MARITIME MOBILE (con.)

kHz	Mode	Call	Location	Service	kW	Remarks
16,876.1	CW	XSM	Xiamen Radio, PRC	FC	6.0	
	CW	FUG3	La Regine (Castelnaudary), France	FC	....	French Navy
	CW	6WW	Dakar Naval Radio, Senegal	FC	10.0	French Navy
16,877.5	CW	JDB	Nagasaki Radio, Japan	FC	15.0	
16,878.0	CW	URB2	Klaipeda Radio, Lithuanian SSR	FC	....	
16,879.0	CW	ICB	Genoa P.T. Radio, Italy	FC	5.0	
16,880.0	CW/RTTY	3SB	Datong Naval Radio, PRC	FC	15.0	PRC Navy
	CW	YIR	Basrah Control, Iraq	FC	1.0	
16,880.9	CW	NMC	COMMSTA San Francisco, CA	FC	10.0	USCG/AMVER
16,881.0	CW	ICB	Genoa P.T. Radio, Italy	FC	5.0	
16,882.5	CW/TOR	GKS6	Portishead Radio, England	FC	15.0	
16,882.9	CW	XSQ7	Guangzhou Radio, PRC	FC	7.0	
16,883.0	CW	JOU	Nagasaki Radio, Japan	FC	15.0	
16,883.2	CW	5BA	Nicosia Radio, Cyprus	FC	....	
16,884.5	CW	ODR	Beirut Radio, Lebanon	FC	1.0	
	CW	XSQ7	Guangzhou Radio, PRC	FC	15.0	
16,885.0	CW	RMP	Rostov Radio, USSR	FC	....	
16,886.0	CW	UJY	Kaliningrad Radio, USSR	FC	15.0	
16,887.3	CW	SPH81	Gdynia Radio, Poland	FC	10.0	[aka SPA8]
	CW	YUR	Rijeka Radio, Yugoslavia	FC	6.0	
16,890.5	CW	UFN	Novorossiysk Radio, USSR	FC	15.0	
16,890.8	CW	ZSC39	Cape Town Radio, RSA	FC	10.0	
16,892.9	CW	Y5M	Ruegen Radio, GDR	FC	15.0	
16,893.5	CW	JFC	Misaki Gyogyo Radio, Japan	FC	1.0	
16,895.3	CW	IAR7	Rome P.T. Radio, Italy	FC	15.0	
16,896.4	CW	XFU	Veracruz Radio, Mexico	FC	0.5	
16,897.4	CW/TOR	OXZ82	Lyngby Radio, Denmark	FC	10.0	
16,897.5	CW	OMC	Bratislava Radio, Czechoslovakia	FC	1.0	
	CW	OMK	Komarno Radio, Czechoslovakia	FC	1.0	
16,902.0	CW	PWZ	Rio de Janeiro Naval Radio, Brazil	FC	10.0	Brazil Navy
	CW	XSC	Fangshan Radio, PRC	FC	10.0	
	CW/RTTY	3SD	Fangshan Naval Radio, PRC	FC	15.0	PRC Navy
	CW	PCH6Ø	Scheveningen Radio, Netherlands	FC	10.0	AMVER
16,902.5	CW	DYP	Cebu Radio, Philippines	FC	0.4	
16,903.0	CW	SVM6	Athens Radio, Greece	FC	5.0	
16,904.0	CW	CWM	Montevideo Armada Radio, Uruguay	FC	2.0	Uruguayan Navy
16,904.8	CW	FUV	Djibouti Naval Radio, Djibouti	FC	20.0	French Navy
16,906.0	CW	YIR	Basrah Control, Iraq	FC	5.0	
16,907.3	CW	Y5M	Ruegen Radio, GDR	FC	15.0	
16,909.0	CW	HWN	Paris (Houilles) Naval R., France	FC	....	French Navy
16,909.5	CW	UJO5	Izmail Radio, Ukrainian SSR	FC	1.5	
16,909.7	CW	TFA	Reykjavik Radio, Iceland	FC	1.5	
	CW	PBC217	Goeree Naval Radio, Netherlands	FC	10.0	Dutch Navy
	CW	NMC	COMMSTA San Francisco, CA	FC	10.0	USCG
	CW	....	numerous	FA/FC/FX	....	USCG [Command Net Calling Frequency]
16,910.0	CW	HLJ	Seoul Radio, South Korea	FC	3.5	
16,912.0	CW	LPA2Ø	Rio Grande Radio, Argentina	FC	3.0	
16,913.0	CW	UKX	Nakhodka Radio, USSR	FC	5.0	
	CW	UKA	Vladivostok Radio, USSR	FC	15.0	
16,914.5	CW	SPH84	Gdynia Radio, Poland	FC	10.0	[aka SPK8]
	CW	UGW	Rey Aleksandrovsk Radio, USSR	FC	5.0	
16,915.0	CW/RTTY	FUX	St. Denis Naval Radio, Reunion	FC	5.0	French Navy
16,916.5	CW	XSG8	Shanghai Radio, PRC	FC	10.0	
16,918.0	CW	PPJ	Juncao Radio, Brazil	FC	1.0	NOAA reports wx broadcasts at 0130, 0730 & 1900
	CW	LZW6	Varna Radio, Bulgaria	FC	1.5	
16,918.8	CW	VHP6	COMMSTA Canberra, ACT, Australia	FC	10.0	RA Navy/NOAA reports wx broadcasts at 1730 & 2130
	CW	GKJ6	Portishead Radio, England	FC	15.0	
	CW	WPA	Port Arthur Radio, TX	FC	5.0	
16,920.0	CW	OXZ81	Lyngby Radio, Denmark	FC	10.0	
	CW	9GX	Tema Radio, Ghana	FC	10.0	
16,921.0	CW	CLS	Havana (Industria Pesquera) R., Cuba	FC	5.0	
16,922.0	CW	UQA4	Kiev Radio, Ukrainian SSR	FC	25.0	

kHz	Mode	Call	Location	Service	kW	Remarks
	CW	UDK2	Murmansk Radio, USSR	FC	10.0	
16,923.8	CW	OFJ5	Helsinki Radio, Finland	FC	10.0	
16,925.0	CW/RTTY	URL	Sevastopol Radio, Ukrainian SSR	FC	15.0	
16,925.4	CW	LSO3	Buenos Aires Radio, Argentina	FC	1.0	Prefectura Naval Reported NAVAREA broadcast at 2200/Reported wx broadcasts at 0210, 1210 & 1810
16,926.5	CW	CFH	Maritime Command Radio, Halifax, NS, Canada (Canadian Forces)	FC	10.0	(C13L) NOAA reports wx broadcasts at 0200, 0630, 1400, 1800, 1930--seasonal ice at 0230 & 1430
16,927.0	CW	UJY	Kaliningrad Radio, USSR	FC	15.0	
16,928.0	CW	LZS35	Sofia Naval Radio, Bulgaria	FC	....	
16,928.4	CW	LFX	Rogaland Radio, Norway	FC	5.0	
16,928.5	CW	LZL6	Bourgas Radio, Bulgaria	FC	1.0	
16,930.0	CW	OSN16	Oostende Naval Radio, Belgium	FC	1.0	Belgium Navy
	CW	UXB6	unknown	FC	....	
16,930.4	CW	5RS	Tamatave Radio, Madagascar	FC	2.0	
16,931.0	CW	SXA3	Spata Attikis Naval R., Greece	FC	5.0	(K13A) Greek Navy/NATO
16,932.0	CW	7TA1Ø	Algiers Radio, Algeria	FC	10.0	
16,932.2	CW	JOS	Nagaski Radio, Japan	FC	3.0	
16,933.2	CW	WCC	Chatham Radio, MA	FC	30.0	
16,934.0	CW	RMP	Rostov Radio, USSR	FC	....	
16,935.0	CW	VWB	Bombay Radio, India	FC	1.0	
16,936.0	CW	CCS	Santiago Naval Radio, Chile	FC	....	Chilean Navy
	CW	SXA4	Spata Attikis Naval Radio, Greece	FC	5.0	Greek Navy/NATO
	CW	FFD	St. Denis Radio, Reunion	FC	2.0	
16,937.2	CW	GYA	Whitehall (London) Naval R. England	FC	5.0	British Navy
16,938.0	FAX	GZZ44	Whitehall (Northwood) Naval R., Eng.	FC	5.0	British Navy/ Meteo [aka GYD6]
16,940.0	CW	XSW	Kaohsiung Radio, Taiwan	FC	2.5	
16,942.0	CW	5OZ23	Port Harcourt Radio, Nigeria	FC	1.0	
	CW	RCV	Moscow Naval Radio, USSR	FC	15.0	Russian Navy
16,942.7	CW	YUR7	Rijeka Radio, Yugoslavia	FC	6.0	
16,942.8	CW	EAT5	Santa Cruz de Tenerife R., Canary Is.	FC	0.8	
16,945.0	CW	VNP	Cape Leveque Radio, WA, Australia	FC	1.0	
16,947.0	CW	UFB	Odessa Radio, Ukrainian SSR	FC	5.0	
	CW	UDH	Riga Radio, Latvian SSR	FC	5.0	
16,947.6	CW	VIP5	Perth Radio, WA, Australia	FC	10.0	AMVER/NOAA reports wx broadcasts at 0100 & 1300
	CW	8PO	Barbados Radio, Barbados	FC	1.0	
	CW	VRT	Bermuda (Hamilton) Radio, Bermuda	FC	1.0	
	CW	CBV2	Valparaiso Naval Radio, Chile	FC	1.0	Chilean Navy/ AMVER
	CW	FFT8	St. Lys Radio, France	FC	10.0	
	CW	8RB	Demerara Radio, Guyana	FC	1.0	
	CW	6YI	Kingston Radio, Jamaica	FC	1.0	
16,947.8	CW	TUA9	Abidjan Radio, Ivory Coast	FC	4.0	
16,948.0	CW	HCG	Guayaquil Radio, Ecuador	FC	0.3	AMVER
	CW	RCV	Moscow Naval Radio, USSR	FC	....	Russian Navy
	CW	RIT	unknown, probably in the USSR	FC	....	
16,948.5	CW	VCS	Halifax CG Radio, NS, Canada	FC	1.0	AMVER
16,949.0	CW	ULZ	Ventspils Radio, Latvian SSR	FC	5.0	
	CW/RTTY	UQK	Riga Radio, Latvian SSR	FC	20.0	
	CW/RTTY	UPW2	Liepaja Radio, Latvian SSR	FC	5.0	
16,950.0	CW	XVG	Haiphong Radio, Vietnam	FC	....	
16,950.1	CW	XSQ4	Guangzhou Radio, PRC	FC	10.0	
16,951.5	CW	FUX	Le Port Naval Radio, Reunion	FC	10.0	French Navy
	CW	6WW	Dakar Naval Radio, Senegal	FC	10.0	French Navy
16,952.2	CW	LFT	Rogaland Radio, Norway	FC	5.0	AMVER
16,954.0	CW	YQI4	Constanta Radio, Rumania	FC	1.0	
16,954.4	CW/FAX	GKC6	Portishead Radio, England	FC	15.0	
16,955.0	CW/RTTY	UQK	Riga Radio, Latvian SSR	FC	20.0	
	CW/RTTY	UPW2	Liepaja Radio, Latvian SSR	FCX	5.0	
16,956.0	CW	PZN26	Paramaribo Radio, Suriname	FC	1.0	
16,956.3	CW	TIM	Limon Radio, Costa Rica	FC	10.0	

MARITIME MOBILE (con.)

kHz	Mode	Call	Location	Service	kW	Remarks
16,958.0	CW/RTTY	FUJ	Noumea Naval Radio, New Caledonia	FC	20.0	French Navy
16,959.0	CW	FUV	Djibouti Naval Radio, Djibouti	FC	10.0	French Navy
16,959.2	CW	CUG	Sao Miguel Radio, Azores	FC	5.0	
	CW	CUB	Madeira Radio, Madeira Islands	FC	5.0	
	CW	CUL22	Lisbon Radio, Portugal	FC	15.0	[aka CUA48]
16,960.0	CW	CKN	Vancouver Forces Radio, BC, Canada	FC	15.0	C. Forces/NAWS
	CW	HZY	Ra's Tannurah Radio, Saudi Arabia	FC	1.0	ARAMCO (NOAA reports wx broadcasts on request)
	CW	6VA7	Dakar Radio, Senegal	FC	1.0	
	CW	UMV	Murmansk Radio, USSR	FC	15.0	
16,961.0	CW	CLA4Ø	Havana (Cojimar) Radio, Cuba	FC	5.0	
16,961.5	CW	FUG	La Regine (Castelnaudry) R., France	FC	10.0	French Navy
	CW	FUF	Fort de France Naval, R., Martinique	FC	10.0	" "
	CW	FUX	Le Port Naval Radio, Reunion	FC	10.0	" "
16,962.0	CW	DXF	Davao Radio, Philippines	FC	0.4	
16,962.5	CW	CBV2	Valparaiso Radio, Chile	FC	....	
16,963.0	CW	5AL	Tobruk Radio, Libya	FC	0.5	
	CW	UJY	Kaliningrad Radio, USSR	FC	15.0	
16,964.2	CW	ZRQ6	Cape (Simonstown) Naval Radio, RSA	FC	15.0	RSA Navy
16,965.0	CW/RTTY	Y5M	Ruegen Radio, GDR	FC	10.0	
	CW	UFW	Vladivostok Radio, USSR	FC	15.0	
16,966.0	CW	ZRQ6	Cape (Simonstown) Naval Radio, RSA	FC	15.0	RSA Navy
16,966.2	CW	SVD6	Athens Radio, Greece	FC	10.0	
16,966.5	CW	9VG58	Singapore Radio, Singapore	FC	5.0	
16,967.0	CW	YKM5	Baniyas Radio, Syria	FC	1.0	
16,967.5	CW	WLO	Mobile Radio, AL	FC	30.0	
16,968.0	CW	PPL	Belem Radio, Brazil	FC	1.0	
16,968.5	CW	WLO	Mobile Radio, AL	FC	30.0	
16,970.0	CW	UQB	Kholmsk Radio, USSR	FC	5.0	Reported wx broadcast at 0025
	CW	URD	Leningrad Radio, USSR	FC	15.0	
16,971.0	FAX	JJC	Tokyo Radio, Japan	FC	5.0	Meteo/Nx
16,972.0	CW	WCC	Chatham Radio, MA	FC	22.5	
16,973.6	CW	XST	Qingdao Radio, PRC	FC	30.0	
16,974.1	CW	SPE81	Szczecin Radio, Poland	FC	10.0	[aka SPE8]
16,974.6	CW	GKD6	Portishead Radio, England	FC	10.0	
16,975.0	CW	VWM	Madras Radio, India	FC	....	
16,976.0	CW	NMN	COMMSTA Portsmouth, VA	FC	10.0	USCG/AMVER
16,976.8	CW	JFH	Hamajima Gyogyo Radio, Japan	FC	1.0	
16,977.0	CW	OMP6	Prague Radio, Czechoslovakia	FC	5.0	
16,978.4	CW	3BM6	Mauritius (Bigara) Radio, Mauritius	FC	5.0	NOAA reports wx broadcasts at 0130, 0430, 0900 & 1600. Seasonal cyclone warnings broadcast at 2030
16,980.0	CW	UNQ	Novorossiysk Radio, USSR	FC	5.0	
16,980.4	CW	DAM	Norddeich Radio, GFR	FC	10.0	TIME SIGNALS
16,981.6	CW	SVG6	Athens Radio, Greece	FC	5.0	
16,983.5	CW	EQZ	Abadan Radio, Iran	FC	0.5	
16,983.2	CW	NMR	COMMSTA San Juan, PR	FC	10.0	USCG/AMVER
	CW	URD	Leningrad Radio, USSR	FC	15.0	
16,984.0	CW	PPR	Rio Radio, Brazil	FC	1.0	TIME SIGNALS/Tfc
	CW	UFB	Odessa Radio, Ukrainian SSR	FC	15.0	
16,986.0	CW	PPL	Belem Radio, Brazil	FC	....	
	CW	CTP97	Oeiras Naval Radio, Portugal	FC	5.0	Port. Navy/NATO
16,987.0	CW	VPS79	Cape d'Aguilar Radio, Hong Kong	FC	3.5	
16,987.2	CW	GYU	Gibraltar Naval Radio, Gibraltar	FC	10.0	British Navy
16,990.0	CW	HLO	Seoul Radio, South Korea	FC	3.5	
	CW/RTTY	UFH	Petropavlovsk Radio, USSR	FC	20.0	
	CW	USW2	Rostov Radio, USSR	FC	....	
16,991.0	CW	FUB	Paris (Houilles) Naval R., France	FC	....	French Navy
16,992.8	CW	UAT	Moscow Radio, USSR	FC	15.0	
16,993.0	CW/RTTY	UBN	Jdanov Radio, Ukrainian SSR	FC	15.0	[425/66N]
16,995.0	CW	SVF6	Athens Radio, Greece	FC	5.0	
	CW	9KK22	Kuwait Radio, Kuwait	FC	5.0	
	CW	EBA	Madrid Radionaval, Spain	FC	10.0	Spanish Navy
	CW	UFN	Novorossiysk Radio, USSR	FC	15.0	

kHz	Mode	Call	Location	Service	kW	Remarks
16,997.0	CW	UDH	Riga Radio, Latvian SSR	FC	5.0	
16,997.6	CW	WSL	Amagansett Radio, NY	FC	10.0	
16,998.5	CW	JDC	Choshi Radio, Japan	FC	15.0	
17,000.0	CW	Y5M	Ruegen Radio, GDR	FC	10.0	
	CW	UEK	Feodosia Radio, USSR	FC	1.0	
17,002.5	CW	CBM2	Magallanes (Punta Arenas) R., Chile	FC	2.0	
	CW	XSG29	Shanghai Radio, PRC	FC	15.0	Reported wx broadcasts at 0000 and every 3rd hr/NAVAREA at 0200
17,004.0	CW	HKB	Barranquilla Radio, Colombia	FC	0.5	
17,004.2	CW/RTTY	ZRH6	Cape (Fisantekraal) Naval R., RSA	FC	15.0	RSA Navy
17,005.0	CW	IAR37	Rome P.T. Radio, Italy	FC	15.0	
17,007.2	CW	PCH61	Scheveningen Radio, Netherlands	FC	10.0	AMVER
	CW	KLB	Seattle, WA	FC	3.5	
17,010.0	CW	URD	Leningrad Radio, USSR	FC	15.0	
17,014.0	CW	UEK	Feodosia Radio, USSR	FC	1.0	
17,015.0	CW	UJQ	Kiev Radio, Ukrainian SSR	FC	15.0	
17,015.3	CW	"D"	unknown	FX	....	Beacon
17,016.0	CW	SPH82	Gydnia Radio, Poland	FC	10.0	[aka SPH8]
	CW/DATA	"C"	unknown	RC	....	Beacon
17,016.5	CW	KPH	San Francisco Radio, CA	FC	30.0	
17,017.0	CW	"F"	unknown	RC	....	Beacon
17,017.1	CW	OST6	Oostende Radio, Belgium	FC	30.0	
17,017.9	CW	EBA	Madrid Radionaval, Spain	FC	20.0	(EBCQ) Spanish Navy/Reported NAVAREA broadcast at 0918, 1248 & 1618
17,018.0	CW	ZSC44	Cape Town Radio, RSA	FC	10.0	TIME SIGNALS NOAA reports wx broadcasts at 0930 & 1730
17,020.0	CW	UDK2	Murmansk Radio, USSR	FC	25.0	
17,020.8	CW	WSL	Amagansett Radio, NY	FC	10.0	
17,021.5	CW	TAH	Istanbul Radio, Turkey	FC	3.0	
17,021.6	CW/RTTY	UGG9	Belomorsk Radio, USSR	FC	5.0	
17,025.0	CW	UPB	Providenia Bukhta Radio, USSR	FC	15.0	
17,026.0	CW	KFS	San Francisco Radio, CA	FC	30.0	
17,027.0	CW	FFL8	St. Lys Radio, France	FC	10.0	
17,028.0	CW	PBC317	Goeree Naval Radio, Netherlands	FC	....	(N13A) Dutch Navy
17,029.0	CW	JMC6	Tokyo Radio, Japan	FC	2.5	NOAA reports wx broadcasts at 0318 & 2118
17,030.5	CW	GYA	Whitehall (London) Naval R., England	FC	10.0	British Navy
17,030.8	CW	GYC6	Whitehall (London) Naval R., England	FC	10.0	British Navy
17,032.0	CW	SXE	Aspropyrgos Radio, Greece	FC	3.5	Greek Coast
17,033.0	CW	UKW3	Korsakov Sakhalinsk Radio, USSR	FC	5.0	
17,036.0	CW	9HD	Malta Radio, Malta	FC	5.0	
	CW	UXN	Arkhangelsk Radio, USSR	FC	15.0	
17,037.0	CW	YQI4	Constanta Radio, Rumania	FC	1.0	
17,038.6	CW/RTTY	URL	Sevastopol Radio, Ukrainian SSR	FC	10.0	[170/66N]
17,040.0	CW	PJK317	Suffisant Dorp Naval Radio, Curacao	FC	10.0	Dutch Navy
17,040.8	CW	FFS8	St. Lys Radio, France	FC	10.0	
	CW	FJA26	Mahina Radio, Tahiti	FC	1.0	
17,043.2	CW	JCU	Choshi Radio, Japan	FC	15.0	
17,045.0	CW	ROT2	Moscow Naval Radio, USSR	FC	15.0	Russian Navy
17,045.6	CW	LPD46	Gen. Pacheco Radio, Argentina	FC	15.0	
	CW	YUR8	Rijeka Radio, Yugoslavia	FC	6.0	
	CW	HKC	Buenaventura Radio, Colombia	FC	0.5	
	CW	4PB	Colombo Radio, Sri Lanka	FC	1.0	
17,048.0	CW	DAF	Norddeich Radio, GFR	FC	15.0	
17,049.0	CW	UFB	Odessa Radio, Ukrainian SSR	FC	15.0	
17,050.0	CW	URB2	Klaipeda Radio, Lithuanian SSR	FC	15.0	(4LI) [aka RWWN]
	CW	UKK3	Nakhodka Radio, USSR	FC	15.0	
17,050.1	CW	4XZ	Haifa Naval Radio, Israel	FC	2.0	Israel Navy
17,050.4	CW	ASK	Karachi Radio, Pakistan	FC	3.5	
17,050.5	CW	S3D	Chittagong Radio, Bangladesh	FC	1.0	
17,052.5	CW	JNA	Tokyo Naval Radio, Japan	FC	5.0	JSDA/Reported NAVAREA broadcasts at 0005, 0220, 0405, 0805 & 1205
17,055.0	CW	LZW62	Varna Radio, Bulgaria	FC	1.5	
	CW	UDK2	Murmansk Radio, USSR	FC	25.0	

MARITIME MOBILE (con.)

kHz	Mode	Call	Location	Service	kW	Remarks
17,055.2	CW	CTV7	Monsanto Naval Radio, Portugal	FC	6.0	Port. Navy
17,055.6	CW	D4A8	Sao Vicente Radio, Cape Verde	FC	3.0	
17,057.2	CW	SAB8	Goteborg Radio, Sweden	FC	5.0	
17,059.0	CW	UAT	Moscow Radio, USSR	FC	10.0	
17,060.0	CW	4XO	Haifa Radio, Israel	FC	2.5	
17,061.3	CW	UFB	Odessa Radio, Ukrainian SSR	FC	15.0	
17,062.0	CW	FUX	Le Port Naval Radio, Reunion	FC	10.0	French Navy
17,064.0	CW	SPH83	Gydnia Radio, Poland	FC	10.0	[aka SPI8]
	CW	UCO	Yalta Radio, Ukrainian SSR	FC	5.0	
17,064.9	CW	KOK	Los Angeles Radio, CA	FC	15.0	NOAA reports wx broadcast at 1650
17,065.2	CW	EDZ6	Aranjuez Radio, Spain	FC	15.0	
17,066.0	CW	UAT	Moscow Radio, USSR	FC	10.0	
17,068.5	CW	OXZ8	Lyngby Radio, Denmark	FC	10.0	
17,069.6	CW	JJC	Tokyo Radio, Japan	FC	15.0	NTM
	FAX	JJC	Tokyo Radio, Japan	FC	2.0	Meteo
17,069.6	CW	XFM2	Manzanillo Radio, Mexico	FC	1.0	
	CW	SPE83	Szczecin Radio, Poland	FC	10.0	[aka SPL8]
	CW/RTTY	UJQ7	Kiev Radio, Ukrainian SSR	FC	25.0	
17,071.0	CW	XFS2	Ciudad Madero Radio, Mexico	FC	1.0	
17,071.4	CW	DZF	Manila (Bacoor) Radio, Philippines	FC	0.4	Reported wx broadcasts at 1600 & 2200
17,072.0	CW	GKG6	Portishead Radio, England	FC	10.0	
	CW	DZE	Mandaluyong Radio, Philippines	FC	0.4	Reported wx broadcast at 1000
17,074.4	CW	PNK	Jayapura Radio, Indonesia	FC	1.0	
	CW	LGX	Rogaland Radio, Norway	FC	5.0	AMVER
17,075.0	CW	UAH	Tallinn Radio, Estonian SSR	FC	5.0	
	CW	ZSD47	Durban Radio, RSA	FC	3.5	
17,079.0	CW	HLF	Seoul Radio, South Korea	FC	10.0	
17,079.4	CW	SAG8	Goteborg Radio, Sweden	FC	5.0	AMVER
17,080.0	CW	XSQ	Guangzhou Radio, PRC	FC	5.0	
17,081.6	CW	JFA	Chuo Gyogyo (Matsudo) Radio, Japan	FC	3.0	
17,082.0	CW	DFR28	Frankfurt (Usingen), GFR	FC	15.0	MARPRESS/Reported GG nx broadcast at 1618
17,083.7	CW	IQX	Trieste P.T. Radio, Italy	FC	6.0	
17,085.0	CW	UBN	Jdanov Radio, Ukrainian SSR	FC	15.0	
17,086.0	CW	JFA	Chuo Gyogyo (Matsudo) Radio, Japan	FC	3.0	
17,088.0	CW	RIW	Khiva Naval Radio, Uzbek SSR	FC	....	Russian Navy
17,088.8	CW/RTTY	CTU7	Monsanto Naval Radio, Portugal	FC	6.0	Port. Navy
17,088.8	CW	KPH	San Francisco Radio, CA	FC	30.0	
17,090.0	CW	UKX	Nakhodka Radio, USSR	FC	25.0	Reported nx broadcast at 0100
17,092.0	CW	GKH6	Portishead Radio, England	FC	10.0	
	CW	UFN	Novorossiysk Radio, USSR	FC	5.0	
17,093.5	CW	AQP7	Karachi Naval Radio, Pakistan	FC	7.0	Pakistan Navy
17,093.6	CW	JOR	Nagasaki Radio, Japan	FC	15.0	
	CW	WMH	Baltimore Radio, MD	FC	10.0	
17,094.8	CW	SVA6	Athens Radio, Greece	FC	10.0	
17,096.0	CW	VPS8Ø	Cape d'Aguilar Radio, Hong Kong	FC	2.5	TIME SIGNALS/ NOAA reports wx broadcasts at 0118, 0430 & 0718
17,098.4	CW	GKA6	Portishead Radio, England	FC	15.0	NOAA reports wx broadcasts at 0930 & 2130/Reported NAVAREA at 0730 & 1730
17,100.0	CW	Y5M	Ruegen Radio, GDR	FC	15.0	
	CW	UFN	Novorossiysk Radio, USSR	FC	15.0	
	CW	UOP	Tuapse Radio, USSR	FC	5.0	
17,103.2	CW	XSG3	Shanghai Radio, PRC	FC	20.0	
17,104.2	CW	PCH62	Scheveningen Radio, Netherlands	FC	15.0	AMVER
17,105.0	CW	WWD	La Jolla (Scripps) Radio, CA	FC	5.0	NOAA
	CW	IRM6	CIRM, Rome Radio, Italy	FC	7.5	AMVER/MEDICO
	CW/RTTY	UFW	Vladivostok Radio, USSR	FC	5.0	
17,107.0	CW	UTA	Tallinn Radio, Estonian SSR	FC	5.0	
17,108.0	CW	FUF	Fort de France Naval R., Martinique	FC	20.0	French Navy

kHz	Mode	Call	Location	Service	kW	Remarks
	CW	3BM7	Mauritius (Bigara) Radio, Mauritius	FC	5.0	
	CW	ZLB	Awarua Radio, New Zealand	FC	5.0	
	CW	UFB	Odessa Radio, Ukrainian SSR	FC	10.0	
	CW	UQK2	Riga Radio, Latvian SSR	FC	5.0	
17,110.0	CW	RIW	Khiva Naval Radio, Uzbek SSR	FC	....	Russian Navy
	CW	UFL	Vladisvostok Radio, USSR	FC	15.0	
17,111.0	CW	UJY	Kaliningrad Radio, USSR	FC	5.0	
17,112.0	CW	DZP	Novaliches Radio, Philippines	FC	1.0	Reported wx broadcasts at 0300 & 1100
17,112.6	CW	JCS	Choshi Radio, Japan	FC	3.0	
17,113.0	CW	GKB6	Portishead Radio, England	FC	15.0	
17,115.0	CW	URD	Leningrad Radio, USSR	FC	15.0	
17,117.6	CW	PBC317	Goeree Naval Radio, Netherlands	FC	10,0	Dutch Navy
	CW	WNU45	Slidell Radio, LA	FC	4.4	NOAA reports wx broadcasts at 0430 & 1630
17,118.0	CW	LZS36	Sofia Naval Radio, Bulgaria	FC	....	
	CW	HLG	Seoul Radio, South Korea	FC	15.0	
17,119.0	CW	UAT	Moscow Radio, USSR	FC	15.0	
17,120.0	CW	PPO	Olinda Radio, Brazil	FC	1.0	
17,122.4	CW	PWZ	Rio de Janeiro Naval Radio, Brazil	FC	10.0	Brazil Navy
17,124.5	CW	VRT	Bermuda (Hamilton) Radio, Bermuda	FC	1.0	
17,125.0	CW	JJF	Tokyo Naval Radio, Japan	FC	3.0	JSDA
17,125.9	CW/RTTY	Y7A68	Berlin, GDR	FX	....	EMBASSY
17,127.2	CW	NRK	Keflavik Naval Radio, Iceland	FC	....	USN
	CW	ZLO6	Irirangi Naval Radio, New Zealand	FC	2.0	RNZ Navy
17,128.5	CW	HPN6Ø	Canal (Puerto Armuelles) R., Panama	FC	1.0	AMVER
17,129.0	CW	XSV	Tianjin Radio, PRC	FC	10.0	
17,129.5	CW	HAR	Budapest Naval Radio, Hungary	FC	5.0	
17,129.7	CW	HLW	Seoul Radio, South Korea	FC	15.5	
17,130.0	CW	ROT	Moscow Naval Radio, USSR	FC	15.0	Russian Navy
17,131.0	CW	UJQ7	Kiev Radio, Ukrainian SSR	FC	15.0	
17,131.1	CW/RTTY	BAC7	Beijing, PRC	FC	80.0	Tfc/Tirana
17,132.0	CW	†VPC	Falkland Islands Radio, Falkland Is.	FC	1.0	
17,132.0	CW	ZSJ6	NAVCOMCEN (Silvermine) Cape R., RSA	FC	5.0	RSA Navy/AMVER
17,135.0	CW	PWZ	Rio de Janeiro Naval Radio, Brazil	FC	2.0	Brazil Navy
	CW	UJQ7	Kiev Radio, Ukrainian SSR	FC	15.0	
17,135.7	CW	GKN6	Portishead Radio, England	FC	15.0	
17,136.0	CW	DZR	Manila Radio, Philippines	FC	2.0	NOAA reports wx broadcast at 0120
17,136.8	CW	GKM6	Portishead Radio, England	FC	15.0	
17,137.0	CW/RTTY	FUG	La Regine (Castelnaudry), France	FC	5.0	French Navy
	CW/RTTY	FUO	Toulon Naval Radio, France	FC	5.0	" "
17,137.7	CW	GKO6	Portishead Radio, England	FC	15.0	
17,138.0	CW	URB2	Klaipeda Radio, Lithuanian SSR	FC	5.0	
17,140.0	CW	XDA	Mexico City Radio, DF, Mexico	FC	15.0	
	CW	UMV	Murmansk Radio, USSR	FC	15.0	
17,141.0	CW	UFN	Novorossiysk Radio, USSR	FC	5.0	
17,141.6	CW	JGC	Yokohama Radio, Japan	FC	1.0	
	CW	UBN	Jdanov Radio, Ukrainian SSR	FC	15.0	
	CW	YVG	La Guaira Radio, Venezuela	FC	0.3	
	CW	YVL	Puerto Cabello Radio, Venezuela	FC	....	
17,143.6	CW	DAN	Norddeich Radio, GFR	FC	15.0	
17,144.0	CW	DZD	Manila (Antipolo) Radio, Philippines	FC	0.4	Reported wx broadcasts at 0300 & 1100
17,145.0	CW	LZW63	Varna Radio, Bulgaria	FC	1.5	
17,146.1	CW	4XO	Haifa Radio, Israel	FC	2.5	
17,146.4	CW	CBA4	Arica Radiomaritima, Chile	FC	4.0	
	CW	NRV	COMMSTA Barrigada, Guam	FC	10.0	USCG/AMVER
	CW	XVS8	Ho Chi Minh-Ville Radio, Vietnam	FC	1.0	
17,147.0	CW/RTTY	URL	Sevastopol Radio, Ukrainian SSR	FC	15.0	
17,147.2	CW	SVB6	Athens Radio, Greece	FC	5.0	
17,149.1	CW	TIM	Limon Radio Costa Rica	FC	....	
17,150.0	CW	URD	Leningrad Radio, USSR	FC	15.0	
17,151.2	CW	GKI6	Portishead Radio, England	FC	15.0	

MARITIME MOBILE (con.)

kHz	Mode	Call	Location	Service	kW	Remarks
	CW	NMC	COMMSTA San Francisco, CA	FC	5.0	USCG/Wx broadcasts at 0030, 1900 & 2100
	FAX	NMC	COMMSTA San Francisco, CA	FC	2.0	USCG/Wx pics at 0000 (Mondays, only), 1700, 2000 & 2300
17,152.0	CW	DZZ	Manila Radio, Philippines	FC	0.1	
	CW	UJY2	Kaliningrad Radio, USSR	FC	15.0	
17,153.0	CW	UQA4	Murmansk Radio, USSR	FC	15.0	
	CW	URL	Sevastopol Radio, Ukrainian SSR	FC	5.0	
17,155.0	CW	ROT	Moscow Naval Radio, USSR	FC	15.0	Russian Navy
	CW/RTTY	UFB	Odessa Radio, Ukrainian SSR	FC	15.0	[170/66N]
	CW	UDH	Riga Radio, Latvian SSR	FC	5.0	
	CW/RTTY	UKA	Vladivostok Radio, USSR	FC	25.0	
17,156.0	CW	NGR	Kato Souli Naval Radio, Greece	FC	30.0	USN
17,159.4	CW	WOE	Lantana Radio, FL	FC	5.0	
17,160.0	CW	PWZ	Rio de Janeiro Naval Radio, Brazil	FC	2.0	Brazil Navy NOAA reports wx broadcasts at 0030, 0630 & 1730
	CW	DZO	Manila (Bulacan) Radio, Philippines	FC	2.0	Reported wx broadcasts at 1030 & 2300
17,160.8	CW	LSA6	Boca Radio, Argentina	FC	3.0	
	CW	IAR27	Rome P.T. Radio, Italy	FC	15.0	NOAA reports wx broadcasts at 0700 & 1900
	CW	WOE	Lantana Radio, FL	FC	5.0	
17,161.3	CW	VIS6	Sydney Radio, NSW, Australia	FC	5.0	AMVER
17,161.9	CW	PPO	Olinda Radio, Brazil	FC	1.0	NOAA reports wx broadcasts at 0400 & 2020
17,163.0	CW	RNO	Moscow Radio, USSR	FC	....	SAAMC
17,164.8	CW	ZSC7	Cape Town Radio, RSA	FC	10.0	
17,165.6	CW	CLA41	Havana (Cojimar) Radio, Cuba	FC	5.0	
	CW	LFF	Rogaland Radio, Norway	FC	5.0	AMVER
17,166.4	CW	JCT	Choshi Radio, Japan	FC	3.0	
17,167.5	CW	GKK2	Portishead Radio, England	FC	10.0	
17,168.0	CW	DZN	Manila (Navotas) Radio, Philippines	FC	3.0	NOAA reports wx broadcasts at 0200 & 1400
17,169.0	CW	A9M	Bahrain Radio, Bahrain	FC	2.0	
17,170.0	CW	PPJ	Juncao Radio, Brazil	FC	1.0	
	CW/RTTY	UDK2	Murmansk Radio, USSR	FC	25.0	
17,170.4	CW	CNP	Casablanca Radio, Morocco	FC	2.0	
	CW	PJC	Curacao (Willemstad) Radio, Curacao	FC	2.5	
	CW	ZLB6	Awarua Radio, New Zealand	FC	5.0	AMVER
	CW	ZLW6	Irirangi Naval Radio, New Zealand	FC	5.0	RNZ Navy
	CW	WPD	Tampa Radio, FL	FC	5.0	NOAA reports wx broadcasts at 1430 & 2020
17,170.5	CW	PPL	Belem Radio, Brazil	FX	1.0	NOAA reports wx broadcasts at 0130, 0800 & 1930
17,172.0	CW	UFN	Novorossiysk Radio, USSR	FC	5.0	
17,172.4	CW	9MG10	Penang Radio, Malaysia	FC	5.0	
	CW	WLO	Mobile Radio, AL	FC	30.0	NOAA reports wx broadcasts at 1300 & 2300
17,173.0	CW	ZAD2	Durres Radio, Albania	FC	1.0	
17,174.8	CW	A9M4	Bahrain Radio, Bahrain	FC	2.0	
17,175.2	CW	VAI	Vancouver CG Radio, BC, Canada	FC	5.0	AMVER/NOAA reports wx broadcasts at 0130, 0530 & 1730
	CW	9GA	Takoradi Radio, Ghana	FC	10.0	
	CW	5ZF4	Mombasa Radio, Kenya	FC	1.5	
	CW	9LL	Freetown Radio, Sierra Leone	FC	1.5	
	CW	EDG5	Aranjuez Radio, Spain	FC	15.0	
	CW	7OA	Aden Radio, Democratic Yemen	FC	2.0	
17,175.3	CW	CLS	Havana (Industria Pesquera) R., Cuba	FC	5.0	
	CW	DZG	Manila (Las Pinas) Radio, Philippines	FC	5.0	AMVER/NOAA re-reports wx broadcasts at 0300 & 1500
17,176.0	CW/RTTY	Y7A71	Berlin, GDR	FX	....	EMBASSY
17,177.0	CW/RTTY	URL	Sevastopol Radio, Ukrainian SSR	FC	15.0	
17,177.6	CW/RTTY	DAL	Norddeich Radio, GFR	FC	15.0	
17,180.0	CW	LOL3	Buenos Aires Radio, Argentina	FC	3.0	Prefectura Naval/ TIME SIGNALS

kHz	Mode	Call	Location	Service	kW	Remarks
	CW	HWN	Paris (Houilles) Naval R., France	FC	2.0	French Navy
	CW	JFE	Naha Gyogyo Radio, Okinawa, Japan	FC	1.0	
17,181.0	CW	UDH	Riga Radio, Latvian SSR	FC	5.0	
17,182.0	CW	LZL63	Bourgas Radio, Bulgaria	FC	1.5	
	CW	ICB	Genoa P.T. Radio, Italy	FC	5.0	
17,183.5	CW	RIW	Khiva Naval Radio, Uzbek SSR	FC	....	Russian Navy
17,184.0	CW	DZK	Manila (Bulacan) Radio, Philippines	FC	1.0	
17,184.7	CW	EAD5	Aranjuez Radio, Spain	FC	15.0	
17,184.8	CW	KFS	San Francisco Radio, CA	FC	10.0	NOAA reports
			wx broadcasts at 0420, 1620, 2200 & 2300			
	CW	PKE	Amboina Radio, Indonesia	FC	1.0	
	CW	PKP	Dumai Radio, Indonesia	FC	1.0	
	CW	PKA	Sabang Radio, Indonesia	FC	1.0	
	CW	9YL	North Post Radio, Trinidad	FC	2.0	NOAA reports
			warning broadcasts at 1200 & 1530			
17,185.0	CW	UFB	Odessa Radio, Ukrainian SSR	FC	5.0	
17,187.0	CW	OST62	Oostende Radio, Belgium	FC	10.0	
17,188.2	CW	SVI6	Athens Radio, Greece	FC	5.0	
17,189.6	CW	D3F	Lobito Radio, Angola	FC	2.0	
	CW	D3E	Luanda Radio, Angola	FC	2.0	
	CW	XYR9	Rangoon Radio, Burma	FC	1.0	
	CW	CLQ	Havana (Cojimar) Radio, Cuba	FC	5.0	
	CW	9HD	Malta Radio, Malta	FC	1.0	
17,190.5	CW	UFA/UHK	Batumi Radio, USSR	FC	5.0	
17,192.0	CW	VRN8Ø	Cape d'Aguilar Radio, Hong Kong	FC	3.0	
17,194.0	CW	C2N	Nauru Radio, Nauru	FC	1.0	
17,194.4	CW	VIS64	Sydney Radio, NSW, Australia	FC	5.0	
	CW	PPR	Rio Radio, Brazil	FC	1.0	TIME SIGNALS
	CW	SVB6	Athens Radio, Greece	FC	5.0	
	CW	C9L4	Beira Radio, Mozambique	FC	0.5	
17,197.5	CW/TOR	9VG82	Singapore Radio, Singapore	FC	10.0	
17,198.0	CW/TOR	GKE6	Portishead Radio, England	FC	15.0	
17,199.0	CW	5OW17	Lagos Radio, Nigeria	FC	1.0	
17,200.0	CW/TOR	VIP35	Perth Radio, WA, Australia	FC	1.0	
17,204.1	CW/TOR	VIS69	Sydney Radio, NSW, Australia	FC	10.0	
17,204.3	CW/TOR	VIS84	Sydney Radio, NSW, Australia	FC	10.0	
17,204.5	CW/TOR	WLO	Mobile Radio, AL	FC	15.0	
17,205.0	CW/TOR	HEC17	Berne Radio, Switzerland	FC	10.0	
17,209.5	CW/TOR	WLO	Mobile Radio, AL	FC	15.0	
17,214.0	CW/RTTY	BZP58	Urumqi, PRC	FX	10.0	Nx/Prague
17,215.2	CW/TOR	GKP6	Portishead Radio, England	FC	15.0	
17,223.0	CW/TOR	LGX3	Rogaland Radio, Norway	FC	5.0	
17,231.0	CW/FAX	GKQ6	Portishead Radio, England	FC	15.0	

17,232.9 - 17,356.9 kHz -- MARITIME COAST USB DUPLEX (Worldwide)

kHz	Mode	Call	Location	Service	kW	Remarks
17,232.9	USB	KQM	Kahuku (Honolulu) Radio, HI	FC	0.8	NOAA CH#1601 (16,460.0)
	USB	LFN2	Rogaland Radio, Norway	FC	10.0	CH#1601 " "
	USB	WOM	Miami Radio, FL	FC	10.0	" "
			[NOAA reports wx broadcasts at 1230]			
17,236.0	USB	VIS	Sydney Radio, NSW, Australia	FC	5.0	CH#1602 (16,436.1)
	USB	GKT62	Portishead Radio, England	FC	10.0	" "
	USB	KMI	San Francisco (Dixon) Radio, CA	FC	10.0	" "
			[NOAA reports wx broadcasts at 0000, 0600 & 1500]			
17,239.1	USB	5BA62	Nicosia Radio, Cyprus	FC	10.0	CH#1603 (16,466.2)
	USB	....	Rome Radio 7, Italy	FC	7.0	" "
	USB	LFN3	Rogaland Radio, Norway	FC	10.0	" "
	USB	KMI	San Francisco (Dixon) Radio, CA	FC	10.0	" "
17,242.2	USB	VCS	Halifax CG Radio, NS, Canada	FC	5.0	CH#1604 (16,469.3)
	USB	FFL83	St. Lys Radio, France	FC	10.0	" "

MARITIME COAST USB DUPLEX (con.)

kHz	Mode	Call	Location	Service	kW	Remarks
	USB	JBO	Tokyo Radio, Japan	FC	10.0	" "
	USB	LFN4	Rogaland Radio, Norway	FC	10.0	" "
17,245.3	USB	FFL86	St. Lys Radio, France	FC	10.0	CH#1605 (16,472.4)
	USB	....	Goteborg Radio, Sweden	FC	5.0	" "
	USB	WOO	New York (Ocean Gate, NJ) Radio, NY	FC	10.0	" "
			[NOAA reports wx broadcasts at 0100, 1300 & 1900]			
17,248.4	USB	....	Rome Radio 7, Italy	FC	7.0	CH#1606 (16,475.5)
	USB	....	Moscow Radio, USSR	FC	10.0	" "
	USB	....	several, around USA coastline (Includes: NMA, Miami, FL; NMC, San Francisco, CA & NMG, New Orleans, LA)	FC	....	USCG/CH#1606 (16,475.5)
17,251.5	USB	LFN6	Rogaland Radio, Norway	FC	10.0	CH#1607 (16,478.6)
	USB	WLO	Mobile Radio, AL	FC	10.0	CH#1607 (16,478.6)
17,254.6	USB	ICB	Genoa Radio 7, Italy	FC	7.0	CH#1608 (16,481.7)
	USB	ZSC28	Cape Town Radio, RSA	FC	10.0	" "
	USB	....	Goteborg Radio, Sweden	FC	5.0	" "
17,257.7	USB	SVN6	Athens Radio, Greece	FC	10.0	CH#1609 (16,484.8)
	USB	JBO	Tokyo Radio, Japan	FC	10.0	" "
	USB	WOM	Miami Radio, FL	FC	10.0	" "
			[NOAA reports wx broadcast at 1330]			
17,260.8	USB	VIS	Sydney Radio, NSW, Australia	FC	5.0	CH#1610 (16,487.9)
	USB	DAP	Norddeich Radio, GFR	FC	10.0	" "
	USB	WOM	Miami Radio, FL	FC	10.0	" "
	USB	WOO	New York (Ocean Gate, NJ) Radio, NY	FC	10.0	" "
17,263.9	USB	OHG2	Helsinki Radio, Finland	FC	10.0	CH#1611 (16,491.0)
	USB	HEB17	Berne Radio, Switzerland	FC	10.0	" "
	USB	WOM	Miami Radio, FL	FC	10.0	" "
			[NOAA reports wx broadcast at 2330]			
	USB	....	Rijeka Radio, Yugoslavia	FC	7.5	CH#1611 (16,491.0)
17,267.0	USB	KVH	Norfolk Marine Center, VA	FC	....	NOAA/CH#1612 (16,494.1)
17,270.1	USB	OSU63	Oostende Radio, Belgium	FC	10.0	CH#1613 (16,497.2)
	USB	4XO	Haifa Radio, Israel	FC	10.0	" "
	USB	9VG68	Singapore Radio, Singapore	FC	10.0	" "
	USB	....	numerous, around USA coastline	FC	....	USCG
17,273.2	USB	ICB	Genoa Radio 7, Italy	FC	7.0	CH#1614 (16,500.3)
	USB	....	Goteborg Radio, Sweden	FC	5.0	" "
17,276.3	USB	CUL37	Lisbon Radio, Portugal	FC	5.0	CH#1615 (16,503.4)
	USB	HEB27	Berne Radio, Switzerland	FC	10.0	" "
17,279.4	USB	DAJ	Norddeich Radio, GFR	FC	10.0	CH#1616 (16,506.5)
	USB	....	Rome Radio 7, Italy	FC	7.0	" "
	USB	KMI	San Francisco (Dixon) Radio, CA	FC	10.0	" "
17,282.5	USB	....	....			CH#1617 (16,509.6)
17,285.6	USB	VRT	Bermuda (Hamilton) Radio, Bermuda	FC	1.0	CH#1618 (16,512.7)
17,288.7	USB	FFL84	St. Lys Radio, France	FC	10.0	CH#1619 (16,515.8)
	USB	LFN23	Rogaland Radio, Norway	FC	10.0	" "
17,291.8	USB	LFN24	Rogaland Radio, Norway	FC	10.0	CH#1620 (16,518.9)
	USB	EHY	Pozuelo del Rey Radio, Spain	FC	10.0	" "

MARITIME COAST USB DUPLEX (con.)

kHz	Mode	Call	Location	Service	kW	Remarks
	USB	WFN	Jeffersonville, IN	FC/MS	0.8	Inland Waterway
	USB	WGK	St. Louis, MO	FC/MS	0.8	" "
	USB	WCM	Pittsburgh, PA	FC/MS	0.8	" "
	USB	WJG	Memphis, TN	FC/MS	0.8	" "
	USB	WOO	New York (Ocean Gate, NJ) Radio, NY	FC	10.0	CH#1620 (16,518.9)
			[NOAA reports wx broadcasts at 0100, 1300 & 1900]			
17,294.9	AM/USB	....	INTERNATIONAL COAST STATION CALLING FREQUENCY			CH#1621 (16,522.0)
17,298.0	USB	....	Lyngby Radio, Denmark	FC	10.0	CH#1622 (16,525.1)
	USB	FFL85	St. Lys Radio France	FC	10.0	" "
17,301.1	USB	PCG63	Scheveningen Radio, Netherlands	FC	5.0	CH#1623 (16,528.2)
	USB	....	Kiev Radio, Ukrainian SSR	FC	5.0	" "
	AM/USB	....	Moscow Radio, USSR	FC	5.0	" "
17,304.2	USB	....	Rome Radio 7, Italy	FC	7.0	CH#1624 (16,531.3)
17,307.3	USB	SVN6	Athens Radio, Greece	FC	5.0	CH#1625 (16,534.4)
	USB	NMC	COMMSTA San Francisco, CA [NOAA reports wx broadcasts at 1630 & 2230]	FC	10.0	USCG CH#1625 (16,534.4)
	USB	NMN	COMMSTA Portsmouth, VA [NOAA reports wx broadcast at 1730]	FC	10.0	USCG CH#1625 (16,534.4)
			This frequency also used by USCG Coast Stations: NMA, Miami, FL; NMG, New Orleans, LA; NMF, Boston, MA; NOJ, Kodiak, AK; NMO, Honolulu, HI and NRV, Barrigada, Guam)			
17,310.4	USB	SVN6	Athens Radio, Greece	FC	5.0	CH#1626 (16,537.5)
	USB	WOO	New York (Ocean Gate, NJ) Radio, NY	FC	10.0	" "
17,313.5	USB	LFN26	Rogaland Radio, Norway	FC	10.0	CH#1627 (16,540.6)
17,316.6	USB	FFL81	St. Lys Radio, France	FC	10.0	CH#1628 (16,543.7)
17,319.7	USB	Y5P	Ruegen Radio, GDR	FC	10.0	CH#1629 (16,546.8)
	USB	LFN27	Rogaland Radio, Norway	FC	10.0	" "
17,322.8	USB	EHY	Pozuelo del Rey Radio, Spain	FC	10.0	CH#1630 (16,546.8)
17,325.9	USB	SPC82	Gdynia Radio, Poland	FC	10.0	CH#1631 (16,553.0)
	USB	HEB37	Berne Radio, Switzerland	FC	10.0	" "
	USB	WOO	New York (Ocean Gate, NJ) Radio, NY	FC	10.0	" "
17,329.0	USB	GKW62	Portishead Radio, England	FC	10.0	CH#1632 (16,556.1)
	USB	JBO	Tokyo Radio, Japan	FC	10.0	" "
	USB	WLO	Mobile Radio, AL	FC	10.0	" "
17,332.1	USB	FFL82	St. Lys Radio, France	FC	10.0	CH#1633 (16,559.2)
	USB	ZSD41	Durban Radio, RSA	FC	10.0	" "
17,335.2	USB	....	Norddeich Radio, GFR.	FC	10.0	CH#1634 (16,562.3)
17,338.3	USB	....	Lyngby Radio, Denmark	FC	10.0	CH#1635 (16,565.4)
17,341.4	USB	PCG61	Scheveningen Radio, Netherlands	FC	10.0	CH#1636/AMVER (16,568.5)
17,344.5	USB	EHY	Pozuelo del Rey Radio, Spain	FC	10.0	CH#1637 (16,571.6)
17,347.6	USB	....	Helsinki Radio, Finland	FC	10.0	CH#1638 (16,574.7)
	USB	NMC	COMMSTA San Francisco, CA	FC	10.0	USCG/CH#1638 (16,574.7)

This frequency is also used by USCG Coast Stations; NMF, Boston, MA; NMN, Portsmouth, VA & NMO, Honolulu, HI

MARITIME COAST USB DUPLEX (con.)

kHz	Mode	Call	Location	Service	kW	Remarks
17,350.7	USB	DAH	Norddeich Radio, GFR	FC	10.0	CH#1639 (16,577.8)
	USB	PCG62	Scheveningen Radio, Netherlands	FC	10.0	" "
	USB	EHY	Pozuelo del Rey Radio, Spain	FC	10.0	" "
17,353.8	USB	SVN6	Athens Radio, Greece	FC	5.0	CH#1640 (16,580.9)
17,356.9	USB	9VG69	Singapore Radio, Singapore	FC	10.0	CH#1641 (16,584.0)
	USB	....	Goteborg Radio, Sweden	FC	5.0	CH#1641 (16,584.0)
	USB	WLO	Mobile Radio, AL [NOAA reports wx broadcasts at 1330, 1730 & 2330]	FC	10.0	" "

17,360.0 - 17,700.0 kHz -- FIXED SERVICE (Worldwide)

kHz	Mode	Call	Location	Service	kW	Remarks
17,362.0	CW/RTTY	Y7G29	unknown, probably in Far East	FX	....	GDR EMBASSY
17,365.0	CW	ZVE	Manaus Aeradio, Brazil	AX	1.5	Aero Tfc
	CW	ZVK	Rio de Janeiro Aeradio, Brazil	AX	1.0	Aero Tfc
	CW	9EQ7Ø	Addis Ababa, Ethiopia	FX	0.5	Tfc/London/Rome
	CW	....	Tel Aviv, Israel	FX	0.2	Philippine EMBASSY
	CW/FAX	5YE3	Nairobi, Kenya	FX	10.0	Meteo (1200)
17,367.0	CW	D2U21	Luanda, Angola	AX	3.0	Tfc/Accra, Lubumbashi, Maputo, etc. [Alt: 17,400.0]
17,368.0	FAX	....	unknown	FX	....	
17,380.0	CW	FUM	Papeete Naval Radio, Tahiti	FX	10.0	French Navy
17,385.0	CW/RTTY	ZEF81	Salisbury, Zimbabwe	FX	1.0	PRC EMBASSY
17,386.0	ISB	TJF73	Douala, Cameroon	FX	20.0	Telcom/Tfc
17,391.5	ISB	KUQ2Ø	Pago Pago, American Samoa	FX	10.0	Telcom
17,393.0	ISB	CMLX	Havana (Bauta), Cuba	FX	15.0	Telcom/Mexico
17,395.0	CW	Y7G34	unknown, probably in Far East	FX	....	GDR EMBASSY
17,402.0	CW	....	Ankara, Turkey	FX	1.0	EMBASSY
17,403.0	CW	....	unknown	FX	....	Coded Tfc
17,404.0	CW	UCA	Odessa, Ukrainian SSR	FX	5.0	Tfc/Havana
17,408.6	CW/FAX	WWD	La Jolla (Scripps), CA	FX	....	NOAA reports wx broadcasts 1700 & 2300--June to October

17,410.0 - 17,550.0 kHz -- FIXED SERVICE (Worldwide--post WARC)

kHz	Mode	Call	Location	Service	kW	Remarks
17,411.0	FAX	....	unknown, probably in USSR	FX	....	
17,412.0	ISB	TYK73	Cotonou, Benin	FX	20.0	Telcom/Tfc
17,415.0	CW	OFJ	Helsinki Radio, Finland	FC	....	
17,417.5	ISB	PCK27	Hilversum (Kootwijk), Netherlands	FX	30.0	Backup feeder for Radio Netherland
17,418.0	USB	OHU2Ø	Helsinki, Finland	FX	1.0	EMBASSY [Alt: 14,954.0 & 17,443.0]
	USB	OHU21	Paris, France	FX	1.0	Finnish EMBASSY [Alt: 10,895.0 & 20,124.0]
17,424.0	CW	CMU967	Santiago Naval Radio, Cuba	FX	....	Russian Navy
17,426.0	CW	KKN44	Monrovia, Liberia	FX	....	USA EMBASSY
17,427.0	USB	SAM	Stockholm, Sweden	FX	1.2	EMBASSY (This frequency is used by: SAM46, Baghdad, Iraq; SAM48; Damascus, Syria; SAM5Ø, Cairo, Egypt; SAM52, Tel Aviv, Israel; SAM54, Vientiane, Laos; SAM58, Jakarta, Indonesia; SAM61, New Delhi, India; SAM64, Pyongyang, N. Korea; SAM65, Seoul, S. Korea & SAM81, Bissau, Guinea-Bissau) [Alt: 14,968.5 & 18,808.0]
17,430.0	LSB	SAM	Stockholm, Sweden	FX	1.2	EMBASSY (This frequency is used by: SAM2Ø, Athens, Greece; SAM36, Budapest, Hungary; SAM37, Bucharest, Rumania; SAM45, Ankara, Turkey; SAM47, Beirut, Lebanon; SAM52, Tel Aviv, Israel; SAM39, Hanoi, Vietnam & SAM74, Monrovia, Liberia) [Alt: 14,971.5]
17,435.0	CW/FAX	Y2V37	Berlin (Nauen), GDR	FX	20.0	Nx
17,436.5	FAX	WFK67	New York, NY	FX	50.0	Meteo (1950-2350)
17,437.0	USB	9UB83	Bujumbura, Burundi	FX	1.5	Telcom
17,440.0	CW	ART	Rawalpindi, Pakistan	FX	1.0	Tfc/Ankara

kHz	Mode	Call	Location	Service	kW	Remarks
17,443.0	USB	OHU2Ø	Helsinki, Finland	FX	1.0	EMBASSY [Alt: 17,418.0]
	USB	OHU25	Beijing, PRC	FX	1.0	Finnish EMBASSY [Alt: 15,870.0]
17,445.0	ISB	†....	Holzkirchen, GFR	FX	10.0	RFE feeder
17,448.5	USB	VCR976	Ottawa, Ont., Canada	FX	1.0	Saudi Arabia EMBASSY [Alt: 14,748.5 & 17,622.5]
17,458.0	CW/RTTY	ONN36	Brussels, Belgium	FX	1.0	Angola EMBASSY
17,470.0	ISB	FTR47	Paris (St. Assise), France	FX	20.0	Telcom
17,480.0	CW/USB	VJM	Casey Base, Antarctica	FX	10.0	ANARE
	CW/USB	VLZ	Davis Base, Antarctica	FX	5.0	"
	CW/USB	VLV	Mawson Base, Antarctica	FX	5.0	"
17,480.0	ISB	†ATJ67	Bombay (Kirkee), India	FX	10.0	Telcom/Tfc
17,486.0	CW	CFH	Maritime Command Radio, Halifax, NS, Canada	FX	10.0	Tfc/London
	USB	VDD	Debert, NS, Canada	FX	10.0	RCAF/Lahr
17,488.6	USB	....	Northwest Pacific LORAN Net	FX	....	USCG (Includes: NRT, "Yokota Monitor", Japan; NRT2, Gesashi, Japan; NRT3, Iwo Jima; NRT9, Hokkaido, Japan; NRV, Barrigada, Guam; NRV6, Marcus Is.; NRV7, Yap Island) [Alt: 15,922.0 & 19,297.1]
17,503.0	USB	Y3H	Berlin (Nauen), GDR	FX	20.0	Telcom
17,503.7	USB	FTR5Ø	Paris (St. Assise), France	FX	20.0	Telcom
17,504.0	CW	RIW	Khiva Naval Radio, Uzbek SSR	FX	....	Russian Navy
17,510.0	FAX	OXI	Skamlebaek Radio, Denmark	FX	20.0	Dannish Marine Meteo (1335-1355)
17,515.0	ISB	†TYK75	Cotonou, Benin	FX	20.0	Telcom/Tfc
17,528.0	CW	RLX	Dublin, Ireland	FX	0.5	USSR EMBASSY [Alt: 14,426.0]
17,530.0	CW	NPN	Barrigada (Agana NAS), Guam	FX	15.0	USN
17,532.0	CW	CYS22	Ottawa, Ont., Canada	FX	....	Polish EMBASSY [Alt: 16,024.0 & 18,355.0]
17,537.9	USB	†FZW75	Pointe a Pitre, Guadeloupe	FX	20.0	Telcom/Paris
17,539.5	USB	HZN	Jeddah, Saudi Arabia	FX	10.0	Meteo

17,550.0 - 17,900.0 kHz -- INTERNATIONAL BROADCASTING (Worldwide--post WARC)

kHz	Mode	Call	Location	Service	kW	Remarks
17,550.0	CW	LQC2Ø	Buenos Aires (Mt. Grande), Argentina	FX	5.0	TIME SIGNALS/ Tfc
	CW/RTTY	UJY2	Kaliningrad, USSR	FX	5.0	Tfc/Havana
17,552.0	CW	KWL9Ø	Tokyo, Japan	FX	....	USA EMBASSY
17,553.5	USB	LOL	Buenos Aires Naval Radio, Argentina	FX	3.0	Prefectura Naval
	USB	YWM	Maracaibo Naval Radio, Venezuela	FX	....	Venezuela Navy
17,556.4	CW/USB	OEC	Vienna, Austria	FX	10.0	EMBASSY
	CW/USB	OEC57	Pretoria, RSA	FX	1.0	Austrian EMBASSY [Alt: 13,617.0 & 20,752.0]
17,559.0	ISB	9NB27	presumed Kathmandu, Nepal	FX	....	Telcom/Tfc
17,560.0	CW/FAX	†CFH	Maritime Command Radio, Halifax, NS, Canada	FX	10.0	Canadian Forces/ Meteo
	LSB	RCF42	Moscow, USSR	FX	15.0	R. Moscow Feeder
17,561.0	USB	OHU2Ø	Helsinki, Finland	FX	1.0	EMBASSY
	USB	OHU23	Tel Aviv, Israel	FX	1.0	Finnish EMBASSY [Alt: 14,661.0]
17,572.0	CW	3BN	Plaisance, Mauritius	AX	2.0	Aero Tfc
17,580.0	LSB	....	Moscow, USSR	FX	....	R. Moscow Feeder
17,590.0	CW	LOL	Buenos Aires Naval Radio, Argentina	FX	3.0	Prefectura Naval
	CW	CPF2	La Paz Naval Radio, Bolivia	FX	....	Bolivian Navy
	CW	PWZ	Rio de Janeiro Naval Radio, Brazil	FX	15.0	Brazil Navy
	CW	CCQ	Iquique Naval Radio, Chile	FX	....	Chilean Navy
	CW	CCS	Santiago Naval Radio, Chile	FX	....	Chilean Navy
	CW	5KM	Bogota Naval Radio, Colombia	FX	....	Colombian Navy
	CW	HDN	Quito Naval Radio, Ecuador	FX	....	Ecuador Navy
	CW	OBC	Callao Naval Radio, Peru	FX	....	Peruvian Navy
	CW	CXR	Montevideo Armada Radio, Uruguay	FX	....	Uruguay Navy
	CW	YWM3	Maracaibo Naval Radio, Venezuela	FX	....	Venezuela Navy
17,605.0	CW	†KKN5Ø	Washington, DC	FX	10.0	EMBASSY

kHz	Mode	Call	Location	Service	kW	Remarks
17,612.5	ISB	†....	Holzkirchen, GFR	FX	10.0	RFE Feeder
17,617.0	USB	AFS	Offutt AFB, NE	FX	0.5	USAF/SAC
				[Alt: 15,962.0 & 18,048.6]		
17,620.0	USB	Y3H2	Berlin (Nauen), GDR	FX	20.0	Telcom
	CW	IDQ	Rome Naval Radio, Italy	FX/FC	15.0	Italian Navy
				(Reported Nx in II at 1800)		
17,622.5	USB	VCR976	Ottawa, Ont., Canada	FX	1.0	Saudi Arabia
			EMBASSY	[Alt: 17,448.5 & 23,098.5]		
17,628.0	CW	CLN	Havana (Bauta), Cuba	FX	20.0	Tfc UJY2
17,630.0	CW	VTK6	Tuticorin Naval Radio, India	FX	....	Indian Navy
17,640.0	ISB	WEK67	New York, NY	FX	80.0	Telcom/Tfc
17,642.0	USB	....	Daventry, England	FX	30.0	BBC Feeder
17,650.0	CW	....	Ankara, Turkey	FX	1.0	EMBASSY
17,655.0	CW	FUJ	Noumea Naval Radio, New Caledonia	FX	2.5	French Navy
17,670.5	FAX	AFA	Washington (Andrews AFB, MD), DC	FX	4.0	USAF/Meteo
17,675.0	USB	....	numerous, throughout Australia	FX	....	Police (For
				list of stations see 13,730.0)		
17,675.0	ISB	FTR67	Paris (St. Assise), France	FX	20.0	Telcom
17,685.0	USB	†PP7A	Paulista, Brazil	FX	1.0	NASA
				[Alt: 20,575.0]		
17,690.0	LSB	†OCK25	Arequipa, Peru	FX	1.0	Smithsonian
17,695.0	ISB	ZFD82	Hamilton, Bermuda	FX	30.0	Emergency
	CW	ZVK	Rio de Janeiro Aeradio, Brazil	AX	1.0	Tfc/Sal
	ISB	JBQ67	Tokyo, Japan	FX	5.0	Telcom/Tfc

17,700.0 - 17,900.0 kHz — INTERNATIONAL BROADCASTING (Worldwide)

17,900.0 - 18,030.0 kHz — AERONAUTICAL MOBILE (Worldwide)

kHz	Mode	Call	Location	Service	kW	Remarks
17,907.0	A3H/USB	....	§SEA moves here from 17,965.0 in 1982			
17,909.0	AM	....	Brazzaville, Congo	FA	5.0	VOLMET in FF
17,917.0	AM/A3H	....	numerous (ICAO SAM)	FA/MA	....	Airports & in-
			flight covering the South American continent			
17,925.0	AM/A3H	....	numerous (ICAO AFI & CAR)	FA/MA	....	Airports & in-
			flight covering the entire African continent, Arabian Sea plus Bombay and Karachi. Also all of the Caribbean area north to New York and south to Bogota & Caracas			
17,933.0	USB	9VA26	Singapore Aeradio, Singapore	FA/MA	6.0	
17,936.0	CW	COL	Havana Aeradio, Cuba	FA/MA	....	Aeroflot
	CW	RFNV	Moscow Aeradio, USSR	FA/MA	....	Aeroflot (Mostly
				in-flight to/from Havana/Moscow)		
17,940.0	USB	9VA3Ø	Singapore Aeradio, Singapore	FA/MA	6.0	
17,941.0	AM/A3H	....	numerous (ICAO NAT)	FA/MA	....	Airports & in-
			flight covering all of the mid and North Atlantic Ocean from New York to Lisbon/Madrid and north to the Arctic			
			§NAT moves to 17,946.0 in 1982			
17,946.0	A3H/USB	....	§NAT moves here from 17,941.0 in 1982			
17,949.0	AM/A3H	....	several (ICAO SAT)	FA/MA	....	Airports & in-
			flight across South Atlantic Ocean from Dakar to Rio de Janeiro			
			§SAT moves to 17,955.0 in 1982			
17,949.0	USB	VIS58	Sydney Aeradio, NSW, Australia	FA	1.5	
17,955.0	A3H/USB	....	§SAT moves here from 17,949.0 in 1982			
17,965.0	AM/A3H	....	numerous (ICAO SEA)	FA/MA	....	Airports & in-
			flight covering all of southeastern Asia from New Delhi to Perth and east to Manila and Biak			
			§SEA moves to 17,904.0 in 1982			
17,972.0	USB	AFA3	Andrews AFB, MD	FA/MA	....	USAF
	USB	AKA5	Elemendorf AFB, AK	FA/MA	....	USAF
17,975.0	USB	....	numerous, worldwide	FA/MA	....	USAF (Usu-
			ally AFH9, March AFB, CA; AJE, Croughton, England; XPH, Thule AB, Greeland & AJG9, Incirlik AB, Turkey) [Has been used for SKY KING broadcasts. Sometimes referred to as the "TANGO" frequency]			
17.982.0	USB	....	Stockholm Aeradio, Sweden	FA/MA	5.0	
	USB	....	several, in Australia & New Zealand	FA/MA	....	RAAF/RNZAF

kHz	Mode	Call	Location	Service	kW	Remarks
	USB	....	numerous, worldwide	FA/MA	....	USN
17,985.0	USB	....	numerous, mostly Atlantic Coast	FA/MA	....	USN
17,992.0	ISB	TFK	Keflavik Airport, Iceland	FA/MA	5.0	USAF
17,993.0	USB	AFA	Andrews AFB, MD	FA/MA	....	USAF
	USB	ZRB2	Pretoria (Waterkloof AFB), RSA	FA/MA	....	RSA Air Force
17,998.0	CW	CAK	Santiago (Los Cerrillos AB), Chile	FA/MA	....	Chilean AF
17,999.0	USB	....	numerous, throughout Australia	FA/MA	....	RAAF
18,002.0	USB	....	numerous, worldwide	FA/MA	1.2	USAF (Usu-
			ally AGA, Hickam AFB, HI; AGA8, Clark AB, Philippines; AFI, McClellan, CA; AFL, Loring AFB, ME; AIF8Ø, Yokota AB, Japan; AFG37, Scott AFB, IL & CUW, Lajes Field, Azores) [Used for "SKY KING" broadcasts. May also be used by AWACS and hurricane hunting a/c]			
18,003.0	USB	....	numerous, throughout Australia	FA	....	RAAF
18,009.0	USB	....	numerous, worldwide	FA/MA	....	USN
18,009.7	FSK	....	unknown	FX	....	Tfc/AA
18,012.0	USB	....	several, in Canada and Europe	FA/MA	....	RCAF
	USB	....	numerous, worldwide	FA/MA	....	USN
18,013.5	CW	CAK	Santiago (Los Cerrillos AB), Chile	FA	0.2	Chilean AF
			(NOAA reports wx broadcasts at 0130, 1330 & 2000)			
18,018.0	USB	....	several, worldwide	FA/MA	....	RAF
18,019.0	USB	....	numerous, worldwide	FA/MA	....	USAF (Usu-
			ally AFE8, MacDill AFB, FL; AFD14, Ascension Island; AHF3, Albrook AFS, Panama; AIF2Ø, Yokota AB, Japan & AGA8, Clark AB, Philippines--may be used for "SKY KING" broadcasts) [Also available to USN]			
18,022.0	USB	A4I	Seeb Aeradio, Oman	FA/MA	1.0	Oman AF
18,025.0	USB	....	several, throughout Australia	FA/MA	....	RAAF [Most fre-
			quently used channel in this band--for list of possible stations see 13,205.0 kHz]			
18,027.0	USB	....	several, worldwide	FA/MA	....	USAF (Parti-
			cularly AFA3, Andrews AFB, MD and "AIR FORCE ONE" when on overseas flights)			

18,030.0 - 19,990.0 kHz -- FIXED SERVICE (Worldwide)

18,030.0 - 18,068.0 kHz -- FIXED SERVICE (Worldwide--post WARC)

kHz	Mode	Call	Location	Service	kW	Remarks
18,032.0	CW	GYA	Whitehall (London) Naval R., England	FX	5.0	British Navy
18,043.4	CW	KKN44	Monrovia, Liberia	FX	....	USA EMBASSY
18,048.6	USB	AFS	Offutt AFB, NE	FX	0.5	USAF/SAC
					[Alt: 17,617.0]	
18,052.5	FAX	....	nknown, probably in Far East	FX	....	Nx
18,060.0	FAX	AXI36	Darwin, NT, Australia	FX	5.0	Meteo (2120-0800)

18,068.0 - 18,168.0 kHz -- FIXED SERVICE & AMATEUR RADIO (Worldwide--shared post WARC)

kHz	Mode	Call	Location	Service	kW	Remarks
18,073.0	USB	†5RY81	Antananarivo, Madagascar	FX	10.0	Telcom
18,087.5	USB	....	several, mostly west coast of USA	FX	....	USN
18,082.5	FAX	....	unknown, probably in Far East	FX	....	Meteo/Nx
18,087.0	CW	8UF75	New Delhi, India	FX	5.0	INTERPOL
18,087.0	CW	JPA24	Tokyo, Japan	FX	5.0	INTERPOL [Far
			East distribution point to Melbourne, Australia; New Delhi, India; Jakarta, Indonesia; Seoul, S. Korea; Hong Kong; Kuala Lumpur, Malaysia; Manila, Philippines; Bang-kok, Thailand; etc. law enforcement agencies]			
18,093.0	FAX	LRO84	Buenos Aires (G. Pacheco), Argentina	FX	5.0	Meteo
18,105.0	USB	....	numerous, throughout RSA	FX	1.0	Governmental
18,109.0	ISB	†IRQ41	Rome (Torrenova), Italy	FX	30.0	Telcom/Tfc
18,120.0	ISB	MKG	London (Stanbridge), England	FX	....	RAF Telcom/Tfc
18,130.0	FAX	JMJ5	Tokyo (Usui), Japan	FX	5.0	Meteo
18,137.5	ISB	....	unknown	FX	....	UN Radio Feeder
18,150.0	CW	†CVM8	Montevideo (Pajas Blancas), Uruguay	FX	14.0	Tfc
	LSB	RTM29	Khabarovsk, USSR	FX	15.0	R. Moscow Feeder

18,168.0 - 18,780.0 kHz -- FIXED SERVICE (Worldwide--post WARC)

kHz	Mode	Call	Location	Service	kW	Remarks
18,170.0	USB	RIF33	Moscow, USSR	FX	20.0	Telcom
18,172.0	ISB	....	Daventry, England	FX	30.0	BBC Feeder
18,175.0	CW	CCV	Valparaiso Naval Radio, Chile	FX	....	Chilean Navy
			NOAA reports wx broadcasts at 0110, 1330 & 1430			
	ISB	TNH81	Brazzaville, Congo	FX	20.0	Telcom/Paris
18,181.00	CW/RTTY	GYU	Gibraltar Naval Radio, Gibraltar	FX	....	British Navy
18,184.0	CW	CLP1	Havana, Cuba	FX	....	EMBASSY
18,187.0	CW	SDQ7	Stockholm (Varberg), Sweden	FX	35.0	Reported nx broadcast at 0815
18,190.0	CW	FSB59	Paris (St. Martin Abbat), France	FX	2.5	INTERPOL
	CW	DEB	Weisbaden, GFR	FX	5.0	"
	CW	PDB2	Utrecht (Bilthoven), Netherlands	FX	1.0	" [Alt: 18,380.0]
	CW	HEP81	Zurich (Waltikon), Switzerland	FX	2.0	INTERPOL
	LSB	†CXL31	Montevideo, Uruguay	FX	2.5	Telcom
18,195.0	LSB	RCI73	Moscow, USSR	FX	20.0	R. Moscow Feeder
18,196.1	USB	....	Atlantic Domestic Emergency Net	FX	....	USCG (Includes: NMA, Miami, FL; NMF, Boston, MA; NMG, New Orleans, LA; NMN, Portsmouth, VA; NMR, San Juan, PR & NOZ, Elizabeth City, NC, plus transportables) [Alt: 15,548.6 & 18,721.1]
	USB	....	Pacific Domestic Emergency Net	FX	....	USCG (Includes: NMC, San Francisco, CA; NMO, Honolulu, HI; NMQ; Long Beach, CA; NOJ, Kodiak, AK & NRV, Barrigada, Guam, plus transportables) [Alt: 15,473.6 & 18,721.1]
18,197.7	ISB	HBO28	Geneva, Switzerland	FX	20.0	Tfc/Pretoria
18,200.0	ISB	†GAW38	London (Rugby), England	FX	30.0	Telcom/Tfc
18,207.5	ISB	†WET48	New York, NY	FX	40.0	Telcom/Tfc
18,220.0	FAX	JMH5	Tokyo, Japan	FX	5.0	Meteo
18,227.0	FAX	ATP38	New Delhi, India	FX	10.0	Meteo (0230-01400)
18,235.0	ISB	4QO81	Colombo (Kotugoda), Sri Lanka	FX	30.0	Telcom/Tfc
18,240.0	FAX	ZRO4	Pretoria, RSA	FX	3.0	Meteo (0545-1745)
18,247.0	CW	VCS838	Ottawa, Ont., Canada	FX	5.0	Yugoslav EMBASSY [Alt: 14,515.0]
18,249.4	USB	FTS25	Paris (St. Assise), France	FX	20.0	Telcom
18,250.0	CW	KNY27	Washington, DC	FX	1.0	Swiss EMBASSY
18,261.0	FAX	GFE24	Bracknell, England	FX	5.0	Meteo
18,275.0	ISB	....	Greenville, NC	FX	50.0	VOA Feeder
18,285.0	USB	HBD41	Pretoria, RSA	FX	0.5	Swiss EMBASSY
18,285.2	USB	RCV28	Moscow, USSR	FX	50.0	R. Moscow Feeder
18,291.8	CW	....	unknown	FX	....	Coded Tfc
18,294.6	USB	TZA218	Bamako, Mali	FX	20.0	Telcom
18,297.5	USB	OMZ3Ø	Prague, Czechoslovakia	FX	4.0	Libyan EMBASSY [Alt: 20,682.5]
18,300.0	CW	†OMZ	Prague, Czechoslovakia	FX	1.0	EMBASSY
18,306.5	CW	†KNY28	Washington, DC	FX	1.0	Algerian EMBASSY
18,312.0	LSB	RYR	Moscow, USSR	FX	15.0	R. Moscow Feeder
18,323.0	CW	OVG	Frederikshaven Naval Radio, Denmark	FX	....	Danish Marine
18,342.8	CW	"K"	unknown	FX	....	Beacon
18,355.0	CW	CYS22	Ottawa, Ont., Canada	FX	....	Polish EMBASSY [Alt: 17,532.0 & 20,030.0]
18,365.0	CW/RTTY	FUB	Paris (Houilles) Naval R., France	FX	10.0	French Navy
18,368.1	CW	CQK55	Sao Tome, Sao Tome e Principe	FX	5.0	Tfc/Lisbon
18,380.0	CW/RTTY	DEB	Weisbaden, GFR	FX	5.0	INTERPOL
	CW/RTTY	PDB2	Utrecht (Bilthoven), Netherlands	FX	1.0	"
	CW/RTTY	HEP83	Zurich (Waltikon), Switzerland	FX	2.0	" [Alt: 18,190.0]
18,395.0	USB	†FTS39	Paris (St. Assise), France	FX	20.0	Telcom/Bangui
18,400.0	CW	†OMZ	Prague, Czechoslovakia	FX	....	EMBASSY
18,400.5	USB	LJA21	Tel Aviv, Israel	FX	0.4	Norwegian EMBASSY [Alt: 18,427.5]
18,402.0	FAX	RUU74	Leningrad, USSR	FX	20.0	Meteo/Nx
18,407.0	CW	ZBP	Pitcairn Island	FX	0.7	
18,423.8	CW	....	unknown, probably in the Far East	FX	....	Hi-speed Tfc

kHz	Mode	Call	Location	Service	kW	Remarks
18,430.0	CW	†KNY21	Washington, DC	FX	1.0	Yugoslav EMBASSY
	CW	†KNY23	Washington, DC	FX	1.0	Czech EMBASSY
18,432.5	ISB	FZM83	Noumea, New Caledonia	FX	20.0	Tfc/Sydney
18,435.0	CW	NPM	Lualualei Naval Radio, HI	FC	15.0	USN
18,436.0	ISB	†TUP84	Abidjan, Ivory Coast	FX	4.0	Telcom/Tfc
18,440.0	ISB	GBB38	London (Rugby), England	FX	30.0	Telcom/Tfc
18,447.5	ISB	†NAW18	Guantanamo, Cuba	FX	....	Telcom/Tfc
18,459.9	CW	KWS78	Athens, Greece	FX	....	USA EMBASSY
18,460.0	USB	RRG29	Moscow, USSR	FX	15.0	R. Moscow Feeder
18,467.0	USB	ORI28	Brussels (Ruiselede), Belgium	FX	30.0	Telcom
18,484.0	CW	HLD89	Seoul, South Korea	FX	....	
18,489.1	CW	RPTN	unknown	FX	....	"VVV's"
18,508.4	CW/FAX	WFK38	New York, NY	FX	50.0	Nx
18,512.0	ISB	ZPZ26	Asuncion, Paraguay	FX	10.0	Telcom/Tfc
18,515.0	USB	FTS51	Paris (St. Assise), France	FX	20.0	Telcom
18,517.5	ISB	FZM85	Noumea, New Caledonia	FX	20.0	Telcom/Papeete
18,520.0	ISB	RRG28	Moscow, USSR	FX	60.0	Telcom/Tfc
18,525.0	CW	KKN5Ø	Washington, DC	FX	10.0	EMBASSY
18,530.0	CW	†ELE28	Harbel, Liberia	FX	15.0	Tfc/Akron
18,535.0	ISB	MKD	Akrotiri, Cyprus	FX	5.0	RAF Telcom/Tfc
18,536.0	USB	†GMO38	London (Ongar), England	FX	20.0	Telcom
18,543.9	CW	KWS78	Athens, Greece	FX	....	USA EMBASSY
18,560.0	CW	RIW	Khiva Naval Radio, Uzbek SSR	FX	....	Russian Navy
18,565.0	ISB	†9XK82	Kigali, Rwanda	FX	10.0	Telcom/Tfc
18,568.0	CW	ZWBE	Belem Aeradio, Brazil	AX	1.0	Tfc/Rio
	CW	ZWRJ	Rio de Janeiro Aeradio, Brazil	AX	1.0	Tfc/Belem
18,578.0	CW	FUM	Papeete Naval Radio, Tahiti	FX	10.0	French Navy
18,585.0	USB	....	Stockholm Aeradio, Sweden	FA	7.0	
18,587.0	USB	OZU25	Copenhagen, Denmark	FX	1.5	EMBASSY (Also
			on this frequency are: OZU3Ø, Cairo, Egypt; OZU33, Tel Aviv, Israel; OZU34, Jeddah, Saudi Arabia; OZU35, Beijing PRC; OZU37, Ankara, Turkey; OZU38, Lagos, Nigeria & OZU39 Nairobi, Kenya) [Alt: 13,490.0 & 26,130.0]			
18,587.5	ISB	5OU88	Lagos, Nigeria	FX	....	Telcom
18,594.0	USB	AFS	Offutt AFB, NE	FX	1.0	USAF/SAC
			[Alt: 23,419.0--sometimes called "ZULU ONE" frequency]			
	USB	AGA	Hickam AFB, HI	FX	0.5	USAF
	USB	AIE	Andersen AFB, Guam	FX	0.5	"
	USB	AIF8Ø	Yokota AB, Japan	FX	0.5	"
	USB	AKE	Eielson AFB, AK	FX	0.5	"
18,607.5	LSB	....	Greenville, NC	FX	50.0	VOA Feeder
18,608.0	CW/RTTY	Y7A73	Berlin, GDR	FX	....	EMBASSY
18,619.0	USB	†PCM88	Hilversum (Kootwijk), Netherlands	FX	30.0	Backup Feeder for Radio Netherland
18,620.0	FAX	NPN	Barrigada (Agana NAS), Guam	FX	15.0	USN Meteo
18,623.8	USB	FZY386	Miquelon, St. Pierre & Miquelon Is.	FX	24.0	Telcom
18,624.0	CW	....	several, unknown, possibly in USSR	FX	....	Coded Tfc
18,630.0	ISB	†IRQ26	Rome (Torrenova), Italy	FX	10.0	Telcom
18,637.0	ISB	BCA2Ø	Shanghai, PRC	FX	15.0	Tfc/Rangoon
18,640.0	USB	YAK	Kabul, Afghanistan	FX	10.0	Telcom
18,653.0	LSB	RBI71	Moscow, USSR	FX	50.0	Telcom
			[May also used as R. Moscow Feeder]			
18,655.0	USB	†FTS66	Paris (St. Assise), France	FX	20.0	Telcom
18,666.0	USB	....	numerous, USA nationwide	FX/MA	....	DEA (Many
			code names preclude positive ID. "ATLAS" appears to be Net Control. Others heard include "CONDOR", "FLINT", "SHARK", "SUNDANCE", etc.) [Alt: 14,686.0]			
18,675.0	USB	†FPS67	Paris (St. Assise), France	FX	20.0	Telcom
18,696.0	CW	RIW	Khiva Naval Radio, Uzbek SSR	FX	....	Russian Navy
18,700.0	CW	KKN5Ø	Washington, DC	FX	....	EMBASSY
18,705.0	ISB	†CXL32	Montevideo, Uruguay	FX	2.5	Telcom/Tfc
18,711.5	ISB	PCW1	The Hague, Netherlands	FX	1.2	EMBASSY (Also
			on this frequency are: PCW2, Jerusalem, Israel; PCW11, Bangkok, Thailand & PCW2Ø, Warsaw, Poland)			

kHz	Mode	Call	Location	Service	kW	Remarks
18,720.2	CW	HIA66	unknown	FX	....	
18,721.1	USB	...	Atlantic Domestic Emergency Net	FX	....	USCG (Includes: NMA, Miami, FL; NMF, Boston, MA; NMG, New Orleans, LA; NMN, Portsmouth, VA; NMR, San Juan, PR & NOZ, Elizabeth City, NC, plus transportables) [Alt: 18,196.1]
	USB	....	Pacific Domestic Emergency Net	FX	....	USCG (Includes: NMC, San Francisco, CA; NMO, Honolulu, HI; NMQ, Long Beach, CA; NOJ, Kodiak, AK & NRV, Barrigada, Guam, plus transportables) [Alt: 18,196.1]
18,722.7	ISB	TLZ87	Bangui, Central African Republic	FX	20.0	Telcom
18,738.0	ISB	5TN244	Nouakchott, Mauritania	FX	....	Telcom/Tfc
18,743.5	USB	CXQ338	Montevideo, Uruguay	FX	5.0	Governmental [Tfc to Rio de Janeiro, Lima, Washington & Caracas]
18,744.4	LSB	5OU93	Lagos (Ikorodu), Nigeria	FX	3.5	Telcom/Paris
18,750.0	ISB	MKG	London (Stanbridge), England	FX	30.0	RAF Telcom/Tfc
18,752.5	ISB	†9KT357	Huban, Kuwait	FX	30.0	Telcom/Tfc
18,760.0	ISB	†....	Lyndhurst, Vic., Australia	FX	....	R. Australia Feeder
18,761.0	USB	OHU2Ø	Helsinki, Finland	FX	1.0	EMBASSY
	USB	OHU25	Beijing, PRC	FX	1.0	Finnish EMBASSY [Alt: 17,443.0 & 18,841.0]
18,771.3	LSB	†FTS77	Paris (St. Assise), France	FX	20.0	Telcom/Tfc

18,780.0 - 18,900.0 kHz — FIXED SERVICE & MARITIME MOBILE (Worldwide—shared post WARC)

kHz	Mode	Call	Location	Service	kW	Remarks
18,783.0	USB	....	Greenville, NC	FX	50.0	VOA Feeder
18,795.0	ISB	ORI48	Brussels (Ruiselede), Belgium	FX	30.0	Telcom
18,808.0	USB	SAM	Stockholm, Sweden	FX	1.2	EMBASSY (Also on this frequency are: SAM45, Ankara, Turkey; SAM46, Baghdad, Iraq; SAM47, Beirut, Lebanon; SAM48, Damascus, Syria; SAM49, Jeddah, Saudi Arabia; SAM5Ø, Cairo, Egypt; SAM51, Teheran, Iran; SAM54, Vientiane, Laos; SAM57, Dacca, Bangladesh; SAM61, New Delhi, India; SAM71, Dar es Salaam, Tanzania; SAM72, Kinshasa, Zaire; SAM73, Lagos, Nigeria; SAM75, Nairobi, Kenya; SAM79, Lusaka, Zambia; SAM8Ø, Gaborone, Botswanna; SAM82, Maputo, Mozambique & SAM89, Havana, Cuba) [Alt: 17,427.0 & 20,010.0]
	CW	RIW	Khiva Naval Radio, Uzbek SSR	FX	....	Russian Navy
	CW	†KNY29	Washington, DC	FX	1.0	British EMBASSY
18,815.0	ISB	FTS81	Paris (St. Assise), France	FX	20.0	Telcom/Kabul
18,823.0	FAX	Y2V38	Berlin (Nauen), GDR	FX	20.0	Nx
18,830.0	LSB	....	Moscow, USSR	FX	....	R. Moscow Feeder
18,833.4	USB	†FTS83	Paris (St. Assise), France	FX	20.0	Telcom
18,840.0	LSB	....	unknown	FX	....	Telcom/FF
18,841.0	USB	OHU25	Beijing, PRC	FX	1.0	Finnish EMBASSY
18,843.5	USB	TLZ88	Bangui, Central African Republic	FX	20.0	Telcom
18,844.0	LSB	ETR84	Addis Ababa, Ethiopia	FX	1.0	Finnish EMBASSY [Alt: 20,176.0]
18,852.5	USB	ZEN89	Cape d'Aguilar, Hong Kong	FX	3.5	Telcom/Nepal
18,853.5	USB	FTS85	Paris (St. Assise), France	FX	20.0	Telcom/Niamey
18,856.7	LSB	5UR88	Niamey, Niger	FX	20.0	Telcom
18,865.0	ISB	FTS86	Paris (St. Assise), France	FX	20.0	Telcom
18,875.8	USB	FTS87	Paris (St. Assise), France	FX	20.0	Telcom/Niamey
18,880.0	USB	VNA2	Sydney, NSW, Australia	FX	1.0	Monitoring Net (Also includes: VNA4, Brisbane; VNA5, Adelaide and VNA6, Perth) [Alt: 9440.0]
	CW	....	unknown	FX	....	Coded Tfc
18,885.0	ISB	MKD	Akrotiri, Cyprus	FX	30.0	RAF Telcom/Tfc
18,894.0	ISB	9XK92	Kigali, Rwanda	FX	....	Telcom

18,900.0 - 19,680.0 kHz — FIXED SERVICE (Worldwide—post WARC)

kHz	Mode	Call	Location	Service	kW	Remarks
18,904.0	ISB	MKG	London (Stanbridge), England	FX	10.0	RAF Telcom/Tfc
18,935.0	USB	FTS93	Paris (St. Assise), France	FX	20.0	Telcom
18,940.0	FAX	XVN47	Hanoi, Vietnam	FX	10.0	

kHz	Mode	Call	Location	Service	kW	Remarks
18,952.0	CW	RIW	Khiva Naval Radio, Uzbek SSR	FX	....	Russian Navy
18,963.0	USB	....	Mediterranean LORAN Net	FX	....	USCG (Includes: AOB5Ø, Estartit, Spain; NCI, Sellia Marina, Italy; NCI3, Lampedusa, Italy & NCI4, Kargabarun, Turkey) [Alt: 15,723.0 & 22,823.0]
18,972.0	CW	KKN5Ø	Washington, DC	FX	10.0	EMBASSY
18,977.2	ISB	5TN3Ø	Nouakchott, Mauritania	FX	20.0	Telcom/Tfc
18,977.8	CW	RVI	unknown	FX	....	
18,984.4	LSB	....	unknown	FX	....	Telcom/FF
18,990.0	ISB	FTS99	Paris (St. Assise), France	FX	20.0	Telcom/Tfc
18,995.0	CW	SOT29	Warsaw, Poland	FX	20.0	Reported PAP nx broadcast at 1400
19,008.2	ISB	GCB39B	London (Rugby), England	FX	30.0	Telcom/Tfc
19,013.0	USB	†KNY24	Washington, DC	FX	1.0	GFR EMBASSY
19,022.5	USB	....	numerous, worldwide	FX	....	US Army MARS
19,030.0	USB	D2J33	Luanda, Angola	FX	10.0	Telcom
	ISB	EHY25	Madrid (Pozuelo del Rey), Spain	FX	4.0	Telcom/Tfc
19,036.0	USB	†70B9Ø	Aden (Hiswa), Democratic Yemen	FX	5.0	Telcom
19,046.0	LSB	ETS8	Addis Ababa, Ethiopia	FX	10.0	Telcom
19,056.0	CW	ULY4	Aleksandrovsk, USSR	FX	....	
19,062.0	CW	PZA2	Paramaribo, Suriname	FX	....	
19,070.0	AM	RRG25	Moscow, USSR	FX	20.0	Telcom [Also used as R. Moscow feeder]
19,089.1	USB	†TTZ9Ø	N'Djamena, Chad	FX	20.0	Telcom
19,090.0	CW	†RIW	Khiva Naval Radio, Uzbek SSR	FX	....	Russian Navy
19,098.0	CW	RCV	Moscow Naval Radio, USSR	FX	....	Russian Navy
19,110.0	USB	†FYT2	Paris (St. Assise), France	FX	20.0	Telcom/Havana
19,115.0	ISB	CUA85	Lisbon (Alfragide), Portugal	FX	10.0	Telcom/Tfc
19,118.0	LSB	....	unknown, probably in Brazil	FX	....	Net
19,123.5	USB	PCW	The Hague, Netherlands	FX	1.2	EMBASSY
19,130.0	CW	DEB	Weisbaden, GFR	FX	5.0	INTERPOL
	CW	OGX	Helsinki, Finland	FX	1.0	"
	CW	8UF75	New Delhi, India	FX	5.0	"
	CW/RTTY	JPA59	Nagoya (Komaki), Japan	FX	10.0	" [Alt: 19,360.0]
19,140.0	FSK	3MA29	Nanjing, PRC	FX	20.0	Tfc/San Francisco
19,143.0	ISB	AFE86	Parham, Antigua	FX	25.0	USAF
19,154.0	ISB	EHY78	Madrid (Pozuelo del Rey), Spain	FX	4.0	Telcom
19,167.5	USB	SUA311	Cairo, Egypt	FX	35.0	Telcom
19,170.0	ISB	RRG24	Moscow, USSR	FX	15.0	Telcom/Tfc
19,178.0	LSB	....	Bonaire, Netherlands Antilles	FX	....	R. Nederland Relay
19,180.0	ISB	CUA86	Lisbon (Alfragide), Portugal	FX	10.0	
19,196.0	ISB	5UR91	Niamey, Niger	FX	20.0	Telcom
19,205.5	LSB	†FTT2Ø	Paris (St. Assise), France	FX	20.0	Telcom
19,225.0	CW	CLP1	Havana, Cuba	FX	...	EMBASSY
19,247.5	USB	†FTT24	Paris (St. Assise), France	FX	20.0	Telcom
19,249.0	LSB	FTT25	Paris (St. Assise), France	FX	20.0	Telcom
19,250.0	ISB	5RY92	Antananarivo, Madagascar	FX	10.0	Telcom/Tfc
	CW/RTTY	ZEF81	Salisbury, Zimbabwe	FX	1.0	PRC EMBASSY
19,255.0	USB	VNJ	Casey Base, Antarctica	FX	10.0	ANARE (Also on this frequency: VLV, Mawson Base; VLZ, Davis Base & VNM, Melbourne)
19,261.5	ISB	....	Bethany, OH	FX	50.0	VOA Feeder
	ISB	†....	Greenville, NC	FX	50.0	VOA Feeder
19,265.0	USB	ZEN91	Cape d'Aguilar, Hong Kong	FX	3.0	Telcom
19,275.0	ISB	CUC23	Lisbon (Alfragide), Portugal	FX	10.0	Tfc
	FAX	RXO74	Khabarovsk, USSR	FX	100.0	Meteo
19,297.1	USB	....	Northwest Pacific LORAN NET	FX	....	USCG (Includes: NRT, "Yokota Monitor", Japan; NRT2, Gesashi, Japan; NRT3, Iwo Jima; NRT9, Hokkaido, Japan; NRV, Barrigada, Guam; NRV6, Marcus Is.; NRV7, Yap Island) [Alt: 17,488.6]
19,303.0	USB	AFE71	"Cape Radio" (Patrick AFB), FL	FX	45.0	USAF/NASA [Alt: 20,191.0]
19,304.8	CW	CLQ	Havana	FX	....	
19,305.0	USB	†FTT3Ø	Paris (St. Assise), France	FX	20.0	Telcom

kHz	Mode	Call	Location	Service	kW	Remarks
19,314.0	USB	....	unknown	FX	....	Telcom/FF
19,320.0	LSB	RRG22	Moscow, USSR	FX	15.0	R. Moscow Feeder
19,324.0	ISB	FZS93	St. Denis (Bel Air), Reunion	FX	20.0	Telcom/Tfc
19,344.5	CW/USB	KUR2Ø	Honolulu, HI	FX	1.0	Tfc/Agana
				[Alt: 20,348.5 & 20,602.5]		
19,360.0	CW	OQX	Helsinki, Finland	FX	1.0	INTERPOL
	CW	FSB	Paris (St. Martin Abbat), France	FX	2.5	"
	CW/RTTY	DEB	Weisbaden, GFR	FX	5.0	"
	CW	8UF75	New Delhi, India	FX	5.0	"
	CW	JPA57	Nagoya (Komaki), Japan	FX	10.0	" (Also
			reported on this frequency are 5BP2, Nicosia, Cyprus; 5OP25, Lagos, Nigeria; 5TP25, Nouakchott, Mauritania & 5YG, Nairobi, Kenya)			
19,379.8	USB	†WEK69	New York, NY	FX	80.0	Telcom
19,385.0	CW	3VA81	Tunis, Tunisia	FX	0.4	
19,405.0	CW/RTTY	FSB63	Paris (St. Martin Abbat), France	FX	2.5	INTERPOL
	CW	TCC2	Ankara, Turkey	FX	....	INTERPOL
19,411.5	ISB	ETS41	Addis Ababa, Ethiopia	FX	10.0	Telcom/Tfc
19,425.0	CW	9ES5Ø	Addis Ababa, Ethiopia	FX	0.5	EMBASSY
19,430.0	ISB	ATS69	Bombay (Kirkee), India	FX	20.0	Telcom/Tfc
	ISB	9RE394	Lubumbashi, Zaire	FX	3.0	Telcom/Tfc
19,430.5	ISB	MKD	Akrotiri, Cyprus	FX	....	RAF Telcom/Tfc
19,449.8	CW	VTK3	Tuticorin Naval Radio, India	FX	....	Indian Navy
19,455.0	ISB	....	Daventry, England	FX	....	BBC Feeder
19,458.0	CW	†KNY2Ø	Washington, DC	FX	1.0	Polish EMBASSY
	CW	†KNY23	Washington, DC	FX	1.0	Czech EMBASSY
19,461.5	USB	ETS43	Addis Ababa, Ethiopia	FX	1.0	Telcom/Lagos
19,465.0	CW	KNY23	Washington, DC	FX	1.0	Czech EMBASSY
19,479.2	ISB	....	Kigali, Rwanda	FX	....	Telcom
19,480.0	ISB	....	Delano, CA	FX	50.0	VOA Feeder
	ISB	....	Greenville, NC	FX	50.0	" "
	ISB	....	Bethany, OH	FX	50.0	" "
19,487.0	CW	TUW22Ø	Abidjan, Ivory Coast	FX	1.0	INTERPOL/Police
			(Bulletins to Central African Republic, Gabon, Mali, Niger, Nigeria, Senegal, Togo & Zaire) [Alt: 14,827.0]			
19,488.2	CW	ZLX31	Wellington (Himatangi), New Zealand	FX	5.0	NOAA reports
			wx broadcasts at 0200, 0500, 0520, 2040 & 2110			
19,500.0	USB	PUZ4	Brasilia, Brazil	AX	1.0	Governmental
	ISB	AFH3	Albrook AFS, Panama	FX	3.0	USAF
					[Alt: 20,600.0]	
19,505.0	ISB	....	Greenville, NC	FX	40.0	VOA Feeder
19,521.5	USB	....	Greenville, NC	FX	40.0	VOA Feeder
19,525.0	ISB	†CXL33	Montevideo, Uruguay	FX	2.5	Telcom
19,528.5	USB	MKT	London (Stanbridge), England	FX	10.0	RAF Telcom/Tfc
19,532.0	ISB	FJY2	Port aux Francais, Kerguelen Is.	FX	35.0	Telcom/Paris
19,532.5	ISB	†....	Greenville, NC	FX	40.0	VOA Feeder
19,554.8	ISB	FZW95	Destrellan, Guadeloupe	FX	6.0	Tfc/Paris
19,565.0	ISB	†HEK6	Berne (Schwarzenburg), Switzerland	FX	4.0	Telcom
19,570.0	ISB	FZF97	Fort de France, Martinique	FX	20.0	Telcom/Tfc
19,577.0	USB	ORI49	Brussels (Ruiselede), Belgium	FX	30.0	Telcom
19,604.0	USB	ATP70	New Delhi (Kirkee), India	FX	10.0	Telcom
19,605.0	CW	YZJ4	Belgrade (Makis), Yugoslavia	FX	30.0	Tfc/Tokyo
					[Alt: 19,865.5]	
19,610.0	ISB	JBU89	Tokyo, Japan	FX	30.0	Telcom/Berne
	USB	ZLO	Irirangi Naval Radio, New Zealand	FX	5.0	RNZ Navy
19,628.5	USB	XYN31	Rangoon, Burma	FX	2.0	Telcom/Poona
19,630.0	ISB	†5YF92	Nairobi, Kenya	FX	30.0	Telcom/Tfc
19,637.5	USB	....	Kabul, Afghanistan	FX	....	R. Afghanistan Feeder
19,645.8	ISB/RTTY	HBO19	Geneva, Swizterland	FX	20.0	Tfc/Rawalpindi
19,665.0	ISB	†CNR34	Rabat, Morocco	FX	20.0	Telcom/Tfc
19,665.1	USB	†5LF19	Monrovia, Liberia	FX	6.0	Telcom/Paris
19,668.4	CW	†SQL47	Islamabad, Pakistan	FX	....	Polish EMBASSY
19,670.1	ISB	ASB33	Karachi, Pakistan	FX	5.0	Telcom/Tfc

kHz	Mode	Call	Location	Service	kW	Remarks
19,675.0	CW	3VA82	Tunis, Tunisia	FX	1.0	
19,678.0	LSB	70B93	Aden (Hiswa), Democratic Yemen	FX	3.5	Telcom

19,680.0 - 19,800.0 kHz -- FIXED SERVICE & MARITIME MOBILE (Worldwide--shared post-WARC)

kHz	Mode	Call	Location	Service	kW	Remarks
19,690.0	FAX	AXM37	Canberra, ACT, Australia	FX	20.0	Meteo
	CW	3MA31	Nanjing, PRC	FX	2.5	Tfc
19,721.8	USB	....	Greenville, NC	FX	50.0	VOA Feeder
19,724.8	USB	RRD	Moscow, USSR	FX	20.0	R. Moscow Feeder
19,728.5	USB	OEC	Vienna, Austria	FX	10.0	EMBASSY
19,745.0	ISB	JBU39	Tokyo, Japan	FX	10.0	Telcom
19,750.0	FAX	6VU79	Dakar, Senegal	AX	5.0	Meteo (0830-2000)

19,800.0 - 19,990.0 kHz -- FIXED SERVICE (Worldwide--post WARC)

kHz	Mode	Call	Location	Service	kW	Remarks
19,805.0	CW	....	unknown, probably in Africa	FX	....	
19,814.5	CW	CUC9	Lisbon (Alfragide), Portugal	FX	10.0	Tfc
19,819.0	CW	....	unknown	FX	....	Coded Tfc
19,845.0	LSB	RWZ74	Moscow, USSR	FX	15.0	R. Moscow Feeder
19,849.2	CW/FAX	WFK39	New York, NY	FX	50.0	Tfc/Nx
19,850.0	ISB	FTT85	Paris (St. Assise), France	FX	20.0	Telcom
19,865.0	ISB	RAD4	Moscow, USSR	FX	15.0	Telcom
19,865.5	CW	YZJ4	Belgrade (Makis), Yugoslavia	FX	30.0	Tfc/Tokyo
					[Alt:	19,605.0]
19,868.0	CW	8KL	unknown	FX	....	Coded Tfc
19,872.9	ISB	ASH	unknown, probably in Pakistan	FX	....	Tfc
19,880.0	ISB	†JBU69	Tokyo, Japan	FX	10.0	Telcom/Tfc
19,885.0	CW/FSK	Y7A79	Berlin, GDR	FX	....	EMBASSY
19,895.0	ISB	MKD	Akrotiri, Cyrpus	FX	10.0	RAF Telcom/Tfc
19,900.0	ISB	†HLA94	Seoul, South Korea	FX	10.0	Telcom/Manila
19,912.0	ISB	....	Delano, CA	FX	40.0	VOA Feeder
19,921.6	LSB	....	unknown	FX	....	VOA Feeder
19,943.0	LSB	5BC94	Nicosia, Cyprus	Fx	6.0	Telcom
19,950.0	USB	†VPC	Port Stanley, Falkland Islands	FX	7.5	Telcom/Tfc
	CW	KNY25	Washington, DC	FX	1.0	Rumanian EMBASSY
19,955.0	FAX	†AFA	Washington (Andrews AFB, MD), DC	FX	3.0	USAF/Meteo
19,960.0	CW	LCK	Oslo (Jeloey), Norway	FX	20.0	Tfc
19,970.0	CW	†PBC5	Goeree Naval Radio, Netherlands	FX	5.0	Dutch Navy
19,985.0	CW	RIW	Khiva Naval Radio, Uzbek SSR	FX	....	Russian Navy

19,990.0 - 20,010.0 kHz -- STANDARD FREQUENCY/TIME & SPACE RESEARCH

kHz	Mode	Call	Location	Service	kW	Remarks
19,993.0	CW	RIW	Khiva Naval Radio, Uzbek SSR	FX	....	Russian Navy
19,995.0	....	....	Infrequent use by USSR Satellites	EH	....	Telemetry
19,996.0	....	....	Infrequent use by PRC Satellites	EH	....	Telemetry
20,000.0	AM	WWV	Fort Collins, CO	SS	10.0	TIME SIGNALS
20,001.0	CW	HGX21	Budapest, Hungary	FX	....	EMBASSY
20,009.0	....	....	Infrequent use by PRC Satellites	EH	....	Telemetry

20,010.0 - 21,000.0 -- FIXED SERVICE (Worldwide)

kHz	Mode	Call	Location	Service	kW	Remarks
20,010.0	USB	SAM	Stockholm, Sweden	FX	1.2	EMBASSY (This frequency is available to all Swedish EMBASSY stations listed on alternative frequencies of 14,971.5 & 18,808.0)
20,011.0	CW/RTTY	†KNY26	Washington, DC	FX	1.0	Hungarian EMBASSY
20,015.0	FAX	NAM	Norfolk Naval Radio, VA	FC	10.0	USN/Meteo
20,022.8	CW	....	unknown	FX	....	Coded Tfc
20,026.0	USB	....	several, USA nationwide	FX	....	FEMA Backup Net (Recently reported have been WGY9Ø7, Kansas City, MO & WGY9Ø8, Santa Rosa, CA) [Alt: 19,970.0]
20,030.0	CW/RTTY	CYS22	Ottawa, Ont., Canada	FX	1.0	Polish EMBASSY [Alt: 18,355.0 & 23,376.0]
20,050.0	LSB	AFA	Andrews AFB, MD	FX	....	USAF
20,060.0	ISB	....	Greenville, NC	FX	50.0	VOA Feeder
20,067.9	CW	RIT	unknown, probably in USSR	FX	....	
20,070.2	ISB/RTTY	4UZ5Ø	Geneva, Switzerland	FX	20.0	UNO

kHz	Mode	Call	Location	Service	kW	Remarks
20,100.0	USB	....	several throughout South Atlantic	FX	....	USN Net (Includes "SKY", Rota NAS, Spain; "DEW", unknown; "SCH", unknown; etc.)
20,124.0	ISB	OHU2Ø	Helsinki, Finland	FX	1.0	EMBASSY
	USB	OHU21	Paris, France	FX	1.0	Finnish EMBASSY [Alt: 20,186.0]
20,125.0	ISB	....	Greenville, NC	FX	40.0	VOA Feeder
20,129.0	CW	....	unknown	FX	....	Coded Tfc
20,130.0	LSB	TKY1	Cotonou, Benin	FX	....	Telcom/Algiers
20,135.0	ISB	SUV66	Cairo, Egypt	FX	10.0	Telcom/Rome
20,152.0	CW	....	unknown, probably in Cuba	FX	....	Coded Tfc
20,160.0	USB	CEC2Ø8	Paine, Chile	FX	3.0	
	USB	CEC224	Arica, Chile	FX	1.0	
20,162.0	USB	ZEN95	Cape d'Aguilar, Hong Kong	FX	10.0	Telcom
20,170.5	CW	Y7A81	Berlin, GDR	FX	....	EMBASSY
20,176.0	USB	OHU2Ø	Helsinki, Finland	FX	1.0	EMBASSY
	USB	OHU25	Beijing, PRC	FX	1.0	Finnish EMBASSY [Alt: 18,841.0 & 20,262.0]
20,185.6	USB	†IRS41	Rome (Torrenova), Italy	FX	30.0	Telcom
20,185.8	USB	....	numerous, Atlantic Ocean area	FX	....	USAF/NASA
20,188.8	USB	....	several, Atlantic Ocean area	FX	....	USAF MARS
20,191.0	LSB	....	numerous, Atlantic Ocean Area (Includes: AFE71, "One" [aka "Cape Radio"], Patrick AFB, FL; "Three", Grand Bahamas Island; "Seven" Grand Turk Island; "Twelve", Ascension Island; "Ninety-nine", Antigua; etc.)	FX	....	USAF/NASA
20,198.0	LSB	....	numerous, Atlantic Ocean area	FX	....	USAF/NASA (Includes same stations as 20,191.0)
20,206.0	USB	3CA67	Banapa, Guinea	FX	4.0	Telcom/Madrid
20,215.0	ISB	....	Holzkirchen, GFR	FX	10.0	RFE Feeder
20,220.0	USB	5VH4Ø2	Tome, Togo	FX	20.0	Telcom/ Paris
20,224.0	LSB	ZEN96	Cape d'Aguilar, Hong Kong	FX	8.0	Telcom
20,225.0	CW	NRK	Keflavik Naval Radio, Iceland	FC	....	USN
	CW	NAM	Norfolk Naval Radio, VA	FC	20.0	(LCMP-2) USN/ NUKO/NOAA reports wx broadcasts at 1230 & 1900/Seasonal ice warnings at 1900 & 2300
20,241.0	ISB	5TA21	Nouadhibou, Mauritania	FX	6.0	Telcom/Paris
20,245.0	CW	SQL49	Islamabad, Pakistan	FX	1.0	Polish EMBASSY
20,247.2	LSB	....	unknown	FX	....	Telcom/FF
20,262.0	USB	OHU2Ø	Helsinki, Finland	FX	1.0	EMBASSY
	USB	OHU25	Beijing, PRC	FX	1.0	Finnish EMBASSY [Alt: 20,176.0]
	USB	OHU36	Baghdad, Iraq	FX	1.0	Finnish EMBASSY
20,262.5	USB	....	Mediterranean LORAN NET	FX	....	USCG (Includes: NCI, Sellia Marina, Spain & ACB5Ø, Estartit, Spain--days)
20,288.0	USB	....	Budapest, Hungary	FX	....	Telcom/Moscow
20,290.0	LSB	MKD	Akrotiri, Cyprus	FX	10.0	RAF Telcom/Tfc
	USB	RCC72	Moscow, USSR	FX	20.0	Telcom
20,295.0	CW	CLN641	Havana (Bauta), Cuba	FX	80.0	Tfc/Moscow
20,320.0	ISB	MKG	London (Stanbridge), England	FX	10.0	RAF Telcom/Tfc
20,327.8	ISB	†6VK221	Dakar, Senegal	FX	20.0	Telcom/Tfc
20,330.0	ISB	MKE	Akrotiri, Cyprus	FX	10.0	RAF Telcom/Tfc
20,332.0	USB	....	unknown	FX	....	Telcom/FF
20,337.8	LSB	D6B4Ø3	Moroni, Republic of the Comores	FX	6.0	Telcom/Tfc
20,344.4	LSB	TYK3	Cotonou, Benin	FX	20.0	Telcom/Paris
20,345.0	ISB	....	Daventry, England	FX	....	BBC Feeder
20,348.5	USB	KOG55	Las Vegas, NV	FX	1.0	
	USB	KUR2Ø	Honolulu, HI	FX	1.0	[Alt: 19,344.5 & 20,602.5]
	USB	KUR5Ø	Agana, Guam	FX	1.0	
20,353.0	CW	KKN44	Monrovia, Liberia	FX	....	USA EMBASSY
20,370.0	ISB	†5VH4Ø3	Lome, Togo	FX	20.0	Telcom/Paris
20,375.0	CW/RTTY	ETT37	Addis Ababa, Ethiopia	FX	1.0	EMBASSY
20,403.5	LSB	D6B4Ø4	Moroni, Republic of the Comores	FX	6.0	Telcom/Paris
20,405.0	ISB	BAZ37	Beijing, PRC	FX	10.0	Telcom/Warsaw

kHz	Mode	Call	Location	Service	kW	Remarks
20,415.5	CW	....	unknown	FX	....	Tfc/SS
20,416.1	USB	DFY	Tel Aviv, Israel	FX	0.1	GFR EMBASSY
20,430.0	ISB/RTTY	IRS24	Rome (Torrenova), Italy	FX	10.0	Telcom/ANSA Nx
20,455.0	LSB	....	Daventry, England	FX	30.0	BBC Feeder
20,455.5	CW	"E"	unknown	RC	....	Beacon
20,458.0	CW	ETT46	Addis Ababa, Ethiopia	FX	5.0	EMBASSY
20,463.0	USB	....	Rio de Janeiro, Brazil	FX	0.5	USA EMBASSY
20,480.0	LSB	OEC	Vienna, Austria	FX	....	EMBASSY
	USB	KGA32	Honolulu, HI	FX	1.0	FAA Control
			(Also includes: KCC97, Diamond Head, HI; KCD72, Agana, Guam & KDM48, Pago Pago, American Samoa)			
	LSB	ZEN97	Cape d'Aguilar, Hong Kong	FX	8.0	Telcom
20,495.4	LSB	OEC	Vienna, Austria	FX	1.0	EMBASSY
	LSB	OEC35	New Delhi, India	FX	1.0	Austrian EMBASSY
20,500.0	ISB	Y4A3	Berlin (Nauen), GDR	FX	20.0	Telcom/Havana
	ISB	RWG	Moscow, USSR	FX	15.0	R. Moscow Feeder
20,505.0	ISB	†....	Holzkirchen, GFR	FX	10.0	RFE Feeder
20,510.0	CW	UJY2	Kaliningrad, USSR	FX	5.0	Tfc/Havana
20,535.0	ISB	ATJ71	Bombay (Kirkee), India	FX	20.0	Telcom
20,535.0	USB/AM	D2J35	Luanda, Angola	FX	10.0	Telcom
20,540.0	CW	GYU	Gibraltar Naval Radio, Gibraltar	FX	....	British Navy
20,545.0	ISB	....	Holzkirchen, GFR	FX	10.0	RFE Feeder
20,558.0	USB	FTU56	Paris (St. Assise), France	FX	20.0	Telcom
20,568.0	CW	KRH5Ø	London, England	FX	....	USA EMBASSY
20,570.0	ISB	EHY23	Madrid (Pozuelo del Rey), Spain	FX	10.0	Telcom
20,575.0	USB	†PP7A	"Observatory" (Paulista), Brazil	FX	1.0	NASA/Smithsonian [Alt: 20,610.0]
20,600.0	ISB	PUZ4	Brasilia, Brazil	FX	1.0	Governmental
			[Bulletins to La Paz, Quito, Asuncion, Lima, Montevideo, Washington & Caracas]			
	ISB	AFH3	Albrook AFS, Panama	FX	3.0	USAF
			[Alt: 15,675.0 & 23,101.0]			
	ISB	MKD	Akrotiri, Cyprus	FX	10.0	RAF Telcom/Tfc
20,600.5	LSB	....	unknown	FX	....	Net/SS
20,602.5	USB	KUR2Ø	Honolulu, HI	FX	1.0	
	USB	KUR5Ø	Agana, Guam	FX	1.0	
			[Alt: 20,348.5 & 23,402.5]			
20,605.0	LSB	OCK25	Arequipa, Peru	FX	....	Smithsonian
	ISB	....	Moscow, USSR	FX	....	R. Moscow Feeder
20,605.2	LSB	....	unknown, possibly Argentina	FX	....	SS Telcom
20,610.0	USB	†PP7A	"Observatory" (Paulista), Brazil	FX	1.0	NASA/Smithsonian [Alt: 20,575.0]
	LSB/RTTY	OCK25	Arequipa, Perua	FX	1.0	NASA/Smithsonian
	USB/RTTY	KCW21	Cambridge, MA	FX	1.0	Tfc
20,610.8	LSB	....	unknown	FX	....	Net/PP
20,615.6	LSB	FTU61	Paris (St. Assise), France	FX	20.0	Telcom
20,619.0	CW	†OMZ	Prague, Czechoslovakia	FX	1.0	EMBASSY
20,623.0	USB	NNNØNRS	Diego Garcia	FX	....	USN MARS
20,630.0	CW	LHK	Oslo (Jeloey), Norway	FX	10.0	Meteo/Nx
20,631.0	USB	....	numerous, worldwide	FX/MA	....	USAF (Usually
			AJE, Croughton, England; AJG9, Incirlik AB, Turkey; AJO, Adana AB, Turkey; etc.--sometimes referred to as "WHISKEY" frequency--sometimes used for SKY KING broadcasts)			
20,640.0	USB	FTU64	Paris (St. Assise), France	FX	20.0	Telcom
20,650.0	USB	†....	Holzkirchen, GFR	FX	10.0	RFE Feeder
20,652.8	ISB	3XF24	Conakry, Guinea	FX	6.0	Telcom/Paris
20,670.0	ISB	EHY22	Madrid (Pozuelo del Rey), Spain	FX	4.0	Telcom/Nouadhibou
20,675.0	ISB	GLF4Ø	London (Rugby), England	FX	30.0	Telcom/Accra
20,682.5	USB	OMZ3Ø	Prague, Czechoslovakia	FX	4.0	Libyan EMBASSY
20,690.0	CW	YWM3	Maracaibo Naval Radio, Venezuela	FX	....	Venezuela Navy
20,695.0	ISB	CUA89	Lisbon (Alfragide), Portugal	FX	10.0	Telcom/Tfc
20,710.0	ISB	†....	Holzkirchen, GFR	FX	10.0	RFE Feeder
20,712.0	USB	....	unknown, probably in Middle East	FX	....	Telcom
20,734.0	CW	4UWG	Geneva, Switzerland	FX	1.5	Tfc/UN
	CW	4UWC	Lubumbashi, Zaire	FX	1.5	Tfc/UN
20,735.0	CW	TJP76	Kousserin, Cameroon	FX	....	UNO

FIXED SERVICE (con.)

kHz	Mode	Call	Location	Service	kW	Remarks
20,745.0	ISB	†....	Holzkirchen, GFR	FX	10.0	RFE Feeder
20,751.4	LSB	OEC64	Lagos, Nigeria	FX	1.0	Austrian EMBASSY
20,751.5	USB	PGA88	The Hague, Netherlands	FX	0.3	INTERNATIONAL
	USB	SAR88	Stockholm, Sweden	FX	...	" RED CROSS
						[Alt: 20,815.0]
20,753.0	CW	EPR88	Teheran, Iran	FX	....	INTERNATIONAL
	CW	ODR88	Beirut, Lebanon	FX	....	" RED CROSS
20,753.5	LSB	OEC	Vienna, Austria	FX	....	EMBASSY
	CW/USB	OEC44	Tel Aviv, Israel	FX	2.0	Austrian EMBASSY
						[Alt: 14,478.5]
	LSB	OEC57	Pretoria, RSA	FX	1.0	Austrian EMBASSY
20,765.0	ISB	CUA88	Lisbon (Alfragide), Portugal	FX	10.0	Telcom/Tfc
20,776.0	CW	NWC	Northwest Cape Naval R., Australia	FX	15.0	USN
20,780.0	USB	VJP	Perth, WA, Australia	FX	....	Telcom/Christmas
20,790.0	ISB	ZEN99	Cape d'Aguilar, Hong Kong	FX	8.0	Telcom/Singapore
20,797.0	USB	PUZ4	Brasilia, Brazil	FX	1.0	Governmental
						[Telcom: London & Paris]
20,798.5	CW/FAX	WFN2Ø	New York, NY	FX	50.0	Tfc/Nx
20,802.0	CW	HGX21	Budapest, Hungary	FX	....	EMBASSY
20,812.0	USB	HBC88	Geneva, Switzerland	FX	1.0	Inter. RED CROSS
						[Alt: 13,997.0 & 20,939.0]
20,815.0	USB	SAR88	Stockholm, Sweden	FX	1.0	Inter. RED Cross
						[Alt: 20,751.5 & 20,942.0]
20,816.4	USB	9TO2Ø	Kinshasa, Zaire	FX	0.2	UN
20,817.6	LSB	D6B4Ø8	Moroni, Republic of the Comores	FX	6.0	Telcom/Tfc
20,835.4	CW/RTTY	CLP1	Havana, Cuba	FX	....	EMBASSY
20,832.0	ISB	TTZ8	N'Djamena, Chad	FX	20.0	Telcom/Tfc
20,837.0	ISB	JBT3Ø	Tokyo, Japan	FX	10.0	Telcom/Tfc
20,845.0	USB	....	Golan Heights, Syria	FX	....	Austrian UN Force
						[Alt: 13,081.0]
20,852.0	USB	....	numerous, USA nationwide	FX	....	FAA Backup Net
20,863.3	LSB	....	unknown	FX	....	Telcom/FF
20,866.5	LSB	....	unknown	FX	....	Telcom/FF
20,870.0	USB	AGB4	Wake Island	FX	1.0	USAF/NASA
						Space Shuttle Support
20,875.0	LSB	XQ8AFI	Cerro Tololo Observatory, Chile	FX	....	wkg KFK92
20,878.5	USB	KFK92	Kitts Peak Obs. (Tucson), AZ	FX	1.0	wkg XQ8AFI
20,885.0	LSB	....	several, in Central & South America	FX	....	USMAAG (Usually heard: AHF3, Albrook AFS, Panama; LQU21, Buenos Aires, Argentina; ZPM261, Asuncion, Paraguay; etc.)
20,890.0	USB	AFG4	Offutt AFB, NE	FX	0.5	USAF/SAC
			[Sometimes referred to as "DELTA ONE" frequency]			[Alt: 14,955.0]
20,897.2	USB	....	unknown	FX	....	Telcom/AA
20,900.0	USB	FZK3	Djibouti, Djibouti	FX	....	Tfc
	CW/RTTY	ETT9Ø	Addis Ababa, Ethiopia	FX	1.0	Czech EMBASSY
	CW	6C2	unknown, probably in Africa	FX	....	EMBASSY Tfc
20,903.2	LSB	FTU9Ø	Paris (St. Assise), France	FX	20.0	Telcom
20,913.6	LSB	ODS91	Beirut, Lebanon	FX	2.0	Telcom/Tfc
20,917.8	USB	...	numerous, worldwide	FX	....	USAF MARS
20,919.6	USB	†FTU92	Paris (St Assise), France	FX	20.0	Telcom
20,928.0	USB	ZLK6	Christchurch, New Zealand	FX	1.0	RNZAF/USAF
20,929.0	CW	KKN44	Monrovia, Liberia	FX	....	USA EMBASSY
20,935.0	ISB	†....	Holzkirchen, GFR	FX	10.0	RFE Feeder
20,936.4	USB	....	numerous, worldwide	FX	....	USN MARS
20,939.0	USB	HBC88	Geneva, Switzerland	FX	1.0	Inter. RED CROSS
						[Alt: 20,812.0 & 20,998.0]
20,940.0	USB	FTU94	Paris (St. Assise), France	FX	20.0	Telcom/Havana
20,942.0	USB	SAR88	Stockholm, Sweden	FX	1.0	Inter. RED CROSS
						[Alt: 20,815.0]
20,946.7	CW	5YC	Nairobi, Kenya	FX	....	Meteo
20,950.0	CW/RTTY	ETT95	Addis Ababa, Ethiopia	FX	1.0	Yugoslav EMBASSY
20,955.0	ISB	†TNI9	Brazzaville, Congo	FX	20.0	Telcom
20,957.0	ISB	CIS9	Petawawa, Ont., Canada	FX	....	Canadian Forces
	CW/RTTY	YZK2	Belgrade (Makis), Yugoslavia	FX	30.0	Tfc/Sydney

FIXED SERVICE (con.)

kHz	Mode	Call	Location	Service	kW	Remarks
20,958.0	USB	SAM	Stockholm, Sweden	FX	1.2	EMBASSY
			(Usually heard Swedish EMBASSIES include: SAM25, Lisbon, Portugal; SAM48, Damascus, Syria; SAM49, Jeddah, Saudi Arabia; SAM52, Tel Aviv, Israel; SAM54, Vientiane, Laos; SAM61, New Delhi, India; SAM64, Pyongyang, N. Korea; SAM65, Seoul, S. Korea; SAM67, Colombo, Sri Lanka; SAM79, Zambia; SAM8Ø, Gaborone, Botswana; SAM81, Bissau, Guinea-Bissua & SAM82, Maputo, Mozambique) [Alt: 14,971.5; 18,808.0; 20,010; 20,958.0 & 23,529.0]			
20,960.0	USB	CPP67	La Paz, Bolovia	FX	0.5	USMAGP
20,970.0	USB	....	several, Middle East & Canada	FX	....	Canadian Forces [Alt: 13,971.0]
	CW/RTTY	ZEF81	Salisbury, Zimbabwe	FX	1.0	PRC EMBASSY
20,972.0	USB	OEC36	Beijing, PRC	FX	1.0	Austrian EMBASSY [Alt: 16,131.5]
20,972.1	ISB	†ODS97	Beirut, Lebanon	FX	7.0	Telcom/Tfc
20,975.3	LSB	....	numerous, worldwide	FX	....	US Army MARS
20,976.0	CW	†FLE23	Paris, France	FX	10.0	Tfc
20,981.2	ISB	MKD	Akrotiri, Cyprus	FX	10.0	RAF Telcom/Tfc
20,980.2	CW	SM5R	unknown	FX	....	Coded Tfc
20,985.0	USB	SAM89	Havana, Cuba	FX	1.0	Swedish EMBASSY
	ISB	....	Holzkirchen, GFR	FX	10.0	RFE Feeder
	USB	SAM	Stockholm, Sweden	FX	1.2	EMBASSY
20,987.4	USB	....	numerous, worldwide	FX	....	USN MARS
20,991.5	CW	"D"	unknown	RC	....	Beacon
20,991.8	CW	"O"	unknown	RC	....	Beacon
20,992.0	LSB	XVX2ØB	Hanoi, Vietnam	FX	....	Telcom
	CW/Data	"C"	unknown	RC	....	Beacon
20,997.0	USB	PGA88	The Hague, Netherlands	FX	1.0	Inter. RED CROSS
	USB	....	several, worldwide	FX	....	USN MARS
20,998.0	USB	SAR88	Stockholm, Sweden	FX	1.0	Inter. RED CROSS [Alt: 20,942.0]
	USB	HBC88	Geneva, Switzerland	FX	1.0	Inter. RED CROSS [Alt: 20,812.0]

21,000.0 - 21,450.0 kHz — AMATEUR RADIO (Worldwide)

21,450.0 - 21,750.0 kHz — INTERNATIONAL BROADCASTING (Worldwide)

21,450.0 - 21,850.0 kHz — INTERNATIONAL BROADCASTING (Worldwide-post WARC)

21,750.0 - 21,870.0 kHz — FIXED SERVICE (Worldwide)

21,850.0 - 21,870.0 kHz — FIXED SERVICE (Worldwide-post WARC)

kHz	Mode	Call	Location	Service	kW	Remarks
21,755.0	CW	9EU76	Addis Ababa, Ethiopia	FX	0.5	Italian EMBASSY
21,760.0	CW	NPN	Barrigada (Agana NAS), Guam	FX	3.0	USN/NOAA reports wx broadcasts at 0400, 1300, 1700 & 2300
	CW	NRV	COMMSTA Barrigada, Guam	FX	....	USCG
	CW	RIF31	Moscow, USSR	FX	20.0	Tfc
21,764.1	CW	RIW	Khiva Naval Radio, Uzbek SSR	FX	....	Russian Navy
21,765.0	CW	RCV	Moscow Naval Radio, USSR	FX	....	Russian Navy
21,773.4	LSB	CUW	Lajes Field, Azores	FX	10.0	USAF
21,784.0	CW	†RIW	Khiva Naval Radio, Uzbek SSR	FX	....	Russian Navy
21,785.0	CW	AYA28	Buenos Aires, Argentina	FX	1.0	INTERPOL
	CW	XJE57	Ottawa (Almonte), Ont., Canada	FX	1.5	INTERPOL
	FAX	NPM	Lualualei Naval Radio, HI	FX	15.0	USN Meteo (1800-0630)
	CW	8UF75	New Delhi, India	FX	5.0	INTERPOL
21,804.0	CW	ZBP	Pitcairn Island	FX	0.7	Tfc
21,804.5	USB	ONN27	Brussels, Belgium	FX	1.0	EMBASSY
21,807.0	CW	TUW21Ø	Abidjan, Ivory Coast	FX	1.0	INTERPOL
21,807.5	CW	FSB67	Paris (St. Martin Abbat), France	FX	2.5	INTERPOL
	CW	8UF75	New Delhi, India	FX	5.0	"
	CW	5OP25	Lagos, Nigeria	FX	1.5	"
21,810.0	CW/RTTY	OMZ	Vienna, Austria	FX	1.0	EMBASSY

kHz	Mode	Call	Location	Service	kW	Remarks
21,814.0	CW	CLN653	Havana (Bauta), Cuba	FX	30.0	Tfc/UJY2
21,820.0	ISB	MKG	London (Stanbridge), England	FX	30.0	RAF Telcom/Tfc
21,824.0	A3H	....	unknown	FX	....	Telcom/FF
21,830.0	FAX	VFE	Edmonton, Alta., Canada	FX	5.0	Meteo
	ISB	FTV83	Paris (St. Assise), France	FX	20.0	Telcom
21,834.0	CW/FAX	ZLK45	Christchurch, New Zealand	FX	15.0	Tfc/McMurdo
21,838.0	CW/RTTY	SOW283	Warsaw, Poland	FX	20.0	Reported PAP nx broadcast at 1400

21,870.0 - 22,000 kHz — AERONAUTICAL FIXED/MOBILE (Worldwide)

kHz	Mode	Call	Location	Service	kW	Remarks
21,886.0	AM/A3H	....	numerous, throughout Africa and Australasia	FA/MA	....	Airports & in-flight
21,925.0	USB	....	Amsterdam Aeradio, Netherlands	FA/MA	6.0	
21,943.5	USB	....	Paris Aeradio (Orly), France	FA/MA	6.0	
21,967.0	USB	DLH	Frankfurt ("Control") Aeradio, GFR	FA/MA	6.0	
21,988.0	USB	HEE92	Berne Aeradio, Switzerland	FA/MA	6.0	
21,996.0	USB	....	numerous, worldwide, operational	FA/MA	....	Airports & in-flight—particularly ARINC New York to Honolulu for telephone patches

22,000.0 - 22,720.0 kHz — MARITIME MOBILE (Worldwide)

22,000.0 - 22,855.0 kHz — MARITIME MOBILE (Worldwide-post WARC)

kHz	Mode	Call	Location	Service	kW	Remarks
22,000.0 – 22,120.9	USB		The band between these two frequencies is divided into 40 channels spaced 3.1 kHz apart and assigned numbers from 2201 to 2240. These are ship transmitting channels and are paired with Coastal USB Stations operating between 22,596.0 and 22,716.9 kHz.			
22,064.4	CW	CCS	Santiago Naval Radio, Chile	FC	....	Chilean Navy
22,071.5	CW/FAX	CCS	Santiago Naval Radio, Chile	FC	....	Chilean Navy
22,124.0	USB	....	Coast & Ship Simplex Frequency	FC/MS	....	Inland & inter-coastal waterway traffic. Employed worldwide by hundreds of base stations and various types of ships. Frequently heard in North America: KHT, Cedar Rapids, IA; KMX, San Diego, CA; KST, Portland, OR; etc.
22,127.1	USB	....	Coast & Ship Simplex Frequency	FC/MS	....	This frequency is also used by USCG Stations: NMA, Miami, FL; NMC, San Francisco, CA; NMF, Boston, MA; NMG, New Orleans, LA; NMN, Portsmouth, VA; NMO, Honolulu, HI; NMQ, Long Beach, CA; NMR, San Juan, PR; NOJ, Kodiak, AK & NRV, Barrigada, Guam
22,130.2	USB	....	Coast & Ship Simplex Frequency	FC/MS	....	
22,133.2	USB	....	Coast & Ship Simplex Frequency	FC/MS	....	Also used by (This frequency is also used by USCG Coast Stations as listed on 22,127.1)
22,136.4	USB	....	Coast & Ship Simplex Frequency	FC/MS	....	
22,227.0 – 22,247.0	CW		The band between these two frequencies is divided into 10 channels spaced 2.0 kHz apart and assigned numbers from 1 to 10. These are ship transmitting channels. The most commonly used channels are 5 and 6, 22,236.0 and 22,238.0 kHz, respectively.			
22,310.5	CW	5OW22	Lagos Radio, Nigeria	FC	....	
22,311.5	CW	A9M	Bahrain Radio, Bahrain	FC	2.0	
22,312.0	CW	FUX	Le Port Naval Radio, Reunion	FC	....	French Navy
22,312.5	CW	XSG3	Shanghai Radio, PRC	FC	10.0	
22,313.0	CW/RTTY	FUF	Fort de France Naval R., Martinique	FC	20.0	French Navy
22,315.0	CW	UJY	Kaliningrad Radio, USSR	FC	5.0	
	CW	URB2	Klaidpeda Radio, Lithuanian SSR	FC	5.0	
22,315.5	CW	VIP6	Perth Radio, WA, Australia	FC	10.0	AMVER
22,318.0	CW	WPA	Port Arthur Radio, TX	FC	5.0	
22,318.5	CW	FFS9	St. Lys Radio, France	FC	10.0	
22,319.5	CW	WLO	Mobile Radio, AL	FC	30.0	NOAA reports wx broadcasts at 1300 & 2300

kHz	Mode	Call	Location	Service	kW	Remarks
22,321.0	CW/RTTY	UJY	Kaliningrad Radio, USSR	FC	15.0	
	CW	USW2	Rostov Radio, USSR	FC	5.0	
22,322.0	CW	A9M	Bahrain Radio, Bahrain	FC	2.0	
22,322.5	CW	UHK	Batumi Radio, USSR	FC	5.0	
22,324.0	CW	XSQ	Guangzhou Radio, PRC	FC	10.0	
22,324.5	CW	PCH7Ø	Scheveningen Radio, Netherlands	FC	10.0	
22,325.0	CW/RTTY	UJY	Kaliningrad, USSR	FC	15.0	
22,325.7	CW	DZJ	Manila (Bulacan), Philippines	FC	1.5	
22,327.5	CW	SVG7	Athens Radio, Greece	FC	5.0	
22,329.5	CW	UMV	Murmansk Radio, USSR	FC	....	
22,330.4	CW	4XZ	Haifa Naval Radio, Israel	FC	2.0	Israel Navy
22,330.5	CW	PBC3	Goeree Naval Radio, Netherlands	FC	10.0	Dutch Navy
22,331.3	CW	D3E81	Luanda Radio, Angola	FC	....	
22,333.0	CW	URB2	Klaidepa Radio, Lithuanian SSR	FC	5.0	(RWWN)
	CW	URD	Leningrad Radio, USSR	FC	....	
22,334.5	CW/RTTY	Y5M	Ruegen Radio, GDR	FC	15.0	
22,335.2	CW	GYU	Gibraltar Naval Radio, Gibraltar	FC	....	British Navy
22,335.5	CW	DZI	Manila (Bacoor), Philippines	FC	1.0	
22,336.0	CW	TIM	Limon Radio, Costa Rica	FC	....	
22,338.0	CW	5RS	Tamatave Radio, Madagascar	FC	....	
22,338.2	CW	YIR	Basrah Control, Iraq	FC	1.0	
22,339.0	CW	PTO	Brasilia Radio, Brazil	FC	....	
22,340.5	CW	DAL	Norddeich Radio, GFR	FC	10.0	
22,342.5	CW	6WW	Dakar Naval Radio, Senegal	FC	10.0	French Navy
22,344.0	CW	LZW7	Varna Radio, Bulgaria	FC	15.0	
22,344.5	CW	HEB	Berne Radio, Switzerland	FC	10.0	
22,346.5	CW	LSA7	Boca (Buenos Aires), Radio, Argentina	FC	3.0	
22,346.5	CW	SVI7	Athens Radio, Greece	FC	5.0	
22,347.5	CW	ZSC2Ø	Cape Town Radio, RSA	FC	10.0	
22,348.6	CW	JFC	Misaki Gyogyo Radio, Japan	FC	1.0	
22,349.5	CW	UFN	Novorossiysk Radio, USSR	FC	15.0	
22,350.0	CW	UFL	Vladivostok Radio, USSR	FC	15.0	Reported wx broadcasts at 0100, 0900 & 1500
22,351.5	CW	OST72	Oostende Radio, Belgium	FC	30.0	
	CW	XSQ	Guangzhou Radio, PRC	FC	10.0	
22,352.4	CW	UFB	Odessa Radio, Ukrainian SSR	FC	15.0	
22,352.5	CW	PPR	Rio Radio, Brazil	FC	1.0	TIME SIGNALS (0125-0130; 1425-1430 & 2125-2130, only)/Tfc
22,354.5	CW	UDK2	Murmansk Radio, USSR	FC	10.0	
22,355.0	CW	UJY	Kaliningrad Radio, USSR	FC	15.0	
22,358.0	CW	UFB	Odessa Radio, Ukrainian SSR	FC	5.0	
22,361.0	CW	DFW36	Frankfurt (Usingen), GFR	FC	15.0	MARPRESS/Reported nx broadcast at 1618
22,363.0	CW	XSG	Shanghai Radio, PRC	FC	5.0	
22,365.0	CW	UBN	Jdanov Radio, Ukrainian SSR	FC	15.0	
22,371.0	CW/RTTY	URL	Sevastopol Radio, Ukrainian SSR	FC	15.0	
22,374.0	CW	LZW72	Varna Radio, Bulgaria	FC	1.5	
22,375.0	CW	XYR24	Rangoon Radio, Burma	FC	1.0	
22,376.0	CW	IAR62	Rome P.T. Radio, Italy	FC	15.0	
22,377.9	CW	VTG9	Bombay Naval Radio, India	FC	1.0	Indian Navy
22,378.0	CW	†IAR32	Rome P.T. Radio, Italy	FC	15.0	
22,383.0	CW/RTTY	URL	Sevastopol Radio, Ukrainian SSR	FC	15.0	
22,384.0	CW	XSE	Qinhuangdao Radio, PRC	FC	6.0	
	FAX	GZZ	Whitehall (Northwood), Naval R., Eng.	FC	5.0	British Navy/Meteo
	CW	EDF6	Aranjuez Radio, Spain	FC	15.0	
22,386.0	CW	JCT	Choshi Radio, Japan	FC	3.0	
22,387.0	CW	UTA	Tallinn Radio, Estonian SSR	FC	5.0	
	CW/RTTY	GKS7	Portishead Radio, England	FC	15.0	
	CW	VTG9	Bombay Naval Radio, India	FC	1.0	Indian Navy
22,387.2	CW	VCS	Halifax CG Radio, NS, Canada	FC	1.0	
22,389.0	CW	HWN	Paris (Houilles) Naval R., France	FC	2.0	French Navy
22,390.1	CW	FUF	Fort de France Naval R., Martinique	FC	10.0	French Navy
22,393.0	CW	Y5M	Ruegen Radio, GDR	FC	15.0	
22,393.2	CW	ZRQ7	Cape (Simonstown) Naval Radio, RSA	FC	10.0	RSA Navy

kHz	Mode	Call	Location	Service	kW	Remarks
22,395.0	CW	UAT	Moscow Radio, USSR	FC	10.0	
22,395.1	CW	HLF	Seoul Radio, South Korea	FC	15.0	
22,396.0	CW	CLA5Ø	Havana (Cojimar) Radio, Cuba	FC	5.0	
	CW	JOS	Nagasaki Radio, Japan	FC	3.0	
	CW	LFL	Rogaland Radio, Norway	FC	5.0	
22,396.5	CW	DHJ59	Sengwarden Naval Radio, GFR	FC	....	GFR Navy/NATO
22,396.7	CW	OFJ9	Helsinki Radio, Finland	FC	10.0	
22,397.5	CW	CFH	Maritime Command Radio, Halifax, NS, Canada	FC (Canadian Forces)	10.0	
22,398.0	CW	JNA	Tokyo Naval Radio, Japan	FC	5.0	JSDA
			Reported NAVAREA 0005, 0405, 0805 & 1205			
22,399.0	CW	SPH92	Gydnia Radio, Poland	FC	10.0	[aka SPH9]
22,401.0	CW	Y5M	Ruegen Radio, GDR	FC	15.0	
22,402.0	CW	XSG	Shanghai Radio, PRC	FC	10.0	
22,403.0	CW	UJY	Kaliningrad Radio, USSR	FC	10.0	
22,404.0	CW	OXZ92	Lyngby Radio, Denmark	FC	10.0	
22,405.0	CW	UKX	Nakhodka Radio, USSR	FC	....	
	CW	UKA	Vladivostok Radio, USSR	FC	15.0	
22,406.0	CW/RTTY	ZRH7	Cape (Fisanteraal) Naval R., RSA	FC	15.0	RSA Navy
	CW	URD	Leningrad Radio, USSR	FC	10.0	
22,407.0	CW	ZLO7	Irirangi Naval Radio, New Zealand	FC	20.0	RNZ Navy
	CW	SPH93	Gydnia Radio, Poland	FC	10.0	[aka SPI9]
	CW	UKW3	Korsakov Sakhalinsk Radio, USSR	FC	....	
22,407.3	CW	GKC7	Portishead Radio, England	FC	5.0	
22,409.0	CW	JOR	Nagasaki Radio, Japan	FC	15.0	
22,410.0	CW	EBA	Madrid Radionaval, Spain	FC	10.0	Spanish Navy
	CW	UKK3	Nakhodka Radio, USSR	FC	....	
	CW/RTTY	UNQ	Novorossiysk Radio, USSR	FC	5.0	
22,410.8	CW	SVB7	Athens Radio, Greece	FC	5.0	
22,411.0	CW/RTTY	UJQ7	Kiev Radio, Ukrainian SSR	FC	25.0	
22,413.0	CW	SAG9	Goteborg Radio, Sweden	FC	5.0	AMVER
	CW	KOK	Los Angeles Radio, CA	FC	15.0	
22,414.0	CW	A9M	Bahrain Radio, Bahrain	FC	2.0	
22,414.9	CW	UFD9	Arkhangelsk Radio, USSR	FC	5.0	[aka RKLM]
22,415.0	CW	DAF	Norddeich Radio, GFR	FC	10.0	
22,417.0	CW	SVA7	Athens Radio, Greece	FC	10.0	
	CW	URB2	Klaipeda Radio, Lithuanian SSR	FC	5.0	
22,419.0	CW	LPD91	Gen. Pacheco Radio, Argentina	FC	15.0	
	CW	OXZ9	Lyngby Radio, Denmark	FC	10.0	
	CW	JCS	Choshi Radio, Japan	FC	3.0	
22,420.0	CW	PPR2	Rio Radio, Brazil	FC	1.0	TIME SIGNALS
			(0125-0130: 1425-1430 & 2125-2130, only)/Tfc			
	CW	UQK	Riga Radio, Latvian SSR	FC	20.0	
22,422.0	CW	Y5M	Ruegen Radio, GDR	FC	20.0	
22,423.0	CW	UAT	Moscow, USSR	FC	15.0	
22,425.0	CW	AQP6	Karachi Naval Radio, Pakistan	FC	20.0	Pakistan Navy
	CW	KFS	San Francisco Radio, CA	FC	30.0	
	FAX	†RHB/RHO	Khabarovsk Radio, USSR	FC	....	Meteo
22,425.5	CW	LGG	Rogaland Radio, Norway	FC	5.0	AMVER
22,428.0	CW	9VG59	Singapore Radio, Singapore	FC	5.0	
22,430.0	CW	URD	Leningrad Radio, USSR	FC	10.0	
	CW	URL3	Sevastopol Radio, Ukrainian SSR	FC	15.0	
22,431.0	CW	PKI	Jakarta Radio, Indonesia	FC	....	
22,432.0	CW	WNU36	Slidell Radio, LA	FC	3.0	NOAA reports wx broadcast at 1630
22,432.2	CW	GKD7	Portishead Radio, England	FC	15.0	
22,435.0	CW	D4U	Sao Vicente Naval Radio, Cape Verde	FC	....	
	CW	URL	Sevastopol Radio, Ukrainian SSR	FC	5.0	
	CW	UFL	Vladivostok Radio, USSR	FC	15.0	
22,437.0	CW	Y5M	Ruegen Radio, GDR	FC	5.0	
22,438.5	CW	SAB9	Goteborg Radio, Sweden	FC	20.0	
22,440.0	CW	JOU	Nagasaki Radio, Japan	FC	15.0	
22,440.2	CW/RTTY	UJQ7	Kiev Radio, Ukrainian SSR	FC	20.0	
22,443.0	CW	YUR9	Rijeka Radio, Yugoslavia	FC	6.0	
22,446.0	CW	EAD6	Aranjuez Radio, Spain	FC	5.0	

kHz	Mode	Call	Location	Service	kW	Remarks
22,448.0	CW	YQI7	Constantza Radio, Rumania	FC	....	
22,448.7	CW	GKB7	Portishead Radio, England	FC	5.0	
22,450.0	CW	ROT	Moscow Naval Radio, USSR	FC	5.0	Russian Navy
22,452.8	CW	XSQ7	Guangzhou Radio, PRC	FC	7.0	
22,454.2	CW	GYA7	Whitehall (London) Naval R., England	FC	10.0	British Navy
22,454.5	CW	ROT	Moscow Naval Radio, USSR	FC	25.0	Russian Navy
22,455.0	CW	XSG	Shanghai Radio, PRC	FC	20.0	
	CW	ZSC4Ø	Cape Town Radio, RSA	FC	10.0	TIME SIGNALS/ NOAA reports wx broadcasts at 0930, 1348 & 1730
	CW	UDH	Riga Radio, Estonian SSR	FC	5.0	
22,457.9	CW	WNU46	Slidell Radio, LA	FC	3.0	
22,458.0	CW	URB2	Klaipeda Radio, Lithuanian SSR	FC	5.0	
22,459.0	CW	XSX	Keelung Radio, Taiwan	FC	....	
22,459.2	CW	OXZ93	Lyngby Radio, Denmark	FC	10.0	
22,460.0	FAX	XSG	Shanghai Radio, PRC	FC	10.0	Meteo
22,461.0	CW	4XO	Haifa Radio, Israel	FC	2.5	
	CW	FUJ	Noumea Naval Radio, New Caledonia	FC	10.0	French Navy
22,463.0	CW	JCU	Choshi Radio, Japan	FC	15.0	
22,464.0	CW	UFB	Odessa Radio, Ukrainian SSR	FC	10.0	
22,465.0	CW	XSG	Shanghai Radio, PRC	FC	5.0	
22,467.0	CW	GKA7	Portishead Radio, England	FC	15.0	NOAA reports wx broadcasts at 0930 & 1230
	CW	KLC	Galveston Radio, TX	FC	1.5	
22,471.5	CW	SVD7	Athens Radio, Greece	FC	5.0	
22,472.0	CW	NMO	COMMSTA Honolulu, HI	FC	5.0	USCG/Wx broadcasts at 0100, 0400, 0700 & 2000
22,473.0	CW	CBV	Valparaiso Radio, Chile	FC	1.0	AMVER
	CW	LFG	Rogaland Radio, Norway	FC	5.0	AMVER
22,474.0	CW	VIS42	Sydney Radio, NSW, Australia	FC	5.0	AMVER
22,475.0	CW	URD	Leningrad Radio, USSR	FC	10.0	
22,476.0	CW	DAM	Norddeich Radio, GFR	FC	15.0	
	CW	NMO	COMMSTA Honolulu, HI	FC	5.0	USCG/NUKO/AMVER
22,479.0	CW	CUL24	Lisbon Radio, Portugal	FC	15.0	
	CW	9VG27	Singapore Radio, Singapore	FC	5.0	
	CW	KPH	San Francisco Radio, CA	FC	30.0	
22,481.0	CW	Y5M	Ruegen Radio, GDR	FC	15.0	
22,482.0	CW	HLG	Seoul Radio, South Korea	FC	10.0	
	CW	DZR	Manila (Bulacan), Philippines	FC	2.0	
	CW	RNI9	Guzar, Uzbek SSR	FX	2.5	Tfc
22,484.0	CW	HAR	Budapest Naval Radio, Hungary	FC	5.0	
22,485.0	CW	VHP7	COMMSTA Canberra, ACT, Australia	FC	20.0	RA Navy
	CW	VIX7	Master Station Canberra, Australia	FC	20.0	NOAA reports wx broadcasts at 0130 & 2130
	CW	UBF2	Leningrad Radio, USSR	FC	....	
	CW/RTTY	UKA	Vladivostok Radio, USSR	FC	25.0	
22,486.0	CW	OMP7	Prague Radio, Czechoslovakia	FC	1.0	
22,486.9	CW	WSL	Amagansett Radio, NY	FC	30.0	
22,490.0	CW	DZZ	Manila Radio, Philippines	FC	....	
	CW/RTTY	UXN	Arkhangelsk Radio, USSR	FC	20.0	
22,491.0	CW	4XO	Haifa Radio, Israel	FC	2.5	
22,493.0	CW	UAT	Moscow Radio, USSR	FC	15.0	
22,494.0	CW	GKK7	Portishead Radio, England	FC	5.0	
22,495.0	CW	SPH91	Gdynia Radio, Poland	FC	10.0	[aka SPA9]
22,497.0	CW	UPW2	Liepaja Radio, Latvian SSR	FC	....	
	CW/RTTY	UDK2	Murmansk Radio, USSR	FC	25.0	
	CW	UQK	Riga Radio, Latvian SSR	FC	20.0	
22,499.4	CW	FUX	Le Port Naval Radio, Reunion	FC	10.0	French Navy
22,500.0	CW	SVF7	Athens Radio, Greece	FC	5.0	
	CW	RIW	Khiva Naval Radio, Uzbek SSR	FC	....	Russian Navy
22,501.0	CW	UFN	Novorossiysk Radio, USSR	FC	....	
22,502.0	CW	DZG	Manila (Las Pinas) Radio, Philippines	FC	5.0	AMVER/NOAA re- wx broadcasts at 0300 & 1500
22,503.0	CW	GKG7	Portishead Radio, England	FC	5.0	
	CW	WOE	Lantana Radio, FL	FC	5.0	
22,504.0	CW	PRS5	Rio de Janeiro Radio, Brazil	FC	....	

kHz	Mode	Call	Location	Service	kW	Remarks
	CW	9KK23	Kuwait Radio, Kuwait	FC	5.0	
22,505.0	CW	SPE91	Szczecin Radio, Poland	FC	10.0	[aka SPE9]
22,506.0	CW	DZI	Manila (Bacoor) Radio, Philippines	FC	....	Reported wx broadcast at 0000
22,508.5	CW	DAM	Norddeich Radio, GFR	FC	5.0	
22,509.0	CW	FFL9	St. Lys Radio, France	FC	10.0	
22,510.0	CW	UAT	Moscow Radio, USSR	FC	15.0	
22,515.0	CW	KFS	San Francisco Radio, CA	FC	30.0	NOAA reports wx broadcasts at 0420, 1620, 2200 & 2300
22,516.0	CW	DAN	Norddeich Radio, GFR	FC	5.0	
22,517.0	CW	EBA	Madrid Radionaval, Spain	FC	5.0	Spanish Navy
22,518.0	CW	WCC	Chatham Radio, MA	FC	15.0	
22,520.0	CW	UPW2	Liepaja Radio, Latvian SSR	FC	....	
	CW	UQK2	Riga Radio, Latvian SSR	FC	5.0	
22,523.0	CW	FUB	Paris (Houilles) Naval R., France	FC	....	French Navy
22,524.0	CW	JFA	Chuo Gyogyo (Matsudo) Radio, Japan	FC	3.0	
22,525.1	CW	IRM	CIRM, Rome Radio, Italy	FC	....	AMVER/Medico
22,525.5	CW	GKH7	Portishead Radio, England	FC	15.0	
22,526.0	CW	XSQ	Guangzhou Radio, PRC	FC	7.0	
	FAX	GKN7	Portishead Radio, England	FC	5.0	Meteo
22,527.0	CW	GKM7	Portishead Radio, England	FC	15.0	
	CW/RTTY	NRV	COMMSTA Barrigada, Guam	FC	10.0	USCG
	CW/RTTY	UAD2	Kholmsk Radio, USSR	FC	25.0	
22,527.9	CW	GKO7	Portishead Radio, England	FC	15.0	
22,528.5	CW	GKI7	Portishead Radio, England	FC	15.0	
22,530.0	CW	PWZ	Rio de Janeiro Naval Radio, Brazil	FC	2.0	Brazil Navy
	CW	URL	Sevastopol Radio, Ukrainian SSR	FC	15.0	
	CW/RTTY	UCP2	Okhotsk Radio, USSR	FC	5.0	
	CW/RTTY	UKA	Vladisvostok Radio, USSR	FC	10.0	
22,532.9	CW	EDZ7	Aranjuez Radio, Spain	FC	5.0	
22,533.0	CW	OST7	Oostende Radio, Belgium	FC	10.0	
	CW	ZLB7	Awarua Radio, New Zealand	FC	5.0	AMVER
22,536.0	CW	VPS22	Cape d'Aguilar Radio, Hong Kong	FC	3.0	TIME SIGNALS/Tfc
22,538.0	CW	DZN	Manila (Navotas) Radio, Philippines	FC	3.0	
22,539.0	CW	PCH79	Scheveningen Radio, Netherlands	FC	3.0	AMVER
	CW	KLB	Seattle Radio, WA	FC	3.3	
22,539.9	CW	PJK322	Suffisant Dorp Naval Radio, Curacao	FC	....	Dutch Navy
22,541.0	FAX	JJC	Tokyo Radio, Japan	FC	15.0	Meteo/Nx
22,543.0	CW	7TA12	Algiers Radio, Algeria	FC	10.0	
22,545.0	CW	XSM	Xiamen Radio, PRC	FC	6.0	
	CW	Y5M	Ruegen Radio, GDR	FC	15.0	
	CW	GKJ7	Portishead Radio, England	FC	5.0	
	CW	....	numerous, USA nationwide	FA/FC/FX	....	USCG/Command Net Calling Frequency for all "Fixed" USCG Coastal Stations [Also used by ships and aircraft]
22,547.5	CW	PJK322	Suffisant Dorp Naval Radio, Curacao	FC	....	Dutch Navy
22,550.0	CW/RTTY	UKX	Nakhodka Radio, USSR	FC	5.0	
22,550.9	CW	VWB	Bombay Radio, India	FC	5.0	
22,551.0	CW	SPE92	Szczecin Radio, Poland	FC	10.0	[aka SPB2]
22,551.0	CW	†CTV2	Monsanto Naval Radio, Portugal	FC	5.0	Port. Navy
22,554.0	CW	OSN	Oostende Naval Radio, Belgium	FC	1.0	Belgium Navy
	CW	LZL7	Bourgas Radio, Bulgaria	FC	1.5	
	CW	LZS39	Sofia Naval Radio, Bulgaria	FC	....	
22,555.0	CW	UFB	Odessa Radio, Ukrainian SSR	FC	5.0	
22,557.0	CW	KPH	San Francisco Radio, CA	FC	30.0	
22,560.0	CW	URL	Sevastopol Radio, Ukrainian SSR	FC	15.0	
22,562.0	CW/TOR	GKE7	Portishead Radio, England	FC	5.0	
22,566.5	CW/TOR	HEC52	Berne Radio, Switzerland	FC	10.0	
22.568.0	CW	RIW	Khiva Naval Radio, Uzbek SSR	FC	....	Russian Navy
22,574.0	CW	RNO	Moscow Radio, USSR	FC	....	SAAM??
22,578.0	CW/TOR	GKP7	Portishead Radio, England	FC	15.0	
22,580.0	CW	UJQ	Kiev Radio, Ukrainian SSR	FC	....	
	CW	RNB27	unknown	FC	....	
22,585.6	CW/TOR	SVU7	Athens Radio, Greece	FC	15.0	
22,587.2	CW/TOR	LGG3	Rogaland Radio, Norway	FC	5.0	[aka LGB]

kHz	Mode	Call	Location	Service	kW	Remarks

22,596.0 - 22,716.9 kHz — MARITIME COAST USB DUPLEX (Worldwide)

kHz	Mode	Call	Location	Service	kW	Remarks
22,596.0	USB	EHY	Pozuelo del Rey Radio, Spain	FC	10.0	CH#2201 (22,000.0)
	USB	WOO	New York (Ocean Gate, NJ) Radio, NY [NOAA reports wx broadcasts at 0100, 1300 & 1900]	FC	10.0	" "
22,599.1	USB	....	Rome Radio 2, Italy	FC	7.0	CH#2202 (22,003.1)
22,602.2	USB	....	Goteborg Radio, Sweden	FC	5.0	CH#2203 (22,006.2)
22,605.3	USB	OHG2	Helsinki Radio, Finland	FC	10.0	CH#2204 (22,009.3)
	USB	FFL92	St. Lys Radio, France	FC	6.0	" "
	USB	ZSC29	Cape Town Radio, RSA	FC	10.0	" "
	USB	....	Rijeka Radio, Yugoslavia	FC	7.5	" "
22,608.4	USB	PCG71	Scheveningen Radio, Netherlands	FC	10.0	CH#2205/AMVER (22,012.4)
	USB	WOO	New York (Ocean Gate, NJ) Radio, NY [NOAA reports wx broadcasts at 0100, 1300 & 1900]	FC	10.0	" "
22,611.5	USB	GKT76	Portishead Radio, England	FC	10.0	CH#2206 (22,015.5)
	USB	....	This frequency also used by USCG Coast Stations: NMA, Miami, FL: NMC, San Francisco, CA; NMF, Boston, MA; NMN, Portsmouth, VA; NMO, Honolulu, HI & NOJ, Kodiak, AK			
22,614.6	USB	DAJ	Norddeich Radio, GFR	FC	10.0	CH#2207 (22,018.6)
	USB	....	Moscow Radio, USSR	FC	10.0	" "
22,617.7	USB	LFN32	Rogaland Radio, Norway	FC	10.0	CH#2208 (22,021.7)
22,620.8	USB	4XO	Haifa Radio, Israel	FC	10.0	CH#2209 (22,024.8)
22,623.9	USB	WOO	New York (Ocean Gate, NJ) Radio, NY [NOAA reports wx broadcasts at 0200, 1400 & 2000]	FC	10.0	CH#2210 (22,027.9)
22,627.0	USB	....	Rome Radio 2, Italy	FC	7.0	CH#2211 (22,031.0)
	USB	....	Goteborg Radio, Sweden	FC	5.0	" "
22,630.1	USB	5BA72	Cyprus Radio, Cyprus	FC	10.0	CH#2212 (22,034.1)
	USB	GKU72	Portishead Radio, England	FC	10.0	" "
22,636.3	USB	HEB52	Berne Radio, Switzerland	FC	10.0	CH#2214 (22,040.3)
22,639.4	USB	FFL95	St. Lys Radio, France	FC	6.0	CH#2215 (22,043.4)
	USB	LFN35	Rogaland Radio, Norway	FC	10.0	" "
	USB	WOM	Miami Radio, FL [NOAA reports wx broadcast at 1230]	FC	10.0	" "
22,642.5	USB	ICB	Genoa Radio 2, Italy	FC	7.0	CH#2216 (22,046.5)
	USB	WOM	Miami Radio, FL [NOAA reports wx broadcast at 1330]	FC	10.0	" "
22,645.6	USB	SVN7	Athens Radio, Greece	FC	5.0	CH#2217 (22,049.6)
22,648.7	AM	....	Izmail Radio, Ukrainian SSR	FC	....	CH#2218 (22,052.7)
	USB	....	This frequency also used by USCG Coast Stations: NMA, Miami, FL; NMC, San Francisco, CA; NMF, Boston, MA; NMN, Portsmouth, VA; NMO, Honolulu, HI & NOJ, Kodiak, AK			
22,651.8	USB	SPO91	Szczecin Radio, Poland	FC	10.0	CH#2219 (22,055.8)
22,654.9	USB	GKU7Ø	Portishead Radio, England	FC	10.0	CH#2220 (22,058.9)
22,658.0	AM/USB	....	INTERNATIONAL COAST STATION CALLING FREQUENCY			CH#2221 (22,062.0)

MARITIME COAST USB DUPLEX (con.)

kHz	Mode	Call	Location	Service	kW	Remarks
22,661.1	USB	DAH	Norddeich Radio, GFR	FC	10.0	CH#2222 (22,065.1)
	USB	WOM	Miami Radio, FL	FC	10.0	" "
22,664.2	USB	....	Rome Radio 2, Italy	FC	7.0	CH#2223 (22,068.2)
22,667.3	USB	SVN7	Athens Radio, Greece	FC	5.0	CH#2224 (22,071.3)
	USB	EHY	Pozuelo del Rey Radio, Spain	FC	10.0	" "
22,670.4	USB	....	....	FC	....	CH#2225 (22,074.4)
22,673.5	USB	FFL91	St. Lys Radio, France	FC	6.0	CH#2226 (22,077.5)
22,676.6	USB	JBO	Tokyo Radio, Japan	FC	10.0	CH#2227 (22,080.6)
22,679.7	USB	....	....	FC	....	CH#2228 (22,083.7)
22,682.8	USB	EHY	Pozuelo del Rey Radio, Spain	FC	10.0	CH#2229 (22,086.8)
22,685.9	USB	....	Goteborg Radio, Sweden	FC	5.0	CH#2230 (22,089.9)
22,689.0	USB	FFL94	St. Lys Radio, France	FC	6.0	CH#2231 (22,093.0)
	USB	SVN7	Athens Radio, Greece	FC	5.0	" "
22,692.1	USB	PCG72	Scheveningen Radio, Netherlands	FC	10.0	CH#2232 (22,096.1)
	USB	HEB62	Berne Radio, Switzerland	FC	10.0	" "
22,695.2	USB	LFN4Ø	Rogaland Radio, Norway	FC	10.0	CH#2233 (22,099.2)
22,698.3	USB	LFN41	Rogaland Radio, Norway	FC	10.0	CH#2234 (22,102.3)
22,701.4	USB	FFL93	St. Lys Radio, France	FC	6.0	CH#2235 (22,105.4)
	USB	SVN7	Athens Radio, Greece	FC	5.0	" "
22,704.5	USB	....	Lyngby Radio, Denmark	FC	10.0	CH#2236 (22,108.5)
	USB	JBO	Tokyo Radio, Japan	FC	10.0	" "
22,707.6	USB	LFN43	Rogaland Radio, Norway	FC	10.0	CH#2237 (22,111.6)
	USB	WLO	Mobile Radio, AL	FC	10.0	" "
			[NOAA reports wx broadcasts at 0000, 1300 & 1800]			
22,710.7	USB	PPR	Rio Radio, Brazil	FC	1.0	CH#2238 (22,114.7)
22,713.8	USB	LFN44	Rogaland Radio, Norway	FC	10.0	CH#2239 (22,117.8)
	USB	....	Rijeka Radio, Yugoslavia	FC	7.5	" "
22,716.9	USB	JBO	Tokyo Radio, Japan	FC	10.0	CH#2240 (22,120.9)
	USB	LFN45	Rogaland Radio, Norway	FC	10.0	" "

22,720.0 - 23,200.0 kHz — FIXED SERVICE (Worldwide)

kHz	Mode	Call	Location	Service	kW	Remarks
22,721.9	CW/RTTY	Y7A83	Berlin (Nauen), GDR	FX	20.0	EMBASSY
22,732.0	ISB/RTTY	Y7A84	Berlin (Nauen), GDR	FX	20.0	EMBASSY
22,740.0	ISB	GBW42	London (Rugby), England	FX	30.0	Telcom/Accra
22,758.5	USB	ETV28	Addis Ababa, Ethiopia	FX	5.0	Telcom/Rome
22,770.0	FAX	JMH6	Tokyo, Japan	FX	5.0	Meteo
	USB	RBK51	Moscow, USSR	FX	20.0	R. Moscow Feeder
22,777.0	FAX	JJC	Tokyo, Japan	FX	....	Meteo
22,783.5	USB	MKT	London (Stanbridge), England	FX	10.0	RAF Telcom/Tfc/ Gibraltar
22,787.2	CW	YZK4	Belgrade (Makis), Yugoslavia	FX	30.0	Tfc/Sydney [Alt: 23,117.0]
22,795.0	ISB	CUA93	Lisbon (Alfragide), Portugal	FX	20.0	
22,800.0	FAX/RTTY	Y2V27	Nauen, GDR	FX	20.0	
22,820.0	CW/RTTY	OMZ	Prague, Czechoslovakia	FX	1.0	EMBASSY

kHz	Mode	Call	Location	Service	kW	Remarks
22,848.0	CW	ONN36	Brussels. Belgium	FX	1.0	Angola EMBASSY
22,851.5	LSB	ZEG25	Salisbury, Zimbabwe	FX	20.0	Telcom

22,855.0 - 23,200.0 — FIXED SERVICE (Worldwide—post WARC)

kHz	Mode	Call	Location	Service	kW	Remarks
22,860.0	CW/RTTY	FUG	La Regine (Castelnaudry), France	FX	2.0	French Navy
22,865.0	FAX	NPO	San Miguel (Capas), Philippines	FX	15.0	USN/Meteo
22,867.0	FAX	5YE	Nairobi, Kenya	FX	30.0	Meteo
22,925.0	FAX	WFN52	New York, NY	FX	50.0	UPI Nx
22,930.0	ISB	....	Daventry, England	FX	30.0	BBC Feeder
22,954.8	ISB	†5UR29	Niamey, Niger	FX	20.0	Telcom/Tfc
22,955.0	ISB	†....	Delano, CA	FX	50.0	VOA Feeder
22,955.0	LSB	....	several, in Atlantic Ocean area	FX	....	USAF/NASA [Alt: 20,191.0 & 23,035.0]
22,960.0	USB	HBD41	Pretoria, RSA	FX	0.1	Swiss EMBASSY
	CW	KNY27	Washington, DC	FX	1.0	Swiss EMBASSY
22,964.0	CW/RTTY	†.....	several, USA nationwide	FX	....	FCC Backup Net [Alt: 23,035.0]
22,970.0	ISB	....	Holzkirchen, GFR	FX	10.0	RFE Feeder
22,984.0	USB	AFE83	Ascension Island	FX	12.5	USAF/NASA
23,009.5	USB	†CXL48	Montevideo, Uruguay	FX	4.0	Telcom
23,030.0	CW	CLP3	unknown	FX	....	Cuban EMBASSY
23,035.0	CW/RTTY	†....	several, USA nationwide	FX	....	FCC Backup Net [Alt: 22,964.0]
23,038.0	ISB	70B23Ø	Aden (Hiswa), Democratic Yemen	FX	3.0	Telcom
23,054.8	CW	4XZ	Haifa Naval Radio, Israel	FX	....	Israel Navy
23,075.0	CW/FAX	WFN23	New York, NY	FX	50.0	AP Nx/Tfc
23,100.0	USB	VCR976	Ottawa, Ont., Canada	FX	1.0	Saudi Arabian EMBASSY [Alt: 17,624.0]
23,101.0	ISB	AFH3	Albrook AFS, Canal Zone	FX	3.0	USAF [Alt: 20,600.0 & 24,860.0]
23,102.5	USB	ETW21	Addis Ababa, Ethiopia	FX	1.0	Nigerian EMBASSY
23,117.0	CW	YZM4	Belgrade (Makis), Yugoslavia	FX	30.0	Tfc/Sydney [Alt: 23,550.0]
23,120.0	USB	....	unknown	FX	....	NUMBERS/YL/GG
23,131.5	USB	....	unknown	FX	....	Telcom/FF
23,137.5	ISB	FPX3	Paris (St. Assise), France	FX	20.0	Telcom/Cotonou
23,143.7	LSB	GLK43	London (Rugby), England	FX	20.0	Telcom/Lagos
23,150.2	USB	FPX5	Paris (St. Assise), France	FX	20.0	Telcom/Bangui
23,165.0	CW	FUM	Papeete Naval Radio, Tahiti	FX	2.0	French Navy
	CW	FUV	Djibouti Naval Radio, Djibouti	FX	2.0	French Navy
23,191.0	ISB	....	Daventry, England	FX	30.0	BBC Feeder

23,200.0 - 23,350.0 kHz — AERONAUTICAL FIXED & MOBILE (Worldwide)

kHz	Mode	Call	Location	Service	kW	Remarks
23,210.0	USB	....	Stockholm Aeradio, Sweden	FA	30.0	
23,221.0	USB	....	"Diego Tower", Diego Garcia	FA/MA	....	USN/USAF
23,227.0	USB	AGA8	Clark AFB, Philippines	FA/MA	....	USAF
23,234.0	CW	FDX	Paris, France	FA	....	French AF
23,263.5	USB	....	several, worldwide	FA/MA	....	USAF
23.265.0	USB	AFA	Andrews AFB, MD	FA/MA	....	USAF
23,281.0	USB	....	numerous, worldwide	FA/MA	....	USN
23,284.0	USB	....	numerous, worldwide	FA/MA	....	USN
23,285.0	USB	HEE91	Berne Aeradio, Switzerland	FA/MA	6.0	
23,287.0	USB	ZRB2	Pretoria (Waterkloof AFB), RSA	FA/MA	5.0	RSA Air Force
	USB	....	numerous, worldwide	FA/MA	....	USAF (Used for SKY KING broadcasts—many code names for bases)
23,298.0	USB	....	numerous, worldwide	FA/MA	....	USN
23,305.0	CW	....	unknown, possibly in Middle East	FA	....	French AF
23,312.0	USB	....	numerous, Pacific Ocean area	FA/MA	....	USN
23,331.4	FAX	KVM7Ø	Honolulu (International), HI	AX	10.0	Meteo (1900-2045)
23,337.0	USB	....	several, worldwide	FA/MA	3.0	USAF (Usually AJE, Croughton, England; AJG9, Incirlik AB, Turkey; XPH, Thule, Greenland; etc.—sometimes referred to as the "UNIFORM" frequency—also used for "SKY KING" broadcasts)

23,350.0 - 25,070.0 kHz -- FIXED SERVICE (Worldwide)

23,350.0 - 24,890.0 kHz -- FIXED SERVICE (Worldwide--post WARC)

kHz	Mode	Call	Location	Service	kW	Remarks
23,365.0	CW	FKO	Djibouti Naval Radio, Djibouti	FX	....	French Navy
23,376.0	CW/RTTY	CYS22	Ottawa, Ont., Canada	FX	....	Polish EMBASSY [Alt: 20,030.0]
23,380.0	ISB	MKD	Akrotiri, Cyprus	FX	30.0	RAF Telcom/Tfc
23,390.0	ISB	9VF239	Singapore (Jurong), Singapore	FX	8.0	Telcom/Tfc
23,402.5	USB	KUR2Ø	Honolulu, HI	FX	1.0	
	USB	KUR5Ø	Agana, Guam	FX	1.0	
				[Alt: 20,602.5]		
23,419.0	USB	AFE7	MacDill AFB, FL	FX	1.0	USAF/NORAD
	USB	AFS	Offutt AFB, NE	FX	1.0	" "
	USB	AFG4Ø	Colorado Springs, CO	FX	1.0	" "
23,420.0	ISB	HEO22	Berne (Schwarzenburg), Switzerland	FXX	4.0	Telcom/Uruguay
23,462.0	ISB	9GN29	Accra, Ghana	FX	8.0	Telcom/London
23,467.1	LSB	5OV234	Lagos (Ikorodu), Nigeria	FX	30.0	Telcom/Tfc
23,480.0	USB	....	unknown	FX	....	NUMBERS/GG/YL
23,510.0	ISB	IRV25	Rome (Torrenova), Italy	FX	20.0	Telcom
23,525.0	CW	RCV	Moscow Naval Radio, USSR	FX	....	Russian Navy
23,529.0	USB	SAM	Stockholm, Sweden	FX	1.2	EMBASSY
			(Usually heard Swedish EMBASSIES include: SAM7Ø, Addis Ababa, Ethiopia; SAM74, Monrovia, Liberia; SAM8Ø, Gaborone, Botswana; SAM81, Bissau, Guinea-Bissau; SAM82, Maputo, Mozambique; etc.) [Alt: 26,660.0]			
23,558.5	USB	PCW	The Hague, Netherlands	FX	1.2	EMBASSY
23,568.5	USB/RTTY	PCW2	Jerusalem, Israel	FX	1.0	Dutch EMBASY
23,590.0	CW	FUM	Papeete Naval Radio, Tahiti	FX	2.0	French Navy
	CW	FUV	Djibouti Naval Radio, Djibouti	FX	2.0	French Navy
23,615.0	ISB	5VH436	Lome, Togo	FX	20.0	Telcom/Paris
23,621.0	FAX	....	unknown	FX	....	
23,654.8	USB	....	unknown, probably in North America	FX	....	
23,668.2	CW	....	unknown	FX	....	VVV's
23,789.4	USB	....	unknown	FX	....	Telcom
23,795.0	ISB	9GN3Ø	Accra, Ghana	FX	8.0	Telcom/Tfc
23,832.0	CW	3ZL3Ø1	King George Island, Antarctica	FX	....	Polish Expedition
23,850.0	USB	ZEG26	Salisbury, Zimbabwe	FX	20.0	Telcom/London
23,860.0	CW/RTTY	CLP1	Havana, Cuba	FX	....	EMBASSY
23,862.5	CW/RTTY	†KKN51	Washington, DC	FX	15.0	EMBASSY
23,880.0	FAX	NPN	Barrigada (Agana NAS), Guam	FX/FC	15.0	USN/Meteo
23,894.0	ISB	FTX89	Paris (St. Assise), France	FX	20.0	Telcom
23,925.0	LSB	5OU239	Lagos, Nigeria	FX	....	Telcom
23,975.0	CW	KKN5Ø	Washington, DC	FX	15.0	EMBASSY
23,982.5	CW	†KKN52	Washington, DC	FX	15.0	EMBASSY
24,005.0	CW	Y7G34	unknown, probably in Far East	FX	....	GDR EMBASSY
24,040.0	USB	....	Hilversum (Kootwijk), Netherlands	FX	....	R. Netherlands Feeder
24,070.0	CW	CSJ28	Lisbon, Portugal	FX	1.0	INTERPOL
24,088.5	USB	....	Mediterranean LORAN Net	FX	....	USCG (Includes:
			ACB5Ø, Estartit, Spain & NCI, Sellia Marina, Italy--days)			
24,108.5	USB	FSB65	Paris (St. Martin Abbat), France	FX	2.5	INTERPOL
24,110.0	CW	AYA29	Buenos Aires, Argentina	FX	1.0	INTERPOL
	CW	XJE57	Ottawa (Almonte) Ont., Canada	FX	1.5	"
	CW	OWS	Copenhagen, Denmark	FX	1.0	"
	CW/RTTY	FSB65	Paris (St. Martin Abbat), France	FX	2.5	"
	CW	8UF75	New Delhi, India	FX	5.0	"
24,145.0	USB	†VPC24	Port Stanley, Falkland Islands	FX	7.5	Telcom/Argentina
24,168.5	USB	MKT	London (Stanbridge), England	FX	10.0	RAF Telcom/Tfc
24,195.0	ISB	CUA96	Lisbon (Alfragide), Portugal	FX	10.0	Telcom/Tfc
24,198.1	USB	....	unknown	FX	....	Telcom
24,226.0	LSB	D6B442	Moroni, Republic of the Comores	FX	6.0	Telcom/Paris
24,274.0	ISB	AFI	McClellan AFB, CA	FX	10.0	USAF

FIXED SERVICE (con.)

kHz	Mode	Call	Location	Service	kW	Remarks
24,862.0	USB	CAA2ØØ	Santiago, Chile	FX	2.5	Telcom/Pt. Arenas
	USB	CAA2Ø1	Punta Arenas, Chile	FX	2.5	Telcom/Santiago
			[Alt: 24,4921.0; 25,202.0; 26,140.0; 26,160.0; etc.]			
24,690.0	ISB	TLZ46	Bangui, Central African Republic	FX	20.0	Telcom/Paris
24,696.4	LSB	....	unknown	FX	....	Telcom/FF
24,769.7	CW	A4I	unknown, not in the Middle East	FX	....	"VVV's"
24,778.5	FAX	....	unknown	FX	....	
24,800.0	CW	Y7D49	unknown, probably in Africa	FX	....	GDR EMBASSY
24,808.2	CW	CLP1	Havana, Cuba	FX	....	EMBASSY
24,860.0	ISB	AFH3	Albrook AFS, Panama	FX	3.0	USAF
			[Alt: 23,101.0]			

24,890.0 - 24,990.0 kHz -- AMATEUR RADIO (Worldwide--post WARC)

kHz	Mode	Call	Location	Service	kW	Remarks
24,920.0	ISB	5RZ49	Antananarive, Madagascar	FX	....	Telcom
24,969.9	ISB	MKG	London (Stanbridge), England	FX	30.0	RAF Telcom/Tfc

24,990.0 - 25,010.0 kHz -- STANDARD FREQUENCY & TIME SIGNALS

25,010.0 - 25,070.0 kHz -- FIXED SERVICE (Worldwide)

kHz	Mode	Call	Location	Service	kW	Remarks
25,052.0	CW	OMP	Prague Radio, Czechoslovakia	FC	1.0	
25,054.0	USB	CXQ342	Montevideo, Uruguay	FX	5.0	Governmental
			(Tfc to Washington; Albrook AFS; Lima; Rio de Janeiro & Caracas) [Alt: 25,350.0]			
25,064.9	CW	OFJ25	Helsinki Radio, Finland	FC	5.0	
25,069.0	CW	Y5M	Reugen Radio, GDR	FC	15.0	

25,070.0 - 25,110.0 kHz -- MARITIME MOBILE (Worldwide)

25,070.0 - 25,210.0 kHz -- MARITIME MOBILE (Worldwide--post WARC)

kHz	Mode	Call	Location	Service	kW	Remarks
25,106.0	CW	....	unknown, probably in Cuba	FC	....	Tfc/SS

25,110.0 - 25,600.0 kHz -- FIXED SERVICE & MARITIME MOBILE (Worldwide)

kHz	Mode	Call	Location	Service	kW	Remarks
25,130.0	CW/RTTY	ROT	Moscow Naval Radio, USSR	FC	25.0	Russian Navy
25,134.2	CW	IDR7	Rome Naval Radio, Italy	FC	....	Italian Navy
25,135.0	CW	OST8	Oostende Radio, Belgium	FC	10.0	
	CW	EBA	Madrid Radionaval, Spain	FC	....	Spanish Navy
25,143.0	CW	OMP	Prague Radio, Czechoslovakia	FC	1.0	
25,160.0	CW	LSA5	Boca Radio, Argentina	FC	3.0	Nx/SS
25,175.0	CW	ROT	Moscow Naval Radio, USSR	FC	25.0	Russian Navy
25,196.0	CW	DAM	Norddeich Radio, GFR	FC	5.0	

25,210.0 - 25,550.0 kHz -- FIXED SERVICE (Worldwide--post WARC)

kHz	Mode	Call	Location	Service	kW	Remarks
25,260.0	USB	9JA28	Lusaka, Zambia	FX	30/0	Telcom/London
25,262.0	CW	OXZ95	Lyngby Radio, Denmark	FC	10.0	
25,277.0	CW	ZSC41	Cape Town Radio, RSA	FC	10.0	
25,308.0	CW	Y5M	Reugen Radio, GDRR	FC	15.0	
	CW	LFR	Rogaland Radio, Norway	FC	5.0	
25,318.00	CW	IDQ15	Rome Naval Radio, Italy	FC	1.0	Italy Navy
25,350.0	LSB	....	Holzkirchen, GFR	FX	10.0	RFE Feeder
	USB	CXQ342	Montevideo, Uruguay	FX	5.0	Governmental
			(Tfc to Washington; Albrook AFS; Lima; Rio de Janeiro & Caracas) [Alt: 25,054.0]			
25,380.0	CW	....	numerous, USA nationwide	FA/FC/FX	....	USCG/Command
			Net Calling Frequency for all "Fixed" USCG Coastal Stations [Also used by ships & a/c]			
25,382.5	CW	LFZ	Rogaland Radio, Norway	FC	5.0	
25,401.0	CW	SVA8	Athens Radio, Greece	FC	5.0	
25,415.5	CW	SAB25	Goteborg Radio, Sweden	FC	5.0	
25,423.0	CW	UFN	Novorossiysk Radio, USSR	FC	....	
25,440.5	USB	OXZ	Lyngby Radio, Denmark	FC	10.0	CH#2501

FIXED SERVICE & MARITIME MOBILE (con.)

kHz	Mode	Call	Location	Service	kW	Remarks
25,443.6	USB	OXZ	Lyngby Radio, Denmark	FC	10.0	CH#2502
25,461.0	CW	SAG25	Goteborg Radio, Sweden	FC	5.0	AMVER
25,470.0	USB	GLJ45	London (Ongar), England	FX	10.0	Telcom/Nairobi
25,493.0	LSB	GAR45	London (Rugby), England	FX	30.0	Telcom
25,495.5	CW	CLQ	Havana (Cojimar) Radio, Cuba	FC	....	
25,535.0	CW	SVD8	Athens Radio, Greece	FC	5.0	

25,550.0 - 25,670.0 kHz -- RADIO ASTRONOMY (Worldwide--post WARC)

kHz	Mode	Call	Location	Service	kW	Remarks
25,578.0	LSB	AFA	Andrews AFB, MD	FX	....	USAF
25,589.5	CW	NRK	Keflavik Naval Radio, Iceland	FC	10.0	USN
25,590.0	CW	NAR	Key West Naval Radio, FL NOAA reports wx broadcasts at 0030, 0630, 1230 & 1900	FC	5.0	USN/NUKO

25,600.0 - 26,100.0 kHz -- INTERNATIONAL BROADCASTING (Worldwide)

25,670.0 - 26,100.0 kHz -- INTERNATIONAL BROADCASTING (Worldwide--post WARC)

26,100.0 - 26,175.0 kHz -- MARITIME MOBILE (Worldwide--post WARC)

kHz	Mode	Call	Location	Service	kW	Remarks
26,108.0	CW	DAN	Norddeich Radio, GFR	FC	15.0	
26,130.0	USB	OZU25	Copenhagen, Denmark	FX	1.5	EMBASSY (This frequency is also available to OZU3Ø, Cairo, Egypt; OZU33, Tel Aviv, Israel; OZU34, Jeddah, Saudi Arabia & OZU35, Beijing, PRC)
26,141.3	USB	LFN47	Rogaland Radio, Norway	FC	10.0	
26,144.7	USB	....	Goteborg Radio, Sweden	FC	10.0	
26,147.5	USB	LFN48	Rogaland Radio, Norway	FC	....	
26,158.5	CW/USB	Z2H27	Salisbury, Zimbabwe	FX	0.5	British EMBASSY

26,175.0 - 27,500.0 kHz -- FIXED SERVICE & MOBILE (Except Aeronautical)

kHz	Mode	Call	Location	Service	kW	Remarks
26,227.5	CW/TOR	DAF	Norddeich Radio, GFR	FC	15.0	
26,440.0	ISB	†....	Daventry, England	FX	30.0	BBC Feeder
26,450.0	CW/RTTY	Y7D35	unknown, probably in Africa	FX	....	GDR EMBASSY
26,455.0	CW	Y5M	Ruegen Radio, GDRR	FC	15.0	
26,490.0	ISB	....	Daventry, England	FX	30.0	BBC Feeder
26,660.0	USB	SAM	Stockholm, Sweden	FX	1.0	EMBASSY (This frequency is also available to SAM58, Jakarta, Indonesia; SAM7Ø, Addis Ababa, Ethiopia; SAM73, Lagos, Nigeria; SAM8Ø, Gaborone, Botswana; SAM81, Bissau, Guinea-Bissau; SAM82, Maputo, Mozambique & SAM83, Santiago, Chile) [Alt: 23,529.0]
26,680.0	ISB	....	Daventry, England	FX	30.0	BBC Feeder
	ISB	†....	Holzkirchen, GFR	FX	10.0	RFE Feeder
26,700.0	USB	VLN51	Sydney (Doonside) NSW, Australia	FX	15.0	Telcom/Poona
26,820.0	ISB	FJY2	Port aux Francais, Kerguelen Is.	FX	....	
27,017.0	CW	OST82	Oostende Radio, Belgium	FC	15.0	
27,998.0	ISB	SAR88	Stockholm, Sweden	FX	1.0	Inter. RED CROSS
	ISB	HBC88	Geneva, Switzerland	FX	1.0	" "

Confidential Callsign List

A reverse listing (by callsign) of active HF non-broadcast and non-radio amateur stations has previously never been been available to the general public. Due to the ease of sorting the basic information in the CONFIDENTIAL FREQUENCY LIST with a word processor, the preparation of a reverse list--complete with operational frequencies--has been greatly simplified. The following 62 pages comprise a callsign listing of the stations that appear in the major CONFIDENTIAL FREQUENCY LIST tabulation (pp. 17-160).

The callsign list enables the CONFIDENTIAL FREQUENCY LIST user to easily locate possible frequency(ies) in use when only the callsign or prefix is known. A prime example of this occurred during the Falkland Islands crisis. International lists identified the one major station in Port Stanley as VPC. Using this reverse list it was a simple task to ascertain possible frequencies--particularly for communications with Buenos Aires (24,145.0 kHz).

Finding the location of a station when only the callsign has been heard (e.g. Coast Station CW signals) is another obvious asset of the reverse listing. Parallel and alternative frequencies may be found in this listing. The list is also of assistance in making tentative identification of new stations operating on new frequencies.

USING THE REVERSE LIST Certain liberties have been taken in assembling this list. To conserve space, details on alternative frequencies have been eliminated from most USCG, Swedish EMBASSY, etc. stations. Usually, the lowest frequency is shown in this listing followed by ",etc." This indicates that reference to the main CONFIDENTIAL FREQUENCY LIST is required to find alternatives--which are listed in ascending order. . . . .ITU SERVICE classification is detailed in this listing. A station operating with the same callsign, but in a different SERVICE is shown in alphabetical order of the classification. . . . .The MODE most frequently monitored is also itemized--with the usual proviso that ISB may be truly ISB, or a station monitored in either LSB or USB at different times.

A FEW FINAL THOUGHTS Although the Soviet Union appears to have relaxed its reticence about identifying the relationship of callsigns and transmitter sites, there is little evidence to confirm that RIW is truly in Khiva, Uzbek SSR. Although the Khiva site is listed here, monitoring of RIW indicates that this call may be signed by any one of 6 or 7 transmitters situated anywhere throughout the eastern third of the USSR. Also, as of this writing, probable Soviet Navy station RIT cannot be cross-correlated with any known USSR Coast Station. . . . .Some inconsistencies between the two lists in this book are due to last-minute additions and corrections.

Call	Location	Service	Mode	kHz
AFA	Andrews AFB, MD	FA	ISB	18,027.0, 23,265.0, 25,578.0
AFA	Washington (Andrews AFB, MD), DC	FX	FAX	4793.0, 10,185.0, 12,201.0, 14,672.0, 15,620.0, 17,670.5, 19,955.0,
		FX	ISB	8052.0, 17,972.0, 18,650.0, 20,050.0, 20,885.0
AFA3	Andrews AFB, MD	FA	USB	8967.0, 11,180.0, 13,201.0, 13,215.0, 13,247.0 15,015.0, 15,048.0
AFC24	Westover AFB, MA	FX	USB	9057.0, 11,483.5, 15,962.0

Call	Location	Service	Mode	kHz
AFC37	Scott AFB, IL	FA	USB	18,002.0
AFD14	Ascension Island	FA	USB	8993.0, 13,244.0, 18,019.0
AFE7	MacDill AFB, FL	FX	USB	23,419.0
AFE8	MacDill AFB, FL	FA	USB	4746.0, 6740.0, 6750.0, 8989.0, 8993.0, 9020.0, 13,201.0, 13,244.0, 15,048.0, 18,019.0
AFE7Ø	"Cape Leader", Patrick AFB, FL	FX	ISB	11,104.0
AFE71	"Cape Radio", Patrick AFB, FL	FX	ISB	4992.0, 5810.0, 6708.0, 9006.0, 9043.0, 10,305.0, 10,780.0, 11,104.0, 11,414.0, 19,303.0
AFE83	Ascension Island	FX	ISB	13,742.0, 15,015.0, 23,984.0
AFE86	Parham, Antigua	FX	ISB	8993.0, 19,143.0
AFF3	Brooks AFB, TX	FX	LSB	13,950.0
AFF4	Bergstrom AFB, TX	FX	LSB	13,950.0
AFF7	Davis-Monthan AFB, AZ	FX	USB	9220.0, 11,483.5
AFF21	Tyler, TX	FX	USB	14,902.0
AFF5Ø	Altus AFB, OK	FX	USB	9057.0
AFG4	Offutt AFB, NE	FX	USB	14,955.0, 20,890.0
AFG8	Malstrom AFB, MT	FX	USB	9220.0
AFG14	Grand Forks AFB, ND	FX	USB	15,962.0
AFG29	Whiteman AFB, MO	FX	USB	9220.0
AFG37	Scott AFB, IL	FA	ISB	6727.0, 9014.0, 9018.0, 11,182.0, 13,247.0,, 15,015.0, 18,002.0
AFG4Ø	Colorado Springs, CO	FX	USB	13,547.0, 23,419.0
AFH3	Albrook AFS, Panama	FA	USB	18,019.0
		FX	ISB	7320.0, 9473.0, 15,675.0, 18,019.0, 19,500.0, 20,600.0, 23,101.0, 24,860.0
AFH9	March AFB, CA	FA	USB	17,975.0
AFH28	Beale AFB, CA	FX	USB	9220.0
AFH39	March AFB, CA	FX	USB	8101.0, 9057.0, 11,407.0, 13,907.0, 15,962.0
AFI	McClellan AFB, CA	FX	USB	24,274.0
AFI2	McClellan AFB, CA	FA	USB	4746.0, 6738.0, 8989.0, 9020.0, 13,201.0, 18,002.0
AFK	Peterson AFB, CO	FX	USB/RTTY	11,607.0
AFL	Loring AFB, ME	FA	USB	9020.0, 11,182.0, 13,201.0, 13,244.0, 18,002.0
AFL2	Loring AFB, ME	FA	USB	9014.0, 13,201.0, 13,244.0, 15,015.0
AFS	Offutt AFB, NE	FX	USB	5026.0, 6826.0, 11,494.0, 11,607.0, 15,962.0, 17,617.0, 18,048.6, 18,594.0, 23,419.0
AFW	March AFB, CA	FX	USB	9057.0
AFX	Barksdale AFB, LA	FX	USB	9057.0, 11,483.5
AF#1	"AIR FORCE ONE"	MA	ISB	6756.0, 11,180.0, 13,201.0, 13,215.0, 15,048., 18,027.0
AF#2	"AIR FORCE TWO"	MA	ISB	6756.0, 9018.0, 11,246.0, 13,210.0, 13,215.0, 15,048.0
AGA	Hickam AFB, HI	FA	USB	18,002.0
		FX	USB	9057.0, 9129.0, 9932.0, 10,430.0, 10,452.0, 11,407.0 18,594.0
AGA1WP	Wright-Patterson AFB, OH	FX	ISB	11,121.0
AGA2	Hickam AFB, HI	FA	USB	6738.0, 8967.0, 11,176.0, 11,179.0, 13,201.0, 13,215.0
		FX	USB	14,744.0
AGA8	Clark AB, Philippines	FA	USB	18,002.0, 18,019.0, 23,227.0
AGB4	Wake Island	FX	USB	10,344.0, 20,870.0
AHF1C	Lojac, Ecuador	FX	LSB	13,950.0
AHF3	Albrook AFS, Panama	FA	USB	9020.0, 15,015.0, 18,019.0
AHF4	Albrook AFS, Panama	FX	LSB	13,950.0
AIC2	Clark AB, Philippines	FA	USB	6738.0, 11,176.0, 13,201.0

Call	Location	Service	Mode	kHz
AID	Misawa AB, Japan	FX	USB	8101.0
AIE	Andersen AFB, Guam	FX	USB	8101.0, 9932.5, 9974.0, 10,344.0, 10,430.0, 11,407.0, 12,129.0, 14,744.0, 18,594.0
AIE2	Andersen AFB, Guam	FA	USB	6738.0, 8967.0, 8989.0, 11,176.0, 13,201.0
AIF2	Yokota AFB, Japan	FA	USB	8967.0, 11,176.0, 11,179.0, 13,201.0, 13,215.0, 15,031.0
AIF2Ø	Yokota AFB, Japan	FA	USB	18,019.0
AIF8Ø	Yokota AB, Japan	FA	USB	6738.0, 8967.0, 11,176.0, 11,179.0, 18,002.0, 18,019.0
		FX	USB	8101.0, 18,594.0
AIR	Andrews AFB, MD	FX	USB	11,121.0
AJE	Croughton (Barford), England	FA	USB	6750.0, 11,176.0, 17,975.0, 23,337.0
		FX	LSB	10,538.0, 16,041.5, 16,454.4
		FX	USB	9242.4, 9921.0, 13,649.0, 20,631.0
AJF7	Rhein Main AB, GFR	FX	USB	4767.0, 6875.0
AJG9	Incirlik AB, Turkey	FA	USB	11,176.0, 13,201.0, 13,215.0, 15,015.0, 17,795.0, 23,337.0
		FX	USB	20,631.0
AJO	Adana AB, Turkey	FA	USB	13,215.0, 15,015.0
AKA	Elemendorf AFB, AK	FX	USB	7938.0
AKA5	Elemendorf AFB, AK	FA	USB	6738.0, 8967.0, 8989.0, 11,176.0, 13,201.0, 13,215.0
AKE	Eielson AFB, AK	FX	USB	18,594.0
AOB5Ø	Estartit, Spain	FX	USB	5066.0, etc.
AOE5Ø	Torrejon AB, Spain	FX	USB	14,423.0, 14,448.0
AOK	Moron de la Frontera, Spain	FX	CW	5918.5, 7706.0
AQI286	Karachi, Pakistan	FX	USB	10,256.4
AQJ	Islamabad Naval Radio, Pakistan	FX	CW	11,585.0
AQP2	Karachi Naval Radio, Pakistan	FC	CW	4325.1, 6390.0
AQP3	Karachi Naval Radio, Pakistan	FC	CW	8490.0
AQP4	Karachi Naval Radio, Pakistan	FC	CW	12,779.5, 13,010.8
AQP6	Karachi Naval Radio, Pakistan	FC	CW	22,425.0
AQP7	Karachi Naval Radio, Pakistan	FC	CW	17,093.5
AQP9	Karachi Naval;Radio, Pakistan	FX	CW	9060.0
AQY286	Karachi, Pakistan	FX	LSB	6976.4
ARH	Gwadar Naval Radio, Pakistan	FC	CW	6410.0, 8495.0
ARL	Karachi Naval Radio, Pakistan	FC	CW	6335.5, 8546.0
ARL71	Karachi Naval Radio, Pakistan	FC	CW	8534.0
ARN	Jiwani Naval Radio, Pakistan	FC	CW	6467.0
ART	Rawalpindi, Pakistan	FX	CW	7539.0, 14,442.5,
		FX	LSB	6796.4, 10,206.4, 11,501.4 17,440.0
ASB33	Karachi, Pakistan	FX	ISB	19,670.1
ASH	Rawalpindi, Pakistan	FX	ISB	11,487.3, 19,872.9
ASK	unknown, probably in Pakistan	FX	CW	10,967.0
ASK	Karachi Radio, Pakistan	FC	CW	6414.1, 8658.0, 8694.0, 13,024.5, 17,050.4
ATA	New Delhi, India	SS	A2	5000.0, 10,000.0, 15,000.0
ATJ67	Bombay (Kirkee), India	FX	ISB	17,480.0
ATJ7Ø	Bombay (Kirkee), India	FX	USB	19,604.0
ATJ71	Bombay (Kirkee), India	FX	ISB	20,535.0
ATP38	New Delhi, India	FX	FAX	18,227.0
ATP57	New Delhi, India	FX	FAX	7405.0
ATS69	Bombay (Kirkee), India	FX	ISB	19,430.0
ATV65	New Delhi, India	FX	FAX	14,842.0
AUY67	Murud, India	FX	LSB	14,541.5
AUY8Ø	Narnaul, India	FX	LSB	14,541.5
AUZ38	Balasore, India	FX	A3H	7811.5
AWB	Bombay Aeradio, India	AX	CW	15,735.0
AXD	"Air Force Laverton, Vic., Australia	FA	USB	4739.0, 8975.0, 11,235.0, 13,205.0, 18,025.0

Call	Location	Service	Mode	kHz
AXF	"Air Force Sydney", NSW, Australia	FA	USB	4739.0, 5718.0, 8975.0, 11,235.0, 13,205.0, 18,025.0
AXH	"Air Force Townsville, Qld., Australia	FA	USB	4739.0, 8975.0, 11,235.0, 13,205.0, 18,025.0
AXI	"Air Force Darwin", NT, Australia	FA	USB	4739.0, 8975.0, 11,235.0, 13,205.0, 18,025.0
AXI32	Darwin, NT, Australia	FX	FAX	5755.0
AXI33	Darwin, NT, Australia	FX	FAX	7535.0
AXI34	Darwin, NT, Australia	FX	FAX	10,555.0
AXI35	Darwin, NT, Australia	FX	FAX	15,615.0
AXI36	Darwin, NT, Australia	FX	FAX	18,060.0
AXJ	"Air Force Perth, WA, Australia	FA	USB	4739.0, 8975.0, 11,235.0, 13,205.0, 18,025.0
AXK	"Air Force Sale", NSW, Australia	FA	USB	4793.0, 8975.0, 11,235.0, 13,205.0, 18,025.0
AXM32	Canberra, ACT, Australia	FX	FAX	5100.0
AXM34	Canberra, ACT, Australia	FX	FAX	11,030.0
AXM35	Canberra, ACT, Australia	FX	FAX	13,920.0
AXM37	Canberra, ACT, Australia	FX	FAX	19,690.0
AXS	"Air Force Learmouth" WA, Australia	FA	USB	4793.0, 8975.0, 11,235.0, 13,205.0, 18,025.0
AXT	"Air Force Edinburgh", SA, Australia	FA	USB	4793.0, 8975.0, 11,235.0, 13,205.0, 18,025.0
AYA	Buenos Aires, Argentina	FX	CW	9200.0
AYA26	Buenos Aires, Argentina	FX	CW	15,738.2
AYA28	Buenos Aires, Argentina	FX	CW	21,785.0
AYA29	Buenos Aires, Argentina	FX	CW	24,110.0
AYA47	Buenos Aires, Argentina	FX	CW	10,390.0
AYA48	Buenos Aires, Argentina	FX	CW	14,817.5
AYA5Ø1	Buenos Aires, Argentina	FX	CW	15,502.5
A2P	Gaborone, Botswana	FX	LSB	6934.5
A4I	Seeb Aeradio, Oman	FA	USB	18,022.0
A4I	unknown, not in Middle East	FX	CW	14,894.1, 24,769.7
A4M	Muscat Radio, Oman	FC	CW	4233.0, 8445.0, 12,675.5, 16,868.0
A9G	Bahrain Aeradio, Bahrain	FA	AM	8828.5
A9M	Bahrain Radio, Bahrain	FC	CW	4284.0, 4302.0, 8448.0, 8454.0, 12,698.0, 12,709.0, 17,169.0, 22,311.5, 22,322.0, 22,414.0
A9M4	Bahrain Radio, Bahrain	FC	CW	17,174.8
A9M43	Hamala, Bahrain	FX	USB	10,845.0
A9M5Ø	Hamala, Bahrain	FX	ISB	11,040.0
A13B	COMMSTA Canberra, ACT, Australia	FC	CW	4286.0 [VHP2]
A13B	COMMSTA Canberra, ACT, Australia	FC	CW	8478.0 [VHP4]
BAC7	Beijing PRC	FC	CW/RTTY	17,131.1
BAF33	Beijing, PRC	FX	FAX	12,110.0
BAF4	Beijing, PRC	FX	FAX	10,115.0
BAF6	Beijing Aeradio, PRC	FA	FAX	5525.0
BAF8	Beijing, PRC	FX	FAX	14,365.0
BAF36	Beijing, PRC	AX	FAX	8120.0
BAZ21	Beijing, PRC	FX	USB	11,402.5
BAZ37	Beijing, PRC	FX	ISB	20,405.0
BAZ53	Beijing, PRC	FX	ISB	12,270.0
BCA2Ø	Shanghai, PRC	FX	ISB	18,637.0
BCO24	Shanghai, PRC	FX	ISB	10,787.0
BEG33	Guangzhou, PRC	FX	USB	10,743.5
BFP99	Xian, PRC	FX	CW	9160.0
BMB	Taipei Naval Radio, Taiwan	FX	CW	5909.0, 8117.0, 13,248.0
BPA	Urumqi, PRC	FA	CW	13,248.0
BPM	Xian, PRC	SS	AM/CW	5000.0, 10,000.0, 15,000.0
		FX	CW	5430.0, 14,961.0
BPV	Shanghai, PRC	FX	CW	9351.0
BSF	Chung-Li, Taiwan	SS	A2	5000.0, 15,000.0
BXM51	unknown	FX	CW	13,537.0
BZP58	unknown	FC	CW	17,214.0

Call	Location	Service	Mode	kHz
CAA2ØØ	Santiago, Chile	FX	USB	24,862.0, etc.
CAA2Ø1	Punta Arenas, Chile	FX	USB	24,862.0, etc.
CAK	Santiago (Los Cerrillos AB), Chile	FA	CW	6745.5, 13,326.0, 13,600.0, 17,998.0, 18,013.5
CAK6H	Santiago (Los Cerrillos AB), Chile	FA	CW	8963.0
CAN6D	Punta Arenas (Montalva), Chile	FX	CW	11,662.5, 15,470.0
CBA3	Arica Radiomaritima, Chile	FS	CW	8461.0
CBA4	Arica Radiomaritima, Chile	FS	CW	17,146.4
CBM2	Magellanes (Punta Arenas) R., Chile	FC	CW	4322.0, 6414.5, 8694.0, 12,826.5, 17,002.5
CBV	Valparaiso Radiomaritima, Chile	FC	CW	4349.0, 12,741.0, 12,747.5 22,473.0
		FX	CW	12,235.9
CBV2	Valparaiso Radiomaritima, Chile	FC	CW	8478.0, 16,947.6, 16,962.5
CBV3	Valparaiso Radiomaritima, Chile	FC	CW	12,714.0
CCM	Magallenes Radionaval, Chile	FC	CW	4256.0, 8509.0
CCP	Puerto Montt Naval Radio, Chile	FS	CW	6405.0
CCQ	Iquique Naval Radio, Chile	FX	CW	17,590.0
CCS	Santiago Naval Radio, Chile	FC	CW	4330.0, 4380.5, 5358.0, 6418.2, 6481.5, 8682.5, 11,395.0, 14,587.0, 16,936.0, 22,064.4
		FS	FAX	4265.0, 8682.5
		FX	CW/FAX	13,525.0, 15,771.4, 17,590.0, 22,071.5
CCV	Valparaiso Naval Radio, Chile	FC	CW	4271.0, 4298.0
		FX	CW	18,175.0
CCV6	Valparaiso Naval Radio, Chile	FC	CW	8558.0, 12,960.0
CEC2Ø8	Paine, Chile	FX	USB	20,160.0
CEC224	Arica, Chile	FX	USB	20,160.0
CFH	Maritime Command Radio, Halifax, NS, Canada	FC	CW	4255.0, 6386.5, 6430.0, 8697.0, 12,726.0, 16,926.5, 22,397.5
		FC	FAX	4271.0, 6329.0
		FC	USB	4363.6, 6509.5
		FX	CW	5097.0, 10,945.0, 13,555.0, 15,920.0, 17,486.0
		FX	CW/FAX	9890.0, 13,510.0, 17,560.0
CFW	Vancouver, BC, Canada	FX	USB	7953.0
CFW	Vancouver Radio, BC, Canada	FC	USB	4410.1
CFW47	Norway House, Man., Canada	FX	USB	4483.0
CFY82	Whitehorse, YU, Canada	FX	USB	5215.0
CFY205	The Pas, Man., Canada	FX	USB	4483.0
CGD2Ø6	Alma, PQ, Canada	FX	USB	5390.0
CGD432	Kenora, Ont., Canada	FX	USB	7400.0
CGF	Halifax, NS, Canada	FC	USB	6509.5
CGK	Vancouver, BC, Canada	FC	USB	6509.5
CGR756	unknown, probably in Canada	FX	USB	10,102.0
CGZ	St. Johns, Nfld., Canada	FX	USB	4616.0, 6509.5
CHF280	Prince Albert, Sask., Canada	FX	USB	4571.0
CHR	"Trenton Military", Ont., Canada	FA	USB	4704.0, 6705.0, 11,209.0, 11,223.0, 13,221.0, 13,254.0
CHU	Ottawa, Ont., Canada	SS	A3H	7335.0, 14,670.0
CIS9	Petawawa, Ont., Canada	FX	ISB	20,957.0
CIW2Ø2	Vancouver, BC, Canada	FX	LSB	14,445.0
CJN911	unknown, in NE Canada	FX	USB	4472.5
CJR68Ø	Sable Island, NS, Canada	FX	USB	5281.5, 7861.0
CJX	"St. Johns Military", Nfld., Canada	FA	USB	4704.0, 6705.0, 6753.0, 11,209.0, 13,221.0, 13,254.9, 15,035.0
CKA214	Inuvik, NWT, Canada	FX	USB	4532.0
CKN	"Vancouver Military", BC, Canada	FA	USB	6705.0
CKN	Vancouver Forces Radio, BC, Canada	FC	CW	4307.0, 6384.5, 6445.5, 6946.0, 8463.0, 8614.0, 12,125.0, 12,702.0, 16,960.0

Call	Location	Service	Mode	kHz
		FC	CW/FAX	4268.0
		FC	USB	4422.5
		FX	CW	15,982.0
		FX	FAX	4497.5, 6946.0, 12,125.0
CLA2	Havana (Cojimar) Radio, Cuba	FC	CW	4235.0
CLA3	Havana (Cojimar) Radio, Cuba	FC	CW	4225.1
CLA4	Havana (Cojimar) Radio, Cuba	FC	CW	6454.0
CLA5	Havana (Cojimar) Radio, Cuba	FC	CW	6337.0
CLA2Ø	Havana (Cojimar) Radio, Cuba	FC	CW	8496.0
CLA21	Havana (Cojimar) Radio, Cuba	FC	CW	8573.0
CLA22	Havana (Cojimar) Radio, Cuba	FC	CW	8701.9
CLA3Ø	Havana (Cojimar) Radio, Cuba	FC	CW	12,748.0
CLA31	Havana (Cojimar) Radio, Cuba	FC	CW	12,792.0
CLA32	Havana (Cojimar) Radio, Cuba	FC	CW	13,062.0
CLA33	Havana (Cojimar) Radio, Cuba	FC	CW	12,673.5
CLA4Ø	Havana (Cojimar) Radio, Cuba	FC	CW	16,961.0
CLA41	Havana (Cojimar) Radio, Cuba	FC	CW	17,165.6
CLA5Ø	Havana (Cojimar) Radio, Cuba	FC	CW	22,396.0
CLN	Havana (Bauta), Cuba	FX	CW	17,628.0
CLN53	Havana (Bauta), Cuba	FX	FAX	6946.2
CLN295	Havana (Bauta), Cuba	FX	ISB	10,853.0
CLN3ØØ	Havana (Bauta), Cuba	FX	ISB	10,788.0
CLN321	Havana (Bauta), Cuba	FX	USB	11,108.0
CLN328	Havana (Bauta), Cuba	FX	ISB	11,675.0
CLN362	Havana (Bauta), Cuba	FX	USB	11,525.0
CLN4Ø4	Havana (Bauta), Cuba	FX	USB	13,474.0
CLN413	Havana (Bauta), Cuba	FX	AM	13,907.0
CLN447	Havana (Bauta), Cuba	FX	ISB	14,670.0
CLN483	Havana (Bauta), Cuba	FX	USB	15,556.0
CLN653	Havana (Bauta), Cuba	FX	CW	21,814.0
CLP1	Havana, Cuba	FX	CW	13,937.7, 13,952.5, 13,970.0, 18,184.0, 23,860.2 24,808.2
		FX	CW/RTTY	20,835.0
CLP3	unknown	FX	CW	23,030.0
CLP6	unknown	FX	CW	8955.0
CLQ	Havana (Cojimar) Radio, Cuba	FC	CW	6435.0, 6477.5, 8480.0, 8690.0 12,748.0, 12,878.0, 17,189.6, 25,495.5
CLS	Havana (Industria Pesuqera) R., Cuba	FC	CW	6360.0, 6392.0, 8489.0, 8510.0, 8516.0, 12,911.0, 16,921.0, 16,921.0
CME9	Havana, Cuba	FX	CW/USB	9062.0
CME3Ø1	Havana, Cuba	FX	CW	9481.0
CME3Ø4	Havana, Cuba	FX	CW	10,484.0
CME31Ø	Havana, Cuba	FX	CW/USB	11,448.0
CME396	Havana, Cuba	FX	ISB	9135.0
CME397	Havana, Cuba	FX	ISB	11,030.0
CME4Ø6	Havana, Cuba	FX	USB	11,508.0
CMH	Santiago Aeradio, Cuba	AX	CW	10,428.0
CMI	Gen. Peraza Aeradio, Cuba	AX	CW	10,428.0
CMLX	Havana (Bauta), Cuba	FX	ISB	17,393.0
CML32	Havana (Bauta), Cuba	FX	ISB	11,650.0
CML49	Havana (Bauta), Cuba	FX	ISB	16,175.0
CMU967	Santiago Naval Radio, Cuba	FX	CW	5258.0, 10,433.5, 10,725.0, 11,114.0, 11,278.0, 11,555.0, 14,792.0, 14,967.0, 15,497.0, 17,424.0
CNL	Rabat, Morocco	FX	CW	14,817.5, 16,238.6
		FX	FAX	15,941.5
CNO	Casablanca, Morocco	FX	CW	5263.0
CNO98	Casablanca, Morocco	AX	CW	11,010.0
CNP	Casablanca Radio, Morocco	FC	CW	8686.0, 12,695.5, 17,170.4
CNR	Rabat, Morocco	FX	USB	9101.0
CNR25	Rabat, Morocco	FX	LSB	9083.5
CNR28	Rabat, Morocco	FX	ISB	13,965.0

Call	Location	Service	Mode	kHz
CNR34	Rabat, Morocco	FX	ISB	19,665.0
CNR39	Rabat, Morocco	FX	USB	11,004.0
CNT	Rabat, Morocco	FX	CW	10,390.0, 14,817.5
COB	Havana Radio, Cuba	FC	USB	4422.5
COB41	Cienfuegos Radio, Cuba	FC	USB	4422.5
COL	Havana Aeradio, Cuba	FA	CW	8842.0, 10, 025.0, 11,312.0, 13,248.0, 15,024.0, 17,936.0
CON3Ø1	Havana, Cuba	FX	USB	4789.0
COY851	Havana Naval Radio, Cuba	FX	CW	13,390.0
COY895	Havana Naval Radio, Cuba	FX	CW	7935.0, 11,406.0, 14,696.5
COZ67	unknown, probably in Cuba	FX	CW	5058.0
CPF2	La Paz Naval Radio, Bolivia	FX	CW	17,590.0
CPP67	La Paz, Bolovia	FX	USB	20,960.0
CQF42	Bissau, Guinea-Bissau	FX	CW	9375.0
CQK	Sao Tome R., Sao Tome & Principe	FC	CW	8634.0
CQK55	Sao Tome, Sao Tome e Principe	FX	CW	18,368.1
CQL	Principe Radio, Sao Tome e Principe	FC	CW	4283.0, 8634.0
CQT	unknown	FC	CW	8784.0
CSF25	Lisbon, Portugal	FX	CW/ISB	9375.0, 11,412.0
CSF36	Ponta Delgada, Azores	FX	CW/ISB	9375.1
CSF37	Funchal, Madeira Islands	FX	CW/ISB	4839.1, 5070.0, 9375.0, 11,412.0
CSF46	Horta, Azores	FX	CW	4839.1
CSF461	Angra do Heroismo, Azores	FX	CW	5070.0
CSF462	Horta, Azores	FX	CW	5070.0
CSJ26	Lisbon, Portugal	FX	CW	10,390.0
CSO	unknown	FX	CW	5192.0
CSR26	Lisbon, Portugal	FX	CW/ISB	12,180.1
CSS414	Ponta Delgada. Azores	FX	ISB	11,562.0
CSY46	Santa Maria Aeradio, Azores	AX	ISB/RTTY	13,364.9, 16,234.0
CSY53	Santa Maria Aeradio, Azores	AX	USB	10,289.0
CSZ69	Lisbon, Portugal	AX	FAX	14,596.0
CTH	Horta Naval Radio, Azores	FC	USB	4394.6
		FX	CW	10,982.2
CTH27	Horta Naval Radio, Azores	FX	USB	7349.5
CTH29	Horta Naval Radio, Azores	FX	USB	12,068.5
CTH38	Horta Naval Radio, Azores	FX	USB	16,041.5
CTH41	Horta Naval Radio, Azores	FC	CW	7351.0
CTH47	Horta Naval Radio, Azores	FC	CW	6334.5
CTH55	Horta Naval Radio, Azores	FC	CW	12,994.0
CTH71	Horta Naval Radio, Azores	FX	USB	11,023.5
CTN37	Apulia Naval Radio, Portugal	FX	CW	14,870.0
CTN81	Apulia Naval Radio, Portugal	FX	USB	11,018,5
CTN84	Apulia Naval Radio, Portugal	FX	USB	16,343.5
CTP	Oeiras Naval Radio, Portugal	FC	CW/RTTY	4235.0, 8451.5, 8674.0
CTP93	Oeiras Naval Radio, Portugal	FC	CW	4278.0
CTP94	Oeiras Naval Radio, Portugal	FC	CW	6389.0
CTP95	Oeiras Naval Radio, Portugal	FC	CW	8551.5
CTP96	Oeiras Naval Radio, Portugal	FC	CW	12,823.5
CTP97	Oeiras Naval Radio, Portugal	FC	CW	16,986.0
CTU2	Monsanto Naval Radio, Portugal	FC	CW	13,001.9
CTU3	Monsanto Naval Radio, Portugal	FC	CW	12,807.5
CTU4	Monsanto Naval Radio, Portugal	FC	CW	4306.0
CTU7	Monsanto Naval Radio, Portugal	FC	CW	17,088.7
CTU28	Monsanto Naval Radio, Portugal	FC	CW	8701.3, 8702.9
CTV	Monsanto Naval Radio, Portugal	FC	CW	4308.0
CTV2	Monsanto Naval Radio, Portugal	FC	CW	13,003.0, 22,551.0
CTV3	Monsanto Naval Radio, Portugal	FC	CW	13,042.0
CTV4	Monsanto Naval Radio, Portugal	FC	CW/FAX	4234.0
CTV6	Monsanto Naval Radio, Portugal	FC	CW	6351.5
CTV7	Monsanto Naval Radio, Portugal	FC	CW	17,055.2
CTV8	Monsanto Naval Radio, Portugal	FC	CW	8641.8
CTV25	Monsanto Naval Radio, Portugal	FX	CW	5016.2
CTV27	Monsanto Naval Radio, Portugal	FC	CW	7353.0

Call	Location	Service	Mode	kHz
CTV28	Monsanto Naval Radio, Portugal	FC	CW	12,810.0
CTV71	Monsanto Naval Radio, Portugal	FX	CW	11,140.0
CTV73	Monsanto Naval Radio, Portugal	FC	CW	8703.0
CTV81	Monsanto Naval Radio, Portugal	FX	CW	11,015.0
CTW8	Monsanto Naval Radio, Portugal	FC	CW/FAX	8525.0
CTW32	Monsanto Naval Radio, Portugal	FX	CW	15,628.5
CUA2Ø	Lisbon (Alfragide), Portugal	FX	ISB	5357.0
CUA23	Lisbon (Alfragide), Portugal	FX	CW/ISB	5765.0
CUA33	Lisbon (Alfragide), Portugal	FX	CW/RTTY	7665.0
CUA37	Lisbon (Alfragide), Portugal	FX	FAX	8010.0
CUA48	Lisbon (Alfragide), Portugal	FX	CW	9419.0
CUA53	Lisbon (Alfragide), Portugal	FX	ISB	10,175.0
CUA54	Lisbon (Alfragide), Portugal	FX	ISB	10,555.0
CUA55	Lisbon (Alfragide), Portugal	FX	ISB	10,905.0
CUA69	Lisbon (Alfragide), Portugal	FX	CW	13,420.0
CUA7Ø	Lisbon (Alfragide), Portugal	FX	ISB	13,970.0
CUA71	Lisbon (Alfragide), Portugal	FX	USB	14,450.0
CUA77	Lisbon (Alfragide), Portugal	FX	ISB	15,957.0
CUA85	Lisbon (Alfragide), Portugal	FX	ISB	19,115.0
CUA86	Lisbon (Alfragide), Portugal	FX	ISB	19,180.0
CUA88	Lisbon (Alfragide), Portugal	FX	ISB	20,765.0
CUA89	Lisbon (Alfragide), Portugal	FX	ISB	20,695.0
CUA93	Lisbon (Alfragide), Portugal	FX	ISB	22,975.0
CUA96	Lisbon (Alfragide), Portugal	FX	ISB	24,195.0
CUB	Madeira Radio, Madeira Islands	FC	CW	8461.0, 8469.0, 12,943.5, 16,959.2
CUB28	Funchal Radio, Maderia Islands	FC	CW	4291.9
CUB29	Madeira Radio, Madeira Islands	FC	CW	6393.5
CUC9	Lisbon (Alfragide), Portugal	FX	CW	19,814.5
CUC23	Lisbon (Alfragide), Portugal	FX	ISB	19,275.0
CUG	Sao Miguel Radio, Azores	FC	CW	8461.0, 16,959.2
		FC	USB	13,119.4
CUG27	Sao Miguel Radio, Azores	FC	CW	12,943.5
CUG3Ø	Lisbon (Alfragide), Portugal	FX	LSB	11,425.0
CUG32	Sao Miguel Radio, Azores	FC	CW	4292.0
CUG33	Sao Miguel Radio, Azores	FC	CW	6393.5
CUG41	Ponta Delgada, Azores	FX	ISB	12,207.5
CUL	Lisbon Radio, Portugal	FC	USB	6509.5, 8722.0
CUL4	Lisbon Radio, Portugal	FC	CW	4292.1
CUL6	Lisbon Radio, Portugal	FC	CW	6393.5
CUL7	Lisbon Radio, Portugal	FC	CW	8489.8
CUL8	Lisbon Radio, Portugal	FC	CW	8469.0
CUL2Ø	Lisbon Radio, Portugal	FC	CW	12,943.5
CUL22	Lisbon Radio, Portugal	FC	CW	16,959.2
CUL24	Lisbon Radio, Portugal	FC	CW	22,479.0
CUL37	Lisbon Radio, Portugal	FC	USB	17,276.3
CUW	Lajes Field, Azores	FA	USB	4746.0, 6750.0, 8967.0, 18,002.0
		FX	ISB	7858.0, 21,773.4
CUW2	Lajes Field, Azores	FA	USB	13,215.0, 13,244.0
CUW3	Lajes Field, Azores	FA	ISB	11,209.0, 15,066.0
CUW2Ø	Lajes Field, Azores	FX	USB	13,927.0, 13,977.0
CVM8	Montevideo (Pajas Blancas), Uruguay	FX	CW	18,150.0
CWA	Cerrito Radio, Uruguay	FC	CW	4363.0, 6435.5, 8602.0, 12,750.0, 12,781.5, 16,871.3
CWM	Montevideo Armada Radio, Uruguay	FC	CW	4280.0, 6376.0, 8578.0, 12,660.0, 16,904.0
CWM25	Isla de Flores Armada R., Uruguay	FC	CW	8578.0
CWO4	Cerrito, Uruguay	FX	CW	13,371.2
CWQ	San Gregorio Tac, Uruguay	FX	CW	10,323.0
CXL21	Montevideo, Uruguay	FX	USB	8037.0
CXL24	Montevideo, Uruguay	FX	ISB	10,770.0
CXL31	Montevideo, Uruguay	FX	LSB	18,190.0
CXL32	Montevideo, Uruguay	FX	ISB	18,705.0
CXL33	Montevideo, Uruguay	FX	ISB	19,525.0

Call	Location	Service	Mode	kHz
CXL48	Montevideo, Uruguay	FX	USB	23,009.5
CXQ338	Montevideo, Uruguay	FX	USB	18,743.5
CXQ342	Montevideo, Uruguay	FX	USB	25,054.0, 25,350.0
CXR	Montevideo Armada Radio, Uruguay	FX	CW	17,590.0
CYS22	Ottawa, Ont., Canada	FX	CW	11,688.0, 14,692.0, 15,582.0, 16,024.0, 17,532.0, 18,355.0, 20,030.0, 23,376.0
CZW	"Halifax Military", NS, Canada	FA	USB	11,249.0
		FX	USB/RTTY	8135.0
C2N	Nauru Radio, Nauru	FC	CW	8686.0, 17,194.0
C5B64	Banjul, Gambia	FX	ISB	13,797.0
C5G	Banjul Radio, Gambia	FC	CW	8441.0, 13,042.0
C6L22	Nassau, Bahamas	FX	USB	5055.5, 8098.5
C6N	Nassau Radio, Bahamas	FC	CW	4220.5, 6376.0, 8441.0, 12,752.0
C6Q51	Nassau, Bahamas	FX	USB	4825.0
C6Q89	Nassau, Bahamas	FX	USB	5217.0
C6Q9Ø	Nassau, Bahamas	FX	USB	8156.0
C6Y42	Nassau, Bahamas	FX	USB	5083.5
C7A	Ocean Station Vessel 62°N & 32°W	MS	CW	5625.0, 8913.5
C7I	Ocean Station Vessel 59°N & 19°W	MS	CW	5625.0, 8913.5
C7J	Ocean Station Vessel 52°N & 20°W	MS	CW	8913.5
C7M	Ocean Station Vessel 66°N & 2°E	MS	CW	5625.0
C7P	Ocean Station Vessel 50°N & 145°W	FA	CW/A3H	5589.0, 8938.0, 13,264.0
C8B53	Maputo, Mozambique	FX	USB	8020.5
C8N27	Maputo, Mozambique	FX	CW/ISB	11,662.5
C9C3	Maputo Naval Radio, Mozambique	FC	CW	4307.0
C9C4	Maputo Naval Station, Mozambique	FX	CW	4358.0
C9C6	Maputo Naval Radio, Mozambique	FC	CW	8642.0
C9C7	Maputo Radionaval, Mozambique	FC	CW	13,042.0
C9E22Ø	Maputo, Mozambique	AX	CW	11,500.0
C9L2	Maputo Radio, Mozambique	FC	CW	4289.0, 6488.0, 8698.0
C9L4	Beira Radio, Mozambique	FC	CW	17,194.4
C9L6	Mozambique Radio, Mozambique	FC	CW	8698.0, 13,056.0
C13E	Vancouver Forces Radio, BC, Canada	FC	CW	4268.0 [CKN]
		FX	CW	6946.0 [CKN]
C13L	Maritime Command R., Halifax, NS, Canada	FC	CW	4255.0, 6430.0, 8697.0 [CFH]
		FX	CW	10,945.0 [CFH]
DAF	Norddeich Radio, GFR	FC	CW	8672.5, 12,832.5, 17,048.0, 22,415.0
		FC	CW/TOR	4220.0, 6363.5, 26,227.0
DAH	Norddeich Radio, GFR	FC	USB	17,350.7, 22,661.1
DAI	Norddeich Radio, GFR	FC	USB	8777.8, 8802.6
DAJ	Norddeich Radio, GFR	FC	USB	8768.5, 13,172.1, 17,279.4, 22,614.6
DAL	Norddeich Radio, GFR	FC	CW	8461.9, 8511.9, 13,027.5, 17,177.6, 22,340.5
		FC	CW/TOR	4244.0, 6456.0,
DAM	Norddeich Radio, GFR	FC	CW	4265.0, 6475.5, 8638.5, 12,763.5, 16,980.4, 22,476.0, 22,508.5, 25,196.0
DAN	Norddeich Radio, GFR	FC	CW	8483.5, 12,898.5, 17,143.6, 22,516.0, 26,108.0
		FC	CW/TOR	4308.2, 6435.5
DAP	Norddeich Radio, GFR	FC	USB	17,260.8
DDH5	Pinneberg, GFR	FX	CW	5876.0
DDH8	Pinneberg, GFR	FX	FAX	13,657.0
DDH9	Pinneberg, GFR	FX	CW	11,039.0
DDJ8	Franfurt (Usingen), GFR	FX	CW/RTTY	13,657.0
DDK2	Quickborn, GFR	FX	CW	13,882.5
DDK9	Pinneberg, GFR	FX	CW	9880.0
DEB	Wiesbaden, GFR	FX	CW	7401.0, 8097.5, 10,390.0,
		FX	CW	12,224.0, 14,817.5, 18,190.0
		FX	CW/RTTY	4837.5, 7532.0, 8038.0, 8045.0, 9200.0, 12,224.0,

Call	Location	Service	Mode	kHz
				18,380.0, 19,130.0, 19,360.0
DEP92	Frankfurt, GFR	FX	ISB	15,920.0
DFE28	Bonames, GFR	FX	FAX	5830.0
DFF41	Frankfurt Radio, GFR	FC	CW	6418.5
DFF97	Frankfurt, GFR	FX	USB	6975.0
DFH43	Frankfurt, GFR	FC	CW	8439.0
DFK92	Frankfurt (Bonames), GFR	FX	USB	10,922.0
DFM95	Frankfurt (Usingen), GFR	FC	CW	12,953.0
DFR28	Frankfurt, GFR	FC	CW	9297.0, 17,082.0
DFW36	Frankfurt Radio, GFR	FC	CW	22,361.0
DFY	Tel Aviv, Israel	FX	USB	20,416.1
DGN62	Quickborn, GFR	FX	FAX	13,627.1
DHJ59	Sengwarden Naval Radio, GFR	FC	CW	4282.7, 8648.0, 12,705.0, 12,750.0, 22,396.5
		FX	CW	4586.0, 16,265.0
DHM94	Lahr Military, GFR	FA	CW	4704.0
DHM95	Lahr Military, GFR	FA	USB	11,209.0, 13,257.0
DML	Sylt, GFR	FX	USB	4857.5, etc.
DUN	Manila Radio, Philippines	FC	CW	12,732.0, 13,032.0
DUN356	Manila, Philippines	FX	CW/RTTY	9285.0
DXD	Manila (Jolo), Philippines	FC	CW	4276.0
DXF	Davao Radio, Philippines	FC	CW	8515.0, 8600.0, 12,900.0, 16,962.0
DXR	Manila Radio, Philippines	FC	CW	8628.0, 12,942.0
DXU	Catabato Radio, Philippines	FC	CW	8614.0
DYM	Cebu Radio, Philippines	FC	CW	4338.0, 6338.0, 8676.0
DYP	Cebu Radio, Philippines	FC	CW	16,902.5
DYV	Manila (Iloilo) Radio, Philippines	FC	CW	4316.0, 6474.0
DZA	Mandaluyong Radio, Philippines	FC	CW	6408.0
DZD	Manila (Antipolo) R., Philippines	FC	CW	6429.0, 8574.0, 12,858.0, 17,144.0
DZE	Mandaluyong Radio, Philippines	FC	CW	4268.0, 6402.0, 8564.0, 12,846.0, 17,072.0
DZF	Manila (Bacoor) Radio, Philippines	FC	CW	4274.0, 6411.0, 8545.8, 12,822.2, 12,828.0, 17,071.4
DZG	Manila (Las Pinas) Radio, Philippines	FC	CW	4294.0, 6440.9, 8588.0, 12,882.0, 17,177.2, 22,502.0
DZH	Manila Radio, Philippines	FC	CW	4324.0, 6486.0, 8648.0, 12,972.0
DZI	Manila (Bacoor) Radio, Philippines	FC	CW	4336.0, 6335.5, 8672.0, 13,008.0, 22,335.5, 22,506.0
DZJ	Manila (Bulacan) Radio, Philippines	FC	CW	4330.0, 6326.5, 6453.0, 8604.2, 8660.0, 12,906.0, 12,990.0, 16,870.4, 22,323.5
DZK	Manila (Bulacan) Radio, Philippines	FC	CW	4296.0, 6444.0, 8595.1, 12,891.0, 17,184.0
DZM	Manila (Bulacan) Radio, Philippines	FC	CW	4328.0, 6492.0, 12,984.0
DZN	Manila (Novotas) Radio, Philippines	FC	CW	4314.0, 6468.0, 8584.0, 12,936.0, 17,168.0, 22,538.0
DZO	Manila (Bulacan) Radio, Philippines	FC	CW	4290.0, 6435.0, 8580.0, 12,870.0, 17,160.0
DZP	Manila (Novaliches) R., Philippines	FC	CW	4278.0, 6417.0, 8555.0, 12,834.5, 17,112.0
DZR	Manila Radio, Philippines	FC	CW	4284.0, 6446.0, 8568.0, 12,852.0, 22,482.0
DZU	Manila (Pasig) Radio, Philippines	FC	CW	4318.0, 6477.0, 8636.0
DZW	Manila Radio, Philippines	FC	CW	4262.0, 6393.0
DZY	Manila (Malabon) Radio, Philippines	FC	CW	4318.0
DZZ	Manila Radio, Philippines	FC	CW	4288.0, 6432.0, 8576.0, 12,864.7, 17,152.0, 22,490.0
D2J32	Luanda, Angola	FX	ISB	16,170.0
D2J35	Luanda, Angola	FX	USB/AM	20,535.0
D2U21	Luanda, Angola	AX	CW	17,367.0
D3E	Luanda Radio, Angola	FC	CW	6369.0, 17,189.6
D3E2	Luanda Radio, Angola	FC	CW	8566.3

Call	Location	Service	Mode	kHz
D3E3	Luanda Radio, Angola	FC	CW	13,023.0
D3E81	Luanda Radio, Angola	FC	CW	22,331.3
D3F	Lobito Radio, Angola	FC	CW	6369.0, 17,189.6
D3G	Mocamedes Radio, Angola	FC	CW	6369.0
D3L	Luanda Radio, Angola	FC	CW	6344.0
D3L25	Luanda Naval Radio, Angola	FC	CW	8642.0
D3L28	Luanda Naval Radio, Angola	FC	CW	4307.0
D3M93	Luanda Naval Radio, Angola	FX	CW	16,090.0
D4A5	Sao Vicente Radio, Cape Verde	FC	CW	6488.0
D4A6	Sao Vicente Radio, Cape Verde	FC	CW	8469.0
D4A7	Sao Vicente Radio, Cape Verde	FC	CW	8642.0, 12,700.0, 12,993.0
D4A8	Sao Vicente Radio, Cape Verde	FC	CW	13,047.0, 17,055.6
D4B44	Sal, Cape Verde	AX	CW	9111.0
D4B62	Sal, Cape Verde	AX	CW	5060.0
D4D	Praia de Cabo Radio, Cape Verde	FC	CW	4292.0, 8469.0, 12,700.0
		FX	CW	9374.6
D4E25	Praia, Cape Verde	FX	ISB	14,464.9
D4E29	Praia, Cape Verde	FX	ISB	16,276.0
D4U	Sao Vicente Naval Radio, Cape Verde	FC	CW	22,435.0
		FX	CW	19,805.0
D6B403	Moroni, Republic of the Comores	FX	LSB	20,337.8
D6B404	Moroni, Republic of the Comores	FX	LSB	20,403.5
D6B408	Moroni, Republic of the Comores	FX	LSB	20,817.6
D6B422	Moroni, Republic of the Comores	FX	LSB	24,226.0
EAC	Cadiz (Tarifa) Radio, Spain	FC	CW	4275.5
EAD	Aranjuez radio, Spain	FC	CW	4349.0, 12,764.5
EAD2	Aranjuez Radio, Spain	FC	CW	6382.0
EAD3	Aranjuez Radio, Spain	FC	CW	8682.0
EAD4	Aranjuez Radio, Spain	FC	CW	13,065.0
EAD5	Aranjuez Radio, Spain	FC	CW	17,184.7
EAD6	Aranjuez Radio, Spain	FC	CW	22,446.0
EAD44	Aranjuez Radio, Spain	FC	CW	12,801.0, 12,887.5
EAT5	Santa Cruz de Tenerife R., Canary Is.	FC	CW	16,942.8
EA3V	unknown	FX	CW	5381.0
EBA	Madrid Radionaval, Spain	FC	CW	6377.7, 6388.0, 6408.5, 8447.0, 8528.5, 12,717.0, 13,052.4, 13,059.0, 17,017.9, 22,410.0, 22,517.0, 25,135.0
		FC	CW/RTTY	4279.0, 4311.3, 8465.0, 8640.0, 12,932.5
		FX	CW	14,640.9
EBB	Ferrol de Caudillo Radionaval, Spain	FX	CW	7920.0
EBC	Cadiz Radionaval, Spain	FC	CW	4.....
EBC	Cadiz Radionaval, Spain	FX	CW	6840.0, 7920.0
EBCQ	Madrid Radionaval, Spain	FC	CW	4311.3 [EBA]
EBJ	Palma Radionaval, Mallorca, Spain	FX	CW	7926.0
EBK	Las Palmas Radionaval, Canary Is.	FC	CW	4303.0, 6377.7
ECA7	Madrid, Spain	AX	FAX	6918.5
EC3Y	unknown	FX	CW	8158.5, 13,582.0
EDF3	Aranjuez Radio, Spain	FC	CW	8473.0
EDF6	Aranjuez Radio, Spain	FC	CW	22,384.0
EDG3	Aranjuez Radio, Spain	FC	CW	8457.0
EDG4	Aranjuez Radio, Spain	FC	CW	13,056.0
EDG5	Aranjuez Radio, Spain	FC	CW	17,175.2
EDZ2	Aranjuez Radio, Spain	FC	CW	6400.5
EDZ4	Aranjuez Radio, Spain	FC	CW	8618.0
EDZ5	Aranjuez Radio, Spain	FC	CW	12,934.5
EDZ6	Aranjuez Radio, Spain	FC	CW	17,065.2
EDZ7	Aranjuez Radio, Spain	FC	CW	22,532.9
EEQ	Madrid, Spain	FX	CW	7374.0, 8072.0, 9796.0, 10,390.0, 12,035.0, 13,440.0
EHY	Pozuelo del Rey Radio, Spain	FC	USB	4372.9, 4376.0, 4388.4, 4403.9, 6504.4, 8728.2, 8746.8, 8771.6, 13,122.5, 13,128.7, 13,175.2, 13,181.4,

Call	Location	Service	Mode	kHz
				17,291.8, 17,322.8, 17,344.5, 17,350.7, 22,596.0, 22,667.0, 22,682.8
EHY22	Madrid (Pozuelo del Rey), Spain	FX	ISB	20,670.0
EHY23	Madrid (Pozuelo del Rey), Spain	FX	ISB	20,570.0
EHY24	Madrid (Pozuelo del Rey), Spain	FX	LSB	16,330.5
EHY25	Madrid (Pozuelo del Rey), Spain	FX	ISB	19,030.0
EHY26	Madrid (Pozuelo del Rey), Spain	FX	AM	14,985.0
EHY78	Madrid (Pozuelo del Rey), Spain	FX	ISB	19,154.0
EHY96	Madrid (Pozuelo del Rey), Spain	FX	ISB	16,056.5
EIP	Shannon Aeradio (Rineanna), Ireland	FA	USB	5533.0, 8833.0, 13,312.0
ELC	Monrovia Radio, Liberia	FC	CW	8518.0, 12,709.0
ELE1Ø	Harbel, Liberia	FX	CW/RTTY	10,310.0
ELE13	Harbel, Liberia	FX	CW/RTTY	13,945.0
ELE14	Harbel, Liberia	FX	CW/RTTY	14,605.0
ELE25	Harbel, Liberia	FX	CW/RTTY	15,940.0
ELE28	Harbel, Liberia	FX	CW	18,530.0
EPC	B. Abbas Naval Radio, Iran	FC	CW	8470.0
EPR88	Teheran, Iran	FX	CW	20,753.0
EQI	Abbas Radio, Iran	FC	CW	8469.0, 13,069.5
EQK	Khoramshahr Radio, Iran	FC	CW	8469.0, 13,069.5
EQZ	Abadan Radio, Iran	FC	CW	4292.0, 6362.0, 8471.0, 13,069.5, 16,983.5
ETC	Assab Radio, Ethiopia	FC	CW	8618.0, 12,673.5
ETM85	Addis Ababa, Ethiopia	FX	CW	13,850.0
ETM99	Addis Ababa, Ethiopia	FX	USB	13,997.0
ETN4Ø	Addis Ababa, Ethiopia	FX	CW/RTTY	14,400.0
ETN52	Addis Ababa, Ethiopia	FX	CW	14,525.0
ETN66	Addis Ababa, Ethiopia	FX	USB	14,661.0
ETR45	Addis Ababa, Ethiopia	FX	CW	16,450.0
ETR84	Addis Ababa, Ethiopia	FX	LSB	18,844.0
ETS	Addis Ababa, Ethiopia	FX	LSB	19,046.0
ETS41	Addis Ababa, Ethiopia	FX	ISB	19,411.5
ETS43	Addis Ababa, Ethiopia	FX	USB	19,461.5
ETT46	Addis Ababa, Ethiopia	FX	CW	20,458.0
ETT9Ø	Addis Ababa, Ethiopia	FX	CW/RTTY	20,900.0
ETT95	Addis Ababa, Ethiopia	FX	CW/RTTY	20,950.0
ETV28	Addis Ababa, Ethiopia	FX	USB	22,758.5
ETW21	Addis Ababa, Ethiopia	FX	USB	23,102.5
FAD58	Tours, France	FX	CW	4897.0
FDA35	Lyon Air, France	FX	CW	11,355.0
FDC	Metz Air, France	FX	CW	4886.0, 4940.0, 6372.0
FDC4	Reims Air, France	FX	CW	9085.9
FDC6	Luxevil Air, France	FX	CW	7930.0
FDC7	Strasbourg Air, France	FX	CW	7407.0, 9085.9
FDC9	Nancy Air, France	FX	CW	4864.5
FDC2Ø	Toul Air, France	FX	CW	4637.5
FDC21	Drackenbronn Air, France	FX	CW	5036.0
FDC22	Contrexeville Air, France	FX	CW	5240.0
FDE	Villacoublay Air, France	FA	CW	4533.5, 4753.0
FDE4	Bourges Air, France	FX	CW	4617.5, 4882.0, 5026.5
FDE8	Doullens Air, France	FX	CW	5021.5
FDG	Bordeaux Air, France	FA	CW	4867.5
FDI	Aix les Milles Air, France	FX	CW	4940.0, 5160.0, 7816.5
FDI21	Nimes Air, France	FX	CW	5461.5
FDI22	Narbonne Air, France	FX	CW	5462.0
FDX	Paris, France	FA	CW	23,234.0
FDY	Orleans Air, France	FX	CW	4942.0, 5010.0, 5123.5, 6768.0, 9342.0, 15,667.7
FDY4	unknown	FX	CW	5010.0, 9085.9
FDY5	Aix les Milles Air, France	FX	CW	5010.0, 7575.0
FEB21	Chichi Jima, Bonin Is., Japan	FX	ISB	10,670.0
FFD	St. Denis Radio, Reunion	FC	CW	4226.0, 16,936.0
FFD28	St. Denis Radio, Reunion	FC	CW	8469.0
FFL2	St. Lys Radio, France	FC	CW	4328.0

Call	Location	Service	Mode	kHz
FFL4	St. Lys Radio, France	FC	CW	8522.5
FFL21	St. Lys Radio, France	FC	USB	4366.7
FFL22	St. Lys Radio, France	FC	USB	4403.9
FFL23	St. Lys Radio, France	FC	USB	4369.8
FFL24	St. Lys Radio, France	FC	USB	4413.2
FFL41	St. Lys Radio, France	FC	USB	8808.8
FFL42	St. Lys Radio, France	FC	USB	8793.3
FFL43	St. Lys Radio, France	FC	USB	8768.5
FFL6	St. Lys Radio, France	FC	CW	12,912.6
FFL8	St. Lys Radio, France	FC	CW	17,027.0
FFL9	St. Lys Radio, France	FC	CW	22,509.0
FFL44	St. Lys Radio, France	FC	USB	8802.6
FFL61	St. Lys Radio, France	FC	USB	13,178.3
FFL62	St. Lys Radio, France	FC	USB	13,187.6
FFL63	St. Lys Radio, France	FC	USB	13,193.8
FFL64	St. Lys Radio, France	FC	USB	13,165.9
FFL81	St. Lys Radio, France	FC	USB	17,316.6
FFL82	St. Lys Radio, France	FC	USB	17,332.1
FFL83	St. Lys Radio, France	FC	USB	17,242.2
FFL84	St. Lys Radio, France	FC	USB	17,288.7
FFL85	St. Lys Radio, France	FC	USB	17,298.0
FFL86	St. Lys Radio, France	FC	USB	17,245.3
FFL91	St. Lys Radio, France	FC	USB	22,673.5
FFL92	St. Lys Radio, France	FC	USB	22,605.3
FFL93	St. Lys Radio, France	FC	USB	22,701.4
FFL94	St. Lys Radio, France	FC	USB	22,689.0
FFL95	St. Lys Radio, France	FC	USB	22,639.4
FFP2	Fort de France Radio, Martinique	FC	CW	4263.0
FFP3	Fort de France Radio, Martinique	FC	CW	8675.2
FFP7	Fort de France Radio, Martinique	FC	CW	12,831.0
FFS3	St. Lys Radio, France	FC	CW	6449.5
FFS4	St. Lys Radio, France	FC	CW	8510.0
FFS6	St. Lys Radio, France	FC	CW	12,678.0
FFS8	St. Lys Radio, France	FC	CW	17,040.8
FFS9	St. Lys Radio, France	FC	CW	22,318.5
FFT4	St. Lys Radio, France	FC	CW	8550.0
FFT6	St. Lys Radio, France	FC	CW	12,655.5
FFT8	St. Lys Radio, France	FC	CW	16,947.6
FIT46	Cahors, France	FX	CW	7565.0
FIT75	Paris, France	FX	CW	7565.0
FJA8	Mahina Radio, Tahiti	FC	CW	8461.0
FJA26	Mahina Radio, Tahiti	FC	CW	17,040.8
FJA41	Mahina Radio, Tahita	FC	CW	4298.0
FJD251	Papeete, Tahiti	FX	LSB	5035.0
FJD291	Papeete, Tahiti	FX	LSB	9205.0
FJP6	Noumea Radio, New Caledonia	FC	CW	6330.0
FJP8	Noumea Radio, New Caledonia	FC	CW	8698.0
FJP23	Noumea Radio, New Caledonia	FC	CW	12,708.0
FJY2	Port aux Francais, Kerguelen Is.	FX	ISB	19,532.0, 26,820.0
FJY4	St. Paul et Amsterdam Radio French Antarctic	FC	CW	4287.0, 8690.0, 12,722.0
FKO	Djibouti Naval Radio, Djibouti	FA	CW	11,208.0
		FX	CW	23,365.0
FLE23	Paris, France	FX	CW	20,976.0
FLK	unknown	FC	CW	4299.7
FLR25	Lyon, France	FX	CW	6913.0
FLR26	Metz, France	FX	CW	5187.0
FNO32	Paris (Orly), France	AX	CW	5415.0
FNO43	Paris (Orly), France	AX	CW	7722.5
FPF29	Paris (Le Vernet), France	FX	USB	5297.6
FPG8Ø	Paris (St. Assise), France	FX	USB	6802.0
FPI88	Paris, France	FX	FAX/ISB	8185.0
FPK5	Paris (St. Assise), France	FX	ISB	10,155.0
FPK9	Paris (St. Assise), France	FX	USB	10,193.5
FPK58	Paris (St. Assise), France	FX	ISB	10,582.5

Call	Location	Service	Mode	kHz
FPM12	Paris (St. Assise), France	FX	ISB	12,101.0
FPP93	Paris (St. Assise), France	FX	ISB	15,935.0
FPS67	Paris (St. Assise), France	FX	USB	18,675.0
FPX3	Paris (St. Assise), France	FX	ISB	23,137.5
FPX5	Paris (St. Assise), France	FX	ISB	23,153.0
FSB	Paris (St. Martin Abbat), France	FX	CW/RTTY	5208.0, 5305.5, 7401.0, 8038.0, 14,817.5
FSB54	Paris (St. Martin Abbat), France	FX	CW	6792.0
FSB57	Paris (St. Martin Abbat), France	FX	CW	10,390.0
FSB57	Paris (St. Martin Abbat), France	FX	USB	10,388.5
FSB59	Paris (St. Martin Abbat), France	FX	CW	18,190.0
FSB62	Paris (St. Martin Abbat), France	FX	CW	19,360.0
FSB63	Paris (St. Martin Abbat), France	FX	CW/RTTY	19,405.0
FSB65	Paris (St. Martin Abbat), France	FX	CW/RTTY	24,110.0
FSB65	Paris (St. Martin Abbat), France	FX	USB	24,108.5
FSB67	Paris (St. Martin Abbat), France	FX	CW	21,807.5
FSB69	Paris (St. Martin Abbat), France	FX	CW	7832.0
FSB7Ø	Paris (St. Martin Abbat), France	FX	CW	7906.0
FSB71	Paris (St. Martin Abbat), France	FX	CW	10,295.0
FSS267	Strasbourg, France	FX	CW	4635.0
FTE4	Paris (St. Assise), France	FX	FAX	4047.5
FTF46	Paris (St. Assise), France	FX	ISB	5462.5
FTF84	Paris (St. Assise), France	FX	ISB	5842.0
FTG83	Paris (St. Assise), France	FX	ISB	6830.0
FTG84	Paris (St. Assise), France	FX	ISB	6847.5
FTH37	Paris (St. Assise), France	FX	ISB	7770.0
FTH42	Paris (St. Assise), France	FX	ISB	7370.0
FTH67	Paris (St. Assise), France	FX	CW	7428.0
FTH68	Paris (St. Assise), France	FX	ISB	7677.0
FTH77	Paris (St. Assise), France	FX	ISB	7688.0
FTI2	Paris (St. Assise), France	FX	ISB/RTTY	8022.5
FTI9	Paris (St. Assise), France	FX	ISB	8075.0
FTJ39	Paris (St. Assise), France	FX	USB	9392.5
FTJ84	Paris (St. Assise), France	FX	LSB	9840.3
FTK46	Paris (St. Assise), France	FX	ISB	10,462.5
FTK56A	Paris (St. Assise), France	FX	CW	10,563.0
FTK69	Paris (St. Assise), France	FX	ISB	10,695.0
FTK76	Paris (St. Assise), France	FX	ISB	10,765.0
FTK77	Paris (St. Assise), France	FX	ISB	10,775.0
FTK82	Paris (St. Assise), France	FX	ISB	10,820.7
FTL3	Paris (St. Assise), France	FX	USB	11,033.9
FTL6	Paris (St. Assise), France	FX	ISB	11,065.0
FTM25	Paris (St. Assise), France	FX	ISB	12,250.0
FTM3Ø	Paris (St. Assise), France	FX	FAX	12,305.0
FTN53	Paris (St. Assise), France	FX	ISB	13,535.0
FTN56	Paris (St. Assise), France	FX	USB	13,565.0
FTN6Ø	Paris (St. Assise), France	FX	LSB	13,603.0
FTN71	Paris (St. Assise), France	FX	ISB	13,712.0
FTN87	Paris (St. Assise), France	FX	CW	13,873.0
FTO35	Paris (St. Assise), France	FX	USB	14,357.0
FTO42	Paris (St. Assise), France	FX	ISB	14,423.0
FTO43	Paris (St. Assise), France	FX	ISB	14,432.0
FTO49	Paris (St. Assise), France	FX	USB	14,495.0
FTO73	Paris (St. Assise), France	FX	USB	14,730.0
FTO77	Paris (St. Assise), France	FX	USB	14,776.2
FTO85	Paris (St. Assise), France	FX	USB	14,855.0
FTO88	Paris (St. Assise), France	FX	LSB	14,891.0
FTO9Ø	Paris (St. Assise), France	FX	USB	14,900.0
FTP6Ø	Paris (St. Assise), France	FX	ISB	15,605.0
FTP68	Paris (St. Assise), France	FX	ISB	15,682.0
FTP8Ø	Paris (St. Assise), France	FX	LSB	15,805.0
FTP83	Paris (St. Assise), France	FX	ISB	21,830.0
FTQ36	Paris (St. Assise), France	FX	USB	16,365.0
FTQ41	Paris (St. Assise), France	FX	USB	16,408.6

Call	Location	Service	Mode	kHz
FTR47	Paris (St. Assise), France	FX	ISB	17,470.0
FTR5Ø	Paris (St. Assise), France	FX	USB	17,503.7
FTR67	Paris (St. Assise), France	FX	ISB	17,675.0
FTS25	Paris (St. Assise), France	FX	USB	18,249.4
FTS39	Paris (St. Assise), France	FX	USB	18,395.0
FTS51	Paris (St. Assise), France	FX	USB	18,515.0
FTS66	Paris (St. Assise), France	FX	USB	18,655.0
FTS77	Paris (St. Assise), France	FX	LSB	18,771.3
FTS81	Paris (St. Assise), France	FX	ISB	18,815.0
FTS83	Paris (St. Assise), France	FX	USB	18,833.4
FTS85	Paris (St. Assise), France	FX	USB	18,853.5
FTS86	Paris (St. Assise), France	FX	ISB	18,865.0
FTS87	Paris (St. Assise), France	FX	USB	18,875.8
FTS93	Paris (St. Assise), France	FX	USB	18,935.0
FTS99	Paris (St. Assise), France	FX	ISB	18,990.0
FTT2Ø	Paris (St. Assise), France	FX	LSB	19,205.5
FTT24	Paris (St. Assise), France	FX	USB	19,247.5
FTT25	Paris (St. Assise), France	FX	LSB	19,249.0
FTT3Ø	Paris (St. Assise), France	FX	USB	19,305.0
FTT85	Paris (St. Assise), France	FX	ISB	19,850.0
FTU56	Paris (St. Assise), France	FX	USB	20,558.0
FTU61	Paris (St. Assise), France	FX	LSB	20,615.6
FTU64	Paris (St. Assise), France	FX	USB	20,640.0
FTU9Ø	Paris (St. Assise), France	FX	LSB	20,903.2
FTU92	Paris (St. Assise), FRance	FX	USB	20,919.6
FTU94	Paris (St. Assise), France	FX	USB	20,940.0
FTV83	Paris (St. Assise), France	FX	ISB	21,830.0
FTX89	Paris (St. Assise), France	FX	ISB	23,894.0
FTY4	Paris (Le Vernet), France	FX	USB	24,040.0
FUB	Paris (Houilles) Naval R., France	FC	CW	4325.0, 12,701.0, 16,991.0
		FX	CW	7605.0, 8000.0, 16,045.0
		FX	CW/RTTY	14,925.0, 18,365.0
FUC	Cherbourg Naval Radio, France	FX	CW	11,690.6
FUE	Brest Naval Radio, France	FC	CW	6352.0
FUF	Fort de France Naval R., Martinique	FC	CW	6387.9, 8478.5, 8554.0, 12,894.0, 13,031.2, 16,961.5, 17,108.0, 22,313.0, 22,390.1
		FX	CW	4645.0, 7895.0, 9935.0,
FUG	La Regine (Castelnaudry) R., France	FC	CW	4312.9, 6352.0, 8478.5 8666.0, 12,875.0, 16,876.1, 16,961.5
		FX	CW/RTTY	18,135.0, 22,860.0
FUJ	Noumea Naval Radio, New Caledonia	FC	CW	8646.1, 12,697.8, 16,958.0, 22,461.0
		FX	CW	5455.0. 7605.0, 9290.0 17,655.0
FUJ2	Noumea Naval Radio, New Caledonia	FC	CW	12,858.0
FUM	Papeete Naval Radio, Tahiti	FC	CW	6264.0, 8625.0, 12,885.0
		FX	CW	5455.0, 7605.0, 9000.0, 9181.0, 17,380.0, 18,578.0, 23,165.0, 23,590.0
FUO	Toulon Naval Radio, France	FC	CW	4289.0, 6352.0, 12,664.5, 12,875.0
FUV	Djibouti Naval Radio, Djibouti	FC	CW	12,858.0, 13,042.6, 16,904.8, 16,959.0
		FX	CW	4645.0, 8567.9, 12,095.0, 23,590.0
FUX	Le Port Naval Radio, Reunion	FA	CW	13,215.5
		FC	CW	6348.0, 6362.0, 8475.5, 8550.4, 12,691.4, 16,915.0, 17,062.0, 22,312.0, 22,499.4
FYJ3	Paris (St. Assise), France	FX	USB	9117.5
		FX	CW/RTTY	14,925.0, 18,365.0
FUC	Cherbourg Naval Radio, France	FX	CW	11,690.6

Call	Location	Service	Mode	kHz
FUE	Brest Naval Radio, France	FC	CW	6352.0
FUF	Fort de France Naval R., Martinique	FC	CW	6387.9, 8478.5, 8554.0, 12,894.0, 13,031.2, 16,961.5, 17,108.0, 22,313.0, 22,390.1
		FX	CW	4645.0, 7895.0, 9935.0,
FUG	La Regine (Castelnaudry) R., France	FC	CW	4312.9, 6352.0, 8478.5 8666.0, 12,875.0, 16,876.1, 16,961.5
		FX	CW/RTTY	18,135.0, 22,860.0
FUJ	Noumea Naval Radio, New Caledonia	FC	CW	8646.1, 12,697.8, 16,958.0, 22,461.0
		FX	CW	5455.0. 7605.0, 9290.0 17,655.0
FUJ2	Noumea Naval Radio, New Caledonia	FC	CW	12,858.0
FUM	Papeete Naval Radio, Tahiti	FC	CW	6264.0, 8625.0, 12,885.0
		FX	CW	5455.0, 7605.0, 9000.0, 9181.0, 17,380.0, 18,578.0, 23,165.0, 23,590.0
FUO	Toulon Naval Radio, France	FC	CW	4289.0, 6352.0, 12,664.5, 12,875.0
FUV	Djibouti Naval Radio, Djibouti	FC	CW	12,858.0, 13,042.6, 16,904.8, 16,959.0
		FX	CW	4645.0, 8567.9, 12,095.0, 23,590.0
FUX	Le Port Naval Radio, Reunion	FA	CW	13,215.5
		FC	CW	6348.0, 6362.0, 8475.5, 8550.4, 12,691.4, 16,915.0, 17,062.0, 22,312.0, 22,499.4
FYJ3	Paris (St. Assise), France	FX	USB	9117.5
FYN87	Paris (St. Assise), France	FX	ISB	13,878.0
FYT2	Paris (St. Assise), France	FX	USB	19,110.0
FZC36	Cayenne, Guiana	FX	ISB	13,685.0
FZE93	Fort de France, Martinique	FX	ISB	9245.0
FZF37	Fort de France, Martinique	FX	ISB	13,760.0
FZF38	Fort de France, Martinique	FX	USB	13,843.0
FZF57	Fort de France, Martinique	FX	LSB	15,741.2
FZF97	Fort de France, Martinique	FX	ISB	19,570.0
FZK3	Djibouti, Djibouti	FX	LSB	20,900.0
FZM83	Noumea, New Caledonia	FX	ISB	18,432.5
FZM85	Noumea, New Caledonia	FX	ISB	18,517.5
FZO9Ø	Papeete, Tahiti	FX	USB	9051.5
FZP64	Papeete (Papenoo), Tahiti	FX	USB	16,384.0
FZS6Ø	St. Denis (Bel Air), Reunion	FX	USB	16,095.0
FZS63	St. Denis (Bel Air), Reunion	FX	FAX	16,335.0
FZS93	St. Denis (Bel Air), Reunion	FX	ISB	19,324.0
FZW75	Pointe a Pitre, Guadeloupe	FX	USB	17,537.9
FZW95	Destrellan, Guadeloupe	FX	ISB	19,554.8
FZX322	Miquelon, St. Pierre & Miquelon Is.	FX	ISB	12,294.7
FZY363	Miquelon, St. Pierre & Miquelon Is.	FX	ISB	16,355.5
FZY386	Miquelon, St. Pierre & Miquelon Is.	FX	USB	18,623.8
F71	unknown, possibly in India	FX	CW	16,293.4
GAB33	London (Rugby), England	FX	USB	13,585.0
GAP36	London (Rugby), England	FX	ISB	16,380.0
GAR45	London (Rugby), England	FX	LSB	25,493.0
GAW38	London (Rugby), England	FX	ISB	18,200.0
GBB38	London (Rugby), England	FX	ISB	18,440.0
GBW42	London (Rugby), England	FX	ISB	22,740.0
GCB39B	London (Rugby), England	FX	ISB	19,008.2
GFA22	Bracknell, England	FX	FAX	4610.0
GFA23	Bracknell, England	FX	FAX	8040.0
GFA24	Bracknell, England	FX	FAX	11,086.0
GFA25	Bracknell, England	FX	FAX	14,582.5
GFE21	Bracknell, England	FX	FAX	4782.0
GFE22	Bracknell, England	FX	FAX	9203.0

Call	Location	Service	Mode	kHz
GFE23	Bracknell, England	FX	FAX	14,436.0
GFE24	Bracknell, England	FX	FAX	18,261.0
GFH	Hong Kong, Hong Kong	FX	CW	11,994.0
GFO	Lossiemouth (Milltown), Scotland	FA	USB	11,193.5, 11,223.0
GFT	Bracknell, England	FX	CW	13,382.0
GFT	Bracknell, England	FX	CW/RTTY	5793.0
GFT27	Bracknell, England	FX	CW	7788.0
GFT29	Bracknell, England	FX	CW	13,365.0
GFT42	Bracknell, England	FX	CW	12,232.0
GIN34	London (Rugby), England	FX	ISB	14,520.0
GIW34	London (Rugby), England	FX	1SB	14,646.0
GKA2	Portishead Radio, England	FC	CW	4286.0
GKA3	Portishead Radio, England	FC	CW/TOR	6368.9
GKA4	Portishead Radio, England	FC	CW	8545.9
GKA5	Portishead Radio, England	FC	CW	12,822.0
GKA6	Portishead Radio, England	FC	CW	17,098.4
GKA7	Portishead Radio, England	FC	CW	22,467.0
GKB2	Portishead Radio, England	FC	CW	4274.0
GKB3	Portishead Radio, England	FC	CW	6379.5, 6382.5
GKB4	Portishead Radio, England	FC	CW	8557.9
GKB5	Portishead Radio, England	FC	CW	12,835.5
GKB6	Portishead Radio, England	FC	CW	17,113.0
GKB7	Portishead Radio, England	FC	CW	22,448.7
GKC2	Portishead Radio, England	FC	CW	4251.6
GKC3	Portishead Radio, England	FC	CW	6407.5
GKC4	Portishead Radio, England	FC	CW	8516.0
GKC5	Portishead Radio, England	FC	CW	13,019.9
GKC6	Portishead Radio, England	FC	CW/TOR	16,954.4
GKC7	Portishead Radio, England	FC	CW	22,407.3
GKD2	Portishead Radio, England	FC	CW	4256.0
GKD4	Portishead Radio, England	FC	CW	8569.0
GKD5	Portishead Radio, England	FC	CW	12,788.5
GKD6	Portishead Radio, England	FC	CW	16,974.6
GKD7	Portishead Radio, England	FC	CW	22,432.2
GKE2	Portsihead Radio, England	FC	CW/TOR	4350.5
GKE4	Portishead Radio, England	FC	CW/TOR	8705.5
GKE5	Portishead Radio, England	FC	CW	13,072.0
GKE6	Portishead Radio, England	FC	CW/TOR	17,198.0
GKE7	Portishead Radio, England	FC	CW/TOR	22,562.0
GKF2	Portishead Radio, England	FC	CW	12,815.0
GKG2	Portishead Radio, England	FC	CW/TOR	4267.0
GKG3	Portishead Radio, England	FC	CW	6469.3
GKG4	Portishead Radio, England	FC	CW	8591.5
GKG5	Portishead Radio, England	FC	CW	12,790.0
GKG6	Portishead Radio, England	FC	CW	17,072.0
GKG7	Portishead Radio, England	FC	CW	22,503.0
GKH3	Portishead Radio, England	FC	CW	6470.8
GKH4	Portishead Radio, England	FC	CW	8604.0
GKH5	Portishead Radio, England	FC	CW	12,791.5
GKH6	Portishead Radio, England	FC	CW	17,092.0
GKH7	Portishead Radio, England	FC	CW	22,525.5
GKI2	Portishead Radio, England	FC	CW	4317.5
GKI3	Portishead Radio, England	FC	CW	6472.3
GKI4	Portishead Radio, England	FC	CW	8606.0
GKI6	Portishead Radio, England	FC	CW	17,151.2
GKI7	Portishead Radio, England	FC	CW	22,528.5
GKJ2	Portishead Radio, England	FC	CW	4326.5
GKJ3	Portishead Radio, England	FC	CW	6477.5
GKJ4	Portishead Radio, England	FC	CW	8684.0
GKJ5	Portishead Radio, England	FC	CW	12.871.5
GKJ6	Portishead Radio, England	FC	CW	16,918.8
GKJ7	Portishead Radio, England	FC	CW	22,545.0
GKK2	Portishead Radio, England	FC	CW	4336.0
GKK2	Portishead Radio, England	FC	CW	17,167.5
GKK3	Portishead Radio, England	FC	CW	6342.0

Call	Location	Service	Mode	kHz
GKK4	Portishead Radio, England	FC	CW	8552.0
GKK5	Portishead Radio, England	FC	CW	13,006.5
GKK7	Portishead Radio, England	FC	CW	22,494.0
GKM2	Portishead Radio, England	FC	CW	4316.0
GKM3	Portishead Radio, England	FC	CW	6397.0
GKM4	Portishead Radio, England	FC	CW	8581.6
GKM5	Portishead Radio, England	FC	CW	12,714.0
GKM6	Portishead Radio, England	FC	CW	17,136.8
GKM7	Portishead Radio, England	FC	CW	22,527.0
GKN3	Portishead Radio, England	FC	CW	6379.9
GKN3	Portishead Radio, England	FC	FAX	6396.0
GKN4	Portishead Radio, England	FC	CW/FAX	8580.5
GKN5	Portishead Radio, England	FC	CW/FAX	12,712.9
GKN6	Portishead Radio, England	FC	CW	17,135.7
GKN7	Portishead Radio, England	FC	FAX	22,526.0
GKO3	Portishead Radio, England	FC	CW	6397.9
GKO4	Portishead Radio, England	FC	CW	8582.5
GKO5	Portishead Radio, England	FC	CW	12,714.9
GKO6	Portishead Radio, England	FC	CW	17,137.7
GKO7	Portishead Radio, England	FC	CW	22,527.9
GKP5	Portishead Radio, England	FC	CW	13,085.0
GKP6	Portishead Radio, England	FC	CW/TOR	17,215.2
GKP7	Portishead Radio, England	FC	CW/TOR	22,578.0
GKQ6	Portishead Radio, England	FC	CW/FAX	17,231.0
GKR	Wick Radio, England	FC	CW	6333.5
GKR3	Wick Radio, England	FC	CW	12,709.0
GKR4	Wick Radio, England	FC	CW	4228.0
GKR8	Wick Radio, England	FC	CW	8468.8
GKS2	Portsihead Radio, England	FC	CW	4344.5
GKS3	Portishead Radio, England	FC	CW	6402.0
GKS4	Portishead Radio, England	FC	CW/TOR	8496.5
GKS6	Portishead Radio, England	FC	CW/TOR	16,882.5
GKS7	Portishead Radio, England	FC	CW	22,387.0
GKT62	Portishead Radio, England	FC	USB	17,236.0
GKT76	Portishead Radio, England	FC	USB	22,611.5
GKU46	Portishead Radio, England	FC	USB	8765.4
GKU49	Portishead Radio, England	FC	USB	8774.7
GKU7Ø	Portishead Radio, England	FC	USB	22,654.9
GKU72	Portishead Radio, England	FC	USB	22,630.1
GKV54	Portishead Radio, England	FC	USB	13,172.1
GKW62	Portishead Radio, England	FC	USB	17,329.0
GLC7	Whitehall (London) Naval R., England	FC	CW	22,454.2
GLF4Ø	London (Rugby), England	FX	ISB	20,675.0
GLJ45	London (Ongar), England	FX	USB	25,470.0
GLJ368	London (Ongar), England	FX	LSB	16,447.0
GLK43	London (Rugby), England	FX	LSB	23,143.7
GLP	Whitehall (London) Naval R., England	FC	CW	4300.2, 6413.7
GLQ3Ø	London (Rugby), England	FX	CW	10,929.0
GMO38	London (Ongar), England	FX	USB	18,536.0
GMP	London (W. Wickham), England	FX	CW	9200.0, 10,390.0
GTK26	Portishead Radio, England	FC	USB	4372.9
GTK51	Portishead Radio, England	FC	USB	13,100.8
GTK52	Portishead Radio, England	FC	USB	13,103.9
GXH	Thurso Naval Radio, Scotland	FC	CW	6487.0, 7505.5, 8459.0, 12, 691.0
GYA	Whitehall (London) Naval R., England	FC	CW	4244.2, 4246.2, 4307.0, 4321.0, 6407.5, 6413.7, 6433.0, 6440.8, 6490.7, 8492.6, 8497.7, 8499.0, 8534.0, 8600.0, 8613.2, 8673.3, 12,740.2, 12,807.7, 12,921.0, 16,937.2, 17,030.5
		FX	CW	18,032.0
GYA3	Whitehall (London) Naval R., England	FC	CW	6362.0
GYC	Whitehall (London) Naval R., England	FC	CW	8613.0

Call	Location	Service	Mode	kHz
GYC3	Whitehall (London) Naval R., England	FC	CW	6414.1
GYC6	Whitehall (London) Naval R., England	FC	CW	17,030.8
GYC7	Whitehall (London) Naval R., England	FC	CW	22,454.2
GYJ3	Whitehall (London) Naval R., England	FC	FAX	6492.3
GYU	Gibraltar Naval Radio, Gibraltar	FC	CW	4221.2, 6434.7, 8646.0, 12,824.2, 12,988.5, 16,987.2, 22,335.2
		FX	CW	14,690.0, 18,180.0, 20,540.0
		FX	ISB	7771.5, 15,760.0
GYU3	Gibraltar Naval Radio, Gibraltar	FC	CW	6371.3
GYU4	Gibraltar Naval Radio, Gibraltar	FC	CW	8625.2
GZO4	Hong Kong Naval Radio, Hong Kong	FC	CW	8554.0
GZO5	Hong Kong Naval Radio, Hong Kong	FC	CW	12,849.0
GZU	Portsmouth (Petersfield), England	FX	CW	11,625.0
		FX	USB	5014.0, 13,345.0, 16,095.0
GZZ	Whitehall (Northwood) Naval R., Eng.	FC	FAX	6435.5, 12,844.5, 22,384.0
GZZ2	Whitehall (Northwood) Naval R., Eng.	FC	FAX	4247.8
GZZ4Ø	Whitehall (Northwood) Naval R., Eng.	FC	FAX	8494.8
GZZ44	Whitehall (Northwood) Naval R., Eng.	FC	FAX	12,741.8, 16,938.0
G23B	Sengwarden Naval Radio, GFR	FC	CW	8648.0 [DHJ59]
		FX	CW	16,265.1 [DHJ59]
HAR	Budapest Naval Radio, Hungary	FC	CW	4239.0, 6375.0, 8451.0, 8458.0, 12,770.0, 12,962.9, 17,129.5, 22,484.0
HAR3	Budapest Naval Radio, Hungary	FC	CW	8466.0, 8616.0
HBC88	Geneva, Switzerland	FX	CW	6998.5
		FX	ISB	13,997.0, 20,812.0, 20,939.0, 20,998.0, 27,998.0
HBD41	Pretoria, RSA	FX	USB	18,285.0, 22,960.0
HBE35	Berne, Switzerland	FX	ISB	14,500.0
HBE53	Berne, Switzerland	FX	ISB	13,593.0
HBO19	Geneva, Swizterland	FX	ISB/RTTY	19,645.8
HBO2Ø	Geneva, Switzerland	FX	ISB	10,144.9
HBO28	Geneva, Switzerland	FX	ISB	18,197.7
HBO42	Geneva, Switzerland	FX	ISB	12,065.7
HBO51	Geneva, Switzerland	FX	ISB	11,465.8
HBO73	Geneva, Switzerland	FX	ISB	13,817.0
HBO75	Geneva, Switzerland	FX	ISB	15,827.4
HBP26	Geneva, Switzerland	FX	ISB	16,053.2
HBX58	Geneva, Switzerland	FX	ISB	8063.5
HCG	Guayaquil Radio, Ecuador	FC	CW	12,711.0, 16,948.0
HCG9	Guayaquil Radio, Ecuador	FC	CW	8476.0
HDN	Quito Naval Radio, Ecuador	FX	CW	17,590.0
HD21ØA	Guayaquil, Ecuador	SS	A2	5000.0, 7600.0
HEB	Berne Radio, Switzerland	FC	CW	4259.0, 8601.5, 13,023.7, 16,863.3, 22,344.5
HEB13	Berne Radio, Switzerland	FC	USB	13,103.9
HEB14	Berne Radio, Switzerland	FC	USB	4379.1
HEB17	Berne Radio, Switzerland	FC	USB	17,263.9
HEB18	Berne Radio, Switzerland	FC	USB	8740.0
HEB23	Berne Radio, Switzerland	FC	USB	13,181.4
HEB24	Berne Radio, Switzerland	FC	USB	4428.7
HEB27	Berne Radio, Switzerland	FC	USB	17,276.3
HEB28	Berne Radio, Switzerland	FC	USB	8790.2
HEB33	Berne Radio, Switzerland	FC	USB	13,190.7
HEB37	Berne Radio, Switzerland	FC	USB	17,325.9
HEB38	Berne Radio, Switzerland	FC	USB	8811.9
HEB52	Berne Radio, Switzerland	FC	USB	22,636.3
HEB62	Berne Radio, Switzerland	FC	USB	22,692.1
HEC13	Berne Radio, Switzerland	FC	CW	13,080.0
HEC17	Berne Radio, Switzerland	FC	CW/TOR	17,205.0
HEC18	Berne Radio, Switzerland	FC	CW/TOR	8709.0
HEC52	Berne Radio, Switzerland	FC	CW/TOR	22,565.5
HEE4Ø	Berne Aeradio, Switzerland	FA	USB	8930.0
HEE61	Berne Aeradio, Switzerland	FA	USB	13,324.0

Call	Location	Service	Mode	kHz
HEE92	Berne Aeradio, Switzerland	FA	USB	21,988.0
HEK6	Berne (Schwarzenburg), Switzerland	FX	ISB	19,565.0
HE022	Berne (Schwarzenburg), Switzerland	FX	ISB	23,420.0
HEP25	Zurich (Waltikon), Switertland	FX	CW	12,224.0
HEP26	Zurich (Waltikon), Switzerland	FX	CW/RTTY	6905.0
HEP39	Zurich (Waltikon), Switzerland	FX	CW	10,390.0
HEP46	Zurich (Waltikon), Switzerland	FX	CW	4632.5
HEP52	Zurich (Waltikon), Switzerland	FX	CW	5208.0
HEP58	Zurick (Waltikon), Switzerland	FX	CW	11,538.0
HEP74	Zurich (Waltikon), Switzerland	FX	CW	7401.0
HEP81	Zurich (Waltikon), Switzerland	FX	CW	18,190.0
HEP83	Zurich (Waltikon), Switzerland	FX	CW	18,380.0
HEP85	Zurich (Waltikon), Switzerland	FX	CW/RTTY	8045.0
HEP88	Zurich (Waltikon), Switzerland	FX	CW	8038.0
HEP92	Zurich (Waltikon), Switzerland	FX	CW	9200.0
HEP96	Zurich (Waltikon), Switzerland	FX	CW	10,295.0
HGX21	Budapest, Hungary	FX	CW	20,001.0, 20,802.0
HGX31	Warsaw, Poland	FX	CW/RTTY	10,400.0
HGX39	New Delhi, India	FX	CW	7937.5
HIA	Domingo Piloto Radio, Dominican Rep.	FC	CW	13,087.2
HIA2	Domingo Piloto Radio, Dominican Rep.	FC	CW	8642.0
HIA66	unknown	FX	CW	18,720.2
HIP2	Puerto Plata R. Dominican Republic	FC	CW	8642.0
HIP491	Santo Domingo, Dominican Republic	FX	USB	10,935.0
HJM273	Bogota, Colombia	FX	USB	6778.5
HKB	Barranquilla Radio, Colombia	FC	CW	4277.1, 6463.4, 8666.0, 8700.4, 12,781.5, 13,087.2, 17,004.0
HKC	Buenaventura Radio, Colombia	FC	CW	4324.8, 12,653.5, 12,853.3, 17,045.6
HKC2	Buenaventura Radio, Columbia	FC	CW	6386.3
HKC3	Buenaventura Radio, Colombia	FC	CW	8574.1
HLA94	Seoul, South Korea	FX	ISB	19,900.0
HLB28	Seoul, South Korea	FX	ISB	13,535.0
HLD89	Seoul, South Korea	FX	CW	18,484.0
HLF	Seoul Radio, South Korea	FC	CW	4273.0, 6344.0, 8484.2, 12,916.5, 17,079.0, 22,395.1
HLG	Seoul Radio, South Korea	FC	CW	4308.0, 6451.0, 8473.0, 12,935.0, 17,118.0, 22,482.0
HLJ	Seoul Radio, South Korea	FC	CW	8497.0, 12,727.0, 12,902.0, 16,910.0
HLL	Seoul, South Korea	FX	CW/AM	5810.0, 11,620.0
HLO	Seoul Radio, South Korea	FC	CW	8577.0, 12,843.0, 16,990.0
HLP2	Pusan Radio, South Korea	FC	CW	4235.0
HLQ62	Seoul, South Korea	FX	ISB	10,425.1
HLW	Seoul Radio, South Korea	FC	CW	8635.8, 13.005.5, 17,129.7
HLW2	Seoul Radio, South Korea	FC	CW	12,923.1
HLW3	Seoul Radio, South Korea	FC	CW	12,712.0
HLX	Seoul Radio, South Korea	FC	CW	13,025.0
HMA22	Seoul, South Korea	FX	CW/RTTY	9285.0, 14,607.5
HME28	Pyongyang, North Korea	FX	CW	13,780.0
HMH56	Pyongyang, North Korea	FX	CW	9480.0
HMK21	Pyongyang, North Korea	FX	CW	9375.0
HMN51	Pyongyang, North Korea	FX	CW	11,430.0
HMN53	Pyongyang, North Korea	FX	CW	11,570.0
HMR23	Pyongyang, North Korea	FX	CW	12,175.0
HMS19	Pyongyang, North Korea	FX	CW/FSK	13,580.0
HMU94	unknown, possibly South Korea	FX	ISB	11,565.0
HPN6Ø	Canal (Puerto Armuelles), R., Panama	FC	CW	4240.0, 6467.0, 8608.0, 12,873.5, 17,128.5
HRW	La Ceiba Radio, Honduras	FC	CW	6350.0
HSA2	Bangkok Radio, Thailand	FC	CW	8686.0
HSA3	Bangkok Radio, Thailand	FC	CW	12,800.0
HSA4	Bangkok Radio, Thailand	FC	CW	8573.5
HSA/HSJ	Bangkok, Thailand	FC	CW	7955.0

Call	Location	Service	Mode	kHz
HSD85	Bangkok, Thailand	AX	ISB	8006.2
HSD93	Bangkok Aeradio, Thailand	AX	ISB	9278.0
HSP27	Bangkok (Laki), Thailand	FX	ISB	15,630.0
HSQ	Bangkok, Thailand	FX	CW/RTTY	9285.0
HSW61	Bangkok, Thailand	FX	FAX	17,520.0
HSW64	Bangkok, Thailand	FX	FAX	7395.0
HSW69	Bangkok, Thailand	FX	FAX	6765.0
HSX	Songkhla Naval Radio, Thailand	FC	CW	6450.0, 6469.0
HSY	Sattahip Naval Radio, Thailand	FC	CW	6450.0, 6469.0
HSY62	Sattahip Naval Radio, Thailand	FC	CW	8466.0
HSZ	Bangkok Naval Radio, Thailand	FC	CW	6450.0, 6469.0
HVC	Vatican City, Vatican	FX	USB	7553.5
HWN	Paris (Houilles) Naval R., France	FC	CW	4232.0, 4338.0, 6348.0, 8453.0, 8477.0, 16,909.0, 17,180.0, 17,190.0, 22,389.0
		FX	CW	13,235.5
HXZ24	Paris (St. Assise), France	FX	CW	14,505.0
HZN	Jeddah, Saudi Arabia	FX	USB	17,539.5
HZQ566	Riyadh, Saudi Arabia	FX	USB	15,665.0
HZW	Khafji Radio, Saudi Arabia	FC	CW	4305.0
HZY	Ra's Tannurah Radio, Saudi Arabia	FC	CW	8480.0, 12,811.3, 16,960.0
HZY473	Dhahran, Saudi Arabia	FX	USB	4753.0
H4H7	Honiara Radio, Solomon Islands	FC	USB	8808.8
H4H29	Honiara Radio, Solomon Islands	FC	CW	12,700.0
IAM	Rome, Italy	SS	CW	5000.0
IAR	Rome Radio 4, Italy	FC	USB	4391.5
IAR3	Rome P.T. Radio, Italy	FC	CW	13,015.5
IAR4	Rome P.T. Radio, Italy	FC	CW	4320.0
IAR6	Rome P.T. Radio, Italy	FC	CW	6409.5
IAR7	Rome P.T. Radio, Italy	FC	CW	16,895.3
IAR8	Rome P.T. Radio, Italy	FC	CW	8669.9
IAR23	Rome P.T. Radio, Italy	FC	CW	13,011.0
IAR24	Rome P.T. Radio, Italy	FC	CW	4292.0
IAR26	Rome P.T. Radio, Italy	FC	CW	6418.2
IAR27	Rome P.T. Radio, Italy	FC	CW	17,160.8
IAR28	Rome P.T. Radio, Italy	FC	CW	8530.0
IAR32	Rome P.T. Radio, Italy	FC	CW	22,378.0
IAR33	Rome P.T. Radio, Italy	FC	CW	12,996.0
IAR37	Rome P.T. Radio, Italy	FC	CW	17,005.0
IAR38	Rome P.T. Radio, Italy	FC	CW	8656.0
IAR62	Rome P.T. Radio, Italy	FC	CW	22,376.0
IBF	Turin, Italy	SS	A2	5000.0
ICB	Genoa P.T. Radio, Italy	FC	CW	4235.0, 6425.0, 8649.5, 12,978.0, 16,879.0, 16,881.0, 17,182.0
ICB	Genoa Radio 2, Italy	FC	USB	22,642.5
ICB	Genoa Radio 7, Italy	FC	USB	17,254.6, 17,273.2
ICB	Genoa Radio 8, Italy	FC	USB	8787.1
ICG	Milan, Italy	FX	USB	14,378.6, 14,388.6
ICT	Taranto Radio, Italy	FC	CW	6435.0
IDQ	Rome Naval Radio, Italy	FC	CW	4280.0
		FX	CW	17,620.0
IDQ3	Rome Naval Radio, Italy	FC	CW	6390.3
IDQ4	Rome Naval Radio, Italy	FC	CW	8486.0
IDQ6	Rome Naval Radio, Italy	FC	CW	12,890.0
IDQ15	Rome Naval Radio, Italy	FC	CW	25,318.0
IDQ26	Rome Naval Radio, Italy	FC	CW	12,760.0
IDR	Rome Naval Radio, Italy	FX	CW	13,657.0
IDR2	Rome Naval Radio, Italy	FX	CW	4615.4, 4762.0
IDR3	Rome Naval Radio, Italy	FC	CW	6435.5, 7302.1
IDR4	Rome Naval Radio, Italy	FX	CW	9090.0
IDR5	Rome Naval Radio, Italy	FX	CW	16,004.0
IDR7	Rome Naval Radio, Italy	FX	CW	25,134.2
IDR2Ø	Rome Naval Radio, Italy	FX	CW	9065.0
IMB31	Rome, Italy	FX	FAX	4777.5

Call	Location	Service	Mode	kHz
IMB54	Rome, Italy	AX	FAX	8146.6
IMB56	Rome, Italy	AX	FAX	13,600.0
IQH	Naples Radio, Italy	FC	CW	6376.0
IQX	Trieste P.T. Radio, Italy	FC	CW	6418.0, 8502.0, 8679.0, 12,975.0, 17,083.7
IRE25	Rome, Italy	FX	ISB	7518.0
IRF4Ø	Rome (Torrenova), Italy	FX	ISB	8037.0
IRH28	Rome (Torrenova), Italy	FX	ISB	10,846.0
IRH31	Rome (Torrenova), Italy	FX	ISB	10,135.0
IRH37	Rome (Torrenova), Italy	FX	ISB	10,715.0
IRK43	Rome (Torrenova), Italy	FX	ISB	13,394.0
IRL27	Rome (Torrenova), Italy	FX	USB	14,736.0
IRM	CIRM, Rome Radio, Italy	FC	CW	22,525.1
IRM2	CIRM, Rome Radio, Italy	FC	CW	4342.5, 4350.5
IRM4	CIRM, Rome Radio, Italy	FC	CW	6365.0
IRM5	CIRM, Rome Radio, Italy	FC	CW	6420.0
IRM6	CIRM, Rome Radio, Italy	FC	CW	8685.0
IRM6	CIRM, Rome Radio, Italy	FC	CW	17,105.0
IRM7	CIRM, Rome Radio, Italy	FC	CW	12,760.0
IRM8	CIRM, Rome Radio, Italy	FC	CW	12,748.0
IRQ26	Rome (Torrenova), Italy	FX	ISB	18,630.0
IRQ41	Rome (Torrenova), Italy	FX	ISB	18,109.0
IRS24	Rome (Torrenova), Italy	FX	ISB/RTTY	20,430.0
IRS41	Rome (Torrenova), Italy	FX	USB	20,185.6
IRV25	Rome (Torrenova), Italy	FX	ISB	23,510.0
ISX57	Rome (Santa Rosa), Italy	FX	FAX	15,724.0
IU481	Rome, Italy	FX	CW	10,390.0
JAE27	Tokyo, Japan	FX	FAX	7327.5
JAL21	Tokyo, Japan	FX	FAX	11,012.5
JAM55	Tokyo, Japan	FX	FAX	15,945.2
JBA44	Tokyo, Japan	FX	ISB	4860.0
JBE3Ø	Tokyo, Japan	FX	ISB	10,128.0
JBE39	Tokyo, Japan	FX	ISB	9112.5
JBE59	Tokyo, Japan	FX	ISB	9295.0
JBE67	Tokyo, Japan	FX	ISB	7630.0
JBE7Ø	Tokyo (Oyamo), Japan	FX	CW/ISB	10,477.5
JBF4Ø	Tokyo, Japan	FX	ISB	10,396.0
JBF5Ø	Tokyo, Japan	FX	ISB	10,857.5
JBF6Ø	Tokyo, Japan	FX	ISB	10,640.0
JBK3	Tokyo (Kemigawa), Japan	FX	FAX	5767.5
JBL54	Tokyo (Oyama), Japan	FX	USB	14,730.0
JBL55	Tokyo, Japan	FX	ISB	15,835.5
JBO	Tokyo Radio, Japan	FC	USB	4376.0, 4431.8, 4434.9, 8746.8, 8753.0, 8777.8, 13,119.4, 13,134.9, 13,153.5, 17,242.2, 17,257.7, 17,329.0, 22,676.6, 22,704.5, 22,716.9
JBQ36	Tokyo, Japan	FX	CW/USB	16,228.5
JBQ67	Tokyo, Japan	FX	ISB	17,695.0
JBT3Ø	Tokyo, Japan	FX	ISB	20,837.0
JBU39	Tokyo, Japan	FX	ISB	19,745.0
JBU69	Tokyo, Japan	FX	ISB	19,880.0
JBU89	Tokyo, Japan	FX	ISB	19,610.0
JCE82	Naha Radio, Okinawa, Japan	FC	CW	12,759.0
JCG	Yokohama Radio, Japan	FC	CW	8534.0
JCK	Kobe Radio, Japan	FC	CW	4250.0, 4332.5
JCS	Choshi Radio, Japan	FC	CW	4349.0, 6467.0, 8653.6, 12,826.5, 17,112.6, 22,419.0
JCT	Choshi Radio, Japan	FC	CW	4225.0, 8686.0, 12,687.0, 17,166.4, 22,386.0
JCU	Choshi Radio, Japan	FC	CW	6485.0, 8479.0, 12,878.0, 17,043.2, 22,463.0
JCX	Naha Radio, Okinawa, Japan	FC	CW	6470.0, 12,667.5
JDB	Nagasaki Radio, Japan	FC	CW	13,063.0, 16,877.5
JDC	Choshi Radio, Japan	FC	CW	8647.5, 13,054.0, 16,998.5

Call	Location	Service	Mode	kHz
JEB2Ø	Choshi, Japan	FX	ISB	11,150.0
JEB21	Chichi Jima, Bonin Is., Japan	FX	ISB	11,674.5
JFA	Chuo Gyogyo (Matsudo) Radio, Japan	FC	CW	4304.0, 6342.0, 6388.0, 8547.5, 8628.0, 12,788.0, 12,847.5, 17,081.6, 17,086.0, 22,524.0
		FC	FAX	4274.2
JFC	Misaki Gyogyo Radio, Japan	FC	CW	16,893.5, 22,348.6
JFE	Naha Gyogyo Radio, Okinawa, Japan	FC	CW	12,880.0, 17,180.0
JFF	Yaizu Gyogyo Radio, Japan	FC	CW	12,858.0
JFH	Hamajima Gyogyo Radio, Japan	FC	CW	4278.5, 12,811.5, 16,976.8
JFQ	Makuraki Gyogyo Radio, Japan	FC	CW	12,704.5
JFS	Hachinohe Gyogyo Radio, Japan	FC	CW	12,858.0
JFU	Ishinomaki Gyogyo Radio, Japan	FC	CW	12,858.0
JFW	Iwaki Gyogyo Radio, Japan	FC	CW	12,858.0
JGC	Yokohama Radio, Japan	FC	CW	8698.0, 12,660.0, 13,015.5, 17,141.6
JHC	Choshi Radio, Japan	FC	CW	6332.0
JJC	Tokyo Radio, Japan	FC	FAX	4316.0, 8467.5, 12,747.0, 16,971.0, 17,069.6, 22,541.0
		FX	FAX	22,777.0
JJC2Ø	Tokyo, Japan	FX	CW	6805.0
		FX	CW/FAX	8017.4, 9273.0
JJD	Tokyo, Japan	FX	CW	10,415.0, 12,000.0
JJD2	Tokyo, Japan	FX	CW	15,950.0
JJF	Tokyo Naval Radio, Japan	FC	CW	6474.0, 12,948.0, 17,125.0
JJG3	Yokosuka Radio, Japan	FC	CW	6488.0
JJG8	Atsugi Ichihar, Japan	FA	CW/USB	5695.5
JJH	Kure Radio, Japan	FC	CW	8662.0
JJI	Sasebo Naval Radio, Japan	FC	CW	8490.0
JJJ	Maizuru Radio, Japan	FC	CW	6481.0
JJK	Ominato Radio, Japan	FC	CW	6390.0
JJY	Tokyo (Sanwa, Ibaraki), Japan	SS	AM/CW	5000.0, 8000.0, 10,000.0, 15,000.0
JKA2	Tokyo (Kemigawa), Japan	FX	FAX	6820.0
JKA3	Tokyo (Kemigawa), Japan	FX	FAX	9260.0
JKB4	Tokyo (Kemigawa), Japan	FX	FAX	9135.0
JKC	Tokyo (Kemigawa), Japan	FX	FAX	5457.0
JKC2	Tokyo (Kemigawa), Japan	FX	FAX	9855.0
JKC3	Tokyo (Kemigawa), Japan	FX	FAX	12,322.0
JKD2	Tokyo, Japan	FX	FAX	6893.0
JKD3	Tokyo, Japan	FX	FAX	9885.0
JKE2	Tokyo, Japan	FX	FAX	5851.0
JKE3	Tokyo, Japan	FX	FAX	8088.0
JKE5	Tokyo (Usui), Japan	FX	FAX	7370.0
JKE6	Tokyo (Usui), Japan	FX	FAX	9410.5
JMA	Narita Aeradio, Japan	FA	A3H	5519.0, 13,344.0
JMB2	Tokyo, Japan	FX	CW	7515.0
JMB3	Tokyo, Japan	FX	CW	14,605.0
JMB4Ø	Tokyo (Fusa), Japan	FX	FAX	4902.0
JMC2	Tokyo Radio, Japan	FC	CW	4298.0
JMC3	Tokyo Radio, Japan	FC	CW	6397.0
JMC4	Tokyo Radio, Japan	FC	CW	8526.0
JMC5	Tokyo Radio, Japan	FC	CW	12,840.0
JMC6	Tokyo Radio, Japan	FC	CW	17,029.0
JMH2	Tokyo, Japan	FX	FAX	7305.0
JMH3	Tokyo, Japan	FX	FAX	9970.0
JMH4	Tokyo, Japan	FX	FAX	13,597.0
JMH5	Tokyo, Japan	FX	FAX	18,220.0
JMH6	Tokyo, Japan	FX	FAX	22,770.0
JMJ2	Tokyo (Usui), Japan	FX	FAX	5405.0
JMJ3	Tokyo (Usui), Japan	FX	FAX	9438.0
JMJ4	Tokyo (Usui), Japan	FX	FAX	14,692.5
JMJ5	Tokyo (Usui), Japan	FX	FAX	18,130.0
JNA	Tokyo Naval Radio, Japan	FC	CW	4276.0, 8492.0, 12,942.0,

Call	Location	Service	Mode	kHz
				17,052.5, 22,398.0
JNE	Hiroshima Radio, Japan	FC	CW	12,660.0
JOR	Nagasaki Radio, Japan	FC	CW	6457.5, 8523.4, 13,008.0, 17,093.6, 22,409.0
JOS	Nagasaki Radio, Japan	FC	CW	4328.0, 6491.5, 8437.0, 13,069.5, 17,093.6, 22,396.0
JOU	Nagasaki Radio, Japan	FC	CW	8463.0, 12,673.6, 16,883.0, 22,440.0
JPA22	Tokyo, Japan	FX	CW/RTTY	8006.5
JPA23	Tokyo, Japan	FX	CW	14,607.5
JPA24	Tokyo, Japan	FX	CW	18,087.0
JPA33	Tokyo, Japan	FX	CW/RTTY	9285.0
JPA34	Tokyo, Japan	FX	CW	14,623.5
JPA35	Tokyo, Japan	FX	CW	14,707.0
JPA55	Nagoya (Komaki), Japan	FX	CW/RTTY	7532.0
JPA56	Nagoya (Komaki), Japan	FX	CW	10,390.0
JPA58	Nagoya (Komaki), Japan	FX	CW	13,820.0
JPA59	Nagoya (Komaki), Japan	FX	CW	19,130.0
JPA6Ø	Nagoya (Komaki), Japan	FX	CW	9200.0
JPA61	Nagoya (Komaki), Japan	FX	CW	13,520.0
JPA62	Nagoya (Komaki), Japan	FX	CW/RTTY	15,684.0
JWT	Stavanger Naval Radio, Norway	FC	CW	8686.0
JWT	Stavanger Naval Radio, Norway	FX	CW	4562.5, 11,007.0
JXL	BØ, Norway	FX	USB	4857.5, 7512.5, 7717,5, 10,337.5, 13,423.0
JXP	Jan Mayen Island	FX	USB	4857.5, 7512.5, 7717.5, 10,337.5, 13,423.0
JXU	BØ NavalRadio, Norway	FX	CW	5226., 5811.0
JYO	Aqaba Radio, Jordan	FC	CW	4236.0, 4326.5, 6390.0 6479.0
J2A4	Djibouti Radio, Djibouti	FC	CW	4262.0
J2A6	Djibouti Radio, Djibouti	FC	CW	6348.0
J2A8	Djibouti Radio, Djibouti	FC	CW	8510.0, 8682.0
J2A9	Djibouti Radio, Djibouti	FC	CW	12,728.0
J3R	unknown	FX	CW	8925.0
J5G	Bissau, Guinea-Bissau	AX	CW	9111.0
KAB8	Anchorage Aeradio, AK	FA	A3H	4668.0
KAC	Woods Hole Radio, MA	FC	USB	6509.1
KAD26Ø	New Orleans, LA	FX	CW/RTTY	9434.0
KAD68Ø	Buffalo, NY	FX	CW/RTTY	9434.0
KAE41	Albuquerque, NM	FX	LSB	5923.0
KAF	Atlantic Highlands Radio, NJ	FC	USB	6509.5
KAG69	Denver, CO	FX	USB	11,448.0
KAG78	Kansas City, MO	FX	USB	11,448.0
KAG98	Omaha, NE	FX	USB	11,448.0
KAI	San Juan Radio, PR	FC	USB	6509.5
KAK7ØØ	Detroit, MI	FX	CW/RTTY	9434.0
KAW63	Laysan Island, HI	FC	USB	6509.5
KBR	Beaumont Radio, NC	FC	USB	6509.5
KCC97	Diamond Head, HI	FX	USB	20,480.0
KCD72	Agana, Guam	FX	USB	20,480.0
KCI95	Cold Bay Radio, AK	FC	USB	4125.0
KCI98	King Salmon Radio, AK	FC	AM	4125.0
KCW21	Cambridge, MA	FX	USB/RTTY	20,610.0
KDM47	Fort Worth, TX	FX	USB	8125.0
KDM48	Pago Pago, American Samoa	FX	USB	20,480.0
KDM5Ø	Atlanta (Hampton), GA	FX	USB	6870.0, 7475.0, 8125.0, 13,630.0, 16,348.0
KDM95	Gulfport, MS	FX	USB	7475.0
KDR3	unknown	FX	CW	10,639.0
KEM8Ø	Washington, DC	FX	USB	5860.0, 13,630.0
KFK92	Kitts Peak Obs. (Tucson), AZ	FX	USB	10.190.5, 15,527.0, 20,878.5
KFS	San Francisco Radio, CA	FC	CW	4228.0, 4274.0, 6348.0, 6365.5, 8444.5, 8558.4, 12,844.5, 17,026.0, 17,184.8,

Call	Location	Service	Mode	kHz
				22,425.0, 22,515.0
KGA32	Honolulu, HI	FX	USB	20,480.0
KGA57	Agana Aeradio, Guam	AX	USB	11,402.5
KGD55	Redding, CA	FX	LSB	5923.0
KGD58	Annette Radio, AK	FC	AM	4125.0
KGD69	Salt Lake City, UT	FX	LSB	5923.0
KGD91	Yakutat Radio, AK	FC	USB	4125.0
KGE33	Washington, DC	FX	USB	10,913.4
KGN	Delcambre Radio, LA	FC	USB	4366.7
KHB24	Billings, MT	FX	LSB	5923.0
KHW	Pascagoula Radio, MS	FC	USB	6509.5
KIS70	Anchorage Aeradio, AK	FA	A3H	5519.0, 8903.0, 13,344.0
KJS	Kings Point Radio, NY	FC	USB	6509.5
KJY74	"Miami Monitor", FL	FA	CW/USB	4668.0, 9020.0, 11,396.0
KKN44	Monrovia, Liberia	FX	CW	4886.0, 5110.0, 7633.9, 7652.0, 7830.0, 11,474.0, 11,520.0, 11,635.0, 15,917.0, 17,426.0, 18,043.4, 20,353.0, 20,929.0
KKN5Ø	Washington, DC	FX	CW	4880.0, 6925.2, 7470.0, 10,637.0, 11,095.0, 12,022.5, 12,111.5, 14,880.5, 15,492.0, 15,540.0, 17,605.0, 18,525.0, 18,700.0, 18,972.0, 20,365.0, 23,975.0
KKN51	Washington, DC	FX	CW/RTTY	23,862.5
KKN52	Washington, DC	FX	CW	23,982.5
KKU4Ø	Kansas City, MO	FX	USB	7475.0
KLB	Seattle Radio, WA	FC	CW	4349.0, 6411.0, 8546.0, 8582.0, 8658.0, 12,907.5, 12,916.5, 17,007.2, 22,539.0
KLC	Galveston Radio, TX	FC	CW	4256.0, 6369.0, 8626.8, 8666.0, 13,038.0, 16,871.4, 22,467.0
KME57	Portland, OR	FX	LSB	5923.0
KMI	San Francisco (Dixon) Radio, CA	FC	USB	4357.4, 4403.9, 4407.0, 8728.2, 8740.0, 8743.7, 13,100.8. 13,187.6, 17,236.0, 17,239.1, 17,279.4
KMV	Agana Radio, Guam	FC	USB	8796.4
KNY2Ø	Washington, DC	FX	CW	15,804.0, 19,458.0
KNY21	Washington, DC	FX	CW/RTTY	11,303.5, 13,377.5, 14,875.0, 15,704.0, 18,430.0
KNY23	Washington, DC	FX	CW	7719.0, 15,704.0, 15,804.0, 18,430.0, 19,458.0, 19,465.0
		FX	CW/RTTY	13,377.5, 14,649.0
KNY24	Washington, DC	FX	USB	19,013.0
KNY25	Washington, DC	FX	CW/RTTY	9040.5, 11,090.0, 16,065.0, 19,950.0
KNY26	Washington, DC	FX	CW/RTTY	9041.5, 10,642.5, 13,379.0, 16,392.0, 20,011.0
KNY27	Washington, DC	FX	CW/RTTY	13,605.0, 18,250.0, 22,960.0
KNY28	Washington, DC	FX	CW	10,100.5, 18,306.5
KNY29	Washington, DC	FX	CW	11,106.0, 18,808.0
KNY34	Washington, DC	FX	CW/RTTY	14,353.5
KNY37	Washington, DC	FX	CW	11,448.0
KOG55	Las Vegas, NV	FX	USB	20,350.0
KOK	Los Angeles Radio, CA	FC	CW	4283.0, 4283.0, 8591.0, 12,933.0, 12,933.0, 17,064.9, 22,413.0
KPA64	Battle Creek, MI	FX	USB	10,493.0
KPA65	Denton, TX	FX	USB	12,216.0
KPA66	Denver, CO	FX	USB	10,493.0
KPH	San Francisco Radio, CA	FC	CW	4247.0, 6477.5, 6489.0,

Call	Location	Service	Mode	kHz
				8618.0, 8642.0, 12,808.5, 13,002.0, 17,016.5, 17,088.8, 22,479.0, 22,557.0
KPH		FC	CW/TOR	4356.1, 6500.5
KPI29	unknown	FX	USB	5167.5
KQM	Kahuku (Honolulu) Radio, HI	FC	USB	4410.1, 8740.0, 13,134.9,
		FC	USB	17,232.9
KRH5Ø	London, England	FX	CW	4626.1, 5425.0, 7724.0, 10,680.0, 11,142.0, 13,815.0, 16,458.0, 20,568.0
KRV	Ponce Radio, PR	FC	USB	6515.7, 8811.9
KSF7Ø	Oakland Aeradio, CA	FA	A3H	5519.0, 8903.0, 13,344.0
KUP61	Medford, OR	FX	LSB	5923.5
KUP65	Dalap, Majuro Atoll	AX	USB	11,402.0
		FX	USB	5205.0
KUP66	Param, Ponape Island	FC	USB	5205.0
KUP67	Truk Aeradio, Caroline Islands	AX	USB	11,401.0
	Moen, Truk Islands	FC	USB	5205.0
KUP68	Korak, Palau Islands	FC	USB	5205.0
KUP69	Nif, Yap Islands	FC	USB	5205.0
KUP71	Kobler, Saipan, Mariana Islands	FC	USB	5205.0
KUP72	Shinaparnu, Rota Is., Mariana Islands	FX	USB	5205.0
KUQ	Pago Pago Radio, American Samoa	FC	CW	6361.0, 8585.0, 12,871.5
KUQ2Ø	Pago Pago, American Samoa	FX	ISB	17,391.5
KUQ2ØA	Pago Pago, American Samoa	FX	LSB	10,708.5
KUQ2ØC	Pago Pago, American Samoa	FX	USB	15,623.5
KUR2Ø	Honolulu, HI	FX	CW/USB	12,138.5, 19,344.5, 20,348.5, 20,602.5, 23,402.5
KUR5Ø	Agana, Guam	FX	CW/USB	12,138.5, 20,348.5, 20,602.5, 23,402.5
KVH	Atlantic Marine Center, Norfolk, VA	FC	USB	6509.5, 13,141.1, 17,267.0
KVJ	Seattle Radio, WA	FC	USB	6509.5
KVK	Miami Radio, FL	FC	USB	4379.1, 6509.5
KVM7Ø	Honolulu (International), HI	AX	FAX	9982.5, 13,627.5, 16,135.0, 23,331.4
		FA	A3H	5519.0, 8903.0, 13,344.0
		FX	FAX	5037.5, 7770.0, 11,090.0
KVR	Detroit Radio, MI	FC	USB	6509.5
KWJ91	Anchorage Radio, AK	FC	USB	4366.7, 8793.3
KWL21	Auke Bay, AK	FC	USB	6509.5
KWL24	Tern Island, HI	FX	USB	5907.5
KWL39	Little Port Walter, AK	FC	USB	6509.5
KWL43	King Salmon, AK	FC	USB	6509.5
KWL47	Bethel, AK	FX	USB	5907.5
KWL9Ø	Tokyo, Japan	FX	CW	5823.0, 6866.5, 7662.0, 9224.0, 10,464.0, 12,210.0, 13,485.1, 14,616.0, 14,782.0, 17,552.0
KWO3	Anchorage Aeradio, AK	FA	A3H	5547.0
		FA	USB	10,057.0
KWS78	Athens Greece	FX	CW	4910.0, 5271.0, 7434.0, 7645.0, 10,255.0, 10,285.0, 14,360.0, 18,459.9, 18,543.9
KWT73	Honolulu Aeradio, HI	FX	USB	11,402.5
KWY43	Kodiak, AK	FX	USB	8067.0
KZU	unknown	FC	CW	8605.0
K13A	Spata Attikis Naval Radio, Greece	FC	CW	8659.5 [SXA4]
LBA2	Stavanger Naval Radio, Norway	FX	CW	5426.8
LBA5	Stavanger Naval Radio, Norway	FX	CW	5426.0
LBA9	Stavanger Naval Radio, Norway	FX	CW	11,454.0
LBJ5	Harstad Naval Radio, Norway	FX	CW	5392.5
LBJ7	Harstad Naval Radio, Norway	CW	FX	7465.0
LCB	Oslo (Jeloey), Norway	FX	CW/ISB	9310.5
LCJ	Oslo (Jeloey), Norway	FX	CW	9980.0
LCK	Oslo (Jeloey), Norway	FX	CW	19,960.1

Call	Location	Service	Mode	kHz
LCMP-2	Key West Naval Radio, FL	FX	CW	5876.0 [NAR]
LCMP-3	Norfolk Naval Radio, VA	FX	CW	5917.0, 8090.0 [NAM]
LCN2	Oslo (Jeloey), Norway	FX	CW/ISB	10,197.4
LCO	Oslo (Jeloey), Norway	FX	CW	13,980.0
LCP	Oslo (Jeloey), Norway	FX	CW/ISB	14,550.0
LDN2	Oslo (Jeloey), Norway	FX	ISB	9166.9
LDR	Oslo (Jeloey), Norway	FX	ISB	12,155.1
LFB	Rogaland Radio, Norway	FC	CW	8678.0
LFB2	Rogaland Radio, Norway	FC	CW	8683.6
LFC	Rogaland Radio, Norway	FC	CW	12,682.5
LFF	Rogaland Radio, Norway	FC	CW	17,165.6
LFG	Rogaland Radio, Norway	FC	CW	22,473.0
LFI	Rogaland Radio, Norway	FC	CW	12,961.5
LFJ	Rogaland Radio, Norway	FC	CW	12,876.0
LFL	Rogaland Radio, Norway	FC	CW	22,396.0
LFL5	Rogaland Radio, Norway	FC	USB	8743.7
LFL6	Rogaland Radio, Norway	FC	USB	8746.8
LFL7	Rogaland Radio, Norway	FC	USB	8749.9
LFL8	Rogaland Radio, Norway	FC	USB	8756.1
LFL31	Rogaland Radio, Norway	FC	USB	13,110.1
LFL34	Rogaland Radio, Norway	FC	USB	13,131.8
LFL37	Rogaland Radio, Norway	FC	USB	13,150.4
LFL4Ø	Rogaland Radio, Norway	FC	USB	13,165.9
LFL42	Rogaland Radio, Norway	FC	USB	13,175.2
LFL44	Rogaland Radio, Norway	FC	USB	13,184.5
LFL45	Rogaland Radio, Norway	FC	USB	13,193.8
LFN	Rogaland Radio, Norway	FC	CW	8529.3
LFN2	Rogaland Radio, Norway	FC	USB	17,232.9
LFN3	Rogaland Radio, Norway	FC	USB	17,239.1
LFN4	Rogaland Radio, Norway	FC	USB	17,242.2
LFN6	Rogaland Radio, Norway	FC	USB	17,251.5
LFN23	Rogaland Radio, Norway	FC	USB	17,288.7
LFN24	Rogaland Radio, Norway	FC	USB	17,291.8
LFN26	Rogaland Radio, Norway	FC	USB	17,313.5
LFN27	Rogaland Radio, Norway	FC	USB	17,319.7
LFN32	Rogaland Radio, Norway	FC	USB	22,617.7
LFN35	Rogaland Radio, Norway	FC	USB	22,639.4
LFN4Ø	Rogaland Radio, Norway	FC	USB	22,695.2
LFN41	Rogaland Radio, Norway	FC	USB	22,698.3
LFN43	Rogaland Radio, Norway	FC	USB	22,707.6
LFN44	Rogaland Radio, Norway	FC	USB	22,713.8
LFN45	Rogaland Radio, Norway	FC	USB	22,716.9
LFN47	Rogaland Radio, Norway	FC	USB	26,141.3
LFN48	Ropaland Radio, Norway	FC	USB	26,147.5
LFP	Rogaland Radio, Norway	FC	CW	9014.0
LFR	Rogaland Radio, Norway	FC	CW	25,308.0
LFT	Rogaland Radio, Norway	FC	CW	16,952.2
LFU	Rogaland Radio, Norway	FC	CW	6467.0
LFW	Rogaland Radio, Norway	FC	CW	4325.0
LFX	Rogaland Radio, Norway	FC	CW	16,928.4
LFZ	Rogaland Radio, Norway	FC	CW	25,382.5
LGB	Rogaland Radio, Norway	FC	CW/TOR	8574.0, 8674.0, 13,097.0, 17,223.0, 22,587.2
LGB2	Rogaland Radio, Norway	FC	CW/TOR	8707.0
LGG	Rogaland Radio, Norway	FC	CW	22,425.5
LGG3	Rogaland Radio, Norway	FC	CW/TOR	22,586.9
LGJ	Rogaland Radio, Norway	FC	CW	12,727.6
LGN4	Rogaland Radio, Norway	FC	USB	4376.0
LGN5	Rogaland Radio, Norway	FC	USB	4382.2
LGU	Rogaland Radio, Norway	FC	CW	6432.0
LGU2	Rogaland Radio, Norway	FC	CW/TOR	6498.0
LGW	Rogaland Radio, Norway	FC	CW	4241.0
LGW2	Rogaland Radio, Norway	FC	CW/TOR	4351.0
LGX	Rogaland Radio, Norway	FC	CW	17,074.4
LGX3	Rogaland Radio, Norway	FC	CW/TOR	17,223.0

Call	Location	Service	Mode	kHz
LHB	Oslo (Jeloey), Norway	FX	CW/RTTY	6915.0
LHK	Oslo (Jeloey), Norway	FX	CW	20,630.0
LJA21	Tel Aviv, Israel	FX	USB	13,386.0, 16,251.0, 18,400.5
LJG4	Rogaland Radio, Norway	FC	CW/TOR	13,097.0
LJN23	Ny Aalesund Naval Radio, Spitzbergen	FX	CW/ISB	9960.0
LJP2Ø	Oslo, Norway	FX	CW	10,390.0
LJP24	Oslo, Norway	FC	CW	4632.5
LJP26	Oslo, Norway	FX	CW	6792.0, 7532.0
LJP29	Oslo, Norway	FX	CW	9200.0
LJP34	Oslo, Norway	FX	CW	14,817.5
LMB	Bergen Radio, Norway	FX	CW	4532.5, 11,165.2
LMB5	Bergen, Norway	FX	CW	5780.0
LMB7	Bergen, Norway	FC	CW	7778.2
LMO	Oslo, Norway	FX	FAX	11,097.0
LMO5	Oslo, Norway	FX	FAX	5945.0
LMO8	Oslo, Norway	FX	FAX	8057.5
LMO34	Olso, Norway	FX	FAX	4642.5
LMT	Tromso Naval Radio, Norway	FX	CW	8132.0
LOK	Orcadas Radio, South Orkney Islands	FC	CW	6464.0
		FC	FAX	4250.0
LOK	Orcadas, South Orkney Islands	FX	CW/FAX	8818.0, 9983., 11,147.0
LOL	Buenos Aires, Argentina	FX	CW	4856.0, 17,590.0, 17,665.0
LOL	Buenos Aires, Argentina	SS	A2	5000.0
LOL	Buenos Aires Radio, Argentina	FX	USB	7398.0, 17,553.5
LOL1	Buenos Aires, Argentina	SS	A2	10,000.0, 15,000.0
LOL3	Buenos Aires Radio, Argentina	FC	CW	17,180.0
LOL3	Buenos Aires, Argentina	FX	CW	8030.0
LPA2Ø	Rio Grande Radio, Argentina	FC	CW	16,912.0
LPC43	Ushuaia Radio, Argentina	FC	CW	4225.0
LPD32	Gen. Pacheco Radio, Argentina	FC	CW	13,002.0
LPD41	Gen. Pacheco Radio, Argentina	FC	CW	6411.0
LPD44	Gen. Pacheco Radio, Argentina	FC	CW	6404.0
LPD46	Gen. Pacheco Radio, Argentina	FC	CW	17,045.6
LPD52	Gen. Pacheco Radio, Argentina	FC	CW	8526.0
LPD62	Gen. Pacheco Radio, Argentina	FC	CW	4262.0
LPD68	Gen. Pacheco Radio, Argentina	FC	CW	4268.0
LPD74	Gen. Pacheco Radio, Argentina	FC	CW	4274.0
LPD76	Gen. Pacheco Radio, Argentina	FC	CW	12,763.5
LPD85	Gen. Pacheco Radio, Argentina	FC	CW	8514.0
LPD86	Gen. Pacheco Radio, Argentina	FC	CW	8646.0
LPD88	Gen. Pacheco Radio, Argentina	FC	CW	12,988.5
LPD91	Gen. Pacheco Radio, Argentina	FC	CW	22,419.0
LPL	Gen. Pacheco Radio, Argentina	FC	USB	13,162.8
LPL3	Gen. Pacheco Radio, Argentina	FC	USB	8759.2
LPL4	Gen. Pacheco Radio, Argentina	FC	USB	13,159.7
LPL3Ø	Gen. Pacheco Radio, Argentina	FC	USB	6512.6
LPW63	Bahia Blanca Radio, Argentina	FC	CW	6358.5
LQB9	Buenos Aires (San Martin), Argentina	FX	CW	8167.5
LQC2Ø	Buenos Aires (Mt. Grande), Argentina	FX	CW	17,550.0
LRB39	Buenos Aires (G. Pacheco), Argentina	FX	CW	10,895.0
LRB72	Buenos Aires (G. Pacheco), Argentina	FX	FAX	10,720.0
LRB91	Buenos Aires (G. Pacheco), Argentina	FX	LSB	9115.0
LRO	Buenos Aires (G. Pacheco), Argentina	FX	FAX	5185.0
LRO84	Buenos Aires (G. Pacheco), Argentina	FX	FAX	18,093.0
LSA2	Boca Radio, Argentina	FC	CW	4334.0
LSA3	Boca Radio, Argentina	FC	CW	6460.0
LSA4	Boca Radio, Argentina	FC	CW	8457.0
LSA5	Boca Radio, Argentina	FC	CW	12,709.0
LSA5	Boca Radio, Argetnina	FC	CW	25,160.0
LSA6	Boca Radio, Argentina	FC	CW	17,160.8
LSA7	Boca Radio, Argentina	FC	CW	22,346.5
LSE	Buenos Aires, Argentina	FX	LSB	14,370.0
LSM397	Buenos Aires, Argentina	FX	USB	7770.0
LSM4ØØ	Buenos Aires, Argentina	FX	USB	16,410.0
LSO3	Buenos Aires Radio, Argentina	FS	CW	8528.0, 16,925.4

Call	Location	Service	Mode	kHz
LSO5	Buenos Aires Radio, Argentina	FS	CW	4304.0, 12,728.0
LST9	Buenos Aires Radio, Argentina	FC	CW	12,707.0
LSU2Ø	Bahai Blanca Radio, Argentina	FC	CW	12,706.5
LTC31	Buenos Aires, Argentina	FX	USB	7651.0
LTU265	La Quiara, Argentina	FX	USB	5755.0
LTY99	Buenos Aires, Argentina	FX	CW	9990.0
LTZ	Buenos Aires, Argentina	FX	USB	10,280.0
LWB	Buenos Aires (Moron), Argentina	AX	CW	9785.0
LXF5Ø	Luxembourg	FX	CW	10,390.0
LZJ2	Sofia, Bulgaria	FX	FAX	5093.0
LZL2	Bourgas Radio, Bulgaria	FC	CW	4262.0
LZL4	Bourgas Radio, Bulgaria	FC	CW	8564.0
LZL5	Bourgas Radio, Bulgaria	FC	CW	12,689.0, 12,709.0
LZL6	Bourgas Radio, Bulgaria	FC	CW	16,928.5
LZL7	Bourgas Radio, Bulgaria	FC	CW	22,554.0
LZL63	Bourgas Radio, Bulgaria	FC	CW	17,182.0
LZS24	Sofia Naval Radio, Bulgaria	FC	CW	6402.5
LZS32	Sofia Naval Radio, Bulgaria	FC	CW	12,870.0
LZS35	Sofia Naval Radio, Bulgaria	FC	CW	16,928.0
LZS36	Sofia Naval Radio, Bulgaria	FC	CW	17,118.0
LZS39	Sofia Naval Radio, Bulgaria	FC	CW	22,554.0
LZW	Varna Radio, Bulgaria	FC	CW	12,880.0
LZW2	Varna Radio, Bulgaria	FC	CW	4220.0, 4250.0
LZW3	Varna Radio, Bulgaria	FC	CW	6336.1, 6405.0, 6448.0
LZW5	Varna Radio, Bulgaria	FC	CW	12,659.0, 12,840.0, 12,940.0, 13,000.0
LZW6	Varna Radio, Bulgaria	FC	CW	16,918.0, 17,055.0
LZW7	Varna Radio, Bulgaria	FC	CW	22,344.0
LZW42	Varna Radio, Bulgaria	FC	CW	8460.0, 8532.0
LZW43	Varna Radio, Bulgaria	FC	CW	8607.0
LZW63	Varna Radio, Bulgaria	FC	CW	17,145.0
LZW72	Varna Radio, Bulgaria	FC	CW	22,374.0
MJV	Falmouth (Culdrose) AB, England	FA	USB	11,243.0
MKA	London (Stanbridge), England	FX	USB	10,210.0
MKD	Akrotiri, Cyprus	FX	ISB	14,670.0, 16,280.0, 18,535.0, 18,885.0, 19,430.5, 19,895.0, 20,290.0, 20,600.0, 20,981.2
MKE	Akrotiri, Cyprus	FX	ISB	7483.5, 10,580.0, 15,755.0, 20,320.0
MKG	London (Stanbridge), England	FX	ISB	6898.0, 6995.0, 10,678.0, 11,430.0, 11,550.0, 13,695.0, 13,829.5, 14,969.9, 15,815.0, 18,120.0, 18,750.0, 18,904.0, 20,320.0, 21,820.7
MKL	Pitreavie Castle, Scotland	FA	CW/USB	5688.0
MKL	Edinburgh, Scotland	FA	USB	6697.0
MKT	London (Stanbridge), England	FX	USB	5113.5, 9223.0, 9236.2, 12,125.0, 18,923.5, 19,528.5, 22,783.5, 24,168.5
MLQ21	Croughton (Barford), England	FA	USB	9017.0
MLU	Gibraltar, Gibraltar	FX	CW	13,382.0
MPU	Luca, Malta	FX	CW	13,382.0
MQD	Plymouth, England	FA	CW	6714.0
MQP	West Drayton (Upavon) AB, England	FA	USB	6690.0
MQY	Pitreavie Castle, Scotland	FA	CW	6714.0, 9025.0
MRX	Buttersworth AB, Malaysia	FA	USB	8975.0, 13,205.0, 18,025.0
MSF	Teddington, England	SS	A2	5000.0, 10,000.0
MTI	Plymouth Naval Radio, England	FC	CW	4305.8, 4310.0
MTO	Rosyth Naval Radio, England	FX	CW	5400.0
MUA2	London (Stanbridge), England	FX	ISB	10,200.3
MVU	West Drayton (Upavon) AB, England	FA	CW/USB	4722.0, 11,200.0, 11,203.3
NAA	Cutler Naval Radio, ME	FX	CW/RTTY	12,135.0
NAF	Newport Naval Radio, RI	FX	USB	6697.0
NAM	Norfolk Naval Radio, VA	FC	CW	13,189.0, 20,225.0
		FX	CW	4839.5, 8090.0, 10,779.0,

Call	Location	Service	Mode	kHz
		FX	CW	14,862.0, 16,180.0, 16,265.0
		FX	FAX	4975.0, 8080.0, 10,861.2,
		FX	FAX	16,410.0, 20,015.0
		FX	USB	4505.0, 5071.0, 6697.0, 6833.6, 12,058.5
NAR	Key West Naval Radio, FL	FC	CW	25,590.0
		FX	CW/RTTY	5870.0, 12,135.0
NAR	Key West NAS, FL	FA	USB	6968.5
NAT	unknown	FX	CW	11,402.6
NAU	San Juan Naval Radio, PR	FC	CW/RTTY	17,012.0
NAW	Guantanamo, Cuba	FC	USB	6968.5, 8188.5, 11,132.0
NAW11	Guantanamo, Cuba	FX	CW/LSB	6885.0
NAW12	Guantanamo, Cuba	FX	LSB	7756.5
NAW18	Guantanamo, Cuba	FX	ISB	18,447.5
NAW25	Guantanamo, Cuba	FX	ISB	6957.5
NAW26	Guantanamo, Cuba	FX	ISB	7430.0
NAW27	Guantanamo, Cuba	FX	ISB	7745.0
NAW31	Guantanamo, Cuba	FX	ISB	13,750.0
NAW32	Guantanamo, Cuba	FX	USB	13,759.0
NAW34	Guantanamo, Cuba	FX	USB	11,130.0
NAW36	Guantanamo, Cuba	FX	USB	13,765.2
NAW38	Guantanamo, Cuba	FX	LSB	15,820.0
NAX	Barber's Point Radio, HI	FC	ISB	5783.5, 6693.0, 6968.5
NBL	New London Naval Radio, CT	FX	CW	9235.0
NBW	Guantanamo Bay Naval Radio, Cuba	FX	USB	13,529.0
NBY	Bird Station, Antarctica	FA	USB	8997.0
NCI	Sellia Marina, Italy	FX	USB	5066.0, etc.
NCI3	Lampedusa, Italy	FX	USB	5066.0, etc.
NCI4	Kargabarum, Turkey	FX	USB	5066.0, etc.
NCI10	Rhodes	FX	USB	5066.0, etc.
NDT	Yokosuka Naval Radio, Japan	FX	CW	7428.0
NDT4	Yokosuka Naval Radio, Japan	FX	CW	12,205.0
NDT6	Totsuka Naval Radio, Japan	FC	CW	7428.0, 10,250.0, 16,218.5
NEL	Lakehurst NAS, NJ	FX	USB	6833.6
NFC	Cape May NAS, NJ	FX	USB	6833.6
NGD	McMurdo Station, Antarctica	FA	USB	8997.0
		FX	CW/RTTY	8090.0
		FX	CW/USB	11,552.0, 12,220.5, 13,678.5, 14,766.0
NGR	Kato Souli Naval Radio, Greece	FC	CW	4623.0, 8578.0, 13,372.5, 17,156.0
		FX	FAX	5206.0, 8100.0, 12,903.0
NGZ	Alameda NAS, NJ	FX	USB	6833.6
NHK	Patuxent Naval Radio, MD	FX	USB	6697.0, 6833.6
NHZ	Brunswick NAS, ME	FX	USB	6833.6
NIK	COMMSTA Boston, MA	FC	FAX	8500.1, 12,748.1
		FX	CW	5320.0, 8502.0, 12,750.0
NKT	Cherry Point NAS, NC	FX	CW	7863.0
NKW	Diego Tower, Diego Garcia	FA	USB	11,176.0, 11,234,0
NMA	COMMSTA Miami, FL	FC	USB	4134.6, etc.
		FX	USB	4048.5, 7528.6, etc.
NMA7	Jupiter CG Radio, FL	FX	USB	4048.5, etc.
NMC	COMMSTA San Francisco, CA	FC	CW	8574.0, 8682.0, 12,730.0, 12,743.0, 16,880.9, 16,909.7, 17,151.2
		FC	FAX	4344.1, 8680.1, 12,730.0, 17,151.2
		FC	USB	4125.0, etc.
		FX	USB	7528.6, etc.
NMF	COMMSTA Boston, MA	FC	USB	4134.6, etc.
		FX	USB	7528.6, etc.
NMF32	Nantucket CG Radio, MA	FX	USB	4048.5, etc.
NMF33	Caribou CG Radio, ME	FX	USB	4048.5, etc.
NMG	COMMSTA New Orleans, LA	FC	USB	4125.0, etc.
		FX	USB	7528.6, etc.

Call	Location	Service	Mode	kHz
NMJ1	COMMSTA Juneau, AK	FA	USB	4585.0
NMJ22	Attu CG LORAN Station, AK	FA	USB	6617.0
NMN	COMMSTA Portsmouth, VA	FC	CW	8465.0, 12,718.5, 16,976.0
		FC	USB	4134.6, etc.
		FX	USB	7528.6, etc.
NMN73	Carolina Beach [Cape Fear] CG, NC	FX	USB	4048.5, etc.
NMO	COMMSTA Honolulu, HI	FC	CW	8650.0, 12,866.0, 12,889.5, 22,472.0, 22,476.0
		FC	USB	4125.0, etc.
		FX	CW	9050.0, 13,655.0, 16,457.5,
		FX	USB	7528.6, etc.
NMQ	COMMSTA Long Beach, CA	FX	USB	7528.6, etc.
NMR	COMMSTA San Juan, PR	FC	CW	8471.0, 16,983.2
		FX	USB	7528.6, etc.
NMS	Shetland Islands CG, Scotland	FX	USB	4857.5
NNNØNRS	Diego Garcia	FX	USB	13,974.0, 20,623.0
NOJ	COMMSTA Kodiak, AK	FC	CW	8628.5
		FC	FAX	4296.0, 8457.0
		FC	USB	4125.0, etc.
		FX	USB	7528.6, etc.
NOL	"Bermuda Monitor", Bermuda	FX	USB	4048.5, etc.
NOX	COMMSTA Adak, AK	FC	USB	4428.7, 6506.4
NOZ	COMMSTA Elizabeth City, NC	FX	USB	7528.6, etc.
NPG	San Francisco Naval Radio, CA	FC	CW	12,966.0
		FX	CW/RTTY	6420.0, 9277.5, 12,135.0, 14,470.0
NPM	Lualualei Naval Radio, HI	FC	CW	4313.0
NPM	Lualualei Naval Radio, HI	FX	CW	9338.0, 12,310.0, 15,524.0
NPM	Lualualei Naval Radio, HI	FX	CW/FAX	13,862.5
NPM	Lualualei Naval Radio, HI	FX	FAX	4802.5, 9440.0, 16,400.0 21,785.0
NPN	Barrigada (Agana NAS), Guam	FA	CW	numerous
		FC	FAX	4975.0, 7645.0, 11,522.5
		FX	CW	17,530.0, 21,760.0
		FX	FAX	10,255.0, 13,807.5, 18,620.0, 23,880.0
NPO	San Miguel (Capas), Philippines	FC	CW	4....., 12,200.0
		FX	CW	10,440.5, 12,804.0
		FX	FAX	10,966.0, 22,865.0
NPX	South Pole Station, Antarctica	FA	USB	8997.0, 13,251.0
		FX	USB	11,552.0
NQM	Sand Island, Midway Islands	FC	CW/USB	5783.5, 6878.5
NQU	Siple Station, Antarctica	FA	USB	8997.0
		FX	CW/USB	11,552.0
NQX	Key West Naval Radio, FL	FX	USB	13,529.0
NRK	Keflavik Naval Radio, Iceland	FC	CW	17,127.2, 25,589.5
		FX	CW	5167.0, 8090.0, 20,225.0
NRO	Johnston Island CG Radio, HI	FX	USB	4050.0, etc.
NRO5	Upolu Point CG Radio, HI	FX	USB	4050.0, etc.
NRO7	Kure [Ocean Is.] Is. CG Radio, HI	FX	USB	4050.0, etc.
NRT	"Yokota Monitor", Japan	FX	USB	4550.1, etc.
NRT2	Gesashi CG Radio, Japan	FX	USB	4550.1, etc.
NRT3	Iwo Jima CG Radio, Japan	FX	USB	4550.1, etc.
NRT9	Hokkaido CG Radio, Japan	FX	USB	4550.1, etc.
NRV	COMMSTA Barrigada, Guam	FC	CW	8570.0, 8582.0, 12,743.0, 12,865.0, 13,380.0, 17,146.0
		FC	CW/RTTY	22,527.0
		FC	USB	13,113.2
		FX	CW	8150.0, 10,440.5, 21,760.0
		FX	USB	4550.1, etc.
NRV6	Marcus Island CG Radio,	FX	USB	4550.1, etc.
NRV7	Yap Island CG Radio	FX	USB	4550.1, etc.
NRW2	St. Paul CG Radio, AK	FX	USB	4531.0, etc.
NRW3	Pt. Clarence CG Radio AK	FX	USB	4531.0, etc.
NSS	Washington Naval Radio, DC	FX	CW/RTTY	12,135.0

Call	Location	Service	Mode	kHz
NWC	Northwest Cape Naval R., Australia	FX	CW	20,776.0
NZCM	Williams Field, New Zealand	FA	USB	8997.0
NZJ	El Toro NAS, CA	FX	USB	4513.0
N13A	Goeree Naval Radio, Netherlands	FC	CW	8514.0 [PBC8]
OAV86	Lima, Peru	FX	CW	15,738.2
OBC	Callao Naval Radio, Peru	FX	CW	17,590.0
OBC3	Callao Radio, Peru	FC	CW	4237.0, 4325.0, 6360.0 8650.0, 12,307.0
OBQ5	Iquitos Radio, Peru	FC	CW	4325.0, 6474.0
OBT	Talara, Peru	AX	CW	10,097.0
OCK25	Arequipa, Peru	FX	LSB	17,690.0, 20,605.0, 20,610.0
OCP	Ouagadougou, Upper Volta	FX	LSB	6918.5
ODF98	Beirut, Lebanon	FX	ISB	7984.0
ODI65	Beirut, Lebanon	FX	ISB	10,650.4
ODL79	Beirut, Lebanon	FX	ISB	13,799.0
ODN22	Damascus, Syria	FX	CW/USB	16,440.8
ODO44	Beirut, Lebanon	FX	CW	16,438.4
ODR	Beirut Radio, Lebanon	FC	CW	16,884.5
ODR3	Beirut Radio, Lebanon	FC	CW	8702.0
ODR4	Beirut Radio, Lebanon	FC	CW	12,682.5, 13,101.0
ODR9	Beirut Radio, Lebanon	FC	CW	4221.0, 4351.0
ODR88	Beirut, Lebanon	FX	CW	20,753.0
ODS91	Beirut, Lebanon	FX	LSB	20,913.6
ODS97	Beirut, Lebanon	FX	ISB	20,972.1
ODW22	Beirut, Lebanon	FX	CW	10,390.0
OEC	Vienna, Austria	FX	CW	16,164.0, 17,556.4
		FX	LSB	10,425.5, 14,483.4, 16,162.5,
		FX	LSB	20,480.0, 20,753.5, 20,495.4
		FX	USB	19,728.5
OEC35	New Delhi, India	FX	LSB	20,480.0, 20,495.4
OEC36	Beijing, PRC	FX	USB	16,131.5, 20,972.0
OEC44	Tel Aviv, Israel	FX	CW/LSB	7880.0, 10,296.6, 13,617.0, 14,478.5, 20,753.5
OEC57	Pretoria, RSA	FX	CW/LSB	10.422.5, 13,617.0, 17,556.4, 20,753.5
OEC61	Rome, Italy	FX	CW/LSB	7894.0, 10,425.5
OEC64	Lagos, Nigeria	FX	CW/LSB	14,482.0, 20,751.4
OEC72	Lisbon, Portugal	FX	CW/LSB	10,422.5, 14,483.4
OEQ35	Vienna, Austria	FX	CW/RTTY	4837.5
OEQ36	Vienna, Austria	FX	CW/RTTY	5208.0
OEQ37	Vienna, Austria	FX	CW/RTTY	5305.5
OEQ38	Vienna, Austria	FX	CW/RTTY	6905.0
OEQ39	Vienna, Austria	FX	CW/RTTY	7401.0
OEQ41	Vienna, Austria	FX	CW/RTTY	8038.0, 8045.0
OEQ43	Vienna, Austria	FX	CW/RTTY	8097.5
OEQ44	Vienna, Austria	FX	CW	10,295.0
OEQ45	Vienna, Austria	FX	CW/RTTY	11,538.0
OEQ46	Vienna, Austria	FX	CW/RTTY	12,224.0
OFB28	Helsinki, Finland	FX	FAX	8018.0
OFB35	Helsinki, Finland	FX	CW	5362.0
OFJ	Helsinki Radio, Finland	FC	CW	17,415.0
OFJ3	Helsinki Radio, Finland	FC	CW	12,669.1
OFJ4	Helsinki Radio, Finland	FC	CW	12,687.0
OFJ5	Helsinki Radio, Finland	FC	CW	16,923.8
OFJ6	Helsinki Radio, Finland	FC	CW	6355.0
OFJ8	Helsinki Radio, Finland	FC	CW	4272.0, 8437.0
OFJ9	Helsinki Radio, Finland	FC	CW	22,396.7
OFJ25	Helsinki Radio, Finland	FC	CW	25,064.9
OFJ82	Helsinki Radio, Finland	FC	CW	8457.0
OGX	Helsinki, Finland	FX	CW	7532.0, 10,390.0, 13,520.0 19,130.0, 19,360.0
		FX	CW/RTTY	9200.0
OHG2	Helsinki Radio, Finland	FC	USB	8805.7, 13,147.3, 13,181.4, 17,263.9, 22,605.3
OHU2Ø	Helsinki, Finland	FX	USB	10,736.0, 10,746.5, 10,885.0,

Call	Location	Service	Mode	kHz
				13,433.0, 13,913.0, 14,661.0, 14,693.0, 14,925.0, 14,954.0, 15,726.0, 15,870.0, 17,418.0, 17,443.0, 17,561.0, 18,761.0, 20,124.0, 20,176.0, 20,262.0
OHU21	Paris, France	FX	USB	10,885.0, 14,693.0, 17,418.0 20,124.0
OHU22	Bucharest, Rumanis	FX	USB	10,746.5
OHU23	Tel Aviv, Israel	FX	USB	10,736.0, 10,746.5, 14,661.0 14,693.0, 17,561.0
OHU25	Beijing, PRC	FX	USB	13,433.0, 13,913.0, 14,693.0, 14,661.0, 14,693.0, 17,443.0, 18,761.0, 18,841.0, 20,176.0, 20,262.0
OHU26	Belgrade, Yugoslavia	FX	USB	10,736.0, 14,661.0, 14,693.0
OHU28	Warsaw, Poland	FX	USB	14,661.0, 14,693.0
OHU36	Baghdad, Iraq	FX	USB	20,176.0, 20,262.0
OKL	Prague Aeradio, Czechoslovakia	FA/MA	AM/CW	5575.0, 11,391.0
OLG4	Prague (Podebrad), Czechoslovakia	FX	CW	10,307.5
OMC	Bratislava Radio, Czechoslovakia	FC	CW	4344.5, 6393.5, 6440.0, 8476.0, 8627.5, 12,859.0, 13,036.0, 13,057.0, 16,897.5
OMC2	Bratislava, Czechoslovakia	FX	CW	8020.0, 9846.0
OMK	Komarno Radio, Czechoslovakia	FC	CW	8627.5, 12,818.0, 13,057.0, 16,897.5
OMP	Prague Radio, Czechoslovakia	FC	CW	6440.0, 25,052.0, 25,143.0
OMP2	Prague Radio, Czechoslovakia	FC	CW	4292.5, 4295.5
OMP4	Prague Radio, Czechoslovakia	FC	CW	8627.5
OMP5	Prague Radio, Czechoslovakia	FC	CW	13,007.0
OMP6	Prague Radio, Czechoslovakia	FC	CW	16,977.0
OMP7	Prague Radio, Czechoslovakia	FC	CW	22,486.0
OMP51	Prague Radio, Czechoslovakia	FC	CW	12,818.0
OMZ	Prague, Czechoslovakia	FX	CW	5204.0, 5250.0, 6843.0 6843.0, 6871.0, 7624.0 11,498.0, 16,100.0, 16,180.0, 16,394.9, 18,300.0, 18,400.0
		FX	CW/RTTY	4900.0, 6840.0, 7960.0, 20,619.0, 21,810.0, 22,820.0
OMZ26	Prague, Czechoslovakia	FX	USB	16,298.5
OMZ28	Prague, Czechoslovakai	FX	CW/RTTY	4840.0, 5868.0, 7647.5, 9078.0, 9784.0
OMZ29	Bratislava, Czechoslovakia	FX	CW/RTTY	8008.0, 9784.0
OMZ3Ø	Prague, Czechoslovakia	FX	USB	14,387.0, 18,297.5
ONA2Ø	Brussels, Belgium	FX	CW/RTTY	7401.0, 8038.0, 8045.0, 8097.5, 9200.0, 10,295.0, 10,390.0, 11,538.0, 12,224.5
ONN27	Brussels, Belgium	FX	USB	8176.5, 15,996.0, 21,804.5
ONN29	Brussels, Belgium	FX	CW/RTTY	12,257.0
ONN3Ø	Brussels, Belgium	FX	CW	7812.5, 11,000.0
		FX	CW/RTTY	5830.0, 5868.0, 6803.5, 6987.0, 7537.0, 7647.5, 9910.0
ONN32	Brussels, Belgium	FX	CW	15,657.0
ONN34	Brussels, Blegium	FX	USB	12,238.5, 16,108.5
ONN36	Brussels, Belgium	FX	CW	9942.0, 17,458.0, 22,848.0
ONN38	Brussels, Belgium	FX	CW	7650.0, 9165.0, 13,695.0
ONY27	Rouveroy, Belgium	FX	CW	7693.0, 7879.0
ONY52	unknown	FX	CW	6999.5
OPBE	Belgrade, Yugoslavia	FA	USB	13,248.0
ORI28	Brussels (Ruiselede), Belgium	FX	USB	18,467.0
ORI48	Brussels (Ruiselede), Belgium	FX	ISB	18,795.0
ORI49	Brussels (Ruiselede), Belgium	FX	USB	19,577.0
OSN	Oostende Naval Radio, Belgium	FC	CW	8702.0, 12,725.0, 22,554.0
OSN6	Oostende Naval Radio, Belgium	FC	CW/RTTY	6493.0
OSN12	Oostende Naval Radio, Belgium	FC	CW	12,985.0

Call	Location	Service	Mode	kHz
OSN16	Oostende Naval Radio, Belgium	FC	CW	16,930.0
OSN42	Oostende Naval Radio, Belgium	FC	CW	12,715.0
OSN44	Oostende Naval Radio, Belgium	FC	CW	4312.0
OSN46	Oostende Naval Radio, Belgium	FC	CW	6338.0
OSN48	Oostende Naval Radio, Belgium	FC	CW	8460.0
OST2	Oostende Radio, Belgium	FC	CW	4298.0
OST3	Oostende Radio, Belgium	FC	CW	6411.0
OST4	Oostende Radio, Belgium	FC	CW	8478.0
OST5	Oostende Radio, Belgium	FC	CW	12,781.5
OST6	Oostende Radio, Belgium	FC	CW	17,017.1
OST7	Oostende Radio, Belgium	FC	CW	22,533.0
OST8	Oostende Radio, Belgium	FC	CW	25,135.0
OST22	Oostende Radio, Belgium	FC	CW	4290.0
OST32	Oostende Radio, Belgium	FC	CW	6328.0, 6496.8
OST42	Oostende Radio, Belgium	FC	CW	8652.0
OST52	Oostende Radio, Belgium	FC	CW	13,067.0
OST62	Oostende Radio, Belgium	FC	CW	17,187.0
OST72	Oostende Radio, Belgium	FC	CW	22,351.5
OST82	Oostende Radio, Belgium	FC	CW	27,017.0
OSU41	Oostende Radio, Belgium	FC	USB	8762.3
OSU45	Oostende Radio, Belgium	FC	USB	8756.1
OSU51	Oostende Radio, Belgium	FC	USB	13,119.4
OSU53	Oostende Radio, Belgium	FC	USB	13,138.0
OSU57	Oostende Radio, Belgium	FC	USB	13,156.6
OSU63	Oostende Radio, Belgium	FC	USB	17,270.1
OUN	Ejde, Denmark	FX	USB	4857.5, etc.
OVC	Groennedal, Greenland	FX	CW	5010.0, 10,765.0, 12,329.0
		FX	CW/USB	8148.0, 11,434.6
OVG	Frederikshaven Naval Radio, Denmark	FX	CW	4020.0, 6393.0, 18,323.0
OVG8	Frederikshaven Naval Radio, Denmark	FX	CW	8148.0
OVG12	Frederikshaven Naval Radio, Denmark	FX	CW	12,329.0
OVG16	Frederikshaven Naval Radio, Denmark	FX	CW	16,323.0
OVY	Angissoq, Greenland	FX	USB	4857.5, etc.
OWS	Copenhagen, Denmark	FX	CW	24,110.0
OWS3	Copenhagen, Denmark	FX	CW	13,520.0, 13,820.0
OWS4	Copenhagen, Denmark	FX	CW	10,390.0
OXI	Godthaab, Greenland	FX	CW	9360.0, 13,855.0
OXT	Copenhagen (Skamleback), Denmark	AX	FAX	5850.0
		FX	FAX	13,855.0, 17,510.0
OXZ	Lyngby Radio, Denamrk	FC	USB	4431.0
OXZ2	Lyngby Radio, Denamrk	FC	CW	4303.0
OXZ3	Lyngby Radio, Denmark	FC	CW	6446.8
OXZ4	Lyngby Radio, Denmark	FC	CW	8560.4, 8598.0, 8635.3
OXZ6	Lyngby Radio, Denmark	FC	CW	12,916.5
OXZ8	Lyngby Radio, Denmark	FC	CW	17,068.5
OXZ9	Lyngby Radio, Denmark	FC	CW	22,419.0
OXZ21	Lyngby Radio, Denmark	FC	CW	4319.0
OXZ31	Lyngby Radio, Denmark	FC	CW	6439.0
OXZ41	Lyngby Radio, Denmark	FC	CW	8626.0
OXZ61	Lyngby Radio, Denmark	FC	CW	13,038.0
OXZ62	Lyngby Radio, Denmark	FC	CW	12,753.5
OXZ81	Lyngby Radio, Denmark	FC	CW	16,920.0
OXZ82	Lyngby Radio, Denmark	FC	CW/TOR	16,897.4
OXZ92	Lyngby Radio, Denmark	FC	CW	22,404.0
OXZ93	Lyngby Radio, Denmark	FC	CW	22,459.2
OXZ95	Lyngby Radio, Denmark	FC	CW	25,262.0
OZL4Ø	Angmagssalik, Greenland	FX	CW	7570.0
OZU25	Copenhagen, Denmark	FX	USB	13,490.0, 16,400.0, 18,587.0, 26,130.0
OZU3Ø	Cairo, Egypt	FX	USB	13,490,0, 18,587.0, 26,130.0
OZU33	Tel Aviv, Israel	FX	USB	13,490.0, 18,587.0, 26,130.0
OZU34	Jeddah, Saudi Arabia	FX	USB	13,490.0, 18,587.0, 26,130.0
OZU35	Beijing, PRC	FX	USB	13,490.0, 16,400.0, 18,587.0, 26,130.0
OZU38	Lagos, Nigeria	FX	USB	13,490.0, 18,587.0

Call	Location	Service	Mode	kHz
OZU39	Nairobi, Kenya	FX	USB	13,490.0, 18,587.0
PBC	Goeree Naval Radio, Netherlands	FC	CW	8460.0
PBC3	Goeree Naval Radio, Netherlands	FC	CW	4280.1, 8439.0, 12,840.5, 22,330.5
PBC5	Goeree Naval Radio, Netherlands	FX	CW	16,030.0, 19,970.0
PBC8	Goeree Naval Radio, Netherlands	FC	CW	8514.0
PBC9	Goeree Naval Radio, Netherlands	FC	CW	12,690.0
PBC92	Goeree Naval Radio, Netherlands	FC	CW/RTTY	6895.0
PBC94	Goeree Naval Radio, Netherlands	FC	CW	4229.3
PBC217	Goeree Naval Radio, Netherlands	FC	CW	16,909.7
PBC317	Goeree Naval Radio, Netherlands	FC	CW	17,028.0, 17,117.6
PBC911	Goeree Naval Radio, Netherlands	FC	CW/RTTY	11,135.0
PBV4	Valkenburg AB, Netherlands	FA	USB	8970.0
PBW	Amsterdam (Schiphol), Netherlands	FA	A3H/USB	8952.0
PCG21	Scheveningen Radio, Netherlands	FC	USB	4369.8
PCG22	Scheveningen Radio, Netherlands	FC	USB	4413.2
PCG23	Scheveningen Radio, Netherlands	FC	USB	4385.3
PCG31	Scheveningen Radio, Netherlands	FC	USB	6509.5
PCG41	Scheveningen Radio, Netherlands	FC	USB	8796.4
PCG42	Scheveningen Radio, Netherlands	FC	USB	8731.3
PCG43	Scheveningen Radio, Netherlands	FC	USB	8734.4
PCG51	Scheveningen Radio, Netherlands	FC	USB	13,138.0
PCG52	Scheveningen Radio, Netherlands	FC	USB	13,119.4
PCG53	Scheveningen Radio, Netherlands	FC	USB	13,156.6
PCG61	Scheveningen Radio, Netherlands	FC	USB	17,341.4
PCG62	Scheveningen Radio, Netherlands	FC	USB	17,350.7
PCG63	Scheveningen Radio, Netherlands	FC	USB	17,301.1
PCG71	Scheveningen Radio, Netherlands	FC	USB	22,608.4
PCG72	Scheveningen Radio, Netherlands	FC	USB	22,692.1
PCH8	Scheveningen Radio, Netherlands	FC	CW	12,966.0
PCH3Ø	Scheveningen Radio, Netherlands	FC	CW	6404.0
PCH4Ø	Scheveningen Radio, Netherlands	FC	CW	8562.0
PCH41	Scheveningen Radio, Netherlands	FC	CW	8622.0
PCH42	Scheveningen Radio, Netherlands	FC	CW	8654.4
PCH5Ø	Scheveningen Radio, Netherlands	FC	CW	12,768.0
PCH51	Scheveningen Radio, Netherlands	FC	CW	12,799.5
PCH52	Scheveningen Radio, Netherlands	FC	CW	12,853.5
PCH6Ø	Scheveningen Radio, Netherlands	FC	CW	16,902.0
PCH61	Scheveningen Radio, Netherlands	FC	CW	17,007.2
PCH62	Scheveningen Radio, Netherlands	FC	CW	17,104.2
PCH7Ø	Scheveningen Radio, Netherlands	FC	CW	22,324.5
PCH79	Scheveningen Radio, Netherlands	FC	CW	22,539.0
PCH95	Scheveningen Radio, Netherlands	FC	CW	4250.0
PCK27	Hilversum (Kootwijk), Netherlands	FX	ISB	17,417.5
PCK93	Hilversum (Kootwijk), Netherlands	FX	ISB	13,807.0
PCL2Ø	Amsterdam (Kootwijk), Netherlands	FX	ISB	10,700.5
PCM88	Hilversum (Kootwijk), Netherlands	FX	USB	18,619.0
PCW	The Hague, Netherlands	FX	USB	7698.5, 19,123.5, 23,558.5
PCW1	The Hague, Netherlands	FX	ISB	18,711.5
PCW2	Jerusalem, Israel	FX	ISB/RTTY	7698.5, 18,711.5, 23,568.5
PCW1Ø	Islamabad, Pakistan	FX	USB	15,966.1
PCW11	Bangkok, Thailand	FX	USB	13,768.5, 18,711.5
PCW2Ø	Warsaw, Poland	FX	ISB	13,768.5, 18,711.5
PCW35	The Hague, Netherlands	FX	USB	25,200.0
PDB2	Utrecht (Bilthoven), Netherlands	FX	CW	9200.0, 10,390.0, 18,190.0, 18,380.0
PGA88	The Hague, Netherlands	FX	USB	13,997.0, 20,751.5, 20,997.0
PHW	Amsterdam Heliport, Netherlands	FA/MA	USB	5647.0
PJC	Curacao (Willemstad) Radio, Curacao	FC	CW	4334.0, 6491.5, 8694.0, 13,042.5, 17,170.4
		FC	USB	6215.5, 8790.2
PJK26	Suffisant Dorp Naval Radio, Curacao	FC	CW	6405.0
PJK32	Suffisant Dorp Naval Radio, Curacao	FC	CW	12,000.0
PJK34	Suffisant Dorp Naval Radio, Curacao	FC	CW	4280.0
PJK38	Suffisant Dorp Naval Radio, Curacao	FC	CW	8540.1

Call	Location	Service	Mode	kHz
PJK38	Suffisant Dorp Naval R., Curacao	FC	CW	12,655.2
PJK317	Suffisant Dorp Naval R., Curacao	FC	CW	17,040.0
PJK322	Suffisant Dorp Naval R., Curacao	FC	CW	22,539.9, 22,547.5
PKA	Sabang Radio, Indonesia	FC	CW	8686.0, 17,184.8
PKB2Ø	Lhok Seumawe Radio, Indonesia	FC	CW	6456.5, 8626.0, 12,980.0
PKB	Belawan Radio, Indonesia	FC	CW	4295.0, 8686.0, 12,970.5, 16,861.7
PKC	Palembang Radio, Indonesia	FC	CW	6491.5, 8437.1
PKC2	Plaju Radio, Indonesia	FC	CW	8626.0
PKD	Surabaja Radio, Indonesia	FC	CW	4237.0, 8460.5, 12,703.6, 16,861.7
PKE5	Ternate Radio, Indonesia	FC	CW	6420.0
PKE	Amboina Radio, Indonesia	FC	CW	4295.0, 8473.3, 12,682.5, 17,184.8
PKF	Makassar Radio, Indonesia	FC	CW	4295.0, 8686.0, 12,682.5
PKG	Banjarmasin Radio, Indonesia	FC	CW	4237.0, 8457.0
PKI	Jakarta Radio, Indonesia	FC	CW	8542.0, 12,970.5, 16861.7, 22,431.0
PKI2	Jakarta Radio, Indonesia	FC	CW	8626.0, 12,890.0
PKK	Kupang Radio, Indonesia	FC	CW	8445.0
PKM	Bitung Radio, Indonesia	FC	CW	8694.0, 12,704.5
PKN	Balikpapan Radio, Indonesia	FC	CW	4237.0, 8437.0
PKN2	Balikpapan Radio, Indonesia	FC	CW	6456.5
PKN7	Bontang Radio, Indonesia	FC	CW	6456.5, 8626.0, 12,980.0
PKO	Tarakan Radio, Indonesia	FC	CW	8445.0
PKP	Dumai Radio, Indonesia	FC	CW	8457.0, 12,682.5, 17,184.8
PKR	Semararang Radio, Indonesia	FC	CW	8461.0
PKR3	Cilacap Radio, Indonesia	FC	CW	8445.0
PKR6	Cilacap Radio, Indonesia	FC	CW	6456.5, 8626.0, 12,980.0
PKS	Pontianak Radio, Indonesia	FC	CW	6355.0, 8473.0
PKT	Dili Radio, Indonesia	FC	CW	8445.0
PKY4	Sorong Radio, Indonesia	FC	CW	6337.0, 8461.0
PKY5	Merauke Radio, Indonesia	FC	CW	8457.0
PKY41	Sorong Radio, Indonesia	FC	CW	6456.5, 8456.5
PLC	Jakarta, Indonesia	FX	CW	11,440.0
PNK	Jayapura Radio, Indonesia	FC	CW	8694.0, 12,682.5, 17,074.0
POB35	Jakarta, Indonesia	FX	CW	11,500.0
PPC	Brasilia, Brazil	FX	CW	15,738.2
PPE	Rio de Janeiro, Brazil	SS	A2	5000.0
PPE	Rio de Janeiro Radio, Brazil	FC	CW	8721.0
PPE2	Rio Radio, Brazil	FC	CW	8476.0
PPJ	Juncao Radio, Brazil	FC	CW	4231.0, 4251.0, 8460.0 12,689.0, 12,840.0, 16,918.0, 17,170.0
		FC	USB	8802.6
PPL	Belem Radio, Brazil	FC	CW	4247.5, 4265.0, 8460.4 8502.0, 12,698.0, 12,979.5, 16,986.0, 17,170.5
		FC	USB	4369.8, 4413.2, 8784.0
PPN9	Brasilia, Brazil	FX	FAX	18,080.0
PPO	Olinda Radio, Brazil	FC	CW	4280.0, 4297.9, 8520.0, 12,840.0, 12,958.5, 17,120.0, 17,161.9
		FC	USB	8790.2
PPQ7	Rio de Janeiro (Santa Cruz), Brazil	FX	LSB	14,605.0
PPR	Rio Radio, Brazil	FC	CW	4244.0, 8492.0, 8634.0, 12,687.0, 12,738.0, 16,984.0, 17,194.4, 22,352.5, 22,710.0
		FC	USB	8808.8, 13,141.1
PPR2	Rio Radio, Brazil	FC	CW	22,420.0
PPR48	Rio de Janeiro (Sepetiba), Brazil	FX	USB	14,500.0
PP7A	"Observatory" (Paulista), Brazil	FX	LSB	17,685.0, 20,610.0
PRS5	Rio de Janeiro Radio, Brazil	FC	CW	22,504.0
PRW863	Belo Horizonte, Brazil	FX	USB	5781.0
PRX347	Brasilia, Brazil	FX	USB	7336.0
PSG351	Salvador, Brazil	FX	USB	5858.0

Call	Location	Service	Mode	kHz
PSR	Rio de Janeiro (Sepetiba), Brazil	FX	USB	14,855.0
PTO	Brasilia Radio, Brazil	FC	CW	22,339.0
PUZ4	Brasilia, Brazil	FX	USB	7318.5, 9473.0, 13,700.0, 20,600.0, 20,797.0
		AX	USB	19,500.0
PVX	Rio de Janeiro (Sepetiba), Brazil	FX	USB	10,640.0
PWB	Belem Naval Radio, Brazil	FC	CW	12,750.0, 12,804.0
PWF	Salvador Naval Radio, Brazil	FC	CW	12,750.0
PWI	Recife Naval Radio, Brazil	FC	CW	8480.0
		FX	CW	11,065.0
PWN	Natal Naval Radio, Brazil	FC	CW	8480.0
PWP	Florianopolis Naval Radio, Brazil	FC	CW	8480.0, 12,750.0
PWZ	Rio de Janeiro Naval Radio, Brazil	FC	CW	4289.0, 6420.0, 6435.5, 6453.0, 8530.0, 8542.0, 8550.0, 8664.0, 12,025.0, 12,754.0, 12,795.0, 12,900.0, 12,966.0, 16,902.0, 17,122.4, 17,135.0, 17, 160.0, 22,530.0
		FX	CW	17,590.0
PZA2	Paramaribo, Suriname	FX	CW	19,062.0
PZN2	Paramaribo Radio, Suriname	FC	CW	6355.0
PZN3	Paramaribo Radio, Suriname	FC	CW	8652.5
PZN4	Paramaribo Radio, Suriname	FC	CW	13,044.9
PZN25	Paramaribo Radio, Suriname	FC	CW	4255.3
PZN26	Paramaribo Radio, Suriname	FC	CW	16,956.0
P2M	Port Moresby Radio, Papua New Guinea	FC	CW	6351.5, 8515.0, 13,042.0
		FC	USB	6515.7, 8731.3
P2R	Rabaul Radio, Papua New Guinea	FC	CW	4247.0, 6351.5, 8515.0, 13,042.0
		FC	USB	4407.0, 8731.3
P6Z	Paris, France	FX	CW	5125.0, 10,732.0, 10,792.5
P13A	Monsanto Naval Radio, Portugal	FC	CW	8702.9 [CTU13]
P28BM	Port Moresby, Papua New Guinea	FX	USB	8149.0
P242PT	Lae, Papua New Guinea	FX	USB	4470.0
RAD4	Moscow, USSR	FX	ISB	19,865.0
RAN	Moscow, USSR	FX	USB	13,977.3
RAN77	Moscow, USSR	FX	FAX	6880.0
RAN78	Moscow, USSR	FX	ISB	6770.0
RAT23	Moscow, USSR	FX	ISB	6987.5
RAT25	Moscow, USSR	FX	USB	6808.0, 6918.5
RAW71	Moscow, USSR	FX	USB	8058.5
RAW78	Moscow, USSR	FX	FAX	7750.0
RBI71	Moscow, USSR	FX	LSB	18,653.0
RBK51	Moscow, USSR	FX	USB	22,770.0
RBK74	Moscow, USSR	FX	ISB	5455.0
RBM76	Moscow, USSR	FX	LSB	15,600.0
RBNF	Moscow Aeradio, USSR	AX	CW	16,065.0
RBQ74	Alma Ata, USSR	FX	FAX	6950.0
RBV76	Tashkent, Uzbek SSR	FX	FAX	14,982.5
RBV78	Tashkent, Uzbek SSR	FX	FAX	5890.0
RBV79	Tashkent, Uzbek SSR	FX	ISB	6825.0
RBW48	Murmansk, USSR	FX	FAX	10,130.0
RCC72	Moscow, USSR	FX	USB	20,290.0
RCC76	Moscow, USSR	FX	FAX	7670.0
RCD33	Moscow, USSR	FX	USB	8003.5
RCD34	Moscow, USSR	FX	LSB	12,175.0
RCF42	Moscow, USSR	FX	LSB	17,560.0
RCF47	Moscow, USSR	FX	ISB	16,215.0
RCG79	Moscow, USSR	FX	ISB	14,625.0
RCH	unknown	FX	AM/USB	5087.0
RCH	Tashkent, Uzbek SSR	SS	CW	10,000.0
RCI73	Moscow, USSR	FX	LSB	18,195.0
RCV	Moscow Naval Radio, USSR	FC	CW	4264.0, 8576.0, 12,723.0, 12,744.0, 16,942.0, 16,948.0
		FX	CW	15,465.0, 19,098.0, 21,765.0,

Call	Location	Service	Mode	kHz
				23,525.0
RCV28	Moscow, USSR	FX	USB	18,285.2
RDD79	Moscow, USSR	FX	FAX	10,980.0
RDT72	Alma Ata, USSR	FX	LSB	12,205.0
RDW76	Khabarovsk, USSR	FX	FAX	10,220.1
RDZ79	Moscow, USSR	FX	USB	8125.0
REN35	Moscow, USSR	FX	ISB	5290.0
RFNV	Moscow Aeradio, USSR	FA	CW	6748.0,8842.0, 10,025.0, 11,312.0, 15,024.0, 17,936.0
RGD22	Moscow, USSR	FX	USB	13,710.0
RGD23	Moscow, USSR	FX	USB	13,590.0
RGG42	Kuybyshev, USSR	FX	CW	10,288.0
RGH77	Irkutsk, USSR	FX	FAX	7759.5
RGI24	Moscow, USSR	FX	USB	10,120.0
RGQ2	unknown, probably in the USSR	FC	CW	8607.2
RGW28	Moscow, USSR	FX	LSB	16,140.0
RHA41	Moscow, USSR	FX	ISB	12,240.0
RHB/RHO	Khabarovsk, USSR	FX	FAX	6516.7, 7457.0, 7475.0, 9230.0, 14,737.0, 22,425.0
RHH	Petropavlovsk Radio, USSR	FC	CW	8659.0
RIC77	Moscow, USSR	FX	ISB	13,720.0
RID	Irkutsk, USSR	SS	CW	15,004.0
RIF31	Moscow, USSR	FX	CW	21,760.0
RIF33	Moscow, USSR	FX	USB	18,170.0
RIF35	Moscow, USSR	FX	LSB	16,193.0
RIQ24	Yuzhno-Sakhalinsk, USSR	FX	LSB	5810.0
RIT	unknown, probably in the USSR	FC	CW	6376.0, 8596.0, 16,870.0, 16,948.0
		FX	CW	20,067.9
RIW	Khiva Naval Radio, Uzbek SSR	FC	CW	6394.0, 8508.0, 8523.0, 12,752.0, 12,824.0, 13,064.0, 17,088.0, 17,110.0, 17,183.5, 22,500.0, 22,568.0
		FX	CW	7577.0, 9145.0, 10,434.0, 10,509.7, 10,912.0, 11,046.0, 11,488.0, 12,056.0, 13,425.0, 14,405.0, 14,510.0, 14,544.4, 14,556.0, 14,792.0, 16,338.0, 16,392.0, 17,504.0, 18,560.0, 18,690.0, 18,808.0, 18,952.0, 18,960.0, 19,090.0, 19,985.0, 19,993.0, 21,764.1, 21,784.0
RKA72	Moscow, USSR	FX	LSB	10,740.0
RKA75	Moscow, USSR	FX	USB	16,330.0
RKA78	Moscow, USSR	FX	FAX	10,230.0
RKB78	Moscow, USSR	FX	USB	12,165.0
RKDF	Parnu Radio, Estonian SSR	FC	CW	12,727.0, 12,756.0, 13,065.0
RKD41	Moscow, USSR	FX	USB	16,281.0
RKD48	Moscow, USSR	FX	USB	10,595.0
RKIC	Moscow, USSR	FX	FAX	7530.0
RKLM	Arkhangelsk Radio, USSR	FC	CW	6371.3, 8580.0, 12,830.0, 22,414.9
RKU29	Moscow, USSR	FX	LSB	5470.0
RLG71	Yuzhno-Sakhalinsk, USSR	FX	LSB	7470.0
RLI9	Kalinkovitch, USSR	FX	CW/FAX	7537.6
RLU8	Tiksi Bukhta, USSR	FX	CW	10,382.5
RLX	Dublin, Ireland	FX	CW/RTTY	4893.0, 5181.0, 5744.0, 6842.0, 9217.0, 9842.0, 11,148.0, 12,211.0, 14,426.0, 17,528.0
RMD53	Mosocw, USSR	FX	CW	10,155.0
RME21	Moscow, USSR	FX	USB	10,338.0
RME2Ø	Moscow, USSR	FX	ISB	9210.0
RMN3	Omsk, USSR	FX	CW	13,385.0
RMP44	Petropavlo Kam, USSR	FX	FAX	7475.0

Call	Location	Service	Mode	kHz
RMP	Rostov Radio, USSR	FC	CW	8680.0, 12,720.0, 16,885.0, 16,934.0
		FX	CW	16,016.0
RNB27	unknown	FC	CW	22,580.0
RND72	Moscow, USSR	FX	USB	5893.5
RND77	Moscow, USSR	FX	FAX	5355.0
RNE39	Moscow, USSR	FX	LSB	12,135.0
RNI9	Guzur, Uzbek SSR	FX	CW	22,482.0
RNN58	Moscow, USSR	FX	USB	7548.5
RNO	Moscow Radio, USSR	FC	CW	6473.0, 8653.0, 12,793.0, 17,163.0, 22,574.0
RNR5	unknown	FC	CW	8640.0
ROF73	Novosibirsk, USSR	FX	FAX	4445.0
ROF76	Novosibirsk, USSR	FX	FAX	5335.0
ROK22	Moscow, USSR	FX	ISB	7925.0
ROK28	Moscow, USSR	FX	ISB	7541.5
ROT	Moscow Naval Radio, USSR	FC	CW/RTTY	8456.0, 12,995.0, 13,045.0, 17,130.0, 17,155.0, 22,450.0, 22,454.5, 25,125.0, 25,130.0
ROT2	Moscow Naval Radio, USSR	FC	CW	6445.0, 17,045.0
ROW23	Moscow, USSR	FX	USB	15,540.0
ROW25	Moscow, USSR	FX	ISB	9148.0
ROW26	Moscow, USSR	FX	ISB	14,850.0
ROW27	Moscow, USSR	FX	LSB	13,960.8
RPC21	Kiev, Ukranian SSR	FX	CW	10,335.0
RPLT	Tuapse Radio, USSR	FC	CW	6414.5, 8660.0 [aka UOP]
RPTN	unknown	FX	CW	18,489.1
RPT31	Tashkent, Uzbek SSR	FX	CW/FAX	10,165.0
RRD	Moscow, USSR	FX	USB	19,724.8
RRG22	Moscow, USSR	FX	LSB	19,320.0
RRG24	Moscow, USSR	FX	ISB	19,170.0
RRG25	Moscow, USSR	FX	AM	19,070.0
RRG28	Moscow, USSR	FX	ISB	18,520.0
RRG29	Moscow, USSR	FX	USB	18,460.0
RRRF	Moscow, USSR	FX	ISB	14,500.0
RRRQ	Novosibirsk, USSR	FX	FAX	5765.0
RSF23	Tbilisi, Georgian SSR	FX	CW	10,861.2
RSGV	Petrozavodsk Radio, USSR	FC	CW	8675.0
RTA	Novosibirsk, USSR	SS	CW	10,000.0, 15,000.0
RTA21	Novosibirsk, USSR	FX	FAX	9060.0
RTM29	Khabarovsk, USSR	FX	LSB	18,150.0
RTU4Ø	Moscow, USSR	FX	USB	12,100.0
RUC27	Kiev, Ukrainian SSR	FX	LSB	15,490.0
RULE	Vostok USSR Base, Antarctica	FX	CW	13,385.0
RUQ	Krasnoyarsk, USSR	FX	CW	9392.0
RUU74	Leningrad, USSR	FX	FAX	18,402.0
RUZU	Molodezhnaya USSR Base, Antarctica	FX	CW	9215.0, 9325.0, 10,140.0, 10,555.0, 13,385.0, 15,904.1
		FX	FAX	6283.0, 9280.0
RVF53	Dushambe, USSR	FX	ISB	7440.0
RVI	unknown	FX	CW	18,977.8
RVL24	Khabarovsk, USSR	FX	USB	9905.0
RVO7Ø	Moscow, USSR	FX	USB	9240.0
RVO72	Moscow, USSR	FX	ISB	5815.0
RVO73	Moscow, USSR	FX	FAX	5150.3
RVW55	Moscow, USSR	FX	USB	13,820.0
RWD52	Moscow, USSR	FX	USB	5830.0
RWG	Moscow, USSR	FX	ISB	20,500.0
RWHC	Magadan Radio, USSR	FC	CW	6470.0 [aka UDQ2]
RWM71	Moscow, USSR	FX	USB	15,779.8
RWN	Moscow, USSR	SS	CW	4996.0, 9996.0, 14,996.0
RWN74	Moscow, USSR	FX	LSB	15,720.0
RWWN	Klaipeda Radio, Lithuanian SSR	FC	CW	8541.0 [aka URB2]
RWZ74	Moscow, USSR	FX	LSB	19,845.0
RXO	Warsaw, Poland	FX	CW/RTTY	11,152.0

Call	Location	Service	Mode	kHz
RXO74	Khabarovsk, USSR	FX	FAX	19,275.0
RYP29	Khabarovsk, USSR	FX	FX	5110.0
RYR	Moscow, USSR	FX	LSB	18,320.0
RZA27	Moscow, USSR	FX	USB	10,660.0
RZH	Tashkent, Uzbek SSR	SS	CW	5000.0
SAB2	Goteborg Radio, Sweden	FC	CW	4299.5
SAB3	Goteborg Radio, Sweden	FC	CW	6451.5
SAB4	Goteborg Radio, Sweden	FC	CW	8646.0
SAB6	Goteborg Radio, Sweden	FC	CW	12,755.5, 12,775.5
SAB8	Goteborg Radio, Sweden	FC	CW	17,057.2
SAB9	Goteborg Radio, Sweden	FC	CW	22,438.5
SAB25	Goteborg Radio, Sweden	FC	CW	25,415.5
SAG	Goteborg Radio, Sweden	FC	USB	4357.4
SAG2	Goteborg Radio, Sweden	FC	CW	4262.0
SAG3	Goteborg Radio, Sweden	FC	CW	6372.0
SAG4	Goteborg Radio, Sweden	FC	CW	8498.0
SAG6	Goteborg Radio, Sweden	FC	CW	12,380.5
SAG8	Goteborg Radio, Sweden	FC	CW	17,079.4
SAG9	Goteborg Radio, Sweden	FC	CW	22,413.0
SAG25	Goteborg Radio, Sweden	FC	CW	25,461.0
SAM	Stockholm, Sweden	FX	LSB	14,971.5, 17,430.0
		FX	USB	7605.0, 10,150.0, 10,164.0, 12,101.0, 12,225.0, 12,304.5, 12,306.0, 14,968.5, 18,808.0, 20,010.0, 20,961.0, 20,985.0, 26,660.0
SAM2Ø	Athens, Greece	FX	LSB	10,150.0, etc.
SAM21	Berlin, GFR	FX	USB	14,968.5
SAM24	Copenhagen, Denmark	FX	USB	12,304.5, etc.
SAM25	Lisbon, Portugal	FX	USB	10,150.0, etc.
SAM26	London, England	FX	LSB	14,971.5
SAM3Ø	Madrid, Spain	FX	USB	10,150.0, etc.
SAM35	Belgrade, Yugoslavia	FX	ISB	10,150.0, etc.
SAM36	Budapest, Hungary	FX	ISB	10,150.0, etc.
SAM37	Bucharest, Rumania	FX	LSB	10,150.0, etc.
SAM38	Moscow, USSR	FX	USB	7605.0, 10,164.0
SAM39	Prague, Czechoslovakia	FX	ISB	10,150.0, etc.
SAM4Ø	Warsaw, Poland	FX	USB	14,968.5, etc.
SAM45	Ankara, Turkey	FX	LSB	14,971.5
SAM46	Baghdad, Iraq	FX	USB	17,427.0
SAM47	Beirut, Lebanon	FX	LSB	17,430.0, etc.
SAM48	Damascus, Syria	FX	USB	12,304.5, etc.
SAM49	Jeddah, Saudi Arabia	FX	ISB	14,971.5, etc.
SAM5Ø	Cairo, Egypt	FX	USB	17,427.0
SAM52	Tel Aviv, Israel	FX	ISB	12,304.0, etc.
SAM54	Vientiane, Laos	FX	USB	14,968.5, etc.
SAM55	Bangkok, Thailand	FX	LSB	14,971.5
SAM57	Dacca, Bangladesh	FX	LSB	14,971.5
SAM58	Jakarta, Indonesia	FX	USB	14,968.5, etc.
SAM59	Hanoi, Vietnam	FX	LSB	14,971.5, etc.
SAM6Ø	Islamabad, Pakistan	FX	LSB	14,971.5, etc.
SAM61	New Delhi, India	FX	USB	14,968.5, etc.
SAM62	Beijing, PRC	FX	USB	14,971.5, etc.
SAM64	Pyongyang, N. Korea	FX	USB	14,968.5, etc.
SAM65	Seoul, S. Korea	FX	USB	14,968.5, etc.
SAM67	Colombo, Sri Lanka	FX	USB	20,958.0
SAM7Ø	Addis Ababa, Ethiopia	FX	USB	20,958.0, etc.
SAM71	Dar es Salaam, Tanzania	FX	LSB	14,971.5, etc.
SAM72	Kinshasa, Zaire	FX	LSB	14,971.5, etc.
SAM73	Nairobi, Kenya	FX	LSB	14,971.5, etc.
SAM74	Monrovia, Liberia	FX	USB	14,971.5, etc.
SAM79	Lusaka, Zambia	FX	USB	14,968.5, etc.
SAM8Ø	Gaborone, Botswana	FX	ISB	14,968.5, etc.
SAM81	Bissau, Guinea-Bissau	FX	USB	14,968.5, etc.

Call	Location	Service	Mode	kHz
SAM82	Maputo, Mozambique	FX	ISB	14,968.5, etc.
SAM83	Santiago, Chile	FX	USB	26,660.0
SAM89	Havana, Cuba	FX	USB	18,808.0, etc.
SAR88	Stockholm, Sweden	FX	ISB	6996.0, 27,998.0
		FX	USB	13,915.0, 13,997.0, 20,751.5, 20,815.0, 20,942.0, 20,998.0
SDQ7	Stockholm (Varberg), Sweden	FX	CW	18,187.0
SHA38	Stockholm Aeradio, Sweden	FA	USB	8972.0
SHX	Stockholm, Sweden	FX	CW	4632.5, 7532.0, 9200.0, 10,295.0, 10,390.0, 11,538.0
		FX	CW/RTTY	4837.5, 6905.0, 7969.0, 8038.0. 12.224.0
SMA4	Stockholm (Norrkoping), Sweden	AX	FAX	4037.5
SMA5	Stockholm (Norrkoping), Sweden	FX	FAX	5172.5
SMA6	Stockholm (Norrkoping), Sweden	FX	FAX	6901.0
SMA7	Stockholm (Norrkoping), Sweden	FX	CW	7732.5
SMA8	Stockholm (Norrkoping), Sweden	FX	FAX	8077.5
SMR5	unknown	FX	CW	20,980.2
SOE263	Warsaw, Poland	FX	CW	4630.0
SOH269	Warsaw, Poland	FX	CW	7694.0
SOH289	Warsaw, Poland	FX	CW	7845.0, 7894.0
SOH299	Warsaw, Poland	FX	CW/RTTY	7997.0
SOI208	Warsaw, Poland	FX	CW	8081.0
SOI213	Warsaw, Poland	FX	CW/RTTY	8133.0
SOI219	Warsaw, Poland	FX	CW/RTTY	8192.0
SOJ249	Warsaw (Radom), Poland	FX	CW	9391.0
SOL244	Warsaw, Poland	FX	CW	11,440.0
SOL249	Warsaw, Poland	FX	CW	11,494.0
SON249	Warsaw, Poland	FX	CW/RTTY	13,499.0
SON261	Warsaw, Poland	FX	CW	13,616.0
SON279	Warsaw, Poland	FX	CW	13,793.0
SOO236	Warsaw, Poland	FX	CW/RTTY	14,362.0
SOP251	Warsaw, Poland	FX	CW	15,510.0
SOP256	Warsaw, Poland	FX	CW	15,564.0
SOP298	Warsaw, Poland	FX	CW/RTTY	15,983.0
SOP299	Warsaw, Poland	AX	CW	15,996.0
SOT29	Warsaw, Poland	FX	CW	18,995.0
SOW283	Warsaw, Poland	FX	CW/RTTY	21,838.0
SPC23	Gydnia Radio, Poland	FC	USB	4360.5
SPC41	Gydnia Radio, Poland	FC	USB	8753.0 [SPC4]
SPC61	Gdynia Radio, Poland	FC	USB	13,147.3 [SPC6]
SPC82	Gdynia Radio, Poland	FC	USB	17,325.9 [SPD6]
SPD42	Gdynia Radio, Poland	FC	USB	8790.2 [SPD4]
SPE21	Szczecin Radio, Poland	FC	CW	4306.0 [SPE2]
SPE31	Szczecin Radio, Poland	FC	CW	6459.0 [SPE3]
SPE41	Szezecin Radio, Poland	FC	CW	8557.0 [SPE4]
SPE61	Szczecin Radio, Poland	FC	CW	12,939.0 [SPE6]
SPE62	Szczecin Radio, Poland	FC	CW	6376.0 [SPB3]
SPE81	Szczecin Radio, Poland	FC	CW	16,974.1 [SPE8]
SPE82	Szczecin Radio, Poland	FC	CW	16,861.7 [SPB8]
SPE91	Szczecin Radio, Poland	FC	CW	22,505.0 [SPE9]
SPE92	Szczecin Radio, Poland	FC	CW	22,551.0 [SPB2]
SPH27	Gydnia Radio, Poland	FC	CW	4337.0 [SPA2]
SPH31	Gdynia Radio, Poland	FC	CW	6383.0 [SPA3]
SPH41	Gydnia Radio, Poland	FC	CW	8482.0 [SPA4]
SPH42	Szczecin Radio, Poland	FC	CW	8650.0 [SPB4]
SPH44	Gydnia Radio, Poland	FC	CW	8678.0 [SPI4]
SPH61	Gdynia Radio, Poland	FC	CW	12,721.0 [SPA6]
SPH91	Gdynia Radio, Poland	FC	CW	22,495.0 [SPA9]
SPE62	Szczecin Radio, Poland	FC	CW	12,872.0 [SPB6]
SPH32	Gdynia Radio, Poland	FC	CW	6398.0 [SPH3]
SPH42	Gydnia Radio, Poland	FC	CW	8634.0 [SPH4]
SPH62	Gydnia Radio, Poland	FC	CW	12,826.5 [SPH6]
SPH82	Gydnia Radio, Poland	FC	CW	17,016.0 [SPH8]
SPH92	Gydnia Radio, Poland	FC	CW	22,399.0 [SPH9]

Call	Location	Service	Mode	kHz
SPH81	Gdynia Radio, Poland	FC	CW	16,887.3 [SPA8]
SPH23	Gydnia Radio, Poland	FC	CW	4319.0 [SPI2]
SPH43	Gydnia Radio, Poland	FC	CW	8666.4 [SPI4]
SPH63	Gydnia Radio, Poland	FC	CW	12,928.0 [SPI6]
SPH83	Gydnia Radio, Poland	FC	CW	17,064.0 [SPI8]
SPH93	Gydnia Radio, Poland	FC	CW	22,407.0 [SPI9]
SPH84	Gdynia Radio, Poland	FC	CW	16,914.5 [SPH8]
SPE63	Szczecin Radio, Poland	FC	CW	13,022.0 [SPL6]
SPE83	Szczecin Radio, Poland	FC	CW	17,069.6 [SPL8]
SP091	Szczecin Radio, Poland	FC	USB	22,651.8 [SP09]
SP061	Szczecin Radio, Poland	FC	USB	13,181.4 [SP06)
SQL47	Islamabad, Pakistan	FX	CW	19,668.4
SQL49	Islamabad, Pakistan	FX	CW	20,245.0
SSF	Port Said Naval Radio, Egypt	FC	CW	6498.5
SSG	Matrouh Naval Radio, Egypt	FC	CW	6498.5
SSJ	Safaga Naval Radio, Egypt	FC	CW	6498.5
SSK	Alexandria Naval Radio, Egypt	FC	CW	6498.5
SSL	Alexandria Naval Radio, Egypt	FX	CW	16,310.0
STP	Port Sudan Radio, Sudan	FC	CW	4226.5, 8652.0
SUA311	Cairo, Egypt	FX	USB	19,167.5
SUC5Ø	Cairo Aeradio, Egypt	FA	AM	8819.0
SUH3	Alexandria Radio, Egypt	FC	CW	8578.0
SUH4	Alexandria Radio, Egypt	FC	CW	12,970.5
SUP	Port Said Radio, Egypt	FC	CW	4325.0, 8471.7
SUU2	Cairo, Egypt	FX	FAX	10,123.0
SUU36	Cairo, Egypt	FX	FAX	4526.0
SUV66	Cairo, Egypt	FX	ISB	20,135.0
SVA3	Athens Radio, Greece	FC	CW	6478.5
SVA4	Athens Radio, Greece	FC	CW	8687.0
SVA5	Athens Radio, Greece	FC	CW	13,046.9
SVA6	Athens Radio, Greece	FC	CW	17,094.8
SVA7	Athens Radio, Greece	FC	CW	22,417.0
SVA8	Athens Radio, Greece	FC	CW	25,401.0
SVB2	Athens Radio, Greece	FC	CW	4343.0
SVB3	Athens Radio, Greece	FC	CW	6344.0
SVB4	Athens Radio, Greece	FC	CW	8704.0
SVB5	Athens Radio, Greece	FC	CW	13,029.0
SVB6	Athens Radio, Greece	FC	CW	17,147.2, 17,194.4
SVB7	Athens Radio, Greece	FC	CW	22,410.8
SVD2	Athens Radio, Greece	FC	CW	4223.0
SVD3	Athens Radio, Greece	FC	CW	6411.0
SVD4	Athens Radio, Greece	FC	CW	8536.5
SVD5	Athens Radio, Greece	FC	CW	12,942.0
SVD6	Athens Radio, Greece	FC	CW	16,966.2
SVD7	Athens Radio, Greece	FC	CW	22,471.5
SVD8	Athens Radio, Greece	FC	CW	25,535.0
SVF3	Athens Radio, Greece	FC	CW	6444.5
SVF4	Athens Radio, Greece	FC	CW	8690.0, 8692.5
SVF5	Athens Radio, Greece	FC	CW	12,833.0, 12,859.0
SVF6	Athens Radio, Greece	FC	CW	16,995.0
SVF7	Athens Radio, Greece	FC	CW	22,500.0
SVG4	Athens Radio, Greece	FC	CW	8454.5
SVG5	Athens Radio, Greece	FC	CW	12,720.0, 12,727.5
SVG6	Athens Radio, Greece	FC	CW	16,981.6
SVG7	Athens Radio, Greece	FC	CW	22,327.5
SVI4	Athens Radio, Greece	FC	CW	8681.0
SVI6	Athens Radio, Greece	FC	CW	17,188.2
SVI7	Athens Radio, Greece	FC	CW	22,346.5
SVJ4	Athens Radio, Greece	FC	CW	8530.0
SVM6	Athens Radio, Greece	FC	CW	16,903.0
SVN4	Athens Radio, Greece	FC	USB	8734.4, 8759.2
SVN5	Athens Radio, Greece	FC	USB	13,134.9, 13,159.7, 13,196.0
SVN6	Athens Radio, Greece	FC	USB	17,257.7, 17,307.3, 17,310.4, 17,353.8
SVN7	Athens Radio, Greece	FC	USB	22,645.6, 22,667.3, 22,689.0,

Call	Location	Service	Mode	kHz
				22,701.4
SVU5	Athens Radio, Greece	FC	CW	8712.6
SVU7	Athens Radio, Greece	FC	CW/TOR	22,585.6
SVZ	Athens Radio, Greece	FC	CW	13,054.0
SXA2	Spata Attikis Naval Radio, Greece	FC	CW	6508.0
SXA3	Spata Attikis Naval Radio, Greece	FC	CW	4226.8, 6385.0, 12,689.0, 16,931.0
SXA4	Spata Attikis Naval Radio, Greece	FC	CW	8659.5, 16,936.0
SXA8	Spata Attikis Naval Radio, Greece	FC	CW	8671.0
SXA24	Spata Attikis Naval Radio, Greece	FC	CW	6470.1
SXA34	Spata Attikis Naval Radio, Greece	FC	CW	4293.9
S3D	Chittagong Radio, Bangladesh	FC	CW	6414.5, 8694.0, 13,056.0, 17,050.5
S3E	Khulna Radio, Bangladesh	FC	CW	8658.0, 13,024.0
S3V	Cox's Bazaar Radio, Bangladesh	FC	CW	12,728.0
S7Q	Seychelles Radio, Seychelles	FC	CW	4349.0, 8445.0, 12,660.0
TAH	Istanbul Radio, Turkey	FC	CW	4253.0, 6491.5, 8611.5, 8662.0, 12,735.8, 17,021.5
TBA2	Izmir Naval Radio, Turkey	FC	CW	4260.0]TCC21]
TBA3	Izmir Naval Radio, Turkey	FC	CW	6395.0, 8195.0
TBA6	Izmir Naval Radio, Turkey	FC	CW	8572.0
TCA21	Izmir Naval Radio, Turkey	FC	CW	6395.0 [aka TBA3]
TBB5	unknown, in Turkey	FC	CW	8555.0
TBO2	Izmir Naval Radio, Turkey	FC	CW	6374.3
TBO3	Izmir Naval Radio, Turkey	FC	CW	8505.0
TCCH	Istanbul Naval Radio, Turkey	FC	CW	8611.5
TCCQ	Tophane Kulesi Naval R., Turkey	FC	CW	8444.5, 8662.0 [aka TCR]
TCC2	Ankara, Turkey	FX	CW	10,390.0. 19,405.
TCC21	Izmir Naval Radio, Turkey	FC	CW	4260.0 [aka TBA2]
TCR	Tophane Kulesi Naval Radio, Turkey	FC	CW	8444.5, 8662.0
TFA	Reykjavik Radio, Iceland	FC	CW	4289.1, 8578.4, 16,909.7
TFA3	Reykjavik Radio, Iceland	FC	CW	8690.0
TFA6	Reykjavik Radio, Iceland	FC	CW	6344.0
TFA13	Reykjavik Radio, Iceland	FC	CW	13,069.5
TFK	Keflavik Airport, Iceland	FA	ISB	17,992.0
		FX	USB	6775.0
		FX	USB/RTTY	7799.0
TFR	Sandur CG Radio, Iceland	FX	USB	4857.5, etc.
TFR2	Reykjavik CG Radio, Iceland	FX	USB	4857.5, etc.
TIM	Limon Radio, Costa Rica	FC	CW	4291.0, 4297.0, 8478.1, 13,000.0, 16,874.2, 16,956.3, 22,336.0
TJC7	Douala Radio, Cameroon	FC	CW	8449.0
TJC8	Douala Radio, Cameroon	FC	CW	8718.0
TJC9	Douala Radio, Cameroon	FC	CW	13,069.5
TJF58	Douala, Cameroon	FX	ISB	15,832.0
TJF69	Douala, Cameroon	FX	USB	6945.0
TJF73	Douala, Cameroon	FX	ISB	17,386.0
TJF78	Douala, Cameroon	FX	ISB	7810.0
TJF9Ø	Douala, Cameroon	FX	USB	9044.0
TJP76	Kousserin, Cameroon	FX	CW	20,735.0
TKY1	Cotonou, Benin	FX	LSB	20,130.0
TLZ2	Bangui, Central African Republic	FX	USB	11,021.3
TLZ46	Bangui, Central African Republic	FX	ISB	24,690.0
TLZ58	Bangui, Central African Republic	FX	ISB	5813.0
TLZ63	Bangui, Central African Republic	FX	ISB	16,386.0
TLZ87	Bangui, Central African Republic	FX	ISB	18,722.7
TLZ88	Bangui, Central African Republic	FX	USB	18,843.5
TLZ93	Bangui, Central African Republic	FX	LSB	9305.0
TNA8	Pointe-Noire Radio, Congo	FC	CW	8453.0
TNA12	Pointe-Noire Radio, Congo	FC	CW	12,682.5
TNH21	Brazzaville, Congo	FX	USB	12,125.0
TNH81	Brazzaville, Congo	FX	ISB	18,175.0
TNI9	Brazzaville, Congo	FX	ISB	20,955.0

Call	Location	Service	Mode	kHz
TNI92	Brazzaville, Congo	FX	ISB	9241.9
TNL	Brazzaville, Congo	AX	CW	16,245.0
TRP8	Banjul Shell Radio, Gambia	FC	CW	13,070.0
TTR1Ø3	N'Djamena, Chad	FX	LSB	10,391.5
TTZ8	N'Djamena, Chad	FX	ISB	20,832.0
TTZ34	N'Djamena, Chad	FX	USB	13,427.0
TTZ9Ø	N'Djamena, Chad	FX	USB	19,089.1
TUA3	Abidjan Radio, Ivory Coast	FC	CW	4343.0
TUA4	Abidjan Radio, Ivory Coast	FC	CW	8465.0
TUA5	Abidjan Radio, Ivory Coast	FC	CW	13,060.5
TUA9	Abidjan Radio, Ivory Coast	FC	CW	16,947.8
TUA47	Abidjan, Ivory Coast	FX	USB	7823.5
TUA211	Abidjan, Ivory Coast	FX	USB	11,560.0
TUP	Abidjan, Ivory Coast	FX	USB	10,147.0
TUP8	Abidjan, Ivory Coast	FX	LSB	10,836.0
TUP38	Abidjan, Ivory Coast	FX	ISB	13,862.5
TUP43	Abidjan, Ivory Coast	FX	USB	4577.0
TUP45	Abidjan, Ivory Coast	FX	USB	4597.2
TUP57	Abidjan, Ivory Coast	FX	USB	5773.5
TUP84	Abidjan, Ivory Coast	FX	ISB	18,436.0
TUW21Ø	Abidjan, Ivory Coast	FX	CW	21,807.0
TUW22Ø	Abidjan, Ivory Coast	FX	CW	14,827.0, 19,487.0
TYA3	Cotonou Radio, Benin	FC	CW	6383.0
TYK3	Cotonou, Benin	FX	LSB	20,344.4
TYK8	Cotonou, Benin	FX	LSB	10,894.5
TYK73	Cotonou, Benin	FX	ISB	17,412.0
TYK75	Cotonou, Benin	FX	ISB	17,515.0
TYK92	Cotonou, Benin	FX	USB	9205.5
TZA218	Bamako, Mali	FX	USB	18,294.6
TZA512	Bamako, Mali	FX	ISB	12,225.0
TZB216	Bamako, Mali	FX	USB	16,425.0
T2AY	Funafuti, Tuvala	FX	CW	4484.0
T2U	Funafuti, Tuvala	FX	USB	6945.0
UAD2	Kholmsk Radio, USSR	FC	CW/FAX	22,527.0
UAH	Tallinn Radio, Estonian SSR	FC	CW	4285.0, 6405.0, 6485.0, 8476.0, 12,723.0, 12,970.0, 13,001.9, 17,075.0
UAI3	Nakhodka Radio, USSR	FC	CW	17,090.0
UAQ	unknown	FX	CW	14,451.0
UAT	Moscow Radio, USSR	FC	CW	6345.0, 6460.0, 6475.0, 8440.6, 8471.0, 8480.0, 8565.0, 8645.0, 12,720.0, 12,738.0, 12,820.0, 12,910.0, 12,980.0, 16,872.5, 16,992.0, 17,059.0, 17,066.0, 17,119.0, 22,395.0, 22,423.0, 22,493.0, 22,512.0
UBE	Petropavlovsk Radio, USSR	FC	CW	6379.5, 13,020.0
UBE2	Petropavlovsk Radio, USSR	FC	CW	4271.0, 6370.0, 8610.0, 12,835.5, 13.000.0
UBF2	Leningrad Radio, USSR	FC	CW	8451.0, 22,485.0
UBF5	Leningrad Radio, USSR	FC	CW	4255.0
UBN	Jdanov Radio, Ukrainian SSR	FC	CW	4075.0, 4265.0, 6355.0, 6379.5, 6477.5, 8495.0, 8540.0, 8607.0, 8620.0, 12,663.5, 12,697.0, 12,711.0, 12,732.0, 17,085.0, 17,141.6 22,365.0
		FC	CW/RTTY	16,993.0
UCA	Odessa Radio, Ukrainian SSR	FX	CW	17,404.0
UCO	Yalta Radio, Ukrainian SSR	FC	CW	8436.0, 17,064.0
UCP2	Okhotsk Radio, USSR	FC	CW/RTTY	6365.0, 22,530.0
UCW4	Leningrad Radio Radio, USSR	FC	CW	4223.0, 4290.0, 6443.0, 8505.0, 12,765.0
UCY2	Astrakhan Radio, USSR	FC	CW	4040.0, 6400.0, 8474.5

Call	Location	Service	Mode	kHz
UDH	Riga Radio, Latvian SSR	FC	CW	4236.0, 4335.0, 6341.5 6410.0, 8505.9, 8510.0 12,671.5, 12,895.0, 13,065.9, 13,070.0, 16,947.0, 16,997.0, 17,155.0, 17,181.0, 22,455.0
UDH2	Riga Radio, Latvian SSR	FC	CW	8515.0
UDK2	Murmansk Radio, USSR	FC	CW	4260.0, 6393.5, 6455.0 8580.0, 8698.0, 8713.0, 12,797.0, 12,803.0, 12,925.0, 13,050.0, 16,922.0, 17,020.0, 17,055.0, 17,170.0, 22,354.5, 22,497.0
UDN	Jdanov Radio, Ukrainian SSR	FC	CW	12,919.0
UDY	Novolassarevskaya USSR Base, Antarctica	FX	CW	13,385.0
UEK	Feodosia Radio, USSR	FC	CW	6340.0, 6371.0, 17,000.0, 17,014.0
UFA/UHK	Batumi Radio, USSR	FC	CW	17,190.5
UFB	Odessa Radio, Ukrainian SSR	FC	CW	4230.0, 6341.5, 6345.0, 6483.0, 8495.5, 8520.0, 8528.5, 8593.0, 8630.0, 11,623.0, 12, 683.0, 12,760.0, 12,947.0, 12,948.5, 12,966.0, 12,997.0, 13,041.0, 16,947.0, 16,984.0, 17,049.0, 17,061.3, 17,108.0, 17,185.0, 22,352.4, 22,358.0, 22,464.0, 22,555.0
			CW/RTTY	17,155.0
UFD9	Arkhangelsk Radio, USSR	FC	CW	6371.5, 12,942.0
UFE	Mirnyy USSR Base, Antarctica	FX	CW	13,385.0
UFH	Petropavlovsk Radio, USSR	FC	CW	4323.0, 6405.0, 6447.0, 8436.0, 8555.0, 8626.0, 12,715.0, 12,825.0
UFJ	Rostov Radio, USSR	FC	CW	6369.0, 8485.0, 13,010.0, 13,060.0
UFL	Vladivostok Radio, USSR	FC	CW	4235.0, 6345.0, 6379.5, 6385.0, 6445.0, 8515.0, 8590.0, 8595.0, 12,729.0, 12,955.0, 17,110.0, 22,350.0, 22,435.0, 22,530.0
UFL6	Vladivostok Radio, USSR	FC	CW	12,710.0
UFM3	Nevelsk Radio, USSR	FC	CW	4345.0, 6334.5, 8446.0, 12,684.5
UFN	Novorossiysk Radio, USSR	FC	CW	4245.0, 4275.0, 6405.0, 6440.0, 8439.0, 8485.0, 8540.0, 8571.0, 8590.0, 8663.0, 12,891.0, 12,920.0, 13,040.0, 16,890.5, 16,995.0, 17,092.0, 17,100.0, 17,141.0, 17,172.0, 22,349.5, 22,501.0
UFW	Vladivostok Radio, USSR	FC	CW	8535.0, 12,854.0
UGAB	Moscow (Tuchino) Aeradio, USSR	FA	CW	11,390.0
UGE2	Bellingshausen USSR Base, Antarctica	FX	CW	7665.0, 9215.0, 9280.0, 10,140.0, 13,385.0
UGG9	Belomorsk Radio, USSR	FC	CW/RTTY	6430.0, 12,860.0, 17,021.0
UGH2	Juzno-Sakhalinsk Radio, USSR	FC	CW	4223.0, 6375.0, 8557.0, 12,684.5
UGK2	Kaliningrad Radio, USSR	FC	CW	6365.5, 8408.0, 12,734.0
UGW	Rey Aleksandrovsk Radio, USSR	FC	CW	16,914.5
UHF3	Yeysk Staro Radio, USSR	FC	CW	8552.0
UHK	Batumi Radio, USSR	FC	CW	4295.0, 6345.0, 6410.0, 8590.0, 8695.1, 12,912.0, 22,322.5
UJE	Moscow Naval Radio, USSR	FC	CW	6890.0, 12,810.0, 12,967.0
UJO3	Izmail Radio, Ukrainian SSR	FC	CW	4295.0
UJO5	Izmail Radio, Ukrainian SSR	FC	CW	4474.0, 6386.4, 8586.0,

Call	Location	Service	Mode	kHz
				12,754.5, 12,963.0, 16,909.5
UJQ	Kiev Radio, Ukrainian SSR	FC	CW	6416.0, 8657.0, 12,656.5, 12,990.0, 17,015.0, 22,580.0
UJQ2	Kiev Radio, Ukrainian SSR	FC	CW	8635.0
UJQ7	Kiev Radio, Ukrainian SSR	FC	CW/RTTY	6420.0, 8663.5, 12,715.0, 12,805.0, 13,054.0, 17,131.0, 17,135.0, 22,411.0, 22,440.2
UJY	Kaliningrad Radio, USSR	FC	CW	6360.0, 6386.5, 6465.0, 8408.0, 8460.0, 8640.0, 12,690.0, 12,788.0 , 12,865.0, 12,877.5, 12,886.0, 12,889.0, 12,970.0, 12,973.0, 16,865.0, 16,927.0, 16,963.0, 17,111.0, 22,315.0, 22,321.0, 22,325.0, 22,355.0, 22,403.0
UJY2	Kaliningrad, USSR	FX	CW/RTTY	9415.0, 14,680.0, 17,152.0, 17,550.0, 20,510.0
UJY7	Kaliningrad Radio, USSR	FC	CW	13,054.0
UKA	Vladivostok Radio, USSR	FC	CW/RTTY	4241.0, 6411.0, 6430.0, 6460.0, 6465.0, 8565.0, 12,664.5, 12,852.0, 12,855.0, 12,870.0, 12,876.0, 12,933.0, 13,047.0, 16,913.0, 17,155.0, 22,405.0, 22,485.0
UKG	Kerch Radio, Ukrainian SSR	FC	CW	12,993.0
UKJ	Astrakhan Radio, USSR	FC	CW	6430.0, 8474.0
UKK3	Nakhodka Radio, USSR	FC	CW	6439.0, 8510.0, 12,990.0, 17,050.0, 22,410.0
UKW3	Korsakov Sakhalinsk Radio, USSR	FC	CW	4283.0, 6487.0, 8520.0, 12,765.0, 16,862.5, 17,033.0, 22,407.0
UKX	Nakhodka Radio, USSR	FC	CW/RTTY	4233.0, 6417.0, 8575.0, 8620.0, 12,872.5, 13,047.0, 13,067.0, 16,913.0, 17,090.0, 22,405.0, 22,550.0
ULY4	Aleksandrovsk Radio, USSR	FC	CW	12,960.0
ULY4	Aleksandrovsk, USSR	FX	CW	19,056.0
ULZ	Ventspils Radio, Latvian SSR	FC	CW	16,949.0
UMS	Moscow Naval Radio, USSR	FX	CW	11,132.0, 11,430.0
UMV	Murmansk Radio, USSR	FC	CW	4255.0, 4307.0, 6400.0, 6480.0, 8515.0, 8620.0, 8690.0, 12,661.5, 12,730.0, 12,905.0, 16,960.0, 17,140.0, 22,329.5
UMV2	Murmansk Radio, USSR	FC	CW	6331.5
UNM2	Klaipeda Radio, Lithuanian SSR	FC	CW	4343.0, 6375.0, 6475.0, 6493.0, 8473.0, 8701.0, 12,703.0, 12,745.0, 12,895.0
UNM3	Klaipeda Radio, Lithuanian SSR	FC	CW	4229.0
UNO2	Severo Kurilsk Radio, USSR	FC	CW	6420.0
UNQ	Novorossiysk Radio, USSR	FC	CW	8438.0, 8580.0, 16,980.0, 22,410.0
UNQ2	Novorossiysk Radio, USSR	FC	CW	12,740.0
UON	Baku Radio, Azerbaijan SSR	FC	CW	6455.0, 12,844.5
UOP	Tuapse Radio, USSR	FC	CW	6414.5, 8590.0, 12,990.0, 17,100.0
UPB	Providenia Bukhta Radio, USSR	FC	CW	6390.0, 8570.0, 12,720.0, 12,980.0, 17,025.0
UPW2	Liepaja Radio, Latvian SSR	FC	CW	12,785.0, 16,949.0, 22,497.0, 22,520.0
UQA4	Kiev Radio, Ukrainian SSR	FC	CW	16,922.0
UQB	Kholmsk Radio, USSR	FC	CW	6380.0, 8675.0, 8679.0, 12,678.0, 13,029.0, 16,970.0
UQD2	Magadan Radio, USSR	FC	CW	6470.0, 8570.5
UQK	Riga Radio, Latvian SSR	FC	CW	6369.0, 8445.0, 16,949.0,

Call	Location	Service	Mode	kHz
				22,420.0, 22,497.0
UQK2	Riga Radio, Latvian SSR	FC	CW	4283.0, 8446.0, 8530.0, 8554.0, 12,688.5, 12,706.0, 12,865.0, 16,955.0, 17,108.0, 22.520.0
UQZC	unknown	FX	CW	7536.9
URB2	Klaipeda Radio, Lithunianian SSR	FC	CW	6493.5, 8541.0, 12,725.0, 12,815.0, 16,865.0, 16,878.0, 17,050.0, 17,138.0, 22,315.0, 22,333.0, 22,417.0, 22,458.0
URD	Leningrad Radio, USSR	FC	CW	4315.0, 6354.0, 6370.0, 6481.0, 8575.0, 8600.0, 8675.0, 8687.6, 12,693.0, 12,901.5, 13,030.0, 16,970.0, 16,983.2, 17,010.0, 17,115.0, 17,150.0, 22,333.0, 22,430.0, 22,475.0
URD2	Leningrad Radio, USSR	FC	CW	8617.5
URL	Sevastopol Radio, Ukrainian SSR	FC	CW/RTTY	4244.0, 6383.0, 8614.0, 8470.0, 8515.0, 12,735.0, 12,876.0, 12,915.0, 12,970.0, 12,977.0, 12,993.0, 16,925.0, 17,038.6, 17,147.0, 17,153.0, 17,177.0, 22,371.0, 22,383.0, 22,435.0, 22,530.0, 22,560.0
URL3	Sevastopol Radio, Ukrainian SSR	FC	CW	8580.0, 22,430.0
USW2	Rostov Radio, USSR	FC	CW	8606.0, 12,983.0, 16,990.0, 22,321.0
UTA	Tallinn Radio, Estonian SSR	FC	CW	6475.0, 8535.0, 8650.0, 12,755.0, 17,107.0, 22,387.0
UUK	unknown	FC	CW	12,780.0
UVD	Magadan 1 Radio, USSR	FC	CW	4243.0
UVD	Magadan Radio, USSR	FC	CW	6365.0, 6405.0
UXB6	unknown	FC	CW	16,930.0
UXN	Arkhangelsk Radio, USSR	FC	CW	4253.0, 4349.0, 6337.0, 6407.0, 6470.0, 8549.0, 8610.0, 8703.0, 12,674.0, 12,700.0, 12,714.0, 12,795.0, 13,047.0, 16,867.5, 16,872.5, 17,036.0, 22,490.0
UZU8	unknown	FX	CW	11,020.0
VAF	"Alert Military", NWT, Canada	FA	USB	11,209.0
VAI	Vancouver CG Radio, BC, Canada	FC	CW	4235.0, 6351.5, 6493.0 8452.9, 12,876.0, 17,175.2
		FC	USB	4385.3, 6518.8
VAP	Churchill CG Radio, Man., Canada	FC	USB	4376.0
VAW	Killinek CG Radio, NWT, Canada	FC	USB	4376.0
VCA77	St. Anthony, Nfld., Canada	FX	USB	4826.0, 6788.5
VCA294	unknown, probably NS, Canada	FX	USB	4601.4
VCA318	Chipman, NB, Canada	FX	USB	4460.0
VCA388	Lower Lahave, NS, Canada	FX	USB	5060.0
VCA737	Battle Harbor, Nfld., Canada	FC	USB	4601.4
VCA744	Cape St. Francis, Nfld., Canada	FX	USB	4601.4
VCB513	Halifax, NS, Canada	FX	USB	4963.0
VCB9Ø1	Fredericton, NB, Canada	FX	USB	4963.0
VCC42	Dartmouth, NS, Canada	FX	USB	4601.4
VCC43	Dartmouth, NS, Canada	FX	USB	5281.5, 7861.0
VCR976	Ottawa, Ont., Canada	FX	CW	17,448.5
		FX	USB	5806.0, 14,748.5, 15,595.5, 17,662.5, 23,100.0
VCS838	Ottawa, Ont., Canada	FX	CW	7445.0, 15,705.0, 18,247.0
		FX	CW/RTTY	8172.0, 10,780.0, 12,295.0, 14,515.0, 16,459.0
VCS	Halifax CG Radio, NS, Canada	FC	CW	4285.0, 6491.5, 8172.0,

Call	Location	Service	Mode	kHz
				8439.9, 16,948.5, 22,387.2
		FC	USB	4394.6, 4410.1, 6518.8, 8787.1, 13,138.0, 17,242.2
VDD	Debert, NS, Canada	FX	USB	17,486.0
VDH9	Alert, NWT, Canada	FX	USB	13,971.0
VEB2	unknown, probably in Canada	FX	AM	4625.0
VET9	Damascus, Syria	FX	USB	13,971.0
VEW	unknown	FX	USB	6785.0
VFA	Inuvik CG Radio, NWT, Canada	FC	CW	6335.5
VFC	Cambridge CG Radio, NWT, Canada	FC	CW	6493.0, 12,761.0
		FC	USB	4363.6
VFE	Edmonton, Alta., Canada	FX	FAX	11,615.5, 15,770.5, 21,830.0
VFF	Frobisher CG Radio, NWT, Canada	FC	CW	4236.5, 8443.0, 12,671.0
		FX	FAX	7710.0, 12,667.0, 15,644.0
		FC	USB	4376.0
VFG	Gander Aeradio, Nfld., Canada	FA	USB	5652.0, 8868.0, 13,272.0
VFR	Resolute Bay CG Radio, NWT, Canada	FC	USB	4376.0
VFU	Coral Harbor CG Radio, NWT, Canada	FC	USB	4376.0
VFU2	Cape Dorset CG Radio, NWT, Canada	FC	USB	4376.0
VFZ	Goose Bay CG Radio, NWT, Canada	FC	USB	4379.1
VGF91	Mount Bertha, BC, Canada	FX	USB	4627.0
VHM5	COMMSTA Darwin, NT, Australia	FC	CW	12,831.0
VHP	COMMSTA Canberra, ACT, Australia	FC	CW	12,831.0
VHP2	COMMSTA Canberra, ACT, Australia	FC	CW	4286.0
VHP4	COMMSTA Canberra, ACT, Australia	FC	CW	8478.0
VHP5	COMMSTA Canberra, ACT, Australia	FC	CW	12,907.5
VHP6	COMMSTA Canberra, ACT, Australia	FC	CW	16,918.8
VHP7	COMMSTA Canberra, ACT, Australia	FC	CW	22,485.0
VHR2	COMMSTA Darwin, NT, Australia	FC	CW	4304.0
VIA	Adelaide Radio, SA, Australia	FC	CW	4272.5, 6463.5
		FC	USB	4428.7
VIB	Brisbane Radio, Qld., Australia	FC	CW	4230.5, 6351.5
		FC	USB	4428.7
VIC	Carnarvon Radio, WA, Australia	FC	CW	4323.0, 4339.0, 6407.5
		FC	USB	4428.7
VID	Darwin Radio, NT, Australia	FC	CW	4272.0, 6463.5, 8478.0
		FC	USB	4428.7, 8762.3, 13,181.4
VIE	Esperance Radio, WA, Australia	FC	CW	4423.6, 6407.5
		FC	USB	4428.7
VIH	Hobart Radio, Tasmania, Australia	FC	USB	4428.7
VII	Thursday Island, Qld., Australia	FC	CW	4228.5, 6333.5
		FC	USB	4428.7, 4620.0
VIM	Melbourne Radio, Vic, Australia	FC	CW	4228.5, 6333.5
		FC	USB	4428.7
VIO	Broome Radio, WA, Australia	FC	CW	6407.5
		FC	USB	4428.7
VIP	Perth Radio, WA, Australia	FC	CW	4229.0, 6512.6
		FC	USB	4428.7, 13,178.3
VIP2	Perth Radio, WA, Australia	FC	CW	6407.5
VIP3	Perth Radio, WA, Australia	FC	CW	8597.0
VIP4	Perth Radio, WA, Australia	FC	CW	12,994.0
VIP5	Perth Radio, WA, Australia	FC	CW	16,947.6
VIP6	Perth Radio, WA, Australia	FC	CW	22,315.5
VIP24	Perth Radio, WA, Australia	FC	CW/TOR	13,074.0
VIP33	Perth Radio, WA, Australia	FC	CW/TOR	8707.4
VIP4Ø	Perth Radio, WA, Australia	FX	CW	13,076.0, 17,200.0
VIR	Rockhampton Radio, Qld., Australia	FC	CW	4255.6, 6333.5
		FC	USB	4428.7
VIS	Sydney Radio, NSW, Australia	FC	USB	4407.0, 4428.7, 6512.6, 8722.0, 8805.7, 13,107.0,
VIS3	Sydney Radio, NSW, Australia	FC	CW	6464.0
VIS4	Sydney Radio, NSW, Australia	FC	CW	8478.0
VIS5	Sydney Radio, NSW, Australia	FC	CW	12,952.5
VIS6	Sydney Radio, NSW, Australia	FC	CW	17,161.3

Call	Location	Service	Mode	kHz
VIS26	Sydney Radio, NSW, Australia	FC	CW	8521.0
VIS28	Sydney Radio, NSW, Australia	FC	CW	8481.0
VIS35	Sydney Radio, NSW, Australia	FC	CW	8452.0
VIS42	Sydney Radio, NSW, Australia	FC	CW	22,474.0
VIS42	Sydney Radio, NSW, Australia	FC	CW	22,474.0
VIS49	Sydney Radio, NSW, Australia	FC	CW	12,979.0
VIS53	Sydney Radio, NSW, Australia	FC	CW	4245.0
VIS54	Sydney Aeradio, NSW, Australia	FA	USB	4696.0
VIS56	Sydney Aeradio, NSW, Australia	FA	USB	10,093.0
VIS58	Sydney Aeradio, NSW, Australia	FA	USB	17,949.0
VIS64	Sydney Radio, NSW, Australia	FC	CW	17,194.4
VIS65	Sydney Radio, NSW, Australia	FC	CW/TOR	8711.6
VIS67	Sydney Radio, NSW, Australia	FC	CW/TOR	13,078.1
VIS69	Sydney Radio, NSW, Australia	FC	CW/TOR	17,204.1
VIS84	Sydney Radio, NSW, Australia	FC	CW	17,204.3
VIT	Townsville Radio, Qld., Australia	FC	CW	4229.5, 6463.5
		FC	USB	4428.7, 8768.5
VIX	Master Station Canberra, Australia	FC	CW	12,907.5
VIX7	Master Station Canberra, Australia	FC	CW	22,485.0
VIY	Christmas Island, Australia	FC	CW	13,065.0
VJB	Derby, WA, Australia	FX	USB	6925.0
VJC	Broken Hill, NSW, Australia	FX	USB	4635.0, 6920.0, 8150.0
VJD	Alice Spring, NT, Australia	FX	USB	6950.0
VJI	Mt. Isa, Qld., Australia	FX	USB	6965.0
VJJ	Charleville, Qld., Australia	FX	USB	6845.0
VJM	Macquarie Island, Antarctica	FX	CW/USB	12,255.0, 14,415.0
VJN	Cairns, Qld., Australia	FX	USB	5145.0, 5300.0, 5865.0 7465.0
VJP	Perth (Gnangara), WA, Australia	FX	ISB	9222.5, 14,535.0, 20,780.0
VJQ	Kalgoorlie, WA, Australia	FX	USB	5865.0, 6825.0
VJT	Carnarvon, WA, Australia	FX	USB	6890.0, 6960.0
VJU	Kingston, Norfolk Island	FX	ISB	7785.0, 10,953.0, 16,330.0
VJY	Darwin, NT, Australia	FX	USB	6840.0, 6975.0
VKA	Adelaide, SA, Australia	FX	USB	4560.0, 5180.0, 7660.0, 10,505.0, 13,730.0
VKC	Melbourne, Vic., Australia	FX	USB	4560.0, 4780.0, 7660.0, 10,505.0, 13,730.0
VKF	Wyndham, WA, Australia	FX	USB	6945.0
VKG	Sydney, NSW, Australia	FX	USB	13,730.0
VKH	Sydney (Doonside), NSW, Australia	FX	USB	12,227.5
VKI	Perth, WA, Australia	FX	USB	7660.0, 15,505.0, 13,730.0
VKJ	Meekatharra, WA, Australia	FX	USB	6960.0
VKL	Port Hedland, WA, Australia	FX	USB	6960.0
VKM	Darwin, NT, Australia	FX	USB	13,730.0
VKO	unknown	FA	CW	11,208.0
VKR	Brisbane, Qld, Australia	FX	USB	4560.0, 6905.0, 7660.0, 10,505.0, 13,730.0
VKT	Hobart, Tasmania, Australia	FX	USB	4560.0, 7660.0, 10,505.0, 13,730.0
VKV1	Cato Is., Qld., Australia	FX	CW	6895.0
VKV2	Frederick, Qld., Australia	FX	CW	6937.0
VKV3	Marion, Qld., Australia	FX	CW	6937.0
VKX	Canberra, ACT, Australia	FX	USB	7660.0, 10,505.0, 13,730.0
VLN	Sydney (Doonside), NSW, Australia	FX	ISB	15,610.0
VLN2Ø	Sydney (Doonside), NSW, Australia	FX	USB	9250.0, 9330.0
VLN32	Sydney (Doonside), NSW, Australia	FX	USB	13,500.0
VLN33	Sydney (Doonside), NSW, Australia	FX	ISB	13,800.0
VLN51	Sydney (Doonside), NSW, Australia	FX	USB	26,700.0
VLN63	Sydney (Doonside), NSW, Australia	FX	USB	13,640.0
VLS	Sydney Aeradio, NSW, Australia	FA	A3H	6680.0, 10,017.0
VLV	Mawson Base, Antarctica	FX	CW/RTTY	7922.5
		FX	CW/USB	12,255.0, 14,415.0, 15,845.0, 17,480.0, 19, 255.0
VLZ	Davis Base, Antarctica	FX	CW/USB	6850.0, 12,255.0, 14,415.0, 15,845.0, 17,480.0. 19,255.0
VL4SX	Birdsville, Qld., Australia	FX	USB	7422.0

Call	Location	Service	Mode	kHz
VL5LM	Alice Springs, NT, Australia	FX	USB	7422.0
VL5MB	Darwin, NT, Australia	FX	ISB	4441.5, 7998.0
VL5MR	Grooteyland, NT, Australia	FX	LSB	8053.0
VL6BA	Hammersley, WA, Australia	FX	USB	7310.0
VL6NX	Wanneroo, WA, Australia	FX	USB	7312.0
VL6OF	Giles, WA, Australia	FX	USB	7422.0
VMH	unknown	FX	CW	5418.9
VNA2	Sydney, NSW, Australia	FX	CW/USB	18,880.0
VNA3	Melbourne, Vic., Australia	FX	CW/USB	4550.0, 6810.0, 9440.0
VNA4	Brisbane, Qld., Australia	FX	CW/USB	4550.0, 6810.0, 9440.0, 18,880.0
VNA5	Adelaide, SA, Australia	FX	CW/USB	4550.0, 6810.0, 9440.0, 18,880.0
VNA6	Perth, WA, Australia	FX	CW/USB	4550.0, 6810.0, 9440.0, 18,880.0
VNA21	Townsville, Qld., Australia	FX	CW/USB	4550.0, 6810.0, 9440.0
VNA7	Hobart, Tasmania, Australia	FX	CW/USB	4550.0, 6810.0, 9440.0
VNG	Lyndhurst, Vic., Australia	SS	A2/AM	4500.0, 7500.0, 12,000.0
VNJ	Casey Base, Antarctica	FX	CW/RTTY	7922.5
		FX	CW/USB	6850.0, 12,225.0, 14,415.0, 15,845.0, 17,480.0, 19,255.0
VNM	Melbourne, Vic., Australia	FX	CW/USB	12,225.0, 14,415.0, 15,845.0, 19,255.0
VNP	Cap Leveque Radio, WA, Australia	FC	CW	16,945.0
VNV	Sydney (Doonside), NSW, Australia	FX	ISB	6970.0, 9475.5
VNV3Ø	Sydney (Doonside), NSW, Australia	FX	ISB	7980.0
VNV49	Sydney (Doonside), NSW, Australia	FX	ISB	11,465.0
VNV52	Sydney (Doonside), NSW, Australia	FX	USB	9046.0
VNV56	Sydney (Doonside), NSW, Australia	FX	ISB	15,800.0
VNV71	Sydney (Doonside), NSW, Australia	FX	USB	14,945.0
VNV113	Sydney (Doonside), NSW, Australia	FX	ISB	14,930.0
VNV13Ø	Sydney (Doonside), NSW, Australia	FX	USB	11,435.2
VNZ	Port Augusta, SA, Australia	FX	USB	6890.0, 8165.0
VOK	Cartwright CG Radio, Nfld., Canada	FC	USB	4376.0
VPC	Falkland Islands Radio, Falkland Is.	FC	CW	6386.5, 17,132.0
VPC	Port Stanley, Falkland Islands	FX	ISB	13,482.0, 19,950.0
VPC24	Port Stanley, Falkland Islands	FX	ISB	24,145.0
VPS22	Cape d'Aguilar Radio, Hong Kong	FC	CW	22,536.0
VPS25	Cape d'Aguilar Radio, Hong Kong	FC	CW	6371.0
VPS35	Cape d'Aguilar Radio, Hong Kong	FC	CW	8539.0
VPS36	Cape d'Aguilar Radio, Hong Kong	FC	CW	8584.0
VPS61	Cape d'Aguilar Radio, Hong Kong	FC	CW	13,044.0
VPS6Ø	Cape d'Aguilar Radio, Hong Kong	FC	CW	13,020.4
VPS79	Cape d'Aguilar Radio, Hong Kong	FC	CW	16,987.0
VPS8Ø	Cape d'Aguilar Radio, Hong Kong	FC	CW	17,096.0
VQJ366	Honiara, Solomon Islands	FX	LSB	6831.5
VRD	Hong Kong, Hong, Kong	FX	CW/RTTY	9285.0
VRN	Cape d'Aguilar Radio, Hong Knog	FC	CW	8628.0
VRN35	Cape d'Aguilar Radio, Hong Kong	FC	CW	8619.0
VRN6Ø	Cape d'Aguilar Radio, Hong Kong	FC	CW	13,031.0
VRN8Ø	Cape d'Aguilar Radio, Hong Kong	FC	CW	17,192.0
VRT	Bermuda (Hamilton) Radio, Bermuda	FC	CW	4....., 6487.5, 8449.4 12,709.2, 16,947.6, 17,124.5
		FC	USB	8768.5, 17,285.6
VSD	Halley Base, Antarctica	FX	CW/USB	9106.0
VTG3	Bombay Naval Radio, India	FC	CW	8534.0
VTG4	Bombay Radio Naval, India	FC	CW	4268.0
VTG5	Bombay Naval Radio, India	FC	CW	6467.0
VTG7	Bombay Naval Radio, India	FC	CW	12,808.5
VTG8	Bombay Naval Radio, India	FC	CW	16,440.0
VTG9	Bombay Naval Radio, India	FC	CW	22,377.9, 22,387.0
VTK	Tuticorin Naval Radio, India	FX	CW	9865.0
VTK3	Tuticorin Naval Radio, India	FX	CW	19,449.8
VTK6	Tuticorin Naval Radio, India	FX	CW	17,630.0
VTO4	Vishakhapatnam Naval Radio, India	FC	CW	8566.0

Call	Location	Service	Mode	kHz
VTP3	Vishakhapatnam Naval Radio, India	FC	CW	8618.0
VTP4	Vishakhapatnam Naval Radio, India	FC	CW	4237.0, 8647.4
VTP5	Vishakhapatnam Naval Radio, India	FC	CW	6418.0
VTP6	Vishakhapatnam Naval Radio, India	FC	CW	8634.0, 8643.9, 12,839.3
VWB	Bombay Radio, India	FC	CW	4....., 8514.0, 8630.0, 12,710.0, 12,966.0, 16,935.0, 22,550.9
VWC	Calcutta Radio, India	FC	CW	4286.0, 8526.0, 12,745.5
VWM	Madras Radio, India	FC	CW	4301.0, 8674.4, 12,718.5, 16,975.0
VXA	"Edmonton Military", Alta, Canada	FA	CW	4704.0, 6753.0
		FA	USB	6705.0, 11,209.0, 13,221.0, 13,254.0
VXN9	Nicosia, Cyprus	FX	USB	13,971.0, 14,445.0
VXV9	Golan Heights, Syria	FX	USB	13,971.0, 14,445.0
VYW61	Montreal Aeradio, PQ, Canada	FA	USB	11,319.0
VZLH	Lord Howe Island, Australia	AX	USB	9815.0
VZSY	Sydney (Llandilo), NSW, Australia	AX	ISB	5940.0
WAK	New Orleans Radio, LA	FC	USB	4369.8, 4419.0
WAR	Washington, DC	FX	CW	5216.5, 14,403.0
WAR42	Mt. Weather, MD	FX	USB	12,216.0
WBH29	Kodiak Radio, AK	FC	USB	4125.0
WBL	Buffalo Radio, NY	FC	USB	4382.2, 4410.1, 8796.4
WBR7Ø	Miami, FL	FX	USB	8125.0
WCC	Chatham Radio, MA	FC	CW	4238.0, 4268.0, 4331.0, 6333.5, 6337.0, 6376.0, 8586.0, 8630.0, 12,925.5, 12,961.5, 13,033.5, 16,933.2, 16,972.0, 22,518.0
WCM	Pittsburgh Radio, PA	FC	USB	4063.0, 4382.2, 4410.1, 6515.7, 8213.6, 8737.5, 12,333.1, 16,518.9
WDB2	Nome Aeradio, AK	FA	USB	4696.0
WEE73	New York, NY	FX	ISB	13,612.0
WEH97	New York, NY	FX	USB	7610.0
WEK34	New York, NY	FX	USB	14,534.0
WEK53	New York, NY	FX	USB	13,427.5
		FX	LSB	13,434.0
WEK54	New York, NY	FX	USB	14,600.0
WEK64	New York, NY	FX	USB	14,609.0
WEK67	New York, NY	FX	ISB	17,640.0
WEK69	New York, NY	FX	USB	19,379.8
WEM2Ø	New York, NY	FX	USB	10,222.6
WET48	New York, NY	FX	ISB	18,207.5
WFA29	New York, NY	FX	FAX	9290.0
WFE34	New York, NY	FX	ISB	14,875.0
WFH29	New York, NY	FX	FAX	9389.5
WFK21	New York, NY	FX	CW/FAX	11,461.0
WFK34	New York, NY	FX	ISB	14,630.0
WFK38	New York, NY	FX	CW/FAX	18,508.4
WFK39	New York, NY	FX	CW/FAX	19,849.2
WFK5Ø	New York, NY	FX	FAX/FSK	10,340 0
WFK67	New York, NY	FX	FAX	17,436.5
WFL25	New York, NY	FX	ISB	15,730.0
WFL35	New York, NY	FX	CW	15,823.5
WFL51	New York, NY	FX	FAX	11,035.0
WFM55	New York, NY	FX	CW	15,778.0
WFM55	New York, NY	FX	FAX	15,785.0
WFN2Ø	New York, NY	FX	CW/FAX	20,798.5
WFN23	New York, NY	FX	CW/FAX	23,075.0
WFN52	New York, NY	FX	FAX	22,925.0
WFN	Louisville, KY	FC	USB	4115.7, 4410.1, 6212.4, 8725.1, 13,103.9, 17,291.8
WGK	St. Louis, MO	FC	USB	4410.1, 6209.3, 6212.4, 8737.5, 13,103.9, 17,291.8

Call	Location	Service	Mode	kHz
WGY9Ø3	Olney, MD	FX	USB	12,216.0
WGY9Ø7	Kansas City, MO	FX	USB	20,026.0
WGY9Ø9	Santa Rosa, CA	FX	USB	10,493.0, 12,216.0, 20,026.0
WIC	Tulsa, OK	FX	LSB	5327.0
WJG	Memphis Radio, TN	FC	USB	4087.8, 4382.2, 6209.3, 6515.7, 8201.0, 8725.1, 12,333.1, 16,518.9, 17,291.8
WKB2Ø	Slidell, LA	FX	CW	10,460.0
WLC	Rogers City, MI	FC	USB	4087.8
WLC	Rogers City Radio, MI	FC	CW	4316.0
		FC	USB	4382.2, 4410.1, 8796.4
WLO	Mobile Radio, AL	FC	CW	4256.5, 6446.0, 8445.0, 8474.5, 12,704.5, 12,885.0, 12,886.5, 16,967.5, 16,968.5, 17,172.4, 22,319.5
		FC	CW/TOR	6501.5, 8707.0, 8711.9, 8717.0, 12,886.5, 13,078.5, 13,083.5, 17,204.5, 17,209.5
		FC	USB	4397.7, 4413.2, 8790.2, 8805.7, 8808.8, 13,134.9, 13,175.2, 13,178.3, 17,251.5, 17,329.0, 17,356.9, 22,707.6
WMA29	San Francisco (Dixon), CA	FX	USB	9170.0
WMC94	San Francisco (Oakland), CA	FX	USB	14,792.5
WMH	Baltimore Radio, MD	FC	CW	4346.0, 6333.5, 6351.5 8610.0, 8686.0, 12,952.9, 17,093.6
WMI	Lorain Radio, OH	FC	USB	4382.2, 4410.1, 8796.4
WNU31	Slidell Radio, LA	FC	CW	4294.0
WNU32	Slidell Radio, LA	FC	CW	6326.5
WNU33	Slidell Radio, LA	FC	CW	8525.0
WNU34	Slidell Radio, LA	FC	CW	12,826.5
WNU35	Slidell Radio, LA	FC	CW	16,861.7
WNU36	Slidell Radio, LA	FC	CW	22,432.0
WNU41	Slidell Radio, LA	FC	CW	4310.0
WNU42	Slidell Radio, LA	FC	CW	6389.6
WNU43	Slidell Radio, LA	FC	CW	8688.0
WNU44	Slidell Radio, LA	FC	CW	12,869.0
WNU45	Slidell Radio, LA	FC	CW	8570.0
WNU46	Slidell Radio, LA	FC	CW	13,011.0
WNU53	Slidell Radio, LA	FC	CW	17,117.6
WNU54	Slidell Radio, LA	FC	CW	22,457.9
WOE	Lantana Radio, FL	FC	CW	4292.0, 6411.4, 8486.0, 12,970.7, 17,159.4, 17,160.8, 22,503.0
WOM	Miami Radio, FL	FC	USB	4363.5, 4391.5, 4407.0, 4425.6, 4428.7, 8722.0, 8731.3, 8746.8, 8793.3, 8811.9, 13,116.3, 13,122.5, 13,125.6, 13,144.2, 17,232.9, 17,257.7, 17,260.8, 17,273.9, 22,639.4, 22,642.5, 22,661.1
WOO	New York (Ocean Gate, NJ), Radio, NY	FC	USB	4385.3, 4388.4, 4403.9, 4422.5, 8740.6, 8749.9, 8762.3, 8796.4, 13,107.0, 13,128.7, 13,131.8, 13,184.5, 13,184.5, 13,190.7, 17,245.3, 17,260.8, 17,291.8, 17,310.4, 17,325.9, 22,596.0, 22,608.4, 22,623.9
WPA	Port Arthur Radio, TX	FC	CW	4322.0, 6436.3, 8550.0, 12,839.6, 16,918.8, 22,318.0
WPD	Tampa Radio, FL	FC	CW	4274.0, 6365.5, 8615.5, 13,051.5, 17,170.4
WQB23	Akron, OH	FX	CW	13,945.0

Call	Location	Service	Mode	kHz
WQB35	Akron, OH	FX	CW	15,780.0
WSL	Amagansett Radio, New York	FC	CW	4342.7, 6414.1, 6418.0, 8514.0, 8658.0, 12,660.0, 12,997.5, 13,024.9, 16,997.0, 17,020.8, 22,486.9
WSY7Ø	New York Aeradio, NY	FA	USB	5652.0, 8868.0, 13,272.0
WUB4	Baltimore, MD	FX	LSB	5015.0
WUD	Omaha, NE	FX	USB	5400.0
WUD2	Buffalo, NY	FX	USB	5400.0
WUD6	St. Paul, MN	FX	USB	5400.0
WUD7	Rock Island, IL	FX	USB	5400.0
WUH	Omaha, NE	FX	USB	5398.5
WUH7	Rock Island, IL	FX	USB	5400.0
WUJ3	Portland, OR	FX	USB	6785.0
WWD	La Jolla (Scripps) Radio, CA	FC	CW	17,105.0
		FC	FAX/USB	8644.1
		FC	USB	4431.8, 13,147.3
		FX	CW/FAX	17,408.6
WWV	Fort Collins, CO	SS	AM/A2	5000.0, 10,000.0, 15,000.0, 20,000.0
WWVH	Kauai, HI	SS	AM/A2	5000.0, 10,000.0, 15,000.0
XBD492	Mexico City, DF, Mexico	FX	USB	13,377.6
XBD959	unknown, probably in Mexico	FX	USB	7330.1
XDA	Mexico City Radio, DF, Mexico	FC	CW	8567.0, 8575.5, 17,140.0
XDA358	Mexico City, DF, Mexico	FX	USB	13,527.0
XDD212	Mexico City, DF, Mexico	FX	LSB	7350.0
XDD229	Mexico City, DF, Mexico	FX	ISB	10,563.0
XFA	Acapulco Radio, Mexico	FC	CW	4292.0, 6414.5, 8514.0, 8690.0, 12,752.0
XFB2	Campeche Radio, Mexico	FC	CW	8690.0, 12,790.0
XFB3	Ciudad del Carmen Radio, Mexico	FC	CW	12,790.0
XFF2	Coatzacoalcos Radio, Mexico	FC	CW	12,790.0
XFK	La Paz Baja California R., Mexico	FC	CW	8505.0, 12,675.0
XFK2	La Paz Baja California R., Mexico	FC	CW	12,790.
XFL	Mazatlan Radio, Mexico	FC	CW	4250.0, 8470.0, 8514.0 12,703.0
XFL2	Mazatlan Radio, Mexico	FC	CW	12,790.0
XFM	Manzanillo Radio, Mexico	FC	CW	4225.2, 6354.0, 8568.3 12,829.5
XFM2	Manzanillo Radio, Mexico	FC	CW	6442.5, 12,790.0, 17,069.6
XFP	Chetumal Radio, Mexico	FC	CW	4292.0, 8595.5, 12,843.0, 12,843.0
XFQ	Salina Cruz Radio, Mexico	FC	CW	6360.0, 8631.0
XFQ2	Frontera (Tabasco) Radio, Mexico	FC	CW	8690.0, 12,790.0
XFQ3	Salina Cruz Radio, Mexico	FC	CW	8690.0
XFS	Tampico Radio, Mexico	FC	CW	4271.0, 6365.0, 8644.0, 12,761.0
XFS2	Ciudad Madero Radio, Mexico	FC	CW	8690.0, 12,790.0, 17,071.0
XFS3	Tampico Radio, Mexico	FC	CW	6442.5, 8690.0
XFU	Veracruz Radio, Mexico	FC	CW	4285.3, 6367.0, 6389.9, 8656.0, 12,775.0, 16,896.4
		FC	USB	13,125.6
XFU2	Veracruz Radio, Mexico	FC	CW	4346.0, 6442.5, 8690.0 12,687.0, 12,790.0
XFY	Guaymas Radio, Mexico	FC	CW	8513.0
XFY2	Guaymas Radio, Mexico	FC	CW	8690.0, 12,790.0
XJD48	Ottawa, Ont., Canada	FX	CW	9105.0
XJE57	Ottawa (Almonte), Ont., Canada	FX	CW	10,390.0, 21,785.0, 24,110.0
XJP26	unknown, probably Nfld., Canada	FX	USB	4888.4
XLI334	Frobisher Bay, NWT, Canada	FX	USB	5031.0, 5033.0
XLN45	Thunder Bay, Ont., Canada	FX	USB	4460.0
XMH356	Dartmouth, NS, Canada	FX	USB	4982.0, 5876.5
XNZ583	St. Johns, Nfld., Canada	FX	USB	4883.0
XNZ584	Tote River, Nfld., Canada	FX	USB	4883.0
XNZ585	Bernard River, Nfld., Canada	FX	USB	4883.0

Call	Location	Service	Mode	kHz
XOP654	Fort Chimo, PQ, Canada	FX	USB	4483.5
XOU5Ø1	Eldorado, Sask., Canada	FX	LSB	5075.0
XPH	Thule AB, Greenland	FA	USB	6738.0, 13,201.0, 17,975.0, 23,337.0
XQ8AFI	Cerro Tololo Observatory, Chile	FX	LSB	20,875.0
XSC	Fangshan Radio, PRC	FC	CW	8698.0
		FC	FAX	12,905.5
XSE	Qinhuangdao Radio, PRC	FC	CW	8484.0, 22,384.0
XSG	Shanghai Radio, PRC	FC	CW	4271.0, 4319.0, 6414.5, 6452.0, 6484.5, 8514.0, 11,156.0, 12,661.5, 12,871.5, 12,898.0, 16,871.3, 22,363.0, 22,401.0, 22,455.0, 22,467.0
		FC	FAX	22,460.0
XSG3	Shanghai Radio, PRC	FC	CW	8502.0, 8665.0, 17,103.2, 22,312.5
XSG4	Shanghai Radio, PRC	FC	CW	8451.0
XSG7	Shanghai Radio, PRC	FC	CW	12,856.0
XSG8	Shanghai Radio, PRC	FC	CW	16,916.5
XSG26	Shanghai Radio, PRC	FC	CW	8487.0
XSG28	Shanghai Radio, PRC	FC	CW	12,953.5
XSG29	Shanghai Radio, PRC	FC	CW	17,002.5
XSH	Basuo Radio, PRC	FC	CW	8634.0, 12,912.0
XSJ	Zhanjiang Radio, PRC	FC	CW	4225.0, 4295.0, 8541.0 8614.0, 12,890.0
XSK4	Haimen Radio, PRC	FC	CW	4256.0, 8640.0
XSL	Fuzhou Radio, PRC	FC	CW	4343.0, 6407.5, 8598.0
XSM	Xiamen Radio, PRC	FC	CW	8460.0, 16,876.1, 22,545.0
XSP	Shantou Radio, PRC	FC	CW	4228.0, 8457.0, 12,709.0
XSQ	Guangzhou Radio, PRC	FC	CW	4288.0, 4340.1, 6390.0, 8624.0, 12,877.0, 13,054.0, 17,080.0, 22,324.0, 22,351.5, 22,526.0
XSQ4	Guangzhou Radio, PRC	FC	CW	8490.0, 8513.0, 12,698.9, 16,950.1
XSQ7	Guangzhou Radio, PRC	FC	CW	16,882.9, 16,884.5, 22,452.8
XSR	Haikow Radio, PRC	FC	CW	4266.0, 8702.0
XST	Qingdao Radio, PRC	FC	CW	6369.0, 6477.5, 8545.8, 12,799.5, 16,973.6
XSU	Yantai Radio, PRC	FC	CW	4274.0, 8574.0, 12,817.5
XSV	Tianjin Radio, PRC	FC	CW	4283.1, 6383.0, 6484.5, 8599.9, 8617.0, 8630.0, 12,822.0, 12,969.0, 16,862.5, 17,129.0
XSW	Kaohsiung Radio, Taiwan	FC	CW	8582.0, 12,727.5, 16,940.0
XSX	Keelung Radio, Taiwan	FC	CW	8445.0, 12,695.2, 22,459.0
XSY	Hualien Radio, Taiwan	FC	CW	8546.0
XSZ	Dairen Radio, PRC	FC	CW	4305.0, 6333,5, 8678.0,
		FC	CW	8694.0, 12,799.5
XTA9	Ouagadougou, Upper Volta	FX	USB	11,095.3
XTA49	Ouagadougou, Upper Volta	FX	ISB	14,921.0
XTA63	Ouagadougou, Upper Volta	FX	ISB	16,386.0
XTA77	Ouagadougou, Upper Volta	FX	ISB	7700.0
XUK2	Kompong Som-Ville Radio, Cambodia	FC	CW	8480.0
XUK3	Kompong Som-Ville Radio, Cambodia	FC	CW	8680.0
XUK4	Kompong Som-Ville Radio, Cambodia	FC	CW	4342.0
XUQC	unknown	FX	CW	7536.9
XVG	Haiphong Radio, Vietnam	FC	CW	16,950.0
XVG9	Haiphong Radio, Vietnam	FC	CW	8470.0
XVG22	Haiphong Radio, Vietnam	FC	CW	4315.0
XVN47	Hanoi, Vietnam	FX	FAX	18,940.0
XVS	Ho Chi Minh-Ville, Vietnam	FX	CW	5263.0, 14,975.0
XVS8	Ho Chi Minh-Ville Radio, Vietnam	FC	CW	8590.0, 17,146.4
XVS9	Ho Chi Minh Ville Radio, Vietnam	FC	CW	13,042.0
XVS37	Ho Chi Minh-Ville, Vietnam	FX	CW	8194.0

Call	Location	Service	Mode	kHz
XVT5	Danang Radio, Vietnam	FC	CW	8570.0
XVX2ØB	Hanoi, Vietnam	FX	LSB	20,992.0
XXG	Macao Radio, Macao	FC	CW	8469.0
XYN31	Rangoon, Burma	FX	USB	19, 628.5
XY084	Rangoon, Burma	FX	ISB	14,794.8
XYQ65	Mandalay, Burma	FX	USB	
XYR6	Rangoon Radio, Burma	FC	CW	4292.0
XYR7	Rangoon Radio, Burma	FC	CW	8441.0
XYR8	Rangoon Radio, Burma	FC	CW	12,867.0
XYR9	Rangoon Radio, Burma	FC	CW	17,189.6
XYR24	Rangoon Radio, Burma	FC	CW	22,375.0
X3Q	unknown	FX	CW	12,282.0
YAB8	Kabul, Afghanistan	FX	USB	16,238.0
YAK	Kabul, Afghanistan	FX	ISB	8098.5, 9093.0, 18,640.0
YCN2	Pontianak, Indonesia	FX	CW	8090.4
YIE35	Baghdad (Abu Ghuraib), Iraq	FX	ISB	13,410.0
YIR	Basrah Control, Iraq	FC	CW	4220.0, 6330.0, 6339.5, 6493.2, 8440.0, 8458.0, 8668.0, 12,659.9, 16,880.0, 16,906.0, 22,338.2
YJM3	Port-Vila Radio, Vanuatu	FC	CW	12,678.0
YJM4	Port-Vila Radio, Vanuatu	FC	CW	4343.0
YJM6	Port-Vila Radio, Vanuatu	FC	CW	6344.0
YJM8	Port Vila Radio, Vanuatu	FC	CW	8502.0
YKI	Tartous Radio, Syria	FC	CW	12,698.0
YKM5	Baniyas Radio, Syria	FC	CW	16,967.0
YKM7	Lattakia Radio, Syria	FC	CW	8625.0, 12,728.0
YKW1Ø7	Damascus, Syria	FX	USB	16,397.0
YKY32	Damascus, Syria	FX	USB	14,747.5
YMA22	Ankara, Turkey	FX	FAX	6790.0
YMA35	Ankara, Turkey	FX	FAX	4560.0
YMH3	Bandirma, Turkey	FX	CW	6965.0
YMN	New Delhi, India	FX	CW	16,455.0
YMN5	Karachi, Pakistan	FX	CW	16,455.0
YMN6	Ankara, Turkey	FX	CW	16,455.0
YMN12	Madrid, Spain	FX	CW	16,455.0
YMS	Warsaw, Poland	FX	CW/RTTY	13,931.0, 14,741.0
YMY	Samsum, Turkey	FX	CW	6965.0
YO99	Bucharest, Rumania	FX	CW	10,390.0
YPO	Giurgiu Radio, Rumania	FC	CW	8663.0
YQBF	unknown	FX	CW	6475.0, 9993.0, 10,125.0, 14,834.7
YQI2	Constanta Radio, Rumania	FC	CW	12,757.0
YQI3	Constanta Radio, Rumania	FC	CW	6473.5
YQI4	Constanta Radio, Rumania	FC	CW	4323.5, 16,954.0, 17,037.0
YQI5	Constanta Radio, Rumania	FC	CW	8459.0
YQI7	Constanta Radio, Rumania	FC	CW	22,448.0
YRA	Bucharest Aeradio, Rumania	FA/MA	CW	11,357.0
YRR4	Bucharest, Rumania	FX	CW	4002.0
YUR	Rijeka Radio, Yugoslavia	FC	CW	4346.0, 4346.0
YUR3	Rijeka Radio, Yugoslavia	FC	CW	8700.0
YUR4	Rijeka Radio, Yugoslavia	FC	CW	8445.0
YUR5	Rijeka Radio, Yugoslavia	FC	CW	12,780.5
YUR6	Rijeka Radio, Yugoslavia	FC	CW	12,907.5
YUR7	Rijeka Radio, Yugoslavia	FC	CW	16,942.7
YUR8	Rijeka Radio, Yugoslavia	FC	CW	17,045.6
YUR9	Rijeka Radio, Yugoslavia	FC	CW	22,443.0
YVG	La Guaira Radio, Venezuela	FC	CW	4352.0, 6351.5, 8460.7, 12,817.5, 17,141.6
YVL	Puerto Cabello Radio, Venezuela	FC	CW	4322.0, 17,141.6
YVL5	Caracas, Venezuela	FX	LSB	7389.5
YVM	Puerto Ordaz Radio, Venezuela	FC	CW	4322.0, 6372.5, 8453.0
YVTO	Caracas, Venezuela	BC	AM	6100.0
YVZ3	Caracas, Venezuela	FX	CW	15,738.2
YWM	Maracaibo Naval Radio, Venezuela	FX	USB	17,553.5

Call	Location	Service	Mode	kHz
YWM3	Maracaibo Naval Radio, Venezuela	FX	CW	17,590.0, 20,690.0
YZC2	Belgrade (Makis), Yugoslavia	FX	CW	14,632.0
YZG6	Belgrade (Makis), Yugoslavia	FX	CW/RTTY	5226.0
YZI4	Belgrade (Makis), Yugoslavia	FX	CW/RTTY	16,343.0
YZJ4	Belgrade (Makis), Yugoslavia	FX	CW	19,605.0, 19,865.5
YZJ7	Belgrade (Makis), Yugoslavia	FX	CW	16,065.0
YZK2	Belgrade (Makis), Yugoslavia	FX	CW/RTTY	20,957.0
YZK4	Belgrade (Makis), Yugoslavia	FX	CW	22,787.2
YZM4	Belgrade (Makis), Yugoslavia	FX	CW	23,117.0
YZZ1	Belgrade, Yugoslavia	FX	FAX	5800.0
Y2V25	Nauen, GDR	FX	FAX	14,825.0
Y2V27	Nauen, GDR	FX	CW/RTTY	22,800.0
Y2V37	Berlin, GDR	FX	CW/FAX	17,435.0
Y2V38	Berlin (Nauen), GDR	FX	FAX	18,823.0
Y2V47	Berlin, GDR	FX	CW/FAX	13,895.0
Y3S	Berlin (Nauen), GDR	FX	CW	4525.0
Y4X	unknown	FX	CW	6377.0
Y5M	Ruegen Radio, GDR	FC	CW	4233.5, 4240.5, 4242.0, 6358.5, 6449.0, 6466.0, 8443.0, 8463.0, 8484.0, 8660.0, 8696.0, 12,681.5, 12,702.0, 12,745.0, 12,860.5, 13,062.5, 16,892.9, 16,907.3, 16,965.0, 17,000.0, 17,100.0, 22,334.5, 22,393.0, 22,401.0, 22,422.0, 22,437.0, 22,481.0, 22,545.0, 25,069.0, 25,308.0, 26,455.0
Y5P	Ruegen Radio, GDR	FC	USB	8722.0, 17,319.7
Y7A24	Berlin, GDR	FX	CW/RTTY	4840.0
Y7A24	Berlin, GDR	FX	CW	6706.0
Y7A36	Berlin, GDR	FX	CW	9062.0
Y7A37	Berlin, GDR	FX	CW	9078.0
Y7A49	Berlin, GDR	FX	CW	11,448.0
Y7A55	Berlin, GDR	FX	CW	13,950.0
Y7A6Ø	Berlin, GDR	FX	CW	14,817.5
Y7A64	Berlin (Nauen), GDR	FX	CW/RTTY	16,243.1
Y7A65	Berlin, GDR	FX	CW/RTTY	16,268.0
Y7A68	Berlin, GDR	FX	CW/RTTY	17,125.9
Y7A71	Berlin, GDR	FX	CW/RTTY	17,176.0
Y7A73	Berlin, GDR	FX	CW/RTTY	18,608.0
Y7A79	Berlin, GDR	FX	CW/FSK	19,885.0
Y7A81	Berlin, GDR	FX	CW	20,170.5
Y7A83	Berlin, GDR	FX	CW/RTTY	22,721.9
Y7A84	Berlin (Nauen), GDR	FX	ISB/RTTY	22,732.0
Y7B32	Belgrade, Yugoslavia	FX	CW	7812.0
Y7B94	unknown, probably in Europe	FX	CW	16,354.0
Y7D35	unknown, probably in Africa	FX	Cw/RTTY	26,450.0
Y7D49	unknown, probably in Africa	FX	CW	24,800.0
Y7F48	unknown, probably in Middle East	FX	CW	16,356.0
Y7F54	Algiers, Algeria	FX	CW	14,410.0
Y7F68	Beirut, Lebanon	FX	CW	14,411.0
Y7G29	unknown, probably in Far East	FX	CW	17,362.0
Y7G34	unknown, probably in Far East	FX	CW	17,362.0, 24,005.0
Y7L36	Havana, Cuba	FX	CW	14,410.0
ZAD2	Durres Radio, Albania	FC	CW	4220.0, 4302.0, 6434.0, 8696.0, 12,690.0, 17,173.0
ZBH	Grytviken, So. Georgia Is.	FX	CW/USB	9106.0
ZBI67	Ascension Island	FX	USB	14,767.0
ZBP	Pitcairn Island	FX	CW	7859.0, 9200.0, 10,264.0, 12,110.0, 15,718.0, 18,407.0, 21,804.0
ZCB4	unknown	FC	CW	4305.0, 8818.6
ZEF81	Salisbury, Zimbabwe	FX	CW/RTTY	16,125.0, 17,385.0, 19,250.0, 20,970.0

Call	Location	Service	Mode	kHz
ZEG25	Salisbury, Zimbabwe	FX	LSB	22,851.5
ZEG26	Salisbury, Zimbabwe	FX	USB	23,850.0
ZEN32	Cape d'Aguilar, Hong Kong	FX	ISB	7427.5
ZEN42	Cape d'Aguilar, Hong Kong	FX	ISB	9331.0
ZEN44	Cape d'Aguilar, Hong Kong	FX	ISB	9442.5
ZEN63	Cape d'Aguilar, Hong Kong	FX	ISB	13,688.0
ZEN69A	Cape d'Aguilar, Hong Kong	FX	CW	14,987.0
ZEN77	Cape d'Aguilar, Hong Kong	FX	ISB	16,274.5
ZEN89	Cape d'Aguilar, Hong Kong	FX	USB	18,852.5
ZEN91	Cape d'Aguilar, Hong Kong	FX	USB	19,265.0
ZEN95	Cape d'Aguilar, Hong Kong	FX	USB	20,162.0
ZEN96	Cape d'Aguilar, Hong Kong	FX	LSB	20,224.0
ZEN97	Cape d'Aguilar, Hong Kong	FX	ISB	20,480.0
ZEN99	Cape d'Aguilar, Hong Kong	FX	ISB	20,790.0
ZEO66	Cape d'Aguilar, Hong Kong	FX	USB	13,941.5
ZFD43	Hamilton, Bermuda	FX	ISB	9106.0
ZFD44	Hamilton, Bermuda	FX	ISB	9909.5
ZFD49	Hamilton, Bermuda	FX	ISB	10,635.9
ZFD62	Hamilton, Bermuda	FX	USB	13,503.5
ZFD82	Hamilton, Bermuda	FX	ISB	17,695.0
ZHF43	Signy Base, South Orkney Islands	FX	USB	9106.0
ZHF44	Faraday Base, Antarctica	FX	CW	9106.0
ZHF45	Rothera Base, Adelaide Island	FX	CW/USB	9106.0, 11,557.0
ZHH	St. Helena Radio, St. Helena Is.	FC	CW	12,963.5
ZKAK	Auckland Aeradio, New Zealand	AX	USB	4940.0
ZKI	Taujunu, Manihiki Atoll, Cook Is.	FX	USB	14,445.0
ZKJ	Omaka, Tongareva Atoll, Cook Is.	FX	USB	14,445.0
ZKP5	Nassau, Cook Islands	FX	USB	14,445.0
ZKR	Rarotonga Radio, Cook Islands	FC	CW	8449.0
		FC	USB	8793.3
ZKRG	Rarotonga Aeradio, Cook Islands	AX	USB	4940.0, 14,500.0
		FA	USB	11.319.0
ZKS	Rarotonga, Cook Islands	AX	USB	7740.0
		FX	USB	4832.0, 5930.0, 7390.0, 9095.0, 9386.0, 14,445.0, 14,900.0
ZKS28	Ratotonga, Cook Islands	FX	ISB	11,100.0
ZKS34	Rarotonga, Cook Islands	FX	ISB	7840.0
ZKW	Christchurch, New Zealand	FA	USB	8975.0, 11,180.5, 13,205.0
ZKW	Wanganui, New Zealand	FA	USB	8975.0
ZKX	Auckland, New Zealand	FA	USB	8975.0, 13,205.0
ZKY	Palmerston North, New Zealand	FA	USB	8975.0
		FX	USB	8055.0, 15,690.0
ZLB	Awarua Radio, New Zealand	FC	USB	4134.6
ZLB2	Awarua Radio, New Zealand	FC	CW	4277.0
ZLB3	Awarua Radio, New Zealand	FC	CW	6393.5
ZLB4	Awarua Radio, New Zealand	FC	CW	8504.0
ZLB5	Awarua Radio, New Zealand	FC	CW	12,740.0
ZLB6	Awarua Radio, New Zealand	FC	CW	17,170.4
ZLB7	Awarua Radio, New Zealand	FC	CW	22,533.0
ZLBC	Campbell Island, New Zealand	FX	USB	11,552.0
ZLBC4	Campbell Island, New Zealand	FX	USB	9950.0
ZLBC5	Campbell Island, New Zealand	FX	USB	4601.0
ZLBC6	Campbell Island, New Zealand	FX	USB	12,152.5
ZLC7	Chatham Island, New Zealand	FX	USB	5254.0
ZLD	Auckland Radio, New Zealand	FC	USB	4134.6
ZLK	Christchurch, New Zealand	FA	USB	8997.0
ZLK6	Christchurch, New Zealand	FX	USB	20,928.0
ZLK36	Christchurch (Weedons), New Zealand	FX	ISB	16,152.0
ZLK43	Christchurch (Weedons), New Zealand	FX	USB	12,118.6
ZLK45	Christchurch, New Zealand	FX	CW/FAX	21,834.0
ZLO	Irirangi Naval Radio, New Zealand	FX	USB	19,610.0
ZLO2	Irirangi Naval Radio, New Zealand	FC	CW	4260.4
ZLO3	Irirangi Naval Radio, New Zealand	FC	CW	6336.8
ZLO4	Irirangi Naval Radio, New Zealand	FC	CW	8598.4, 8678.0

Call	Location	Service	Mode	kHz
ZLO5	Irirangi Naval Radio, New Zealand	FC	CW	12,716.9
ZLO6	Irirangi Naval Radio, New Zealand	FC	CW	16,874.5, 17,127.2
ZLO7	Irirangi Naval Radio, New Zealand	FC	CW	22,407.0
ZLP2	Irirangi Naval Radio, New Zealand	FC	CW	4250.0
ZLP3	Irirangi Naval Radio, New Zealand	FC	CW	6435.5
ZLP5	Irirangi Naval Radio, New Zealand	FC	CW	12,943.5
ZLQ8	Scott Base, Antarctica	FX	USB	14,655.0
ZLS6	Irirangi Naval Radio, New Zealand	FC	USB	16,873.5
ZLW	Wellington Radio, New Zealand	FC	USB	4134.6, 6506.4, 8737.5, 13,125.6
		FX	USB	5866.0
ZLX22	Wellington (Himatangi), New Zealand	FX	CW	11,130.0
ZLX31	Wellington (Himatangi), New Zealand	FX	CW	19,488.2
ZLX37	Wellington (Himatangi), New Zealand	FX	CW	14,850.0
ZLX82	Wellington (Himatangi), New Zealand	FX	USB	13,570.0
ZLZ2Ø	Wellington (Himatangi), New Zealand	FC	CW	5915.0
ZLZ22	Wellington (Himatangi), New Zealand	FX	CW	7600.0
ZME3	Raoul (Sunday) Is., Kermandec Is.	FX	USB	9950.0
ZME4	Raoul (Sunday) Is., Kermandec Is.	FX	USB	12,152.5
ZPB74	Asuncion, Paraguay	FX	ISB	7430.0
ZPG14	Asuncion, Paraguay	FX	ISB	11,427.0
ZPG4	Asuncion, Paraguay	FX	ISB	10,415.0
ZPG36	Asuncion, Paraguay	FX	AM	13,605.0
ZPG45	Asuncion, Paraguay	FX	USB	14,580.0
ZPG57	Asuncion, Paraguay	FX	CW	15,780.0
ZPM261	Asuncion, Paraguay	FX	LSB	20,885.0
ZPZ	Asuncion, Paraguay	FX	CW	15,738.2
ZPZ26	Asuncion, Paraguay	FX	ISB	18,512.0
ZRB2	Pretoria (Waterkloof AFB), RSA	FA	USB	15,056.0, 17,993.0, 23,287.0
ZRH	Cape (Fisantekraal) Naval R., RSA	FC	CW/RTTY	16,964.3
ZRH2	Cape (Fisantekraal) Naval R., RSA	FC	CW/RTTY	4246.5
ZRH4	Cape (Fisantekraal) Naval R., RSA	FC	CW/RTTY	8605.0
ZRH5	Cape (Fisantekraal) Naval R., RSA	FC	CW/RTTY	12,947.3
ZRH6	Cape (Fisantekraal) Naval R., RSA	FC	CW/RTTY	17,004.2
ZRH7	Cape (Fisantekraal) Naval R., RSA	FC	CW/RTTY	22,406.0
ZRO2	Pretoria, RSA	FX	FAX	7508.0
ZRO3	Pretoria, RSA	FX	FAX	13,773.0
ZRO4	Pretoria, RSA	FX	FAX	18,240.0
ZRQ2	Cape (Simonstown) Naval Radio, RSA	FC	CW	4223.9, 4352.2
ZRQ3	Cape (Simonstown) Naval Radio, RSA	FC	CW	6336.5
ZRQ4	Cape (Simonstown) Naval Radio, RSA	FC	CW	8470.0
ZRQ5	Cape (Simonstown) Naval Radio, RSA	FC	CW	12,692.3
ZRQ6	Cape (Simonstown) Naval Radio, RSA	FC	CW	16,962.2, 16,964.2
ZRQ7	Cape (Simonstown) Naval Radio, RSA	FC	CW	22,393.2
ZSC4	Cape Town Radio, RSA	FC	CW	8461.0
ZSC6	Cape Town Radio, RSA	FC	CW	8688.5
ZSC7	Cape Town Radio, RSA	FC	CW	17,164.8
ZSC9	Cape Town Radio, RSA	FC	CW	12,698.0
ZSC2Ø	Cape Town Radio, RSA	FC	CW	22,347.5
ZSC21	Cape Town Radio, RSA	FC	CW	6379.5
ZSC22	Cape Town Radio, RSA	FC	CW	8449.0
ZSC23	Cape Town Radio, RSA	FC	CW	12,709.0
ZSC25	Cape Town Radio, RSA	FC	USB	4369.8
ZSC26	Cape Town Radio, RSA	FC	USB	8731.3
ZSC28	Cape Town Radio, RSA	FC	USB	17,254.6
ZSC29	Cape Town Radio, RSA	FC	USB	22,605.3
ZSC33	Cape Town Radio, RSA	FC	CW	4317.0
ZSC34	Cape Town Radio, RSA	FC	CW	6478.0
ZSC36	Cape Town Radio, RSA	FC	CW	6467.0
ZSC37	Cape Town Radio, RSA	FC	CW	8502.0
ZSC38	Cape Town Radio, RSA	FC	CW	12,772.5
ZSC39	Cape Town Radio, RSA	FC	CW	16,890.8
ZSC4Ø	Cape Town Radio, RSA	FC	CW	22,455.0
ZSC41	Cape Town Radio, RSA	FC	CW	25,277.0
ZSC43	Cape Town Radio, RSA	FC	CW	12,724.0

Call	Location	Service	Mode	kHz
ZSC44	Cape Town Radio, RSA	FC	CW	17,018.0
ZSC45	Cape Town Radio, RSA	FC	CW	4261.0
ZSC46	Cape Town Radio, RSA	FC	CW	4291.0
ZSD4	Durban Radio, RSA	FC	CW	4242.5
ZSD37	Durban Radio, RSA	FC	USB	4376.0
ZSD38	Durban Radio, RSA	FC	A3H	8740.6
ZSD39	Durban Radio, RSA	FC	USB	13,172.1
ZSD41	Durban Radio, RSA	FC	USB	17,332.1
ZSD43	Durban Radio, RSA	FC	CW	4240.0
ZSD44	Durban Radio, RSA	FC	CW	6372.0
ZSD45	Durban Radio, RSA	FC	CW	8576.6
ZSD46	Durban Radio, RSA	FC	CW	13,028.0
ZSD47	Durban Radio, RSA	FC	CW	17,075.0
ZSD5Ø	Durban Radio, RSA	FC	CW	8436.5
ZSD51	Durban Radio, RSA	FC	CW	12,663.0
ZSD52	Durban Radio, RSA	FC	CW	16,874.2
ZSJ2	NAVCOMCEN (Silvermine) Cape R., RSA	FC	CW	4283.0
ZSJ3	NAVCOMCEN (Silvermine) Cape R., RSA	FC	CW	6386.3
ZSJ4	NAVCOMCEN (Silvermine) Cape R., RSA	FC	CW	8566.0
ZSJ5	NAVCOMCEN (Silvermine) Cape R., RSA	FC	CW	12,849.0
ZSJ6	NAVCOMCEN (Silvermine) Cape R., RSA	FC	CW	17,132.0
ZSV8	Walvis Bay Radio, RSA	FC	USB	4357.4
ZUD49	Olifantsfontein Radio, RSA	FC	CW	8705.0
ZUD76	Olifantsfontein Radio, RSA	FC	USB	6504.0
ZUD81	Olifantsfontein Radio, RSA	FC	CW/TOR	13,073.2
ZUD214	Olifantsfontein, RSA	FX	LSB	5175.0
ZUJ	Johannesburg (Jan Smuts), RSA	AX	CW	17,571.8
		FA	USB	5688.0
ZUO	Pretoria, RSA	SS	A2	5000.0
ZVE	Manaus Aeradio, Brazil	AX	CW	17,365.0
ZVK	Rio de Janeiro Aeradio, Brazil	AX	CW	17,365.0, 17,695.0
ZVN3	Belem Aeradio, Brazil	AX	CW	5085.0
ZWBE	Belem Aeradio, Brazil	AX	CW	18,568.0
ZWRJ	Rio de Janeiro Aeradio, Brazil	AX	CW	18,568.0
Z2H27	Salisbury, Zimbabwe	FX	CW/USB	16,270.5, 26,158.5
2K6	unknown	FX	CW	14,718.0
2Z7	unknown	FX	CW	13,685.0
3AC8	Monaco Radio, Monaco	FC	USB	8728.2
3AC9	Monaco Radio, Monaco	FC	USB	8743.7
3AD1	Monaco, Monaco	FX	USB	12,294.0, 12,305.0
3BA3	Mauritius (Bigara) Radio, Mauritius	FC	CW	6393.5
3BA4	Mauritius (Bigara) Radio, Mauritius	FC	CW	8554.0
3BA5	Mauritius (Bigara) Radio, Mauritius	FC	CW	12,831.0
3BM3	Mauritius (Bigara) Radio, Mauritius	FC	CW	6351.5
3BM5	Mauritius (Bigara) Radio, Mauritius	FC	CW	12,988.5
3BM6	Mauritius (Bigara) Radio, Mauritius	FC	CW	16,978.4
3BM7	Mauritius (Bigara) Radio, Mauritius	FC	CW	17,108.0
3BN	Plaisance, Mauritius	AX	CW	17,572.0
3BT4	Bigara, Mauritius	FX	CW	15,955.0
3CA67	Banapa, Guinea	FX	USB	20,206.0
3DP	Suva Radio, Fiji	FC	CW	12,700.0
		FC	USB	8746.8
		FC	AM	8794.0
3DP3	Suva Radio, Fiji	FC	CW	8690.0
3DV31	Suva Radio, Fiji	FC	USB	4454.0
3MA29	Nanjing, PRC	FX	FSK	19,140.0
3MA31	Nanjing, PRC	FX	CW	19,690.0
3SB	Datong Naval Radio, PRC	FC	CW/RTTY	4224.0, 8472.0, 12,808.0
				16,880.0
3SD	Fangshan Naval Radio, PRC	FC	CW/RTTY	16,902.0
3VA3Ø	Tunis, Tunisia	FX	CW	13,945.0
3VA81	Tunis, Tunisia	FX	CW	19,385.0
3VA82	Tunis, Tunisia	FX	CW	19,675.0
3VP	La Skhirra Radio, Tunesia	FC	CW	4251.0, 8534.0
3VP8	La Skhirra Radio, Tunisia	FC	CW	12,806.0

Call	Location	Service	Mode	kHz
3XC2	Conakry Radio, Guinea	FC	CW	6383.0
3XF23	Conakry, Guinea	FX	USB	5315.0
3XF24	Conakry, Guinea	FX	ISB	20,652.8
3XF26	Conakry, Guinea	FX	ISB	15,839.9
3ZL3Ø1	King George Island, Antarctica	FX	CW	23,832.0
4KA	Odessa Radio, Ukrainian SSR	FC	CW	8495.5 [UFB]
4KT	Nakhodka Radio, USSR	FC	CW	8510.0 [UKK3]
4LA	Kiev Radio, Ukrainian SSR	FC	CW	8634.0 [UJQ2]
4LI	Klaipeda Radio, Lithuanian SSR	FC	CW	8541.0 [URB2]
4LN	Riga Radio, Latvian SSR	FC	CW	8530.0 [UQK2]
4LS	Murmansk Radio, USSR	FC	CW	8580.0 [UQA4]
4LY	Arkhangelsk Radio, USSR	FC	CW	8580.0 [RKLM]
4MJ4	Caracas, Venezuela	FX	A3H	15,575.0
4MO	unknown	FX	CW	7819.9
4NX7	Belgrade, Yogoslvia	FX	CW	10,390.0
4OC3	Belgrade (Makis), Yugoslavia	FX	CW/RTTY	5112.0
4PB	Colombo Radio, Sri Lanka	FC	CW	8475.0, 12,927.0, 17,045.6
4Q029	Colombo (Kotugoda), Sri Lanka	FX	USB	5215.0
4Q081	Colombo (Kotugoda), Sri Lanka	FX	ISB	18,235.0
4UWC	Lumbumbashi, Zaire	FX	CW	20,734.0
4UWG	Geneva, switzerland	FX	CW	20,734.0
4UZ5Ø	Geneva, Switzerland	FX	ISB/RTTY	20,070.2
4WA49	Ghuraff, Yemen, A.R.	FX	USB	10,892.0
4WD3	Hodeidah Port Radio, Yemen, A.R.	FC	CW	8588.0, 12,855.8
4XI	Lod, Israel	AX	CW	16,245.0
4XL	Ben Gurion Aeradio, Israel	FA	AM	5575.0, 11,391.0
4XO	Haifa Radio, Israel	FC	CW	4237.0, 6430.0, 6470.5, 8485.0, 8694.4, 12,860.0, 13,051.5, 17,060.0, 17,146.1, 22,461.0, 22,491.0
		FC	USB	8799.5, 13,119.4, 13,144.2, 17,270.1, 22,620.8
4XP41	Tel Aviv, Israel	FX	CW	10,390.0
4XZ	Haifa Naval Radio, Israel	FC	CW	4241.4, 4289.0, 6379.0, 8437.2, 8448.0, 8518.1, 8705.9, 12,984.2, 17,050.0, 22,330.4
		FX	CW	22,330.4
5AL	Tobruk Radio, Libya	FC	CW	4241.0, 8480.0, 16,963.0
5AT	Tripoli Radio, Libya	FC	CW	8515.0
5BA	Nicosia Radio, Cyprus	FC	CW	4347.5, 8465.0, 12,666.5, 16,883.2
5BA42	Nicosia Radio, Cyprus	FC	USB	8737.5
5BA44	Nicosia Radio, Cyprus	FC	USB	8771.6
5BA54	Nicosia Radio, Cyprus	FC	USB	13,122.5
5BA62	Nicosia Radio, Cyprus	FC	USB	17,239.1
5BA72	Nicosia Radio, Cyprus	FC	USB	22,630.1
5BP2	Nicosia, Cuprus	FX	CW	19,360.0
5BP3	Nicosia, Cyprus	FX	CW	4632.5
5BP5	Nicosia, Cyprus	FX	CW	9200.0
5BP6	Nicosia, Cyprus	FX	CW	10,390.0
5BC67	Nicosia, Cyprus	FX	USB	14,476.0
5BC94	Nicosia, Cyprus	FX	LSB	19,943.0
5KM	Bogota Naval Radio, Colombia	FX	CW	17,590.0
5LF9	Monrovia, Liberia	FX	LSB	9894.8
5LF1Ø	Monrovia, Liberia	FX	USB	10,430.0
5LF19	Monrovia, Liberia	FX	USB	19,665.1
5OP25	Lagos, Nigeria	FX	CW	6792.0, 9200.0, 19,360.0
5OU88	Lagos, Nigeria	FX	ISB	18,587.5
5OU93	Lagos (Ikorodu), Nigeria	FX	LSB	18,744.3
5OU239	Lagos, Nigeria	FX	LSB	23,925.0
5OV234	Lagos (Ikorodu), Nigeria	FX	LSB	23,467.1
5OW12	Lagos Radio, Nigeria	FC	CW	12,658.0
5OW22	Lagos Radio, Nigeria	FC	CW	22,310.5

Call	Location	Service	Mode	kHz
5OZ23	Port Harcourt Radio, Nigeria	FC	CW	8461.5, 8490.0, 8496.0, 12,696.5, 12,723.0, 12,935.5, 16,942.0
5RS	Tamatave Radio, Madagascar	FC	CW	6348.0, 12,697.8, 12,860.8, 16,930.4, 22,338.0
5RS4	Tamatave Radio, Madagascar	FC	CW	4232.5
5RS8	Tamatave Radio, Madagascar	FC	CW	8465.0
5RY5	Antananarivo, Madagascar	FX	ISB	10,515.0
5RY81	Antananarivo, Madagascar	FX	USB	18,073.0
5RY92	Antananarivo, Madagascar	FX	ISB	19,250.0
5RZ49	Antananarivo, Madagascar	FX	ISB	24,920.0
5ST81	Antananarivo, Madagscar	AX	CW	16,245.0
5TA	Nouadhibou Radio, Mauritania	FC	CW	8572.0
5TA17	Nouadhibou, Mauritania	FX	ISB	11,165.0
5TA14	Nouadhibou, Mauritania	FX	ISB	7306.5
5TA21	Nouadhibou, Mauritania	FX	ISB	20,241.0
5TN22	Nouakchott, Mauritania	FX	ISB	5735.1
5TN27	Nouakchott, Mauritania	FX	USB	10,917.0
5TN29	Nouakchott, Mauritania	FX	ISB	14,366.0
5TN3Ø	Nouakchott, Mauritania	FX	ISB	18,977.2
5TN58	Nouakchott, Mauritania	FX	USB	5398.5
5TN2Ø5	Nouakchott, Mauritania	FX	USB	15,660.0
5TN228	Nouakchott, Mauritania	FX	LSB	7498.5
5TN231	Nouakchott, Mauritania	FX	ISB	4529.0
5TN244	Nouakchott, Mauritania	FX	ISB	18,738.0
5TN277	Nouakchott, Mauritania	FX	ISB	15,635.0
5TP25	Nouakchott, Mauritania	FX	CW	19,360.0
5TP294	Nouakchott, Mauritania	FX	USB	16,041.4
5UR8	Niamey, Niger	FX	USB	10,844.5
5UR29	Niamey, Niger	FX	ISB	22,954.8
5UR8Ø	Niamey, Niger	FX	ISB	5081.0
5UR88	Niamey, Niger	FX	LSB	18,856.7
5UR91	Niamey, Niger	FX	ISB	19,196.0
5VA	Lome, Togo	FX	CW	5265.0
5VH8Ø	Lome, Togo	FX	ISB	8015.0
5VH3Ø5	Lome, Togo	FX	ISB	10,526.3
5VH3Ø8	Lome, Togo	FX	USB	10,799.5
5VH346	Lome, Togo	FX	ISB	14,613.5
5VH347	Lome, Togo	FX	ISB	14,777.0
5VH4Ø2	Tome, Togo	FX	USB	20,220.0
5VH4Ø3	Lome, Togo	FX	ISB	20,370.0
5VH436	Lome, Togo	FX	ISB	23,615.0
5YC	Nairobi, Kenya	FX	CW	20,946.7
5YE	Nairobi, Kenya	FX	CW/FAX	9043.0, 16,315.0, 22,867.0
		FX	USB	9086.0
5YD	Nairobi, Kenya	AX	CW	16,245.0
5YE3	Nairobi, Kenya	FX	CW/FAX	17,365.0
5YE4	Nairobi, Kenya	FX	FAX	5127.0
5YF52	Nairobi, Kenya	FX	ISB	11,495.0
5YG	Nairobi, Kenya	FX	CW	19,360.0
5ZF2	Mombasa Radio, Kenya	FC	CW	8441.4
5ZF3	Mombasa Radio, Kenya	FC	CW	13,065.0
5ZF4	Mombasa Radio, Kenya	FC	CW	17,175.2
6C2	unknown	FX	CW	20,900.0
6OA24	Mogadiscio, Somalia	FX	ISB	9250.0
6VA3	Dakar Radio, Senegal	FC	CW	4295.0
6VA4	Dakar Radio, Senegal	FC	CW	6386.0
6VA5	Dakar Radio, Senegal	FC	CW	8690.0
6VA6	Dakar Radio, Senegal	FC	CW	12,653.5, 12,660.0,
6VA7	Dakar Radio, Senegal	FC	CW	16,960.0
6VK27	Dakar, Senegal	FX	USB	7558.5
6VK49	Dakar, Senegal	FX	LSB	9918.5
6VK221	Dakar, Senegal	FX	ISB	20,327.8
6VK317	Dakar, Senegal	FX	USB	16,117.0
6VK311	Dakar, Senegal	FX	LSB	11,116.2
6VK413	Dakar, Senegal	FX	ISB	13,603.0

Call	Location	Service	Mode	kHz
6VK815	Dakar, Senegal	FX	ISB	15,735.0
6VU73	Dakar, Senegal	FX	FAX	13,667.5
6VU79	Dakar, Senegal	AX	FAX	19,750.0
6VY41	Dakar, Senegal	FX	FAX	7587.5
6WW	Dakar Naval Radio, Senegal	FA	CW	8992.9
		FC	CW	4232.0, 4305.2, 8665.0,
		FC	CW	12,894.0, 16,876.1, 16,951.5, 22,342.5
		FX	CW	13,410.0, 16,135.0
6XH57	Antananarivo, Madagascar	FX	CW	13,660.0
6YF21	Kingston (Coopers Hill), Jamaica	FX	USB	7307.0, 15,462.5
6YI	Kingston Radio, Jamaica	FC	CW	6470.5, 8465.0, 13,065.0, 16,947.6
7CB	unknown	FC	CW	6364.7
7OA	Aden Radio, Democratic Yemen	FC	CW	8441.2, 13,060.5, 17,175.2
7OB9Ø	Aden (Hiswa), Democratic Yemen	FX	USB	19,036.0
7OB93	Aden (Hiswa), Democratic Yemen	FX	LSB	19,680.0
7OB23Ø	Aden (Hiswa), Democratic Yemen	FX	ISB	23,038.0
7OB452	Aden (Khormaksar), Democratic Yemen	FX	USB	11,125.0, 11,985.0
7RA2Ø	Algiers, Algeria	FX	CW	10,390.0
7RV5Ø	Islamabad, Pakistan	FX	CW	11,005.0
7RV7Ø	Warsaw, Poland	FX	CW	10,996.0
7TA2	Algiers Radio, Algeria	FC	CW	4288.0
7TA4	Algiers Radio, Algeria	FC	CW	6415.0
7TA6	Algiers Radio, Algeria	FC	CW	8437.0
7TA8	Algiers Radio, Algeria	FC	CW	12,662.0
7TA1Ø	Algiers Radio, Algeria	FC	CW	16,932.0
7TA12	Algiers Radio, Algeria	FC	CW	22,543.0
7UP16	Algiers, Algeria	FX	USB	16,165.0
8BB39	Jakarta, Indonesia	FX	CW	16,200.0
8CJ	unknown	FX	ISB	12,017.0
8KL	unknown	FX	CW	19,868.0
8PO	Barbados Radio, Barbados	FC	CW	6379.5, 8450.2, 12,709.0, 16,947.6
8PX45	Bridgetown, Barbados	FX	CW	10,115.0
8RB	Demerara Radio, Guyana	FC	CW	8449.0, 12,709.0, 16,947.6
8RB78	Georgetown, Guyana	FX	USB	16,452.5
8UF75	New Delhi, India	FX	CW	7532.0, 8006.5, 9285.0, 10,390.0, 13,520.0, 13,820.0, 14,607.0, 14,817.5, 15,684.0, 18,087.0, 19,130.0, 21,785.0, 21,807.5, 24,110.0
		FX	CW/RTTY	9285.0
9EQ7Ø	Addis Ababa, Ethiopia	FX	CW	17,365.0
9ES5Ø	Addis Ababa, Ethiopia	FX	CW	19,425.0
9EU76	Addis Ababa, Ethiopia	FX	CW	21,755.0
9GA	Takoradi Radio, Ghana	FC	CW	4262.0, 8542.0, 12,669.0, 13,074.0, 171,75.2
9GN29	Accra, Ghana	FX	ISB	23,462.0
9GN3Ø	Accra, Ghana	FX	ISB	23,795.0
9GX	Tema Radio, Ghana	FC	CW	4340.0, 6480.0, 8696.0, 12,765.0, 16,920.0
9HC32	Malta, Malta	FX	USB	7532.0
9HC35	Malta, Malta	FX	ISB	7739.9
9HC72	Malta, Malta	FX	ISB	15,582.1
9HD	Malta Radio, Malta	FC	CW	4223.5, 6333.5, 8441.3, 12,709.2, 17,189.6
9JA28	Lusaka, Zambia	FX	USB	25,260.0
9KK2	Kuwait Radio, Kuwait	FC	CW	4299.0
9KK4	Kuwait Radio, Kuwait	FC	CW	6381.0
9KK6	Kuwait Radio, Kuwait	FC	CW	8525.0
9KK8	Kuwait Radio, Kuwait	FC	CW	12,895.0
9KK2Ø	Kuwait Radio, Kuwait	FC	CW	12,925.0
9KK22	Kuwait Radio, Kuawit	FC	CW	16,995.0
9KK23	Kuwait Radio, Kuwait	FC	CW	22,504.0

Call	Location	Service	Mode	kHz
9KT357	Huban, Kuwait	FX	ISB	18,752.5
9LL	Freetown Radio, Sierra Leone	FC	CW	6411.0, 8710.0, 13,042.0,
		FC	CW	17,175.2
9MB2	Penang Naval Radio, Malaysia	FC	CW	4335.6
9MB3	Penang Naval Radio, Malaysia	FC	CW	6478.0, 8630.5
9MB4	Penang Naval Radio, Malaysia	FC	CW	8626.4
9MG2	Penang Radio, Malaysia	FC	CW	8698.0, 12,677.9
9MG8	Penang Radio, Malaysia	FC	CW	8492.0
9MG42	Penang Radio, Malaysia	FC	USB	4400.8
9NB24	presumed Kathmandu, Nepal	FX	ISB	14,676.0
9NB27	presumed Kathmandu, Nepal	FX	ISB	17,559.0
9PA5	Banana Radio, Zaire	FC	CW	8548.0, 8714.0
9PM7	Matadi Radio, Zaire	FC	CW	8546.0
9RE77	Lubumbashi, Zaire	FX	ISB	7740.0
9RE394	Lubumbashi, Zaire	FX	ISB	19,430.0
9TK21	Kinshasa, Zaire	FX	AM	10,390.0
9TO2Ø	Kinshasa, Zaire	FX	USB	20,816.4
9UB69	Bujumbura, Burundi	FX	ISB	14,424.9
9UB83	Bujumbura, Burundi	FX	USB	17,437.0
9VA25	Singapore Aeradio, Singapore	FA	USB	13,324.0
9VA26	Singapore Aeradio, Singapore	FA	USB	17,933.0
9VA27	Singapore Aeradio, Singapore	FA	USB	8930.0
9VA28	Singapore Aeradio, Singapore	FA	USB	10,078.0, 10,093.0
9VA3Ø	Singapore Aeradio, Singapore	FA	USB	17,940.0
9VA4Ø	Singapore Aeradio, Singapore	FA	AM	6680.0
9VA43	Singapore Aeradio, Singapore	FA	A3H	10,017.0
9VE27	Singapore, Singapore	AX	ISB/RTTY	8035.0
9VE55	Singapore, Singapore	AX	ISB	9815.0
9VF6	Singapore, Singapore	FX	ISB	5072.5
9VF68	Singapore, Singapore	FX	ISB	11,065.0
9VF251	Singapore, Singapore	FX	ISB	10,275.0
9VF284	Singapore, Singapore	FX	ISB	10,961.0
9VG5	Singapore Radio, Singapore	FC	CW	6412.0
9VG26	Singapore Radio, Singapore	FC	CW	12,817.4
9VG27	Singapore Radio, Singapore	FC	CW	22,479.0
9VG33	Singapore Radio, Singapore	FC	CW	4313.0
9VG34	Singapore Radio, Singapore	FC	CW	12,707.0
9VG35	Singapore Radio, Singapore	FC	CW	8530.0
9VG36	Singapore Radio, Singapore	FC	CW	8688.0
9VG37	Singapore Radio, Singapore	FC	CW	12,659.4
9VG9	Singapore Radio, Singapore	FC	CW	6340.5
9VG53	Singapore Radio, Singapore	FC	CW	16,868.5
9VG54	Singapore Radio, Singapore	FC	CW	4322.0
9VG56	Singapore Radio, Singapore	FC	CW	8476.0
9VG57	Singapore Radio, Singapore	FC	CW	12,724.0
9VG58	Singapore Radio, Singapore	FC	CW	16,966.5
9VG59	Singapore Radio, Singapore	FC	CW	22,428.0
9VG6Ø	Singapore Radio, Singapore	FC	USB	4369.8
9VG63	Singapore Radio, Singapore	FC	USB	8728.2
9VG64	Singapore Radio, Singapore	FC	USB	8762.3
9VG65	Singapore Radio, Singapore	FC	USB	8790.2
9VG66	Singapore Radio, Singapore	FC	USB	13,147.3
9VG68	Singapore Radio, Singapore	FC	USB	17,270.1
9VG69	Singapore Radio, Singapore	FC	USB	17,356.9
9VG73	Singapore Radio, Singapore	FC	CW	8609.5
9VG74	Singapore Radio, Singapore	FC	CW/TOR	4350.0
9VG78	Singapore Radio, Singapore	FC	CW/TOR	8709.0
9VG8Ø	Singapore Radio, Singapore	FC	CW/TOR	13,071.5
9VG82	Singapore Radio, Singapore	FC	CW/TOR	17,197.5
9VG89	Singapore Radio, Singapore	FC	USB	8780.9
9VT5	Singapore Radio, Singapore	FC	CW	12,949.0
9WW2Ø	Kuching Radio, Malaysia	FC	CW	8522.0
9XK82	Kigali, Rwanda	FX	ISB	18,565.0
9XK92	Kigali, Rwanda	FX	ISB	18,894.1
9YL	North Post Radio, Trinidad	FC	CW	6470.5, 8441.0, 12,885.0,
				17,184.8

Call	Location	Service	Mode	kHz
				17,184.8
"CP"	unknown	FX	USB	7960.0
"C"	unknown	RC	CW/Data	17,016.0, 20,992.0
"D"	unknown	RC	CW	5305.0, 10,643.5, 17,015.3 20,991.5
"E"	unknown	RC	CW	8703.2, 10,446.1, 20,455.5
"F"	unknown	RC	CW	5307.0, 8647.0, 10,614.0, 10,645.0, 13,637.0, 17,017.0
"KPA2"	unknown	FX	USB	8925.0, 12,747.0, 13,150.0
"K"	unknown	RC	CW	4005.0, 5794.7, 5889.8, 5919.9, 7954.0, 8144.1, 8752.0, 9042.9, 10,570.3, 10,638.0, 10,646.0, 11,155.5, 12,151.0, 14,477.2, 14,587.0, 14,967.0, 18,342.8
"MIW2"	unknown	FX	USB	12,747.0
"O"	unknown	RC	CW	20,991.8
"P"	unknown	RC	CW	6203.0
"R"	unknown	RC	CW	10,645.9
"U"	unknown	RC	CW	8136.5, 9057.5, 10,211.0, 12,185.0, 12,329.0, 13,328.0, 15,655.6, 15,705.0
"W"	unknown	RC	CW	7656.0, 15,700.0
"XS"	unknown	FX	USB	13,431.2

Abbreviations

AB	Air Base
ACT	Australian Capitol Territory
AF	Air Force
AFI	African Intercontinental AMS
AFRTS	Armed Forces Radio-Television Service
AFS	Air Force Station
AK	Alaska, USA
aka	also known as
AL	Alabama, USA
Alta.	Alberta, Canada
AMS	Aeronautical Mobile Service
AMVER	Automated Mutual-assistance VEssel Rescue System
ANARE	Australian National Antarctic Research Expedition
ANSA	Italian News Agency
AP	Associated Press
ARINC	Aeronautical Radio, Inc.
AS	Air Station
AX	Aeronautical Fixed Station
AZ	Arizona, USA
A3H	LSB or USB with 6 dB Suppressed Carrier
BBC	British Broadcasting Corp.
BC	British Columbia, Canada
BC	Broadcast Station
CA	California, USA
CAP	Civil Air Patrol
CAR	Caribbean AMS
CEP	Central Eastern Pacific AMS
CH	Channel
CIRM	Centro Internationale Radio Medico, Rome, Italy
CO	Colorado, USA
cont.	Continuous
con.	Continued
CRS	Compagnies Republicaines de Securite (France)
CT	Connecticut, USA
CW	Continuous Wave (Morse Code)
CWP	Central Western Pacific AMS
DC	District of Columbia, USA
DEA	US Drug Enforcement Agency
DF	District Federal (Mexico)
DOE	US Department of Energy
DW	Deutsche Welle (GFR)
EE	English Language